Toxicological Risk Assessment of Chemicals

A Practical Guide

Toxicological Risk Assessment of Chemicals

A Practical Guide

Elsa Nielsen • Grete Østergaard • John Christian Larsen

informa
healthcare

New York   London

Cover: Theophrastus Phillippus Aureolus Bombastus von Hohenheim, known as "Paracelsus" 1493–1541 ("the equal of Celsus", an early Roman physician). Paracelsus is often called "the father of toxicology." His famous quotation forms the central dogma of regulatory toxicology:

"All substances are poisons; there is none that is not a poison. The right dose differentiates a poison and a remedy."—Paracelsus

Informa Healthcare USA, Inc.
52 Vanderbilt Avenue
New York, NY 10017

© 2008 by Informa Healthcare USA, Inc.
Informa Healthcare is an Informa business

Library of Congress Cataloging-in-Publication Data

Ostergaard, Grete.
    Toxicological Risk Assessment of Chemicals : a practical guide / Elsa Nielsen, Grete Ostergaard, and John Christian Larsen.
        p. ; cm.
    Includes bibliographical references and index.
    ISBN-13: 978-0-8493-7265-0 (hardcover : alk. paper)
    ISBN-10: 0-8493-7265-8 (hardcover : alk. paper)
    1. Toxicology. 2. Toxicity testing. 3. Health risk assessment. 4. Environmental risk assessment. I. Nielsen, Elsa. II. Larsen, John Christian. III. Title.
    [DNLM: 1. Hazardous Substances--toxicity. 2. Environmental Exposure--prevention & control. 3. Risk Assessment--methods. 4. Risk Assessment--standards. WA 670 O845p 2008]

RA1190.O88 2008
615.9--dc22   2007034227

Visit the Informa Web site at
www.informa.com

and the Informa Healthcare Web site at
www.informahealthcare.com

Contents

xiv

Preface

Risk assessment of chemical substances is an ever-developing discipline. Transparent and accurate risk assessments are necessary for decision-makers to make wise risk management decisions. The outcome of risk assessments may have enormous economical consequences, in addition to the consequences for human health and the environment. Globalization is a fact, with huge possibilities for economic and social prosperity. Food and consumer products are produced in one part of the world and put on the market in another. Fair competition rests upon global similarity of regulations. Environmental chemical pollution does not stay within the borders of a country, but moves to even the most remote and isolated places on Earth. Therefore, regulation of chemicals cannot be kept within individual states, and consequently the demand for international agreement on regulation is increasing. Toxicology and chemical risk assessments are important parts of this movement from a national to an international world, and must assist in the development of tools to ensure human safety while free trade is not compromised. The key is science, harmonization, mutual agreement, and acceptance.

Our aim with this book is to provide the reader with a useful guide and a valuable tool for working as a toxicologist and risk assessor in the field of human health risk assessments. The book has been written with emphasis on international harmonization through the United Nations and the Organisation for Economic Co-operation and Development, and taking major federal bodies such as the United States and the European Union into consideration. National programs and methods are mentioned, but only to a very limited extent. The book is mainly concerned with industrial chemicals, but other chemical use categories, for example pesticides, food additives, veterinary drugs, etc., are also mentioned in recognition of the wide use of chemicals for different purposes.

The focus of this book is on the description of the existing risk assessment methodologies for human health. New developments in these methodologies are also mentioned. The major subjects of the book include a description of the various institutions, agencies, and programs involved in chemicals regulation; data used for hazard assessment; the various toxicological endpoints and the associated test methods used in hazard assessment; the process of standard setting for threshold and non-threshold effects with focus on the assessment factors applied for threshold effects; the exposure assessment, the risk characterization, regulatory standards by various bodies; and the risk assessment of chemicals in mixture.

We have attempted to present the state of the art and scientific consensus in an unbiased way; however, what is believed to be true today may not hold tomorrow - politics may change, and science may progress. Fortunately, while methods and techniques are continuing to improve, the basic concepts of toxicological risk assessment remain stable and form the core in this book.

The book has been based primarily on our experiences achieved through many years of practical work as risk assessors within different areas of toxicological risk assessments of chemicals. We have also consulted essential key references for the state of the art as well as relevant Web sites. Links to Web sites were checked right before we forwarded the manuscript to the publisher (1 July 2007); however, as Web sites are undergoing continuing development, links might have changed since then.

Elsa Nielsen and Grete Østergaard are the authors of Chapters 1 through 9, while John Christian Larsen is the author of Chapter 10.

The book is primarily intended for students in health and environmental sciences as well as for risk assessors who are involved in toxicological risk assessments of chemicals. We hope that this book will serve as a fundamental basis for students in their toxicological training and as a useful guide and a valuable tool for risk assessors in their daily work.

Acknowledgment

Very special thanks go to Junior Advisor Krestine Greve, National Food Institute, Technical University of Denmark. We are deeply grateful to you, Krestine, for reading the entire manuscript and providing us your very valuable comments; thank you very much.

Thanks also to Senior Advisor Christine Nellemann, National Food Institute, Technical University of Denmark, for your contributions regarding the alternative methods to experimental animal studies.

Finally, we thank our families who, for quite a long time, only saw us in front of the computer.

Authors

Elsa Nielsen received her master's degree from the Royal School of Pharmacy, Copenhagen (now the Faculty of Pharmaceutical Sciences at the University of Copenhagen), in 1980. Subsequently she undertook research at the school's Department of Biochemistry and received her PhD in 1987. For a number of years, she worked as a chemist in the field of control of chemical substances in drinking water, waste water, and soil. In 1991, she obtained a position at the Institute of Toxicology at the National Food Agency of Denmark (now the National Food Institute at the Technical University of Denmark), where she still holds a position as a risk assessor and senior advisor, primarily as a consultant for the Danish Environmental Protection Agency in the field of toxicological risk assessment of industrial chemicals in relation to national and EU regulations.

Grete Østergaard received her master's degree in veterinary medicine from the Royal Agricultural and Veterinary College in Copenhagen in 1983. After graduation, she worked briefly in veterinary practice and with clinical trials of pharmaceuticals. In 1985, she joined the Institute of Toxicology at the National Food Agency of Denmark, where she did experimental work and toxicological risk assessment of chemicals. In 1997, she received her PhD. In 2003, she obtained a position at the University of Copenhagen's Faculty of Health in the Department of Experimental Medicine, and in 2006 received her master's degree in laboratory animal science from the Royal Agricultural and Veterinary College.

John Christian Larsen received his master's degree from the Royal School of Pharmacy, Copenhagen, in 1969. Following a short appointment in the pharmaceutical industry, he joined the Institute of Toxicology at the National Food Agency of Denmark in 1971, where his research work was concerned with xenobiotic metabolism and biochemical toxicology, including genetic toxicology in vitro and in vivo. During 1987–2004, he was the head of Division for Biochemical and Molecular Toxicology. In 2004, he became a consultant for the institute. John Christian Larsen has participated in numerous national, Nordic, and international expert groups. He is a member of the WHO Expert Advisory Panel on Food Safety and has participated for a number of years in the FAO/WHO Joint Expert Group on Food Additives (JECFA). He is currently also a member of the European Food Safety Authority's (EFSA) Panel on Food Additives, Flavorings, Processing Aids, and Materials in Contact with Food.

Abbreviations and Acronyms

ACGIH   American Conference on Governmental Industrial Hygienists
ADD   Average Daily Dose
ADI   Acceptable Daily Intake
ADME   Absorption, Distribution, Metabolism, Excretion
AF   Assessment Factor
ALARA   As Low as Reasonably Achievable
AOEL   Acceptable Operator Exposure Level
AR   Androgen Receptor
ARfD   Acute Reference Dose
ARI   Aggregate Risk Index
ATC   Acute Toxic Class
ATRA   Air Toxics Risk Assessment
ATSDR   Agency for Toxic Substances and Disease Registry
AUC   Area Under the (blood/plasma concentration versus time) Curve
BEI   Biological Exposure Indices
BIAC   Business and Industry Advisory Committee
BMD   Benchmark Dose
BMDL   Benchmark Dose Lower Limit
BMR   Benchmark Response
BUA   (Beratergremium für Altstoffe) Advisory Committee on Existing Chemicals
           of Environmental Relevance
CA   Competent Authority
CAA   Clean Air Act
CAFE   Clean Air For Europe
CAS   Chemical Abstract Service
CCOHS   Canadian Centre for Occupational Health and Safety
CCPR   Codex Committee on Pesticide Residues
CDC   Centers for Disease Control and Prevention
CDER   Centre for Drug Evaluation and Research
CED   Critical Effect Dose
CEFIC   European Chemical Industry Council
CEM TF   Consumer Exposure Modelling Task Force
CERCLA   Comprehensive Emergency Response, Compensation, and Liability Act
CES   Critical Effect Size
CF   Conversion Factor
CG/HCCS   Coordinating Group for the Harmonization of Chemical Classification Systems
CHAD   Consolidated Human Activity Database
CHMP   Committee for Medicinal Products for Human Use
CICAD   Concise International Chemical Assessment Document
CNS   Central Nervous System
CPDB   Carcinogenic Potency Database
CPF   Carcinogen Potency Factor
CPSC   Consumer Product Safety Commission
CRI   Cumulative Risk Index
CSAF   Chemical-Specific Adjustment Factor

CSSTT	Collaborative Study on the Assessment and Validation of Short-Term Tests for Genotoxicity and Carcinogenicity
CVMP	Committee for Medicinal Products for Veterinary Use
CYP	Cytochrome P450
DAF	Dosimetric Adjustment Factor
DFE	Design for the Environment
DG	Directorate-general
DG SANCO	Health and Consumer Protection Directorate-general
DNEL	Derived No-Effect Level
DRP	Detailed Review Paper
EAP	Environment Action Programme
ECB	European Chemicals Bureau
ECETOC	European Centre for Ecotoxicology and Toxicology of Chemicals
ECHA	European Chemicals Agency
ECVAM	European Center for the Validation of Alternative Methods
ED	Endocrine Disruptor
EDSP	Endocrine Disruptor Screening Program
EDSTAC	Endocrine Disruptor Screening and Advisory Committee
EDTA	Endocrine Disrupter Testing and Assessment
EFSA	European Food Safety Authority
EFTA	European Free Trade Association
EHC	Environmental Health Criteria
EINECS	European Inventory of Existing Commercial Chemical Substances
ELINCS	European List of Notified Chemical Substances
ELISA	Enzyme-Linked Immunosorbent Assay
EMEA	European Medicines Agency
EPCRA	Emergency Planning and Community Right to Know Act
ER	Estrogen Receptor
ESAC	ECVAM Scientific Advisory Committee
ESD	Emission Scenario Document
ESR	Existing Substances Regulation
EU	European Union
EUSES	European Union System for the Evaluation of Substances
FAO	Food and Agriculture Organization of the United Nations
FCA	Freund's Complete Adjuvant
FDA	Food and Drug Administration
FIFRA	Federal Insecticide, Fungicide, and Rodenticide Act
FISH	Fluorescence In Situ Hybridization
FQPA	Food Quality Protection Act
GDCh	Gesellschaft Deutscher Chemiker (German Chemical Society)
GEV	Generic Exposure Value
GHS	Globally Harmonized System of Classification and Labeling of Chemicals
GLEV	Generic Lowest Exposure Value
GLP	Good Laboratory Practice
GM	Geometric Mean
GPMT	Guinea Pig Maximization Test
GSD	Geometric Standard Deviation
HAP	Hazardous Air Pollutant
HBORV	Health-Based Occupational Reference Values
HEC	Human Equivalent Concentration
HED	Human Equivalent Dose

HEDSET	Harmonised Electronic Dataset
HI	Hazard Index
HPV	High Production Volume
HPVC	High Production Volume Chemicals
HQ	Hazard Quotient
HR	Heart Rate
HSDB	Hazardous Substances Data Bank
HSG	Health and Safety Guide
IARC	International Agency for Research on Cancer
ICCVAM	United States Interagency Coordinating Committee on the Validation of Alternative Methods
ICSC	International Chemical Safety Cards
IHCP	Institute for Health and Consumer Protection
ILO	International Labour Organization
ILSI	International Life Sciences Institute
INCHEM	WHO/IPCS global database with evaluated information on chemicals
IOMC	Inter-Organizational Programme for the Sound Management of Chemicals
IPCS	International Programme on Chemical Safety
IR	Inhalation rate
IRIS	Integrated Risk Information System
IUCLID	International Uniform Chemical Information Database
IUR	Inventory Update Rule
JECFA	Joint FAO/WHO Expert Committee on Food Additives
JEMRA	Joint FAO/WHO Expert Meetings on Microbiological Risk Assessment
JMPR	Joint FAO/WHO Meeting on Pesticide Residues
JRC	Joint Research Centre
LADD	Lifetime Average Daily Dose
LAEL	(true) Lowest-Adverse-Effect Level
LCL	Lower Confidence Limit
LED	Lowest Effective Dose
LEL	(true) Lowest-Effect Level
LLNA	Local Lymph Node Assay
LMS	Linearized Multistage
LOAEC	Lowest-Observed-Adverse-Effect Concentration
LOAEL	Lowest-Observed-Adverse-Effect Level
LOEL	Lowest-Observed-Effect Level
LPVC	Low Production Volume Chemicals
MACT	Maximum Achievable Control Technology
MAD	Mutual Acceptance of Data
MBSL	Mouse Biochemical Specific Locus Test
MCL	Maximum Contaminant Level
MCLG	Maximum Contaminant Level Goal
MEST	Mouse Ear Swelling Test
MF	Modifying Factor
MLE	Maximum Likelihood Estimate
MNCL	Mononuclear Cell Leukemia
MOE	Margin of Exposure
MOS	Margin of Safety
MOSref	Reference MOS
MRL	Maximum Residue Level
MSLT	Mouse Visible Specific Locus Test

NAEL	(true) No-Adverse-Effect Level
NCEA	National Center for Environmental Assessment
NCEH	National Center for Environmental Health
NCI	National Cancer Institute
NCOD	National Contaminant Occurrence Database
NCP	National Contingency Plan
NCTR	National Center for Toxicological Research
NEG	Nordic Expert Group for Criteria Documentation of Health Risks from Chemicals
NEL	(true) No-Effect Level
NGO	Nongovernmental Organization
NICEATM	National Toxicology Program Interagency Center for the Evaluation of Alternative Toxicological Methods
NIEHS	National Institute of Environmental Health Sciences
NIH	National Institutes of Health
NIOSH	National Institute for Occupational Safety and Health
NIWL	Swedish National Institute for Working Life
NLM	National Library of Medicine
NOAEC	No-Observed-Adverse-Effect Concentration
NOAEL	No-Observed-Adverse-Effect Level
NOEL	No-Observed-Effect Level
NPL	National Priorities List
NTE	Neuropathy Target Esterase
NTP	National Toxicology Program
OECD	Organisation for Economic Co-operation and Development
OEL	Occupational Exposure Limit
OJ	Official Journal of the European Communities
OP	Organophosphate
OPPT	Office of Pollution Prevention and Toxics
OSHA	Occupational Safety and Health Administration
PAH	Polycyclic Aromatic Hydrocarbons
PBPK	Physiologically Based Pharmacokinetics
PBTK	Physiologically Based Toxicokinetics
PCB	Polychlorinated Biphenyl
PDS	Pesticide Safety Data Sheets
PFC	Plaque Forming Cell
PIM	Poisons Information Monograph
PMN	Premanufacture notice
POD	Point of Departure
PODF	Point of Departure Fractions
PODI	Point of Departure Index
POP	Persistent Organic Pollutant
PPE	Personal Protective Equipment
PPP	Plant Protection Product
PSD	Pesticide Safety Directorate
PT	Physiological Toxicokinetics
PTWI	Provisional Tolerable Weekly Intake
QSAR	Quantitative Structure–Activity Relationship
RACB	Reproductive assessment by continuous breeding
RAF	Risk Assessment Forum
RCRA	Resource Conservation and Recovery Act

RfC	Reference Concentration
RfD	Reference Dose
REACH	Registration, Evaluation and Authorisation of CHemicals
RIP	REACH Implementation Project
RIVM	Dutch National Institute of Public Health and the Environment (Rijksinstituut voor Volksgezondheid en Milieu)
RSA	Response Surface Analysis
SAR	Structure–Activity Relationship
SARA	Superfund Amendments and Reauthorization Act
SCCP	Scientific Committee on Consumer Products
SCE	Sister Chromatid Exchange
SCENIHR	Scientific Committee on Emerging and Newly-Identified Health Risks
SCHER	Scientific Committee on Health and Environmental Risks
SDWA	Safe Drinking Water Act
SIAM	SIDS Initial Assessment Meeting
SIAP	SIDS Initial Assessment Profile
SIAR	SIDS Initial Assessment Report
SIDS	Screening Information Data Set
SIP	State Implementation Plan
SIS	Division of Specialized Information Services
SLRL	Sex-Linked Recessive Lethal
SRBC	Sheep Red Blood Cells
SSL	Soil Screening Level
TBG	Thyroxine-Binding Globulin
TC C&L	Technical Committee on Classification and Labelling of Dangerous Substances
TCNES	Technical Committee for New and Existing Substances
TD	Tumorigenic Dose
TDI	Tolerable Daily Intake
TEQ	Total Equivalent
TEF	Toxic Equivalency Factor
TEHIP	Toxicology and Environmental Health Information Program
TER	Transcutaneous Electrical Resistance
TG	Test Guideline
TGD	Technical Guidance Document
TLV	Threshold Limit Values
TNsG	Technical Notes for Guidance
TNO	Nederlandse Organisatie voor Toegepast-Natuurwetenschappelijk Onderzoek TNO. (The Netherlands Organisation for Applied Scientific Research)
TRH	Thyroid-Releasing Hormone
TSCA	Toxic Substances Control Act
TSH	Thyroid Stimulating Hormone
TT	Treatment Technique
TTC	Threshold of Toxicological Concern
TTD	Target-organ Toxicity dose
UDS	Unscheduled DNA Synthesis
UF	Uncertainty Factor
UN	United Nations
UNCETDG/GHS	United Nations Committee of Experts on the Transport of Dangerous Goods and on the Globally Harmonized System of Classification and Labeling of Chemicals

UNDP	United Nations Development Programme
UNEP	United Nations Environment Programme
UNIDO	United Nations Industrial Development Organization
UNITAR	United Nations Institute for Training and Research
UNSCETDG	United Nations Sub-Committee of Experts on the Transport of Dangerous Goods
US-EPA	Unites States Environmental Protection Agency
UVCB	Chemical Substances of Unknown, or Variable Composition, Complex Reaction Products, and Biological Materials
VOC	Volatile Organic Compound
VR	Ventilation Rate
VSD	Virtually Safe Dose
WGP	Working Group on Pesticides
WHO	World Health Organization
WOE	Weight Of Evidence
WSSD	World Summit on Sustainable Development

1 Introduction

Chemical substances make a vital contribution to almost all aspects of the lives of human beings. Chemicals bring about improvements in the health of the population and our quality of life through new developments in, for example, treatment of disease, food safety, or crop protection. They also make a vital contribution to the economic and social well-being of citizens in terms of trade and employment. The global production of chemicals has increased from about 1 million tons in 1930 to more than 400 million tons today, and more than 100,000 different chemical substances are now on the market.

On the other hand, chemicals may also present risks to human health and the environment. For example, toxic compounds may pollute air, water, soil, and food, and accumulate in the human body following exposure to such a compound as illustrated in Figure 1.1.

Though many of such chemicals have now been totally banned or subjected to other controls, measures were not taken until after the damage was done, because knowledge about the adverse impacts of these chemicals was not available before they were used in large quantities. Consequently, if chemicals are to provide benefits, their adverse impacts must be sufficiently well known in order to prevent or minimize their possible risks to human health and the environment.

Government agencies, for example in the United States as well as the European Union, have recently launched initiatives to acquire testing data for large numbers of chemicals currently on their markets in high volumes on which little is known about their adverse impacts. For chemicals being developed nowadays, various national or regional programs are in force to ensure the generation of data on harmful effects before these substances are put on the market.

Chemicals policy must ensure a high level of protection of human health and the environment both for the present generation as well as for future generations, while also ensuring benefits to the overall quality of life of the population as well as the efficient functioning of the market. Fundamental to achieving these objectives is the "Precautionary Principle." According to this principle, whenever reliable scientific evidence is available that a chemical may have an adverse impact on human health and the environment but there is still scientific uncertainty about the precise nature or the magnitude of the potential damage, decision making must be based on precaution in order to prevent damage to human health and the environment. Another important objective for chemicals policy is to encourage the substitution of dangerous by less dangerous substances where suitable alternatives are available. In addition, policy for chemicals should provide incentives for technical innovation and development of safer chemicals. Recent experience has shown that innovation (e.g., development of new and often safer chemicals) has been hindered by the burdens of the present regulations for introducing new chemicals on the market. Ecological, economic, and social aspects of development have to be taken into account in an integrated and balanced manner in order to reach the goal of "sustainability", the concept of meeting the needs of the present without compromising the ability of future generations to meet their needs.

Fortunately, well-developed methods exist for assessing the risks posed by chemicals to humans and the environment. The focus of this book is on the description of the existing risk assessment methodologies for human health. New developments in these methodologies will also be described. Throughout the book, the emphasis will be on international harmonization, and the application of test methods and guidance documents developed by international expert groups. The book only

1

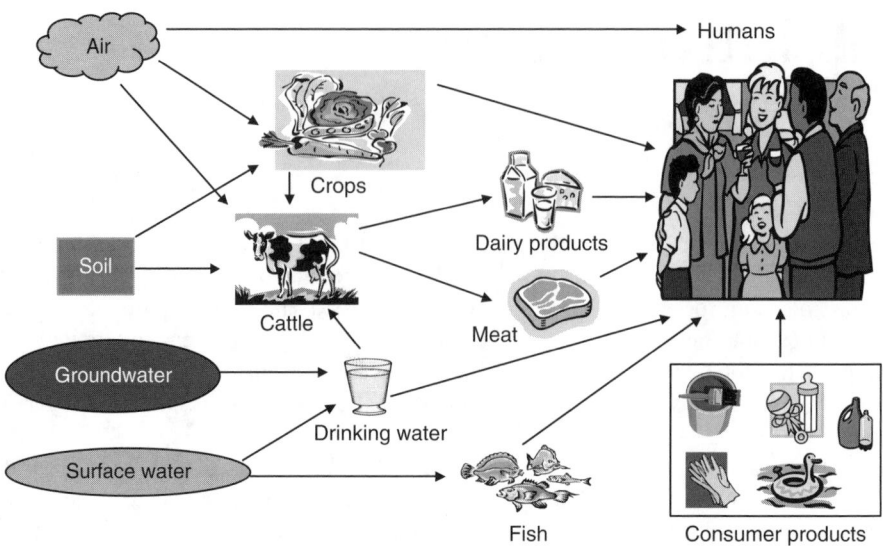

FIGURE 1.1 Human exposure to chemical substances from environmental media and sources.

deals with the risks posed by chemicals to humans, as risks posed by chemicals to the organisms living in the environment, i.e., ecotoxicology, is an entire subject in its own right.

The following are the major subjects of the book: the various institutions, agencies, and programs involved in chemicals regulation (Chapter 2). The data for hazard assessment (Chapter 3) and the hazard assessment process, i.e., identification and characterization of the various toxicological effects and the associated test methods (Chapter 4). Standard setting for threshold effects (Chapter 5) and non-threshold effects (Chapter 6). Exposure assessment (Chapter 7) and risk characterization (Chapter 8). Regulatory standards set by various bodies (Chapter 9) and combined actions of chemicals in mixture (Chapter 10).

Risk assessment is a process by which regulatory and scientific principles are applied in a systematic fashion in order to describe the hazard associated with human exposure to chemical substances. The information provided by a risk assessment may then be used to regulate the use of the substance, or may not, depending on political, social, economic, and technical considerations in the process of risk management.

The definition of risk assessment provided by the Organisation for Economic Co-operation and Development (OECD 2003) is as follows: "A process intended to calculate or estimate the risk to a given target organism, system or (sub)population, including the identification of attendant uncertainties, following exposure to a particular agent, taking into account the inherent characteristics of the agent of concern as well as the characteristics of the specific target system. The risk assessment process includes four steps: hazard identification, hazard characterization (related term: dose–response assessment), exposure assessment, and risk characterization. It is the first component in a risk analysis process." The Risk Assessment process, from data collection to risk characterization, is illustrated in Figure 1.2.

Chemical risk assessments are carried out by many national and international bodies, including major actors such as the World Health Organization (WHO) - particularly the International Programme on Chemical Safety (IPCS), the OECD, the Unites States, and the European Union (EU). The efforts undertaken by these major acting bodies in terms of chemical risk assessments are reflected in this book.

Over the past decades national and international organizations and programs have been faced with the problem of misunderstandings concerning terms used in the hazard/risk assessment of

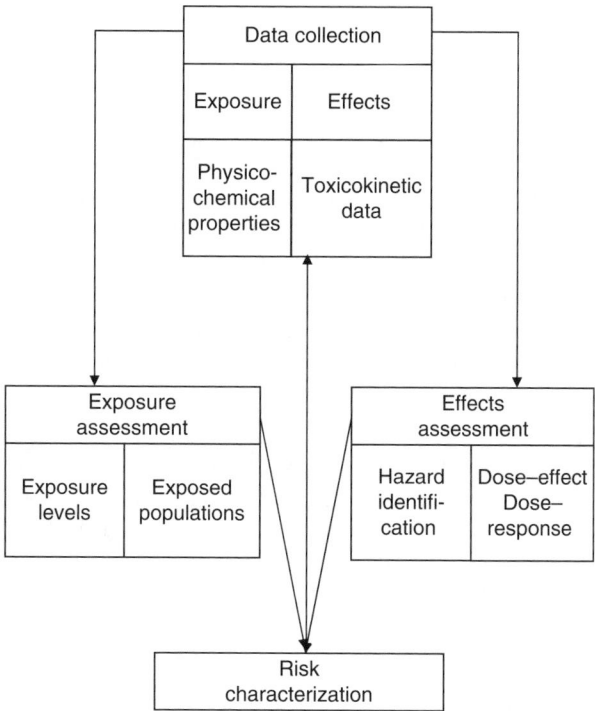

FIGURE 1.2 The risk assessment process from data collection to risk characterization.

chemicals. Due to the lack of an internationally agreed upon glossary of chemical hazard/risk assessment terms, almost every single organization/program has developed for practical reasons its own "working terminology." Therefore, a joint OECD/IPCS project was initiated with the objective to develop internationally harmonized generic terms used in chemical hazard/risk assessment in order to help facilitate the mutual use and acceptance of the assessment of chemicals between countries, saving resources for both governments and industry (OECD 2003). The alphabetical list of selected generic terms in chemical hazard and risk assessment and their descriptions are provided, followed by a compilation of remarks and background notes to each of the terms included in the list. The alphabetical list of the selected generic terms is presented in Annexure 1 of this book.

Throughout this book, the OECD/IPCS terms and definitions in chemical hazard and risk assessment will be used unless otherwise stated. The OECD/IPCS terms are generally concordant with those used by the US-EPA and the EU.

REFERENCE

OECD. 2003. *Descriptions of Selected Key Generic Terms Used in Chemical Hazard/Risk Assessment.* Joint Project with IPCS on the Harmonisation of Hazard/Risk Assessment Terminology. OECD Series on Testing and Assessment No. 44. Environment Directorate, Joint Meeting of the Chemicals Committee and the Working Party on Chemicals, Pesticides and Biotechnology. ENV/JM/MONO (2003)15. Paris: OECD.

ANNEXURE 1
Alphabetical List of Selected Generic Terms in Hazard and Risk Assessment and Their Definitions

Term	Description
Acceptable Daily Intake	Estimated maximum amount of an agent, expressed on a body mass basis, to which an individual in a (sub)population may be exposed daily over its lifetime without appreciable health risk. Related terms: *Reference Dose, Tolerable Daily Intake.*
Acceptable Risk	This is a risk management term. The acceptability of the risk depends on scientific data, social, economic, and political factors, and on the perceived benefits arising from exposure to an agent.
Adverse Effect	Change in the morphology, physiology, growth, development, reproduction or life span of an organism, system, or (sub)population that results in an impairment of functional capacity, an impairment of the capacity to compensate for additional stress, or an increase in susceptibility to other influences.
Analysis	Detailed examination of anything complex, made in order to understand its nature or to determine its essential features.
Assessment	Evaluation or appraisal of an analysis of facts and the inference of possible consequences concerning a particular object or process.
Assessment Endpoint	Qualitative/quantitative expression of a specific factor with which a risk may be associated as determined through an appropriate risk assessment.
Assessment Factor	Numerical adjustment used to extrapolate from experimentally determined (dose–response) relationships to estimate the agent exposure below which an adverse effect is not likely to occur. Related terms: *Safety Factor, Uncertainty Factor.*
Concentration	Amount of a material or agent dissolved or contained in unit quantity in a given medium or system.
Concentration–Effect Relationship	Relationship between the exposure, expressed in concentration, of a given organism, system, or (sub)population to an agent in a specific pattern during a given time and the magnitude of a continuously graded effect to that organism, system, or (sub)population. Related terms: *Effect Assessment, Dose–Response Relationship.*
Dose	Total amount of an agent administered to, taken up, or absorbed by an organism, system, or (sub)population.
Dose–Effect Relationship	Relationship between the total amount of an agent administered to, taken up, or absorbed by an organism, system, or (sub)population and the magnitude of a continuously graded effect to that organism, system, or (sub)population. Related terms: *Effect Assessment, Dose–Response Relationship, Concentration–Effect Relationship.*
Dose-Related Effect	Any effect to an organism, system, or (sub)population as a result of the quantity of an agent administered to, taken up, or absorbed by that organism, system, or (sub)population.

ANNEXURE 1 (continued)

Alphabetical List of Selected Generic Terms in Hazard and Risk Assessment and Their Definitions

Term	Description
Dose Response	Relationship between the amount of an agent administered to, taken up, or absorbed by an organism, system, or (sub)population and the change developed in that organism, system, or (sub)population in reaction to the agent. Synonymous with dose–response relationship. Related terms: *Dose–Effect Relationship, Effect Assessment, Concentration–Effect Relationship.*
Dose–Response Assessment	Analysis of the relationship between the total amount of an agent administered to, taken up, or absorbed by an organism, system, or (sub)population and the changes developed in that organism, system, or (sub)population in reaction to that agent, and inferences derived from such an analysis with respect to the entire population. Dose–response assessment is the second of four steps in risk assessment. Related terms: *Hazard Characterization, Dose–Effect Relationship, Effect Assessment, Dose–Response Relationship, Concentration–Effect Relationship.*
Dose–Response Curve	Graphical presentation of a dose–response relationship.
Dose–Response Relationship	Relationship between the amount of an agent administered to, taken up, or absorbed by an organism, system, or (sub)population and the change developed in that organism, system, or (sub)population in reaction to the agent. Related terms: *Dose–Effect Relationship, Effect Assessment, Concentration–Effect Relationship.*
Effect	Change in the state or dynamics of an organism, system, or (sub)population caused by the exposure to an agent.
Effect Assessment	Combination of analysis and inference of possible consequences of the exposure to a particular agent based on knowledge of the dose–effect relationship associated with that agent in a specific target organism, system, or (sub)population.
Expert Judgment	Opinion of an authoritative person on a particular subject.
Exposure	Concentration or amount of a particular agent that reaches a target organism, system, or (sub)population in a specific frequency for a defined duration.
Exposure Assessment	Evaluation of the exposure of an organism, system, or (sub)population to an agent (and its derivatives). Exposure assessment is the third step in the process of risk assessment.
Exposure Scenario	A set of conditions or assumptions about sources, exposure pathways, amount, or concentrations of agent(s) involved, and exposed organism, system, or (sub)population (i.e., numbers, characteristics, habits) used to aid in the evaluation and quantification of exposure(s) in a given situation.
Fate	Pattern of distribution of an agent, its derivatives, or metabolites in an organism, system, compartment, or (sub)population of concern as a result of transport, partitioning, transformation, or degradation.

(continued)

ANNEXURE 1 (continued)
Alphabetical List of Selected Generic Terms in Hazard and Risk Assessment and Their Definitions

Term	Description
Guidance Value	Value, such as concentration in air or water, which is derived after allocation of the reference dose among the different possible media (routes) of exposure. The aim of the guidance value is to provide quantitative information from risk assessment to the risk managers to enable them to make decisions. (See also: reference dose.)
Hazard	Inherent property of an agent or situation having the potential to cause adverse effects when an organism, system, or (sub)population is exposed to that agent.
Hazard Assessment	A process designed to determine the possible adverse effects of an agent or situation to which an organism, system, or (sub)population could be exposed. The process includes hazard identification and hazard characterization. The process focuses on the hazard in contrast to risk assessment where exposure assessment is a distinct additional step.
Hazard Characterization	The qualitative and, wherever possible, quantitative description of the inherent properties of an agent or situation having the potential to cause adverse effects. This should, where possible, include a dose–response assessment and its attendant uncertainties. Hazard characterization is the second stage in the process of hazard assessment, and the second step in risk assessment. Related terms: *Dose–Effect Relationship, Effect Assessment, Dose–Response Relationship, Concentration–Effect Relationship.*
Hazard Identification	The identification of the type and nature of adverse effects that an agent has as inherent capacity to cause in an organism, system, or (sub) population. Hazard identification is the first stage in hazard assessment and the first step in the process of risk assessment.
Margin of Exposure	Ratio of the No-Observed-Adverse-Effect Level (NOAEL) for the critical effect to the theoretical, predicted, or estimated exposure dose or concentration. Related term: *Margin of Safety.*
Margin of Safety	For some experts the margin of safety has the same meaning as the margin of exposure, while for others, the margin of safety means the margin between the reference dose and the actual exposure dose or concentration. Related term: *Margin of Exposure.*
Measurement Endpoint	Measurable (ecological) characteristic that is related to the valued characteristic chosen as an assessment point.
Reference Dose	An estimate of the daily exposure dose that is likely to be without deleterious effect even if continued exposure occurs over a lifetime. Related term: *Acceptable Daily Intake.*

ANNEXURE 1 (continued)
Alphabetical List of Selected Generic Terms in Hazard and Risk Assessment and Their Definitions

Term	Description
Response	Change developed in the state or dynamics of an organism, system, or (sub)population in reaction to exposure to an agent.
Risk	The probability of an adverse effect in an organism, system, or (sub) population caused under specified circumstances by exposure to an agent.
Risk Analysis	A process for controlling situations where an organism, system, or (sub) population could be exposed to a hazard. The Risk Analysis process consists of three components: risk assessment, risk management, and risk communication.
Risk Assessment	A process intended to calculate or estimate the risk to a given target organism, system, or (sub)population, including the identification of attendant uncertainties, following exposure to a particular agent, taking into account the inherent characteristics of the agent of concern as well as the characteristics of the specific target system. The Risk Assessment process includes four steps: hazard identification, hazard characterization (related term: dose–response assessment), exposure assessment, and risk characterization. It is the first component in a risk analysis process.
Risk Characterization	The qualitative and, wherever possible, quantitative determination, including attendant uncertainties, of the probability of occurrence of known and potential adverse effects of an agent in a given organism, system, or (sub) population, under defined exposure conditions. Risk Characterization is the fourth step in the Risk Assessment process.
Risk Communication	Interactive exchange of information about (health or environmental) risks among risk assessors, managers, news media, interested groups, and the general public.
Risk Estimation	Quantification of the probability, including attendant uncertainties, that specific adverse effects will occur in an organism, system, or (sub) population due to actual or predicted exposure.
Risk Evaluation	Establishment of a qualitative or quantitative relationship between risks and benefits of exposure to an agent, involving the complex process of determining the significance of the identified hazards and estimated risks to the system concerned or affected by the exposure, as well as the significance of the benefits brought about by the agent. It is an element of risk management. Risk Evaluation is synonymous with Risk–Benefit evaluation.
Risk Management	Decision-making process involving considerations of political, social, economic, and technical factors with relevant risk assessment information relating to a hazard so as to develop, analyze, and compare regulatory and non-regulatory options and to select and implement appropriate regulatory response to that hazard. Risk management comprises three elements: risk evaluation, emission and exposure control, and risk monitoring.

(*continued*)

ANNEXURE 1 (continued)
Alphabetical List of Selected Generic Terms in Hazard and Risk Assessment and Their Definitions

Term	Description
Risk Monitoring	Process of following up the decisions and actions within risk management in order to ascertain that risk containment or reduction with respect to a particular hazard is assured. Risk monitoring is an element of risk management.
Safety	Practical certainty that adverse effects will not result from exposure to an agent under defined circumstances. It is the reciprocal of risk.
Safety Factor	Composite (reductive) factor by which an observed or estimated No-Observed-Adverse-Effect Level (NOAEL) is divided to arrive at a criterion or standard that is considered safe or without appreciable risk. Related terms: *Assessment Factor, Uncertainty Factor.*
Threshold	Dose or exposure concentration of an agent below which a stated effect is not observed or expected to occur.
Tolerable Daily Intake	Analogous to acceptable daily intake. The term "Tolerable" is used for agents, which are not deliberately added such as contaminants in food.
Tolerable Intake	Estimated maximum amount of an agent, expressed on a body mass basis, to which each individual in a (sub)population may be exposed over a specified period without appreciable risk.
Toxicity	Inherent property of an agent to cause an adverse biological effect.
Uncertainty	Imperfect knowledge concerning the present or future state of an organism, system, or (sub)population under consideration.
Uncertainty Factor	Reductive factor by which an observed or estimated No-Observed-Adverse-Effect Level (NOAEL) is divided to arrive at a criterion or standard that is considered safe or without appreciable risk. Related terms: *Assessment Factor, Safety Factor.*
Validation	Process by which the reliability and relevance of a particular approach, method, process, or assessment is established for a defined purpose. Different parties define "Reliability" as establishing the reproducibility of the outcome of the approach, method, process, or assessment over time. "Relevance" is defined as establishing the meaningfulness and usefulness of the approach, method, process, or assessment for the defined purpose.

Source: Adapted from OECD, *Descriptions of Selected Key Generic Terms Used in Chemical Hazard/Risk Assessment,* ENV/JM/MONO (2003) 15, Paris, 2003.

2 International and Federal Bodies Involved in Risk Assessment of Chemical Substances

The United Nations (UN) is the superior international organization, which, among many other activities, participates in chemical risk assessment. The UN chemical risk assessment activities are carried out via specialized agencies, see Figure 2.1.

2.1 WORLD HEALTH ORGANIZATION

The World Health Organization (WHO) is the UN's specialized agency for health. It was established in 1948. WHO's objective, as set out in its constitution, is "the attainment by all people of the highest possible level of health." Health is defined in WHO's constitution as "a state of complete physical, mental and social well-being" and not merely as the absence of disease or infirmity.

WHO is governed by the UN's 192 Member States through the World Health Assembly. The Health Assembly is composed of representatives from WHO's Member States. The main tasks of the World Health Assembly are to approve the WHO program and the budget for the following biennium and to decide major policy questions (WHO 2006).

2.1.1 INTERNATIONAL PROGRAMME ON CHEMICAL SAFETY

The International Programme on Chemical Safety (IPCS) was developed and structured on the basis of recommendations of the United Nations Conference on the Human Environment (1972). It is a cooperative venture between WHO, United Nations Environment Programme (UNEP), and International Labour Organization (ILO).

UNEP, established in 1972, is the UN agency for the promotion of wise use and sustainable development of the global environment (UNEP 2006).

ILO, founded in 1919, is the UN specialized agency for the promotion of social justice and internationally recognized human and labor rights.

IPCS was established in 1980 and is responsible for implementation of activities related to chemical safety. WHO is the Executing Agency of the IPCS.

The two main roles of the IPCS are to establish the scientific basis for the safe use of chemicals and to strengthen national capabilities and capacities for chemical safety.

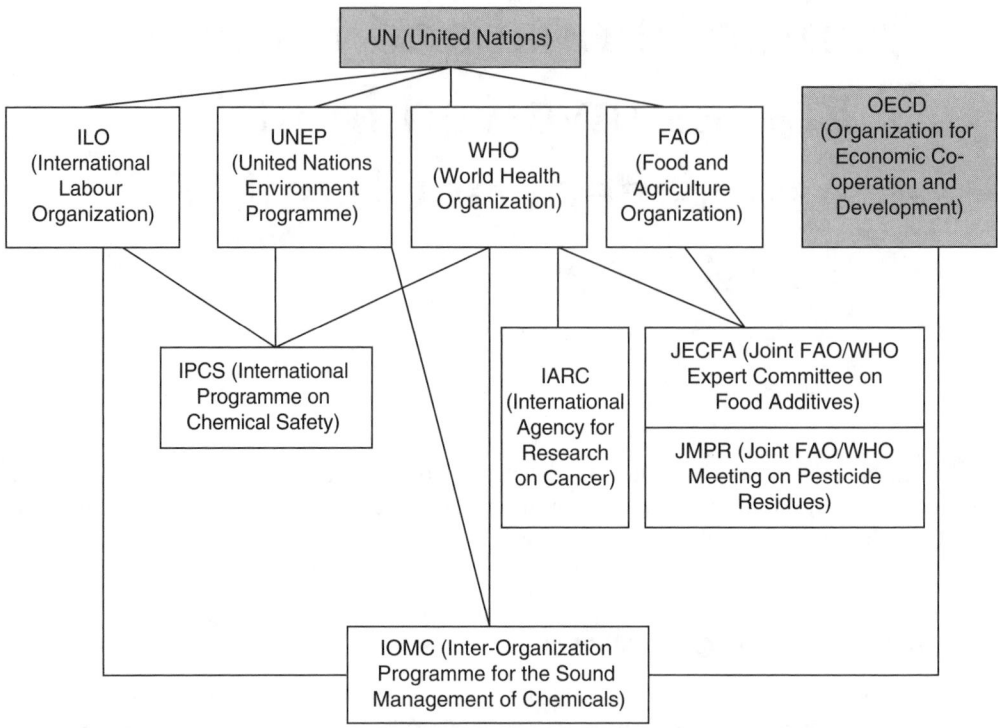

FIGURE 2.1 International bodies involved in chemical risk assessments and their relations.

There are four main elements of work:

1. Evaluation of chemical risks to human health, including:
   * Preparation and publication of chemicals assessments
   * Development and harmonization of scientifically sound methods for chemicals assessment
   * Evaluating the safety of food components, constituents, additives and residues of pesticides and veterinary drugs
2. Poisons information, prevention, and management
3. Chemicals incidents and emergencies
4. Capacity building

IPCS carries out its work in cooperation with national governments, nongovernmental organizations, and through nominated National Focal Points and a network of participating, scientific institutions around the world. One example is the cooperation between IPCS and the Organisation for Economic Co-operation and Development (OECD), under the auspices of the Inter-Organization Programme for the Sound Management of Chemicals (IOMC). Another important relation for IPCS is with the FAO (Food and Agriculture Organization of the United Nations) (IPCS 2006).

2.1.2 FOOD AND AGRICULTURE ORGANIZATION OF THE UNITED NATIONS

The Food and Agriculture Organization of the United Nations (FAO) was founded in 1945 and leads international efforts to defeat hunger. It acts as a neutral forum for agreement negotiation and policy debate and is also a source of knowledge and information. FAO helps developing countries and countries in transition to modernize and improve agriculture, forestry, and fisheries practices and ensure good nutrition for all (FAO 2006).

2.1.3 JOINT FAO/WHO EXPERT COMMITTEE ON FOOD ADDITIVES

The Joint FAO/WHO Expert Committee on Food Additives (JECFA) is an international expert scientific committee that is administered jointly by FAO and WHO. It has been meeting regularly since 1956, initially to evaluate the safety of food additives. Its work now also includes the evaluation of contaminants, naturally occurring toxicants, and residues of veterinary drugs in food. These evaluations are collected in a report published after each JECFA meeting (Section 3.6.1.3) (JECFA 2006).

2.1.4 JOINT FAO/WHO MEETING ON PESTICIDE RESIDUES

The Joint FAO/WHO Meeting on Pesticide Residues (JMPR) is an international expert scientific group that is administered jointly by FAO and WHO. JMPR, which consists of the FAO Panel of Experts on Pesticide Residues in Food and the Environment and the WHO Core Assessment Group, has been meeting regularly since 1963. During the Meetings, the FAO Panel of Experts is responsible for reviewing residue and analytical aspects of the pesticides under consideration, including data on their metabolism, fate in the environment, and use patterns, and for estimating the maximum residue levels (MRLs) that might occur as a result of the use of the pesticides according to good agricultural practices. The WHO Core Assessment Group is responsible for reviewing toxicological and related data and for estimating, where possible, acceptable daily intakes (ADIs) for humans of the pesticides under consideration. These evaluations are collected in a report published after each JMPR meeting (Section 3.6.1.3) (JMPR 2006).

2.1.5 INTER-ORGANIZATION PROGRAMME FOR THE SOUND MANAGEMENT OF CHEMICALS

The Inter-Organization Programme for the Sound Management of Chemicals (IOMC) was established in 1995 to strengthen cooperation and increase coordination in the field of chemical safety. The vision stated by the IOMC is to be the preeminent mechanism for initiating, facilitating, and coordinating international action to achieve the WSSD 2002 (WSSD: World Summit of Sustainable Development, organized by the UN Commission on Sustainable Development) goal for sound management of chemicals.

The seven Participating Organizations of the IOMC are FAO, ILO, OECD, UNEP, the United Nations Industrial Development Organization (UNIDO), the United Nations Institute for Training and Research (UNITAR), and WHO. In addition, two observer organizations are also participating in the IOMC, namely the United Nations Development Programme (UNDP) and the World Bank (IOMC 2006).

2.1.6 INTERNATIONAL AGENCY FOR RESEARCH ON CANCER

The International Agency for Research on Cancer (IARC) was established in May 1965, through a resolution of the 18th World Health Assembly, as an extension of the WHO (IARC 2006).

IARC's founding members were the Federal Republic of Germany, France, Italy, the United Kingdom, and the United States of America. Today, IARC's membership has grown to 18 countries (founding States plus Australia, Belgium, Canada, Denmark, Finland, India, Japan, Norway, the Netherlands, Republic of Korea, Spain, Sweden, and Switzerland). The Agency's headquarter is located in Lyon, France.

IARC's general policy is directed by a Governing Council, composed of the Representatives of Participating States and of the Director-General of the WHO. Its research program is regularly reviewed by a Scientific Council.

IARC activities are mainly funded by the regular budgetary contributions paid by its Participating States. A number of other projects are also funded by extra-budgetary sources, both national and international.

IARC's mission is to coordinate and conduct research on the causes of human cancer, the mechanisms of carcinogenesis, and to develop scientific strategies for cancer control. The Agency is involved in both epidemiological and laboratory research and disseminates scientific information through publications (Section 3.6.1.2), meetings, courses, and fellowships.

The main emphasis of research is on epidemiology, environmental carcinogenesis, and research training. IARC is thus different from all other research institutes, in that it focuses its studies on human cancer and the relationships of man and his environment.

2.1.7 IPCS: CHEMICALS ASSESSMENT

The objective of chemicals assessment is to provide a consensus scientific description of the risks of chemical exposures. These descriptions are published in assessment reports and other related documents so that governments and international and national organizations can use them as the basis for taking preventive actions against adverse health and environmental impacts. For example, the documents are often used as the basis for establishing guidelines and standards for the use of chemicals, and for standards for drinking water, and can assist with the implementation of international agreements such as the Globally Harmonized System of Classification and Labeling of Chemicals (the GHS) (Section 2.5).

Risk assessment activities of IPCS, specifically the production of Concise International Chemical Assessment Documents (CICADs), are supported financially by:

- The Department of Health and Department of Environment, Food and Rural Affairs, United Kingdom
- The Environmental Protection Agency, Food and Drug Administration and National Institute of Environmental Health Sciences, USA
- The European Commission
- The German Federal Ministry of Environment, Nature Conservation and Nuclear Safety
- Health Canada
- The Japanese Ministry of Health, Labour and Welfare
- The Swiss Agency of Environment, Forests and Landscape

The IPCS risk assessment series, available to the public at the INCHEM Web site (INCHEM 2006), include:

- Environmental Health Criteria (EHC)
- Concise International Chemical Assessment Documents (CICADs)
- Health and Safety Guides (HSGs)
- International Chemical Safety Cards (ICSCs)
- Pesticide Safety Data Sheets (PDSs)
- The WHO Recommended Classification of Pesticides by Hazard

The various types of the IPCS risk assessments are described in the following text.

2.1.7.1 Environmental Health Criteria

Two different series of Environmental Health Criteria (EHC) documents are available (Figure 2.2): (i) on specific chemicals, combinations of chemicals, physical factors (e.g., magnetic fields), or biological agents (e.g., *Bacillus thuringiensis*) and (ii) on target organ and tissue toxicity

FIGURE 2.2 Examples of Environmental Health Criteria (EHC) documents published by the WHO's International Programme on Chemical Safety (IPCS).

(e.g., Neurotoxicity Risk Assessment for Human Health: Principles and Approaches) or risk assessment methodologies (e.g., Principles for the Assessment of Risk to Human Health from Exposure to Chemicals). The documents provide international, critical reviews on the above-mentioned issues (see also Section 3.6.1.1).

2.1.7.2   Concise International Chemical Assessment Document

The Concise International Chemical Assessment Documents (CICADs) (see Figure 2.3) are similar to the EHC documents in providing internationally accepted reviews on the effects on human health and the environment of chemicals or combinations of chemicals. They aim to characterize the hazard and dose–response of exposure to chemicals and to provide examples of exposure estimation and risk characterizations for application at the national or local level. They summarize the information considered critical for risk characterization in sufficient detail to allow independent assessment, but are concise, i.e., not repeating all the information available on a particular chemical. For more detail, readers of individual CICADs are referred to the original source document for the CICAD (either a national or regional chemical evaluation document) or an existing EHC (chemicals series).

2.1.7.3   Health and Safety Guide

Health and Safety Guides (HSGs) (see Figure 2.3) provide concise information in nontechnical language, for decision-makers on risks from exposure to chemicals, together with practical advice on medical and administrative issues.

The purpose of an HSG is to facilitate the application of the EHC guidelines for setting exposure limits in national chemical safety programs.

The Guides are intended for occupational health services, and persons in ministries, governmental agencies, industry, and trade unions who are involved in the safe use of chemicals and the avoidance of environmental health hazards.

FIGURE 2.3 Examples of Concise International Chemical Assessment Documents (CICADs) and Health and Safety Guide (HSG) documents published by the WHO's International Programme on Chemical Safety (IPCS).

The first three sections of an HSG highlight the relevant technical information in the corresponding EHC. Section 4 includes advice on preventive and protective measures and emergency action. Within the Guide is a Summary of Chemical Safety Information, which should be readily available, and should be clearly explained, to all who could come into contact with the chemical. The section on regulatory information has been extracted from the legal file of the International Register of Potentially Toxic Chemicals (IRPTC) and from other UN sources.

2.1.7.4   International Chemical Safety Card

The International Chemical Safety Cards (ICSCs) provide essential health and safety information on chemicals to promote their safe use. They are intended for use at the "shop floor" level by workers, but also by other interested parties in factories, agriculture, construction, and other workplaces and often form part of education and training activities.

ICSCs provide information on the intrinsic hazards of specific chemicals together with first aid and firefighting measures, and information about precautions for spillage, disposal, storage, packaging, labeling, and transport. ICSCs have no legal status and may not reflect in all cases the detailed requirements included in national legislation.

ICSCs are prepared by various IPCS Participating Institutions and are finalized in a peer-review meeting.

The information in an ICSC is expressed as far as possible using standard phrases thereby helping to reinforce education and training activities as well as enabling use of computer-aided translation into different languages. A Compiler's Guide is available for download, which provides the collection of standard phrases and criteria for their use.

ICSCs are a potential tool for the implementation of the Globally Harmonized System (GHS) for the labeling and classification of chemicals as they contain the necessary hazard data for this purpose.

2.1.7.5   Pesticide Safety Data Sheet

The Pesticide Safety Data Sheets (PDSs) contain basic information for safe use of pesticides. The PDSs are prepared by WHO in collaboration with FAO and give basic toxicological information on individual pesticides. Priority for development of PDSs is given to substances having a wide use in public health programs and/or in agriculture, or having a high or an unusual toxicity record. The data sheets are prepared by scientific experts and peer reviewed. The comments of industry are provided through the industrial association, Groupement International des Associations Nationales de Fabricants de Produits Agrochemiques (GIFAP). The data sheets are revised from time to time as required.

2.1.7.6   The WHO Recommended Classification of Pesticides by Hazard

This document sets out a classification system to distinguish between the more and the less hazardous forms of selected pesticides based on acute risk to human health (i.e., the risk of single or multiple exposures over a relatively short period of time). It takes into consideration the toxicity of the technical compound and its common formulations.

The document lists common technical grade pesticides and recommended classifications together with a listing of active ingredients believed to be obsolete or discontinued for use as pesticides, pesticides subject to the prior informed consent procedure, limitations to trade because of the Persistent Organic Pollutants (POPs) convention (UN 2001), and gaseous or volatile fumigants not classified under these recommendations.

The present document is regularly updated with new editions. The present edition is from 2004. The present edition, and those before it, complies with original guidelines approved by the World Health Assembly in 1975. In December 2002, the United Nations Committee of Experts on the Transport of Dangerous Goods and on the Globally Harmonized System of Classification and Labeling of Chemicals (UNCETDG/GHS) approved a document called "The Globally Harmonized System of Classification and Labeling of Chemicals" (GHS, Section 2.5) with the intent to provide a globally harmonized system to address classification of chemicals, labels, and safety data sheets. Classification and labeling based on acute toxicity form a part of the GHS, and there are some differences between the GHS and the WHO traditional classification of pesticides by hazard. WHO is in the process of adjusting the Pesticide Classification document to conform to the GHS; the results of this process are expected to be available in the next edition.

2.2   ORGANISATION FOR ECONOMIC CO-OPERATION AND DEVELOPMENT

The Organisation for Economic Co-operation and Development (OECD) was formed to achieve the highest sustainable economic growth and employment and a rising standard of living in Member countries, while maintaining financial stability, and thus to contribute to the development of the world economy; to contribute to sound economic expansion in Member as well as Non-Member countries in the process of economic development; and to contribute to the expansion of world trade on a multilateral, nondiscriminatory basis in accordance with international obligations (OECD 1960).

Twenty countries originally signed the Convention on the Organisation for Economic Co-operation and Development. Since then a further 10 countries have become members of the Organisation. In addition the OECD has active relationships with some 70 other countries, NGOs (nongovernmental organizations), and civil society, and therefore has a global reach. The OECD work covers economic and social issues from macroeconomics to trade, education, development, and science and innovation (OECD 2006a).

The OECD is not a supranational organization, but rather a forum for discussion where governments express their points of view, share their experiences, and search for common ground. More binding decisions can be made through the constitution of a formal OECD Council Act, which is agreed at the highest level of OECD, the Council.

In general, there are two types of Council Acts. A Council Decision, which is legally binding on OECD Member countries, and a Council Recommendation, which is a strong expression of political will. In the area of chemicals, for example, there is a Council Act relating to the Mutual Acceptance of Data (MAD, see below).

Further information can be found at the OECD Web site (OECD 2006a).

2.2.1 THE OECD CHEMICALS PROGRAM

This program works on the development and coordination of environment health and safety activities internationally.

Today, 30 Member countries and the OECD Secretariat work together to develop and coordinate chemical and pesticide related activities on an international basis. Part of working together includes MAD.

The main objectives of the Chemicals Program are to assist OECD Member countries' efforts to protect human health and the environment through improving chemical safety, to make chemical control policies more transparent and efficient and save resources for government and industry, and to prevent unnecessary distortions in the trade of chemicals and chemical products.

While an important focus of the work is on the production, processing, and use of industrial chemicals, some aspects include work on pesticides, chemical accidents, and biotechnology. The work is overseen by the "Joint Meeting" and is undertaken by many subsidiary bodies.

The Chemicals Program produces documents on all aspects of its work. These publications are available to the general public at the OECD Web site (OECD 2006a).

2.2.2 MUTUAL ACCEPTANCE OF DATA

The basis for mutual acceptance of data (MAD) is agreement on the test methods by which to test a chemical, and agreement on a system to ensure high quality and reliability of the data that are generated.

The OECD Council adopted a Decision in 1981 (OECD 1981) stating that data generated in a Member country in accordance with OECD Test Guidelines and Principles of Good Laboratory Practice (GLP) shall be accepted in other Member countries for assessment purposes and other uses relating to the protection of human health and the environment.

The 1981 Council Decision (OECD 1981) sets the policy context agreed by all OECD Member countries which established that safety data developed in one Member country will be accepted for use by the relevant registration authorities in assessing the chemical or product in another OECD country, i.e., the data do not have to be generated a second time for the purposes of safety assessment.

A further Council Act was adopted in 1989 (OECD 1989) to provide safeguards for assurance that the data are indeed developed in compliance with the Principles of GLP. This Council Decision–Recommendation on Compliance with GLP establishes procedures for monitoring GLP compliance through government inspections and study audits as well as a framework for international liaison among monitoring and data-receiving authorities (see also Section 3.3.3). A 1997 Council Decision on the Adherence of Non-Member countries to the Council Acts related to the Mutual Acceptance of Data in the Assessment of Chemicals (OECD 1997) sets out a stepwise procedure for non-OECD countries with a significant chemical industry input to take part as full members in this system.

Many other OECD activities on hazard/risk assessment are undertaken within programs such as Existing Chemicals, New Chemicals, and Pesticides and Biocides, which deal with specific types of chemicals. The work on exposure assessment methods is undertaken by the Task Force on Environmental Exposure Assessment, consisting of experts. Most of the outcome of this work is published in the Series on Testing and Assessment or in Emission Scenario Documents, which are available at the OECD Web site (OECD 2006a).

2.2.3 THE OECD EXISTING CHEMICALS PROGRAM

Since 1988, OECD existing chemicals activities in OECD have centered primarily on the investigation of high production volume (HPV) chemicals. Through a 1990 OECD Council Decision, Member countries decided to undertake the investigation of HPV chemicals in a cooperative way. These HPV chemicals include all chemicals reported to be produced or imported at levels greater than 1000 tons per year in at least one Member country or in the European Union region. The Decision means that Member countries will cooperatively select the chemicals to be investigated; collect characterization, effects, and exposure information from government and public sources; encourage industry to provide information from their files; complete the agreed dossier for the Screening Information Data Set (SIDS) by testing; and make an initial assessment of the potential hazard of each chemical investigated. Member countries share the burden of investigating the chemicals. Each country investigates a proportion of the HPV chemicals. Any chemical on the OECD List of HPV Chemicals can be "sponsored" by an OECD Member country (which thereby becomes the Sponsor country) and thus, all Sponsor countries share the burden of the work on HPV chemicals.

When a full SIDS Dossier on a chemical is available, an initial assessment of the information is undertaken and conclusions are drawn on the potential hazard(s) posed by the chemical and recommendations are made on the need for further work. The conclusions present a summary of the hazards of the chemical, written with sufficient detail and clarity as to be informative and to assist countries with classification work and other hazard-based national decision making; and exposure information to put the hazard information into context (e.g., on use in the Sponsor country). The recommendation, based on these conclusions, can be either that the chemical is currently of low priority for further work or that it is a candidate for further work to clarify its potential risk (e.g., that further information is required to clarify concerns identified in the SIDS process, and that post-SIDS testing is recommended).

The following text explains the process in further detail, see also Table 2.1.

TABLE 2.1

The SIDS Process of the OECD Existing Chemicals Program

Member Country (in cooperation with industry if appropriate)	**OECD**
1. Selection of chemicals from HPV chemicals list	
2. Collection of data	
3. Review of the quality of data, identification of data gaps, preparation of SIDS Dossiers including Robust Study Summaries and SIDS Testing Plans	
4. SIDS Testing as Appropriate	
5. Preparation of SIDS Initial Assessment Report and Profile with Conclusions and Recommendations (SIAR and SIAP)	
	6. SIDS initial assessment and agreement on conclusions on hazard and recommendations at SIDS Initial Assessment Meeting (SIAM)
7. Recommendations: (a) low priority for further work (b) candidate for further work (post-SIDS work)	
	8. Task Force on Existing Chemicals
	9. Joint Meeting
	10. UNEP Chemicals (publication)/IOMC

Once all the data elements of the SIDS for the sponsored chemical (a full SIDS chemical) have been obtained, the SIDS Initial Assessment Report (SIAR) is prepared based on the information in the full SIDS Dossier, including the robust study summaries (Section 2.3.1.4). The SIAR draws conclusions on the potential hazard(s) and recommendations on the need for further work.

The SIAR, which includes evaluations, conclusions, and recommendations and a SIDS Initial Assessment Profile (SIAP, see below), is discussed at the SIDS Initial Assessment Meeting (SIAM), resulting in an internationally agreed assessment for each chemical with agreed conclusions and recommendations.

The SIAM participants include representatives of OECD Member countries, the European Commission, experts nominated by various interested parties, industry, and secretariat staff from OECD, IPCS, and UNEP Chemicals.

The SIAP is a 1–2 page executive summary of the SIAR, which summarizes the rationale for the conclusions and clarifies the recommendations further, if appropriate. The SIAP is used to transmit the conclusions and recommendations of the SIAM on a chemical to the Joint Meeting for endorsement.

In the policy bodies of OECD, Member countries discuss and agree on any follow-up actions on chemicals for which further work is recommended, and indeed, discuss and confirm all conclusions and recommendations made on all chemicals which have undergone an SIDS initial assessment. When full SIDS Dossiers and initial assessment reports are finalized, the results are made available worldwide through the United Nations Environment Programme (UNEP) on Chemicals. A database program, the IUCLID (International Uniform ChemicaL Information Database) software, is the preferred tool for entering data and developing an SIDS Dossier for HPV chemicals in the OECD HPV Chemicals Program. Please see Section 2.4.1.6 and the IUCLID Web site for further information (IUCLID 2006). The SIAPs are available on the publicly accessible OECD HPV database (OECD 2006b).

Work undertaken on a chemical in the OECD HPV Chemicals Program as a follow-up to conclusions and recommendations by SIAM is considered as post-SIDS work, see also Table 2.2. This can include national/regional exposure information gathering and assessment as well as testing of endpoints beyond SIDS to assess a concern identified by SIAM. The Task Force on Existing Chemicals monitors post-SIDS work and can take decisions related to further work to be carried out in OECD in a concerted manner.

The chemical industry supports the OECD activities on HPV chemicals because this work eliminates duplication of efforts to test chemicals to fulfill various national and regional requirements

TABLE 2.2

The Post-SIDS Process of the OECD Existing Chemicals Program

Member Country	OECD

1. Post-SIDS exposure assessment:
  - National/regional priority setting
  - Review of national/regional exposure situations
    if national/regional priority
2. Information exchange on national follow-up or decision
  not to follow-up (monitored by the Secretariat/Task Force)	

	3. Decision of the Task Force
	4. Post-SIDS International Exposure Assessment, Post-SIDS Testing
	5. Post-SIDS Assessment Report
	6. SIAM
	7. Joint IPCS/OECD Risk Assessment, if appropriate

and international commitments. Procedures have been established for close cooperation with the industry in various stages of the Program, which is undertaken in coordination with national, regional, and other international existing chemicals programs.

Risk assessments resulting from the US-EPA and EU programs on chemical risk assessment are fed into the OECD HPV program. The United States is responsible for 25% of almost 300 international HPV chemicals that are now active in the OECD/SIDS testing program, while the other OECD Member countries (including Japan) are contributing the remaining 75% of the work (US-EPA 2006a).

A comprehensive Manual for Investigation of HPV Chemicals is available (OECD 2004). The Manual describes procedures, including the use of electronic discussion groups and the online HPV database; data gathering and testing: SIDS, the SIDS plan, and the SIDS Dossier; data evaluation; initial assessment of data (guidance for assessing the hazards of chemical substances to man and the environment); preparation of the SIAR and SIAP; and post-SIDS work.

For further details, please consult the description of OECD work on investigation of HPV chemicals prepared by the OECD Secretariat based on the agreements reached in the OECD Existing Chemicals Program up to October 2004 (OECD 2006c).

2.2.4 THE OECD NEW CHEMICALS PROGRAM

New industrial chemicals notification and assessment schemes have been established in the majority of OECD Member countries, creating a range of notification and assessment requirements. The OECD has created a *New Industrial Chemicals Information Directory*, which provides readily accessible information about notification and assessment schemes, contact points, sources of information on notification and assessment schemes, and sources of information on specific chemicals for a number of countries.

The *OECD Database on Chemical Risk Assessment Models* includes information on models (computerized or capable of being computerized) that are used by OECD Member governments and industry to predict health or environmental effects (e.g., QSARs), exposure potential, and possible risks. The methods described have not been evaluated or validated by OECD.

2.2.5 THE OECD PESTICIDES AND BIOCIDES PROGRAM

The OECD Pesticides Program was created in 1992 to improve the efficiency and effectiveness of pesticide regulation. The program deals with both chemical pesticides (insecticides, herbicides, fungicides, etc.) and biological pesticides (bacteria, viruses, and predatory insects) used in agriculture. The OECD work concerning antimicrobial pesticides or nonagricultural pesticides (e.g., disinfectants, antifoulants) is managed by the Biocides Program. The Pesticides Program supports three objectives, to:

- Help OECD governments share the work of pesticide registration and reregistration, as the same pesticides are often used in many countries
- Harmonize the data and methods used to test and assess pesticide risks, so as to help governments work together and improve the quality of the data and the rigor of the assessments
- Help OECD governments reduce the risks associated with pesticide use, through a variety of actions to supplement pesticide registration and further reduce the risks that may result even when registered pesticides are used properly

The program is directed by the Working Group on Pesticides (WGP), which is composed primarily of representatives of the 30 OECD governments but also includes representatives of the European Commission and other international organizations, the pesticide industry, and the environmental community. Broad oversight and coordination with other parts of the OECD Chemicals Program is carried out by the Joint Meeting of the Chemicals Committee and the Working Party on Chemicals, Pesticides and Biotechnology.

The WGP is supported by the OECD Secretariat in Paris and three Steering Groups: Registration, Risk Reduction, and Biological Pesticides (OECD 2006a).

2.2.6 OECD TEST GUIDELINES

The OECD Guidelines for the Testing of Chemicals are a collection of the most relevant internationally agreed testing methods used by government, industry, and independent laboratories to characterize potential hazards of new and existing chemical substances and chemical preparations/mixtures. They cover tests for the physico-chemical properties of chemicals, human health effects, environmental effects, and degradation and accumulation in the environment. The test guidelines for human health effects are addressed in more detail in Chapter 4. The adopted test guidelines are listed in Table 2.3 and the draft test guidelines in Tables 2.4 and 2.5.

TABLE 2.3
OECD Test Guidelines, Adopted 2006

401 Acute Oral Toxicity (Deleted Guideline, date of deletion: 20 December 2002)
402 Acute Dermal Toxicity (Updated Guideline, adopted 24 February 1987)
403 Acute Inhalation Toxicity (Original Guideline, adopted 12 May 1981)
404 Acute Dermal Irritation/Corrosion (Updated Guideline, adopted 24 April 2002)
405 Acute Eye Irritation/Corrosion (Updated Guideline, adopted 24 April 2002)
406 Skin Sensitisation (Updated Guideline, adopted 17 July 1992)
407 Repeated Dose 28-day Oral Toxicity Study in Rodents (Updated Guideline, adopted 27 July 1995)
408 Repeated Dose 90-day Oral Toxicity Study in Rodents (Updated Guideline, adopted 21 September 1998)
409 Repeated Dose 90-day Oral Toxicity Study in Non-Rodents (Updated Guideline, adopted 21 September 1998)
410 Repeated Dose Dermal Toxicity: 21/28-Day Study (Original Guideline, adopted 12 May 1981)
411 Subchronic Dermal Toxicity: 90-Day Study (Original Guideline, adopted 12 May 1981)
412 Repeated Dose Inhalation Toxicity: 28-Day or 14-Day Study (Original Guideline, adopted 12 May 1981)
413 Subchronic Inhalation Toxicity: 90-Day Study (Original Guideline, adopted 12 May 1981)
414 Prenatal Developmental Toxicity Study (Updated Guideline, adopted 22 January 2001)
415 One-Generation Reproduction Toxicity Study (Original Guideline, adopted 26 May 1983)
416 Two-Generation Reproduction Toxicity Study (Updated Guideline, adopted 22 January 2001)
417 Toxicokinetics (Updated Guideline, adopted 4 April 1984)
418 Delayed Neurotoxicity of Organophosphorus Substances Following Acute Exposure
     (Updated Guideline, adopted 27 July 1995)
419 Delayed Neurotoxicity of Organophosphorus Substances: 28-Day Repeated Dose Study
     (Updated Guideline, adopted 27 July 1995)
420 Acute Oral Toxicity - Fixed Dose Method (Updated Guideline, adopted 20 December 2001)
421 Reproduction/Developmental Toxicity Screening Test (Original Guideline, adopted 27 July 1995)
422 Combined Repeated Dose Toxicity Study with the Reproduction/Developmental Toxicity Screening Test
     (Original Guideline, adopted 22 March 1996)
423 Acute Oral Toxicity - Acute Toxic Class Method (Updated Guideline, adopted 20 December 2001)
424 Neurotoxicity Study in Rodents (Original Guideline, adopted 21 July 1997)
425 Acute Oral Toxicity: Up-and-Down Procedure (Updated Guideline, adopted 20 December 2001) (Statistical software)
427 Skin Absorption: In Vivo Method (Original Guideline, adopted 13 April 2004)
428 Skin Absorption: In Vitro Method (Original Guideline, adopted 13 April 2004)
429 Skin Sensitisation: Local Lymph Node Assay (Updated Guideline, adopted 24 April 2002)

TABLE 2.3 (continued)
OECD Test Guidelines, Adopted 2006

430 In Vitro Skin Corrosion: Transcutaneous Electrical Resistance Test (TER) (Original Guideline, adopted 13 April 2004)
431 In Vitro Skin Corrosion: Human Skin Model Test (Original Guideline, adopted 13 April 2004)
432 In Vitro 3T3 NRU Phototoxicity Test (Original Guideline, adopted 13 April 2004) (Software: Phototox Version 2.0)
451 Carcinogenicity Studies (Original Guideline, adopted 12 May 1981)
452 Chronic Toxicity Studies (Original Guideline, adopted 12 May 1981)
453 Combined Chronic Toxicity/Carcinogenicity Studies (Original Guideline, adopted 12 May 1981)
471 Bacterial Reverse Mutation Test (Updated Guideline, adopted 21 July 1997)
473 In Vitro Mammalian Chromosomal Aberration Test (Updated Guideline, adopted 21 July 1997)
474 Mammalian Erythrocyte Micronucleus Test (Updated Guideline, adopted 21 July 1997)
475 Mammalian Bone Marrow Chromosomal Aberration Test (Updated Guideline, adopted 21 July 1997)
476 In Vitro Mammalian Cell Gene Mutation Test (Updated Guideline, adopted 21 July 1997)
477 Genetic Toxicology: Sex-Linked Recessive Lethal Test in *Drosophila melanogaster* (Updated Guideline, adopted 4 April 1984)
478 Genetic Toxicology: Rodent Dominant Lethal Test (Updated Guideline, adopted 4 April 1984)
479 Genetic Toxicology: In Vitro Sister Chromatid Exchange Assay in Mammalian Cells (Original Guideline, adopted 23 October 1986)
480 Genetic Toxicology: *Saccharomyces cerevisiae*, Gene Mutation Assay (Original Guideline, adopted 23 October 1986)
481 Genetic Toxicology: *Saccharomyces cerevisiae*, Mitotic Recombination Assay (Original Guideline, adopted 23 October 1986)
482 Genetic Toxicology: DNA Damage and Repair, Unscheduled DNA Synthesis in Mammalian Cells In Vitro (Original Guideline, adopted 23 October 1986)
483 Mammalian Spermatogonial Chromosome Aberration Test (Original Guideline, adopted 21 July 1997)
484 Genetic Toxicology: Mouse Spot Test (Original Guideline, adopted 23 October 1986)
485 Genetic Toxicology: Mouse Heritable Translocation Assay (Original Guideline, adopted 23 October 1986)
486 Unscheduled DNA Synthesis (UDS) Test with Mammalian Liver Cells In Vivo (Original Guideline, adopted 21 July 1997)

Source: Available at the OECD Web site: http://www.oecd.org/document/40/0,3343,en_2649_34377_37051368_1_1_1_1,00.html

Note: Date for the latest revision is given in brackets (). Available to the public free of charge and can be downloaded from the OECD Web site.

2.3   UNITED STATES OF AMERICA

The U.S. Federal bodies involved in chemical risk assessments and their relations are presented in Figure 2.4.

Production and distribution of commercial and industrial chemicals in the United States is covered by the Toxic Substances Control Act (TSCA) of 1976. TSCA was enacted by the U.S. Congress to give the U.S. Environmental Protection Agency (US-EPA) the ability to track the 75,000 industrial chemicals currently produced or imported into the United States. Under TSCA,

TABLE 2.4
OECD Draft Test Guidelines 2006

412 Repeated Dose Inhalation Toxicity: 28-Day or 14-Day Study, Draft Revised Guideline (December 2005)

Source: Available   at   the   OECD   Web   site: http://www.oecd.org/document/40/0,3343.en_2649_34377_37051368_1_1_1_1,00.html

Note: Commenting round; available to the public free of charge and can be downloaded from the OECD Web site.

TABLE 2.5

OECD Draft Test Guidelines 2006

426 Developmental Neurotoxicity Study, Draft New Guideline (September 2003)

433 Acute Inhalation Toxicity, Draft New Guideline (October 2002)

433 Acute Inhalation Toxicity - Fixed Dose Procedure, Draft Revised Guideline (June 2004)

434 Acute Dermal Toxicity - Fixed Dose Procedure, Draft New Guideline (May 2004)

435 In Vitro Membrane Barrier Test Method for Skin Corrosion, Draft New Guideline (May 2004)

436 Acute Inhalation Toxicity - Acute Toxic Class (ATC) Method, Draft New Guideline (December 2004)

487 In Vitro Micronucleus Test, Draft New Guideline (June 2004)

Source: Available at the OECD Web site: http://www.oecd.org/document/40/0,3343.en_2649_34377_37051368_1_1_1_1,00.html

Note: Commenting period has expired, being revised, or finalized; available to the public free of charge and can be downloaded from the OECD Web site.

US-EPA classifies chemical substances as either "existing" chemicals or "new" chemicals. US-EPA repeatedly screens these chemicals and can require reporting or testing of those that may pose an environmental or human health hazard. US-EPA can ban the manufacture and import of those chemicals that pose an unreasonable risk.

Also, US-EPA has mechanisms in place to track the thousands of new chemicals that industry develops each year with either unknown or dangerous characteristics. US-EPA then can control these chemicals as necessary to protect human health and the environment. TSCA supplements

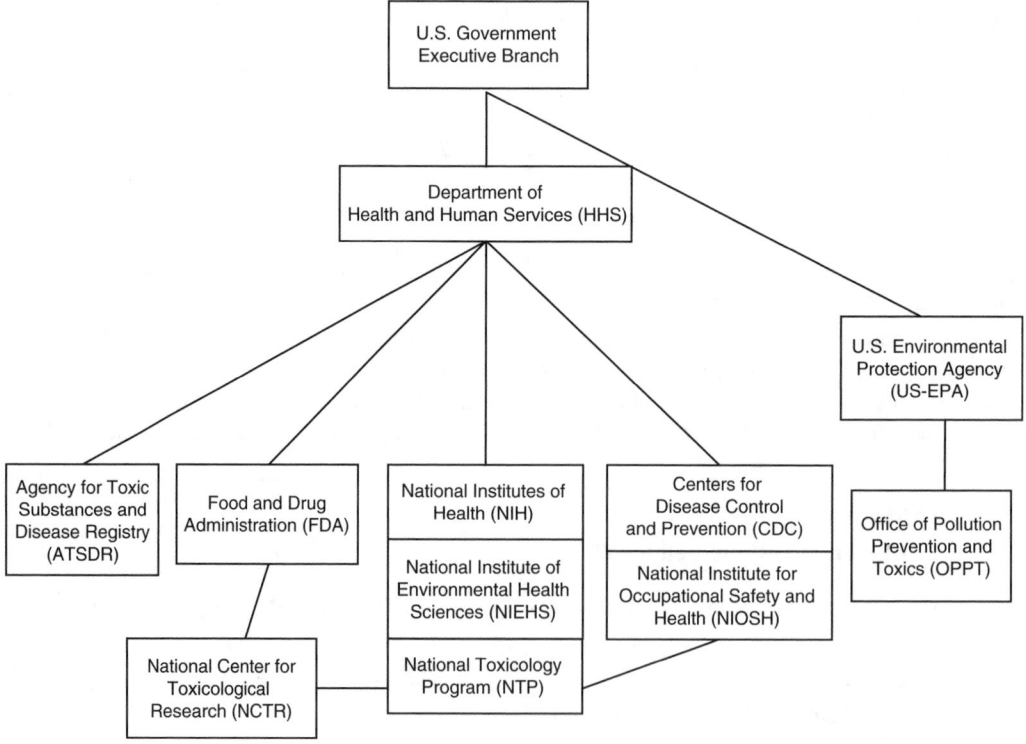

FIGURE 2.4 The U.S. Federal bodies involved in chemical risk assessments and their relations.

other Federal statutes, including the Clean Air Act (Section 9.2.2.1) and the Toxic Release Inventory under the Emergency Planning & Community Right to Know Act (EPCRA).

The TSCA program is federally managed, run by US-EPA and is not delegated to any state agency. The law is overseen by the US-EPA Office of Pollution Prevention and Toxics (OPPT).

2.3.1 UNITED STATES ENVIRONMENTAL PROTECTION AGENCY

The United States Environmental Protection Agency (US-EPA) was established in 1970. The US-EPA is responsible for regulation of chemicals in air, water, and soil (US-EPA 2006a).

2.3.1.1 Office of Pollution Prevention and Toxics

The Office of Pollution Prevention and Toxics (OPPT) was formed in 1977 with the primary responsibility for administering the TSCA. OPPT has the responsibility for assuring that chemicals made available for sale and use in the United States do not pose any adverse risks to human health or to the environment. OPPT's responsibility was expanded still further with the passing of the Pollution Prevention Act of 1990. This act established pollution prevention as the national policy for controlling industrial pollution at its source. In addition, OPPT manages the new Chemical Right-To-Know Initiative, the Design for the Environment (DFE), Green Chemistry programs, and the Lead, Asbestos and Polychlorinated Biphenyls (PCBs) programs (US-EPA 2006b).

2.3.1.2 Toxic Substances Control Act

The U.S. Congress enacted the Toxic Substances Control Act (TSCA) (TSCA, Public Law [Pub. L.] 94–469) in 1976, to become effective 1 January 1977. The act authorizes US-EPA to secure information on all new and existing chemical substances and to control any of these substances that could cause an unreasonable risk to public health or the environment. Drugs, cosmetics, foods, food additives, pesticides, and nuclear materials are exempt from TSCA (US-EPA 2006a).

Under earlier laws US-EPA had the authority to control toxic substances only after damage had occurred. The earlier laws did not require the screening of toxic substances before they entered the marketplace. TSCA closed the gap in the earlier laws by requiring that the health and environmental effects of all new chemicals be reviewed before they are manufactured for commercial purposes.

TSCA has four titles:

Title I - Control of Toxic Substances - includes provisions for testing chemical substances and mixtures, manufacturing and processing notices, regulating hazardous chemical substances and mixtures, managing imminent hazards, and reporting and retaining information.

Title II - Asbestos Hazard Emergency, Title III - Indoor Radon Abatement, and Title IV - Lead Exposure Reduction, deal with regulation of these specific substances.

US-EPA has published a document, which provides an overview of the programs run by the OPPT (US-EPA 2003).

2.3.1.3 Existing Chemicals

The TSCA Inventory provides an overall picture of the organic, inorganic, polymers, and UVCB (chemical substances of Unknown, or Variable Composition, Complex Reaction Products, and Biological Materials) chemicals produced, processed, or imported for commercial purposes in the United States. The Inventory is not a list of chemicals based on toxic or hazardous characteristics, since toxicity/hazard is not a criterion for inclusion in the list. The Inventory includes chemical substances of any commercial use in the United States since 1979 under the Environmental Protection Act, and is prepared by the US-EPA. The current TSCA Inventory contains approximately 81,600 chemicals. Currently, OPPT is focusing on a subset of approximately 3,000 HPV chemicals, which are produced and/or imported in annual volumes of 1 million pounds or more across all U.S. companies. In parallel, the Existing Chemicals Program also focuses on chemicals of concern because of hazardous properties. (US-EPA 2003).

2.3.1.4 High Production Volume Challenge Program

The U.S. HPV chemicals are those which are manufactured in or imported into the United States in amounts equal to or greater than 1 million pounds per year. The U.S. HPV chemicals were identified through information collected under the TSCA Inventory Update Rule (IUR). Organic chemicals that are manufactured in, or imported into, the United States in amounts equal to or exceeding 10,000 pounds per year are subject to reporting under the TSCA IUR. Reporting is required every 4 years.

In 1998, US-EPA's Administrator invited the Chief Executive Officers of more than 900 chemical companies that account for most of the U.S. manufacture and import of HPV chemicals to participate in the HPV Challenge Program. The program encourages chemical manufacturers to voluntarily test chemicals for which little or no health or environmental effects data were publicly available.

A company participates in the HPV Challenge Program by sponsoring the assessment of one or more HPV chemicals. The sponsor company sends a letter to US-EPA announcing its commitment to participate and its willingness to adhere to the Program's procedures. The letter must also specify the names and CAS numbers of the chemicals it will sponsor and the year of the Program in which the sponsor will begin the assessment of each chemical by submitting its robust summary of existing data and its test plan showing how it plans to fill data gaps. The term "robust summary" is used to describe the technical content in electronic summaries of full studies for a particular endpoint. Robust study summaries are intended to provide sufficient information to allow a technically qualified person to make an independent assessment of a given study report without having to go back to the full study report, and to also allow evaluation of the proposed test plan. A "robust study summary" therefore reflects the objectives, methods, results, and conclusions of the full study report, which can either be an experiment or in some cases an estimation or prediction method.

The output of the HPV program is fed into the OECD SIDS program (see Section 2.2.3).

2.3.1.5 New Chemicals

The US-EPA's New Chemicals Program, located in the OPPT, was established to help manage the potential risk from chemicals new to the marketplace. The program is mandated by Section 5 of TSCA.

Chemicals not on the TSCA Inventory are considered "new chemicals" and must go through a review process before they can be marketed. Certain genetically modified microorganisms also come under the definition of new chemicals.

Anyone who plans to manufacture or import a new chemical substance for a nonexempt commercial purpose is required to provide US-EPA with notice before initiating the activity. This premanufacture notice, or PMN, must be submitted at least 90 days prior to the manufacture or import of the chemical. US-EPA can place restrictions up to and including a ban on production, to be placed on the use of a new chemical before it is entered into commerce (US-EPA 2003, 2006c).

2.3.1.6 Integrated Risk Information System

The Integrated Risk Information System (IRIS), prepared and maintained by US-EPA, is an electronic database containing information on human health effects that may result from exposure to various chemicals in the environment. IRIS was initially developed for US-EPA staff in response to a growing demand for consistent information on chemical substances for use in risk assessments, decision making, and regulatory activities. The information in IRIS is intended for those without extensive training in toxicology, but with some knowledge of health sciences (IRIS 2006).

2.3.1.7   Harmonization of US-EPA Risk Assessment

US-EPA has formed the Risk Assessment Forum, a standing committee of senior US-EPA scientists, to promote US-EPA consensus on difficult and controversial risk assessment issues and to ensure that this consensus is incorporated into appropriate US-EPA risk assessment guidance. To fulfill this purpose, the Forum assembles US-EPA risk assessment experts in a formal process to study and report on issues from a US-EPA scientific perspective. Major Forum guidance documents are developed in accordance with the US-EPA's regulatory and policy development process and become US-EPA policy upon approval by the Administrator or the Deputy Administrator. The Risk Assessment Forum products include: risk assessment guidelines, technical panel reports on special risk assessment issues, and peer consultation and peer review workshops addressing controversial risk assessment topics.

The Forum focuses on generic issues fundamental to the risk assessment process and related science policy issues. Forum consensus building activities may include: developing science policy on technical issues, definitive risk assessment guidance, and risk assessment methodology for use in ongoing and prospective US-EPA actions; conducting scientific and technical analysis upon which to base risk assessment positions; and sponsoring colloquia and workshops to foster consensus on risk assessment issues. The Forum does not provide peer review and quality assurance of risk assessments developed by other US-EPA organizations or review nonscientific risk management issues; however, the Forum may undertake special projects at the request of the Deputy Administrator (or delegates) (RAF 2006).

2.3.1.8   US-EPA Risk Assessment Guidelines

US-EPA's Risk Assessment Guidelines set forth recommended principles and procedures to guide US-EPA scientists in assessing the risks from chemicals or other agents in the environment. They also inform US-EPA decision-makers and the public about these procedures.

US-EPA published an initial set of five risk assessment guidelines (relating to cancer, mutagenic effects, developmental effects, exposure assessment, and chemical mixtures) in 1986 as recommended by the National Academy of Sciences. US-EPA continues to revise its risk assessment guidelines and to develop new guidelines as experience and scientific understanding evolve. In 2006 the following health effect guidelines were announced on the Web site of US-EPA's National Center for Environmental Assessment (NCEA): carcinogen risk assessment; chemical mixtures risk assessment; neurotoxicity risk assessment; reproductive toxicity risk assessment; exposure assessment; developmental toxicity risk assessment; and mutagenicity risk assessment. See Table 2.6 for more details (NCEA 2006).

TABLE 2.6
US-EPA Risk Assessment Guidelines

Guidelines for Carcinogen Risk Assessment	2005 Cancer Guidelines and Supplemental Guidance
Guidelines for Chemical Mixtures Risk Assessment	2000 Supplementary guidance and 1986 guidelines, Federal Register 51 (185) 34014–34025, 24 September 1986
Guidelines for Ecological Risk Assessment	Federal Register 63 (93) 26846–26924, 14 May 1998
Guidelines for Neurotoxicity Risk Assessment	Federal Register 63 (93) 26926–26954, 14 May 1998
Guidelines for Reproductive Toxicity Risk Assessment	Federal Register 61 (212) 56274–56322, 31 October 1996
Guidelines for Exposure Assessment	Federal Register 57 (104) 22888–22938, 29 May 1992
Guidelines for Developmental Toxicity Risk Assessment	Federal Register 56 (234) 63798–63826, 5 December 1991
Guidelines for Mutagenicity Risk Assessment	Federal Register 51 (185) 34006–34012, 24 September 1986

Source: National Center for Environmental Assessment (NCEA), 2006. Available at http://cfpub.epa.gov/ncea/index.cfm

The US-EPA staff paper from 2004 titled "An Examination of EPA Risk Assessment Principles and Practices" (US-EPA 2004) is a product of a US-EPA staff review of how risk assessment is conducted at the Agency. US-EPA assembled a group of risk assessment professionals from across US-EPA to examine US-EPA's risk assessment principles and practices and to prepare the paper. The staff paper presents an analysis of US-EPA's general risk assessment practices and provides comprehensive and detailed information on the practices employed.

2.3.1.9 US-EPA Test Guidelines

OPPT has developed a series of test guidelines called the Harmonized Test Guidelines. For health effects, the relevant guidelines are Series 870, Health Effects Test Guidelines. These guidelines are addressed in more detail in Chapter 4. The adopted test guidelines are listed in Table 2.7 and the draft test guidelines are listed in Table 2.8.

TABLE 2.7

US-EPA (OPPT) Harmonized Test Guidelines, Series 870, Health Effects Test Guidelines, Final Guidelines

870.1000 Acute toxicity testing - Background (August 1998)
870.1100 Acute oral toxicity (August 1998)
870.1200 Acute dermal toxicity (August 1998)
870.1300 Acute inhalation toxicity (August 1998)
870.2400 Acute eye irritation (August 1998)
870.2500 Acute dermal irritation (August 1998)
870.2600 Skin sensitization (August 1998)
870.3050 Repeated dose 28-day oral toxicity study in rodents (July 2000)
870.3100 90-Day oral toxicity in rodents (August 1998)
870.3150 90-Day oral toxicity in non-rodent (August 1998)
870.3200 21/28-Day dermal toxicity (August 1998)
870.3250 90-Day dermal toxicity (August 1998)
870.3465 90-Day inhalation toxicity (August 1998)
870.3550 Reproduction/developmental toxicity screening test (July 2000)
870.3650 Combined repeated dose toxicity study with the reproduction/developmental toxicity screening test (July 2000)
870.3700 Prenatal developmental toxicity study (August 1998)
870.3800 Reproduction and fertility effects (August 1998)
870.4100 Chronic toxicity (August 1998)
870.4200 Carcinogenicity (August 1998)
870.4300 Combined chronic toxicity/carcinogenicity (August 1998)
870.5100 Bacterial reverse mutation test (August 1998)
870.5140 Gene mutation in *Aspergillus nidulans* (August 1998)
870.5195 Mouse biochemical specific locus test (August 1998)
870.5200 Mouse visible specific locus test (August 1998)
870.5250 Gene mutation in *Neurospora crassa* (August 1998)
870.5275 Sex-linked recessive lethal test in *Drosophila melanogaster* (August 1998)
870.5300 In vitro mammalian cell gene mutation test (August 1998)
870.5375 In vitro mammalian chromosome aberration test (August 1998)
870.5380 Mammalian spermatogonial chromosomal aberration test (August 1998)
870.5385 Mammalian bone marrow chromosomal aberration test (August 1998)
870.5395 Mammalian erythrocyte micronucleus test (August 1998)
870.5450 Rodent dominant lethal assay (August 1998)
870.5460 Rodent heritable translocation assays (August 1998)
870.5500 Bacterial DNA damage or repair tests (August 1998)

TABLE 2.7 (continued)

US-EPA (OPPT) Harmonized Test Guidelines, Series 870, Health Effects Test Guidelines, Final Guidelines

870.5550 Unscheduled DNA synthesis in mammalian cells in culture (August 1998)

870.5575 Mitotic gene conversion in *Saccharomyces cerevisiae* (August 1998)

870.5900 In vitro sister chromatid exchange assay (August 1998)

870.5915 In vivo sister chromatid exchange assay (August 1998)

870.6100 Acute and 28-day delayed neurotoxicity of organophosphorus substances (August 1998)

870.6200 Neurotoxicity screening battery (August 1998)

870.6300 Developmental neurotoxicity study (August 1998)

870.6500 Schedule-controlled operant behavior (August 1998)

870.6850 Peripheral nerve function (August 1998)

870.6855 Neurophysiology sensory evoked potentials (August 1998)

870.7200 Companion animal safety (August 1998)

870.7485 Metabolism and pharmacokinetics (August 1998)

870.7600 Dermal penetration (August 1998)

870.7800 Immunotoxicity (August 1998)

870.8355 Combined chronic toxicity/carcinogenicity testing of respirable fibrous particles (July 2001)

Source: Available at the US-EPA Web site: http://www.epa.gov/opptsfrs/publications/OPPTS_Harmonized/870_Health_Effects_Test_Guidelines/Series/.

Note: The month for the latest revision is given in brackets ().

The harmonized test guidelines have been developed for use in the testing of pesticides and toxic substances, and the development of test data that must be submitted to the US-EPA for review under Federal regulations. The OPPT guidelines are based on the OECD test guidelines (Section 2.2.6); however, minor deviations may occur. The guidelines are available for download at the OPPT Web site (US-EPA 2006d).

TABLE 2.8

US-EPA (OPPT) Harmonized Test Guidelines, Series 870, Health Effects Test Guidelines - Public Drafts

870.3500 Preliminary developmental toxicity screen (June 1996)

870.3600 Inhalation developmental toxicity study (June 1996)

870.8223 Pharmacokinetic (June 1996)

870.8245 Dermal pharmacokinetics of DGBE and DGBA (June 1996)

870.8300 Dermal absorption for compounds that are volatile and metabolized to carbon dioxide (June 1996)

870.8320 Oral/dermal pharmacokinetics (June 1996)

870.8340 Oral and inhalation pharmacokinetic test (June 1996)

870.8360 Pharmacokinetics of isopropanol (June1 996)

870.8380 Inhalation and dermal pharmacokinetics of commercial hexane (June 1996)

870.8500 Toxicokinetic test (June 1996)

870.8600 Developmental neurotoxicity screen (June 1996)

870.8700 Subchronic oral toxicity test (June 1996)

870.8800 Morphologic transformation of cells in culture (June 1996)

Note: The list contains the guidelines that have not been finalized but only exist as drafts.

2.3.2 Agency for Toxic Substances and Disease Registry

The Agency for Toxic Substances and Disease Registry (ATSDR), based in Atlanta, Georgia, is a federal public health agency of the U.S. Department of Health and Human Services. ATSDR's functions include public health assessments of waste sites, health consultations concerning specific hazardous substances, health surveillance and registries, response to emergency releases of hazardous substances, applied research in support of public health assessments, information development and dissemination, and education and training concerning hazardous substances. ATSDR produces "Toxicological Profiles" for hazardous substances found at National Priorities List (NPL) sites. These hazardous substances are ranked based on frequency of occurrence at NPL sites, toxicity, and potential for human exposure. The Toxicological Profiles (Section 3.6.1.6) are developed from a priority list of 275 substances. (ATSDR 2006).

2.3.3 National Toxicology Program

The National Toxicology Program (NTP) was established in 1978. The program was created as a cooperative effort to coordinate toxicology testing programs within the U.S. Federal government, strengthen the science base in toxicology, develop and validate improved testing methods, provide information about potentially toxic chemicals to health, regulatory, and research agencies, scientific and medical communities, and the public. The NTP is an interagency program whose mission is to evaluate agents of public health concern (NTP 2006).

Three agencies form the NTP:

- National Institute of Environmental Health Sciences of the National Institutes of Health (NIEHS/NIH)
- National Institute for Occupational Safety and Health of the Centers for Disease Control and Prevention (NIOSH/CDC)
- National Center for Toxicological Research of the Food and Drug Administration (NCTR/FDA)

The NTP is located administratively at the NIEHS/NIH and receives advice on its activities from three external groups: The NTP Executive Committee, the NTP Board of Scientific Counselors, and the Scientific Advisory Committee on Alternative Toxicological Methods. The NTP also uses special emphasis panels for independent scientific peer review and advice on targeted issues.

The NTP Executive Committee contains members from ATSDR, CPSC (Consumer Product Safety Commission), US-EPA, US-FDA (Food and Drug Administration), NCEH (National Center for Environmental Health)/CDC, NCI/NIH, NIH, NIOSH/CDC, and OSHA (Occupational Safety & Health Administration).

The NTP designs and conducts studies to characterize and evaluate the toxicological potential of selected chemicals in laboratory animals (usually two species, rats and mice). Chemicals selected for NTP toxicology and carcinogenesis studies are chosen primarily on the basis of human exposure, level of production, and chemical structure. NTP studies include long-term toxicology and carcinogenesis studies, short-term toxicity studies, genetically modified model studies, immunology toxicity studies, developmental toxicity studies, special drinking water studies of water disinfection by-products, and studies of reproductive assessment by continuous breeding (RACB assay).

2.3.4 United States Food and Drug Administration

The United States Food and Drug Administration (US-FDA) originated in June 1906, when President Roosevelt signed the Food and Drugs Act, which was to be implemented by the Bureau of Chemistry of the U.S. Department of Agriculture. The Bureau, the oldest U.S. consumer protection office, eventually became the FDA. The US-FDA is an agency of the U.S. Department of Health and Human Services, and is responsible for regulating food, drugs, medical devices, biologics, animal feed and drugs, cosmetics, radiation-emitting products, and combination products in the United States. The US-FDA derives its authority and jurisdiction from various Congressional acts, the main source being the Federal Food, Drug, and Cosmetic Act. US-FDA employs some 9000 people who work in locations around the United States (US-FDA 2006a).

US-FDA performs premarket review of laboratory, animal, and human clinical testing for new human drugs and biologics, complex medical devices, food and color additives, and animal drugs for safety and effectiveness. Infant formulas are reviewed for required nutrient content and safety in manufacturing practices.

The centers of US-FDA publish guidance documents, which represent the Agency's current thinking on a particular subject without committing the US-FDA in any legally binding way. The guidance documents are available from the Web sites of the individual centers of the US-FDA, i.e., the comprehensive list of current Food and Cosmetic Guidance Documents is available from the home page of the Centre for Food Safety and Applied Nutrition, while the list of drug Guidance Documents is available from the Web site of the Centre for Drug Evaluation and Research (CDER).

With respect to toxicity testing of food additives, guidance is given in the so-called "*Redbook*." The *Redbook* is the popular name for the guidance document "Toxicological Principles for the Safety Assessment of Direct Food Additives and Color Additives Used in Food." The original *Redbook* was published in 1982 (*Redbook I*) and subsequently revised in 1993 (draft *Redbook II*). The current version is "*Redbook 2000*" (US-FDA 2006b).

2.4 THE EUROPEAN UNION

The European Union (EU) consists of 25 European countries, committed to working together for peace and prosperity (EU 2006a). Its political system is founded on a series of treaties. Under these treaties, the Member States of the EU delegate some of their national sovereignty to institutions they share, which represent not only national interests of individual Member States, but also their collective interest. The treaties constitute the so-called "primary" legislation. A large body of "secondary" legislation is derived from the treaties, which mainly consists of regulations, directives, and recommendations. These laws, along with the EU policies in general, are the result of decisions taken by three main institutions:

- The Council of the European Union (the Council), representing the Member States, is the main decision-making institution and consists of one minister from the national governments of each EU country. Which minister depends on which issues are on the agenda for the respective meetings. The Council shares the legislative power as well as the responsibility for the budget with the Parliament. The Council also concludes on international agreements that have been negotiated by the Commission.
- The European Parliament is the elected body that represents the EU's citizens and is, together with the Council, responsible for the legislative process and for the budget. The present parliament, elected in 2004, has 732 members.
- The European Commission is one of the EU's key institutions and represents and upholds the interests of Europe as a whole (EU 2006b). The Commission acts with complete political independence. It has 25 members, the so-called commissioners - one from each EU country, and each commissioner has responsibility for a particular EU policy area. The Commission must ensure that the legislations adopted by the Council and Parliament are being put into effect. The Commission is also the only institution that has the right to propose new EU legislation. The Commission is assisted by a civil service made up of 36 Directorates-General (DGs) and services.

The "secondary" legislation consists mainly of regulations, directives, and recommendations. The purpose of the regulations is the unification of the law and thus, encroach furthest on the national legal systems as they are mandatory, i.e., apply in full in all Member States, and have direct applicability, i.e., do not have to be implemented into national law.

In contrast to the regulations, the purpose of the directives is harmonization of the law. A directive is binding on the Member States as regards the objective to be achieved but leaves it to the national authorities to decide on how the agreed objective is to be incorporated into their national legal systems. That means that a directive does not supersede the laws of the Member States but places the Member States under an obligation to adapt their national law in line with the EU rules.

Recommendations enable the EU institutions to express a view to Member States, which is not binding and does not place any legal obligation on the Member States.

Further information can be found at the EU Web site (EU 2006a,b).

2.4.1 EU CHEMICALS PROGRAM

The EU policy on chemical substances aims to provide an appropriate balance between protection of human health and environment from undesirable exposure to chemicals and the need for continued chemical innovation so that the citizens of the EU can realize the potential benefits (EU 2006c).

Important definitions used in EU legislations on chemicals include:

*   Chemicals: A general term that covers both substances and preparations (EU 2001).
*   Substances: Substances are chemical elements and their compounds in the natural state or obtained by any production process, including any additive necessary to preserve the stability of the product and any impurity deriving from the process used, but excluding any solvent which may be separated without affecting the stability of the substance or changing its composition. While ingredients of pesticides, biocides, pharmaceuticals, or cosmetics might be included in this definition, intentional mixtures or preparations of them for final use would not (EU 2001).
*   Preparations: Intentional mixtures or solutions composed of two or more substances (EU 2001). They are governed by Directive 88/379/EEC (EEC 1988), recently replaced by Directive 1999/45/EC (EC 1999).
*   Dangerous: Substances and preparations are "dangerous" if they are explosive, oxidizing, flammable, toxic, harmful, corrosive, or irritant (EEC 1967) (see Section 2.4.1.8. for details).
*   Existing substances: Substances which are in use within the EU on or before 18 September 1981 and listed in EINECS (EU 2001) (see Section 2.4.1.4 for details).
*   New substances: Substances which are not in use in the EU on or before 18 September 1981 and so not in EINECS (EU 2001) (see Section 2.4.1.5 for details).

2.4.1.1 Actors Involved in Chemical Legislation

Several Commission DGs as well as the Joint Research Center and the European Chemicals Bureau are actors involved in the EU chemical legislation, see Figure 2.5.

2.4.1.1.1 Commission DGs Involved in Chemicals Legislation
There are several Commission DGs and services with different objectives involved in the field of chemicals legislation in general. The main players are the following five DGs:

The Environment Directorate-General (DG Environment) is responsible for the legislation regarding marketing and use of chemical substances in general (EU 2006d). The following areas fall under the management of DG Environment: (i) the new chemical legislation REACH (Registration, Evaluation, and Authorisation of CHemicals) (Section 2.4.1.3), (ii) classification of dangerous substances (Section 2.4.1.8), (iii) existing substances (Section 2.4.1.4), (iv) plant protection

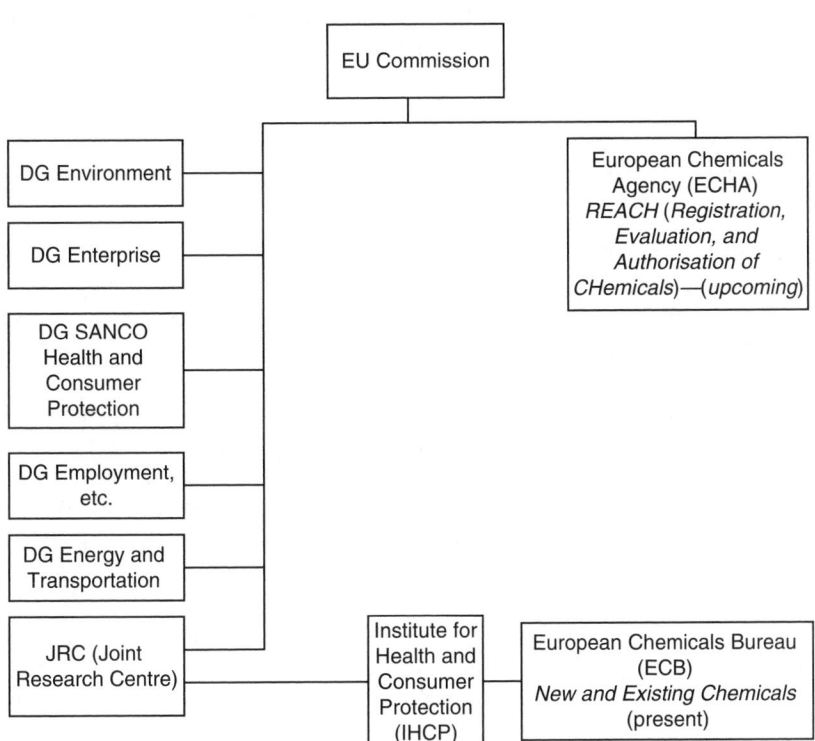

FIGURE 2.5 The European Union institutions involved in chemicals regulation.

products (Section 2.4.1.7.2), (v) biocidal products (Section 2.4.1.7.1), (vi) dioxin exposure and health, (vii) endocrine disrupters, (viii) volatile organic compounds, (ix) chemical accidents, and (x) persistent organic pollutants (POPs).

The Enterprise and Industry Directorate-General (DG Enterprise) is responsible for the legislations in the field of industrial products (EU 2006e). The following areas fall under the management of DG Enterprise: (i) restrictions on the marketing and use of certain dangerous substances and preparations, (ii) classification of dangerous preparations, (iii) detergents, (iv) fertilizers, (v) drug precursors, (vi) explosives for civil uses, (vii) cosmetics, and (viii) pharmaceuticals. DG Enterprise provides the "*Pink Book*," which is a structured list of the legislation for which the DG Enterprise and Industry is responsible. The *Pink Book* is updated twice a year, the most recent edition covers the period up to 31 December 2005 (EU 2005).

The Health and Consumer Protection Directorate-General (DG SANCO) is responsible for the legislations regarding chemicals in food and consumer products (EU 2006f).

The Employment, Social Affairs and Equal Opportunities Directorate-General is responsible for the legislations regarding health and safety at work and thus on controlling chemicals in the workplace (EU 2006g).

The Energy and Transport Directorate-General is responsible for the legislations regarding transport of dangerous goods including transport of dangerous chemicals (EU 2006h).

2.4.1.1.2 Joint Research Centre

The Joint Research Centre (JRC), also a Commission DG, provides the Commission and its policy-making DGs, as well as the Council, the European Parliament, and the Member States, with independent scientific and technical advice (JRC 2006).

The work program is organized into four core areas, three of which relate to vertical sectors: (i) food, chemical products and health, (ii) environment and sustainability, and (iii) nuclear activities, while the fourth covers more general horizontal activities.

JRC is structured into seven institutes, one of these being the Institute for Health and Consumer Protection (IHCP). The IHCP activities are related to genetically modified organisms, biotechnology, chemicals and risk assessment, nano-biotechnology, exposure to environmental stressors, food contact materials and consumer products, and alternative methods to animal testing (IHCP 2006).

2.4.1.1.3   European Chemicals Bureau

IHCP has established five scientific units, one of these being the European Chemicals Bureau (ECB). The ECB is the focal point for collecting information on new and existing chemicals (ECB 2006). Main partners of the ECB are the authorities of the EU Member States and Norway, Commission services such as DG Environment and DG Enterprise, the chemical industry, and NGOs.

The ECB provides scientific and technical support to the conception, development, implementation, and monitoring of EU policies on dangerous chemicals. The ECB coordinates the EU risk assessment programs covering the risks posed by new and existing substances to workers, consumers, and the environment, and supports the legal classification and labeling of substances, the notification of new substances, the authorization of biocides, the development and harmonization of testing methods and (Q)SARs models, and the information exchange on import and export of dangerous substances.

In addition, the ECB is currently providing support to the implementation of the new chemical legislation REACH. In particular, the ECB is coordinating the process of developing guidance documents, software tools, and infrastructure, enabling a smooth start of REACH for all stakeholders.

2.4.1.2   Current EU Regulatory Framework for Chemicals

Many different directives and regulations set up the current EU regulatory framework for chemicals. The most important ones are mentioned below.

In 1967, Council Directive 67/548/EEC (EEC 1967), the so-called 67-Directive, was adopted in order to approximate the national provisions relating to dangerous substances. The Directive introduced common provisions on the classification of dangerous chemical substances to ensure the establishment of a common market in the field of such substances and a high level of protection of human health. The 67-Directive is permanently updated to take account of the scientific and technical progress in the field of dangerous substances. Until today it has been amended 9 times and adapted to technical progress 29 times.

Protecting the environment from the effects of dangerous chemical substances was introduced with the 6th amendment, adopted in 1979 (EEC 1979).

The 6th amendment also introduced a notification system for "new" substances (Section 2.4.1.5) and, consequently, required the establishment of the list of "existing" substances called EINECS (Section 2.4.1.4).

The 7th amendment (EEC 1992) essentially required that the principles of risk assessment for "new" substances be laid down (Section 2.4.1.5).

In addition to the 67-Directive relating to dangerous substance, Council Directive 88/379/EEC (EEC 1988) was adopted in 1988 in order to approximate the national provisions relating to dangerous preparations, recently replaced by Directive 1999/45/EC (EC 1999).

In the 1980s, a Community Action Programme underlined the need for a legislative instrument, which would provide a comprehensive structure for the evaluation of the risks posed by existing substances (Section 2.4.1.4). The Commission proposed a series of legal instruments, one of these being the Council Regulation (EEC) 793/93 on the evaluation and control of the risks of existing substances (EEC 1993a), which came into force on 4 June 1993.

2.4.1.3 New EU Regulatory Framework for Chemicals

The present EU legislative framework for chemical substances is a patchwork of many different Directives and Regulations, which has developed historically and there are different rules for existing and new chemicals (EU 2006i). The "White Paper on the Strategy for a Future Chemicals Policy" (EU 2001), adopted on 13 February 2001, addressed the shortcomings of the current system. The most important drawbacks identified were (i) 100,106 existing substances can be used without testing, (ii) burden of proof on public authorities, (iii) no efficient instrument to ensure safe use of the most problematic substances, and (iv) lack of incentives for innovation, in particular of less hazardous substitutes. The White Paper presented the Commission's proposals for a strategy on future chemicals policy in the Community with the overriding goal of sustainable development.

On 29 October 2003, the Commission adopted a proposal for a new EU regulatory framework for chemicals, the so-called REACH (Registration, Evaluation, and Authorisation of CHemicals). This new regulatory framework achieved all the objectives identified in the White Paper (EU 2001) and thus represents a model of sustainable development by pursuing its three main goals: economic (industrial competitiveness), social (health protection and jobs), and environmental.

Following 2 years of negotiation on the Commission's original proposal, and following the European Parliament's first reading opinion adopted on 17 November 2005, the Council reached a common position on 27 June 2006.

The REACH Regulation was formally adopted on 18 December 2006 by the Council of Ministers following the vote in second reading of the European Parliament on 13 December 2006. REACH entered into force on 1 June 2007.

To enable a smooth transition from the existing chemicals legislation to REACH, the Commission has developed an interim strategy. Within the interim strategy a number of REACH Implementation Projects (RIPs) have been initiated covering the following tasks: (i) setting up the new European Chemicals Agency (ECHA), (ii) preparing the information technology support for the ECHA, and (iii) preparing detailed technical guidance documents for industry and Member States on different REACH elements.

The planned timescale for the implementation of REACH is as follows:

- 2004–June 2007 Interim period: During this period different preparatory actions will take place in order to be able to effectively and efficiently administer the new chemicals legislation when it enters into force.
- June 2007–June 2008 ECHA start-up phase: During this period the ECHA will be set up in Helsinki and will start its preparations for the period when the actual registrations will come in. The activities and committees of the ECB will be transferred to the ECHA and ECHA staff will be trained to take up their duties.
- June 2008–2017 ECHA fully operational: According to the Regulation the ECHA will be at the center of running the REACH system.

REACH mainly refers to four important legal instruments on chemicals currently in force in the EU:

- Dangerous substances - Council Directive 67/548/EEC (EEC 1967) relating to the classification, packaging, and labeling of dangerous substances
- Dangerous preparations - Council Directive 88/379/EEC (EEC 1988) relating to the classification, packaging, and labeling of dangerous preparations
- Existing substances - Council Regulation 793/93/EEC (EEC 1993a) relating to the evaluation and control of the risks of existing substances
- Council Directive 76/769/EEC (EEC 1976) relating to restriction on the marketing and use of certain dangerous substances and preparations

The two most important aims of REACH are to improve the protection of human health and the environment from the risks of chemicals while enhancing the innovative capability and competitiveness of the EU chemicals industry.

The REACH proposal gives greater responsibility to industry to manage the risks from chemicals and to provide safety information on the substances. Industry will be required to gather information on the properties of their substances, which will help industry manage them safely, and to register the information in a central database. Thus, the responsibility for testing and assessing the risks of chemicals will be transferred from the competent authorities to the industry, i.e., the manufacturers, importers, and downstream users.

REACH is very wide in its scope covering all substances (see definition of "substances" in Section 2.4.1) whether manufactured, imported, used as intermediates or placed on the market, either on their own, in preparations or in articles, unless they are radioactive, subject to customs supervision, or are nonisolated intermediates. Waste is specifically exempted. Food is not subject to REACH, as it is not a substance, preparation, or article. Member States may exempt substances used in the interests of defense. Other substances are exempted from parts of REACH, where other equivalent legislation applies; for further details refer EU (2006i).

The REACH system is a single system for both existing and new substances. Its basic elements are:

- All substances are covered by the regulation unless they are explicitly exempted from its scope.
- Registration requires manufacturers and importers of chemicals to obtain relevant information on their substances and to use that data to manage them safely.
- To reduce testing on vertebrate animals, data sharing is required for studies on such animals. For other tests, data sharing is required on request.
- Better information on hazards and risks and how to manage them will be passed down and up the supply chain.
- Downstream users (i.e., producers and importers of products containing a given chemical substance) are brought into the system.
- Evaluation is undertaken by the ECHA (see below) to evaluate testing proposals made by industry or to check compliance with the registration requirements. The ECHA will also coordinate substance evaluation by the authorities to investigate chemicals with perceived risks. This assessment may be used later to prepare proposals for restrictions or authorization.
- Substances with properties of very high concern will be made subject to authorization; the ECHA will publish a list containing such candidate substances. Applicants will have to demonstrate that risks associated with uses of these substances are adequately controlled or that the socioeconomic benefits of their use outweigh the risks and there are no suitable alternative substitute substances or technologies.
- Restrictions provide a procedure to regulate that the manufacture, placing on the market, or use of certain dangerous substances shall be either subject to conditions or otherwise adequately controlled.
- The European Chemicals Agency (ECHA) will manage the technical, scientific, and administrative aspects of the REACH system at Community level, aiming to ensure that REACH functions well and has credibility with all stakeholders.
- A classification and labeling inventory of dangerous substances will help promote agreement within industry on classification of a substance. For some substances of high concern there may be a Community-wide harmonization of classification by the authorities.
- Access to information rules combines a system of publicly available information over the Internet, the current system of requests for access to information, and REACH specific rules on the protection of confidential business information.

More information on REACH can be obtained from a publication "REACH in brief" by DG Enterprise and DG Environment available at the Web site of DG Environment (EU 2006i).

2.4.1.4 Existing Substances

Existing substances are substances deemed to be on the EU Market between 1 January 1971 and 18 September 1981 and listed in EINECS, the European INventory of Existing Commercial Chemical Substances (EU 2006j, ECB 2006). The substances placed on the market for the first time after this target date are "new" (Section 2.4.1.5).

EINECS is a closed list containing 100,106 entries and counts for about 99% of the chemicals' volume on the market. EINECS include chemical substances produced from natural products by chemical modifications or purification, such as metals, minerals, cement, refined oil, and gas; substances produced from animals and plants; active substances of pesticides, medicaments, fertilizers, and cosmetic products; food additives; a few natural polymers; and some waste and by-products. They can be mixtures of different chemicals occurring naturally or as an unintentional result of the production process.

Existing substances do not include: Synthetic polymers (which are registered in EINECS under their building block monomers); medical, cosmetic, and pesticide preparations as intentional mixtures; food; feedstuffs; alloys, such as stainless steel (but individual components of alloys are included); and most naturally occurring raw materials, including coal and most ores.

Chemical substances are treated differently in the Community, depending on when they were introduced on the market. In contrast to new substances (Section 2.4.1.5), there are at present no routine testing requirements for existing substances.

The concern regarding the potential risks of existing substances was already a policy priority in the late 1980s. In 1993, the Council adopted Council Regulation (EEC) 793/93 or the ESR (Existing Substances Regulation) (EEC 1993a) thereby introducing a comprehensive framework for the evaluation and control of existing substances.

According to the ESR (EEC 1993a), evaluation and control of the risks posed by existing substances is carried out in four steps: (i) data collection, (ii) priority setting, (iii) risk assessment, and (iv) risk reduction.

The Existing Chemicals Work Area of the ECB is responsible for the scientific and technical support to the ESR (EEC 1993a) regarding the first three steps.

2.4.1.4.1 Data Collection

For existing substances, the data collection consists of three phases. The ESR was initially concerned with the so-called HPVCs (High Production Volume Chemicals). HPVCs are those substances, which are covered by the data collection phases I and II of the ESR, i.e., those substances which have been imported or produced in quantities exceeding 1000 tons per year and produced/imported between March 23 1990 and March 23 1994. During phase I, 1884 substances were extracted from EINECS - referred to as the HPVC list; these substances are listed in Annex I of ESR. The total list of substances reported under phases I and II of the Regulation is now referred to as the EU-HPVC list.

In phase III of the data collection step, companies which produce or import existing substances in quantities between 10 and 1000 tons per year (LPVCs or Low Production Volume Chemicals) were required to submit a reduced data set by 4 June 1998.

All the data have to be submitted in a specific electronic format HEDSET (Harmonised Electronic DataSET) and is managed by the IUCLID (Section 2.4.1.6). All companies, which have submitted a data set in any of the three data collection phases are required to update the information, at least once every 3 years.

2.4.1.4.2   Priority Setting

The Commission, in consultation with Member States, is required to regularly draw up priority lists of substances or groups of substances requiring immediate attention because of their potential effects on man or the environment. Since 1994, four priority lists have been adopted containing 141 substances.

2.4.1.4.3   Risk Assessment

Substances on the priority lists must undergo an in-depth risk assessment covering the risks posed by the priority chemical to man (covering workers, consumers, and man exposed via the environment) and the environment (covering the terrestrial, aquatic, and atmospheric eco-systems and accumulation through the food chain). Emissions and environmental impact and human exposures at each stage of the life cycle of a chemical (from production, through processing, formulation, and use, to recycling and disposal) are normally covered in the risk assessments.

This risk assessment follows the framework set out in the Commission Regulation (EC) 1488/94 (EC 1994) and implemented in the detailed Technical Guidance Documents (TGD) on Risk Assessment for New and Existing Substances (EC 2003).

For both human health and the environment, the risk assessment process includes (i) an exposure assessment, (ii) an effect assessment (hazard assessment and hazard characterization - addressed in detail in Chapter 4), and (iii) a risk characterization (addressed in detail in Chapter 8). As a part of the effect assessment, classification and labeling of the substance according to the criteria laid down in Directive 67/548/EEC (EEC 1967) is also addressed (Section 2.4.1.8).

The risk characterization for a toxicological endpoint with an effect threshold generally relies on the dose or the concentration below which adverse effects are unlikely to occur. Usually the No-Observed-Adverse-Effect Level (NOAEL) serves as a direct measure for the threshold dose or concentration. The risk characterization is carried out by quantitatively comparing the outcome of the effects assessment to the outcome of the exposure assessment. The ratio resulting from this comparison is called a Margin of Safety (MOS): MOS is equal to NO(A)EL/Exposure. The next step is to derive and apply a so-called "reference MOS" (MOSref), which is an overall assessment factor addressing differences between experimental effect assessment data (usually from animal studies) and the real human exposure situation, taking into account variability and uncertainty. For all relevant combinations of exposed human (sub)populations and toxicological endpoints, scenario-specific MOS values should be evaluated using the corresponding MOSref values.

The possible conclusions of a risk assessment for existing substances are (i) there is need for further information and/or testing, (ii) there is at present no need for further information and/or testing and no need for risk reduction measures beyond those, which are being applied already, or (iii) there is a need for limiting the risks (risk reduction measures, which are already being applied shall be taken into account).

The first draft of the risk assessment reports are written by the Member States, which act as "rapporteurs". Generally, one Member State acts as rapporteur for a prioritized substance or group of substances; however, for some prioritized substances, more than one Member State can act as rapporteurs. The risk assessment process is coordinated by the ECB. Stakeholders are involved in the process through the Technical Committee for New and Existing Substances (TC NES). The Commission mediates the meetings, which attempt to reach consensus on the conclusions of the risk assessments. During the risk assessment process, the Scientific Committee on Health and Environmental Risks (SCHER) is requested to provide an opinion.

After adoption of the risk assessment, three publications are produced: (i) the comprehensive risk assessment report (as a book, on the ECB Web site and in IUCLID), (ii) a summary thereof (as an EUR report and on the ECB Web site), and (iii) a listing of the conclusions in the *Official Journal of the European Communities*.

As a part of the risk assessment, classification and labeling of the substance according to the criteria laid down in Directive 67/548/EEC (EEC 1967) is also addressed (Section 2.4.1.8).

2.4.1.4.4   Risk Reduction

When the outcome of a risk assessment is conclusion (iii), a risk reduction strategy must be developed. The elaboration of such a strategy is done utilizing the Technical Guidance Documents on Risk Reduction (EC 2003). A risk reduction strategy might involve restrictions on marketing and use of dangerous substances and preparations or other relevant existing Community instruments. Directive 76/769/EEC on the restrictions in marketing and use of dangerous substances (EEC 1976), is one of the legal frameworks, which could be invoked to manage the risks identified by the risk assessment. Also other tools for risk reduction may be used, such as voluntary agreements or economic instruments. With regard to regulatory control through other relevant existing Community instruments, a large number of directives are available as tools for the reduction of risks posed by chemicals.

More information on existing substances can be obtained from the ECB Web site (ECB 2006) under the heading "Existing Chemicals."

2.4.1.5   New Substances

New substances are substances, which were not on the EU Market on or before 18 September 1981 and therefore not listed in EINECS (ECB 2006, EU 2006j).

New substances are required to be tested and notified before marketing in volumes above 10 kg. On the basis of the information, they are assessed on their risks to human health and the environment.

2.4.1.5.1   Notification

A harmonized pan-European notification system, i.e., covering the entire EU, was introduced for new substances as part of the 6th Amendment to the 67-Directive (EEC 1967), which was adopted in September 1979 (Directive 79/831/EEC - EEC 1979) and came into force in all Member States on 18 September 1981. Over 6800 notifications in total, representing more than 4300 substances, have been submitted since 1981. Inherent of legislation are principles for notification, including criteria for exemption. Exemption categories include consumer products pertaining to pharmaceuticals, cosmetics, and foodstuffs. The Directive is not applicable to pesticides, radioactive materials, wastes, and substances used in scientific research (ECB 2006).

Besides details on the notifier/manufacturer and the identity of the chemical, a technical dossier for a new substance notification should provide information on the substance as such (e.g., production process, proposed uses) and results from analysis of physico-chemical properties as well as test reports from toxicological and ecotoxicological assays. Proposals for classification and labeling should be submitted, including recommended precautions relating to safety. Requisite dossier detail increases according to substance quantity to be marketed: 10, 100, 1000 kg per year per manufacturer (Annexes VIIC, VIIB, VIIA (base-set), respectively, to Directive 92/32/EEC - EEC 1992), i.e., further toxicological and ecotoxicological testing is required when the quantity of a substance to be marketed exceeds 100 and 1000 tons per year per manufacturer (Annex VIII to Directive 92/32/EEC).

New substances are registered in ELINCS (European LIst of Notified Chemical Substances). ELINCS is published in the *Official Journal of the European Union*. A 5th edition of ELINCS, compiling substances notified until 30 June 1995, is the last available update published in the Official Journal. A 6th edition, comprehensive of substances notified until 30 June 1998, and published in all EU languages as a Commission document, is available online via Europa EUR-Lex Web site: http://eur-lex.europa.eu/, located under "Documents of Public Interest" with reference: COM(2003) 642, dated 29.10.2003. ELINCS is currently maintained in English only, and published only as an Internet document on the ECB Web site (ECB 2006), where the latest update replaces previous versions.

2.4.1.5.2   Risk Assessment

The 7th Amendment (Directive 92/32/EEC - EEC 1992) to the 67-Directive (EEC 1967), adopted in April 1992 with effect from November 1993, introduced a risk assessment for new notified substances. Legal principles of risk assessment of new chemical substances are laid down in Directive 93/67/EEC (EEC 1993b). Detailed guidance, in relation to both human health and environment, is given in the Technical Guidance Document (TGD) on Risk Assessment for New and Existing Substances (EC 2003). The risk assessment process for new substances is similar to that for existing substances (Section 2.4.1.4.3).

For every endpoint investigated, the risk assessment for new substances assigns one of four available conclusions: (i) the substance is of no immediate concern and need not be considered again until further information is made available in accordance with the requirements of Directive 67/548/EEC, (ii) the substance is of concern and the Competent Authority will define information required to refine the assessment and request that it is supplied when the quantity of the substance placed on the market reaches the next supply threshold, (iii) the substance is of concern and the Competent Authority will request that defined information is supplied without further delay, and (iv) the substance is of concern and the Competent Authority will immediately make recommendations for risk reduction.

2.4.1.5.3   Risk Reduction

When the outcome of a risk assessment is conclusion (iv), a risk reduction strategy must be developed. The elaboration of such a strategy for new substances is similar to that for existing substances (Section 2.4.1.4.4).

More information on new substances can be obtained from the ECB Web site (ECB 2006) under the heading "New Chemicals."

2.4.1.6   International Uniform Chemical Information Database

IUCLID (International Uniform Chemical Database) is the basic tool for data collection and evaluation within the EU risk assessment program for existing substances (IUCLID 2006).

IUCLID includes all data sets submitted by industry according to the ESR (EEC 1993a) (Section 2.4.1.4). IUCLID is operational in all EU Member State Authorities involved in the ESR Program and is used to generate the so-called HEDSET (Harmonised Electronic DataSET) with the aim to prepare risk assessments. The data structure has been designed to describe the effects of substances on human health and the environment, in close collaboration between Member States, Industry, and the ECB.

At present, IUCLID at the ECB contains 30,000 dossiers for approximately 10,500 different substances. In addition, a nonconfidential IUCLID CD-ROM is available, a first version in 1996 and a revised version in 2000.

In 1999, IUCLID has also been accepted by the OECD as the data exchange tool under the OECD Existing Chemicals Program (Section 2.2.3). Within the OECD Program, IUCLID is used to generate the so-called Screening Information Data Sets (SIDS), with the aim to prepare hazard assessments for the substances under investigation.

2.4.1.7   Biocides and Plant Protection Products

2.4.1.7.1   Biocides Legislation

A biocide is generally speaking a chemical substance capable of killing different forms of living organisms used in fields such as medicine, agriculture, forestry, and mosquito control. A biocide can be (i) a pesticide, which includes fungicides, herbicides, insecticides, algicides, moluscicides, miticides, and rodenticides or (ii) an antimicrobial, which includes germicides, antibiotics, antibacterials, antivirals, antifungals, antiprotoas, and antiparasites.

The European Parliament and the Council adopted Directive 98/8/EC (EC 1998) in 1998 concerning the placing of biocidal products on the EU Market (Biocidal Product Directive, BPD) (EU 2006d, ECB 2006).

According to Article 2.1(a), biocidal products are defined as "active substances and preparations containing one or more active substances, put up in the form in which they are supplied to the user, intended to destroy, deter, render harmless, prevent the action of, or otherwise exert a controlling effect on any harmful organism by chemical or biological means." And an active substance is defined as "a substance or micro-organism including a virus or a fungus having general or specific action on or against harmful organisms."

The Directive covers 23 different product types specified in Annex V of the Directive. These include disinfectants used in different areas, chemicals used for preservation of products and materials, nonagricultural pesticides, and antifouling products used on hulls of vessels. The Directive will not apply to certain product types already covered by other community legislation, such as plant protection products (PPPs) (Section 2.4.1.7.2), medicines, and cosmetics.

The Directive requires an authorization process for biocidal products containing active substances listed in "positive lists" (Annexes I, IA, and IB of the Directive 98/8/EC). In relation to biocides, existing active substances are substances, which have been on the EU Market for biocidal purposes before 14 May 2000. An active substance, which was not on the market before 14 May 2000 is regarded as a new active substance and has to be approved by the Member States before it will be entered on the positive lists.

The notification procedure of existing active substances, which had started in 2000, was finalized on 31 January 2003. After having succeeded the notification procedure, the names of existing active substances have been published in the *Official Journal of the European Communities* (OJ). At the next step, the full dossier including all test reports should be submitted to the Rapporteur Member State in agreed Data Formats including the risk assessment of the active substance. At the last step, the Rapporteur Member State in cooperation with all other Member States will decide whether the active substance will be entered onto the positive lists or not.

More information on biocides can be obtained from the ECB Web site under the heading "Biocides" (ECB 2006).

2.4.1.7.2 Plant Protection Products Legislation

The European Community has developed a very comprehensive regulatory framework, which regulates the marketing and use of PPPs and their residues in food (EU 2006d,f). Essentially, PPPs are the pesticides, e.g., fungicides, herbicides, and insecticides, aimed at protecting plants.

Directive 91/414/EEC (EEC 1991) concerning the placing of PPPs on the market defines strict rules for the authorization of PPPs.

According to Article 2, PPPs are defined as active substances and preparations containing one or more "active substances, put up in the form in which they are supplied to the user, intended to (i) protect plants or plant products against all harmful organisms or prevent the action of such organisms, insofar as such substances or preparations are not otherwise defined, (ii) influence the life processes of plants, other than as a nutrient (e.g., growth regulators), (iii) preserve plant products, insofar as such substances or products are not subject to special Council of Commission provisions on preservatives, (iv) destroy undesired plants, or (v) destroy parts of plants, check, or prevent undesired growth of plants." And an active substance is defined as "a substance or micro-organism including viruses, having general or specific action (i) against harmful organisms or (ii) on plants, parts of plants, or plant products."

According to the Directive, an active substance cannot be used in a PPP unless it is included in an EU positive list. The Directive also requires very extensive risk assessments for effects on health and environment to be carried out, before a PPP can be placed on the market and used. In 1992, the European Commission started a Community-wide review process for all active substances used in PPPs within the EU. Based on scientific assessments, each applicant had to prove that a substance could be used safely regarding human health, the environment, ecotoxicology and residues in the food chain. This program will be completed by 2008. From the end of 2003, the European Food Safety Authority (EFSA) deals with risk assessment issues and the European Commission retains the risk management decision. Further details are available at the DG SANCO Web site (EU 2006f) under the heading "Food and Feed Safety/Plant Health/Plant Protection/Evaluation & Authorisation."

Pesticides residues in food are regulated by four Council Directives and a Regulation, which consolidates and amends the four Directives. The legislation covers the setting, monitoring, and control of pesticides residues in products of plant and animal origin that may arise from their use in plant protection. Further details are available at the DG SANCO Web site (EU 2006f) under the heading "Food and Feed Safety/Plant Health/Plant Protection/Pesticides Residues."

2.4.1.7.3   Risk Assessment

For biocides, the Technical Guidance Document (EC 2003) as well as the Technical Notes for Guidance (TNsG) should be used as a basis for the risk assessment. Further details on these Guidance Documents are available at the ECB Web site (ECB 2006) under the heading "Biocides."

For PPPs, the standards of the risk assessment are addressed in a number of guidance documents. Further details are available at the DG SANCO Web site (EU 2006f) under the heading "Food and Feed Safety/Plant Health/Plant Protection/Evaluation & Authorisation."

The principles for the risk assessment of active substances intended for use in biocides or in PPPs as laid down in Directive 98/8/EC concerning biocides (EC 1998) or in Directive 91/414/EEC concerning PPPs (EEC 1991) are very similar.

The risk assessment comprises an effect assessment (hazard identification and hazard characterization) and an exposure assessment. The principles for the effect assessment of the active substances are in principle similar to those for existing and new chemicals and are addressed in detail in Chapter 4. Based on the outcome of the effect assessment, an Acceptable Daily Intake (ADI) and an Acceptable Operator Exposure Level (AOEL) are derived, usually from the NOAEL by applying an overall assessment factor addressing differences between experimental effect assessment data (usually from animal studies) and the real human exposure situation, taking into account variability and uncertainty; for further details the reader is referred to Chapter 5. As a part of the effect assessment, classification and labeling of the active substance according to the criteria laid down in Directive 67/548/EEC (EEC 1967) is also addressed (Section 2.4.1.8).

2.4.1.8   Classification and Labeling of Chemical Substances

Classification and labeling involves an evaluation of the hazard of a substance or preparation according to Directive 67/548/EEC (substances) (EEC 1967) or 1999/45/EC (preparations) (EC 1999) and a communication of that hazard via the label (EU 2006c, ECB 2006). Such an evaluation must be made for any substance (new and existing chemical, pesticide, and biocide) or preparation manufactured within or imported into the EU and placed on the EU Market, irrespective of the quantity placed on the market. The evaluation may result in a classification of the substance or preparation as dangerous for one or several endpoints concerning physico-chemical properties, health, or environmental effects. The labeling is the first and often the only information on the hazards of a chemical that reaches the user. Classification and labeling is therefore a useful tool for risk management of chemicals. In addition the classification has a large number of downstream consequences within the EU legislation (for details, see the ECB Web site (ECB 2006) under the heading "Classification & Labelling").

Currently there are 15 classes of danger in Directive 67/548/EEC, such as "explosive," "very toxic," "carcinogenic," or "dangerous for the environment." The Directive includes a list of substances classified as dangerous (Annex I), danger symbols such as a skull with crossed bones underneath (Annex II), standard phrases on the nature of special risks from substances (R-phrases) (Annex III), and the wording of safety precaution phrases (S-phrases) relating to the handling and use of dangerous substances (Annex IV). Annex V contains testing methods to determine the dangerous properties of substances, Annex VI provides detailed criteria on the proper choice of the class of danger and on how to assign the danger symbols, R-, and S-phrases to a tested substance.

The Classification and Labeling Work Area of the ECB provides technical and scientific support to Member State Authorities and Commission Services (mainly DG Environment - Chemicals) concerning the hazard classification and labeling of chemical substances. ECB is responsible for the Technical Committee on Classification and Labelling of Dangerous Substances (TC C&L). In this Committee also experts from the concerned industry, from the trade unions, and from Norway (as a member of the European Area Agreement) are participating. The Committee discusses classification and labeling of substances of special concern, which are then proposed for entry to the list of harmonized classifications of substances, i.e., Annex I of Directive 67/548/EEC, which is legally binding. In addition, the Committee develops the classification criteria for substances and preparations as set up in Annex VI of the Directive.

Annex I, the published list of substances with a harmonized classification and labeling at present contains approximately 2700 existing and 1100 new substance entries (covering approximately 8000 substances) for which classification and labeling have been agreed at Community level. If a dangerous substance not yet included in Annex I is put on the market (as such or contained in a preparation), manufacturers/importers/distributors have to self-classify it according to the criteria in Annex VI. The classification and labeling of preparations subject to Directive 1999/45/EC also uses the criteria set up in Annex VI.

An important element of classification and labeling is the specific concentration limits displayed for a range of substances in Annex I to Directive 67/548/EEC. They relate to dangerous preparations. If a preparation contains a dangerous substance in a concentration that exceeds the specific limit in Annex I, the preparation is classified as dangerous. If no specific limit is given, the "general" limits laid down in the Directive are applicable. The TC C&L has considered that for carcinogens the establishment of these specific concentration limits should also take account of the potency (or power) of a carcinogen to trigger cancer. This element is not yet included in Directive 67/548/EEC.

Current work on the new EU legislation for chemicals, REACH, as well as the implementation of a "Globally Harmonized System of Classification and Labeling of Chemicals" (GHS) (Section 2.5) within the EU will lead to a major revision of the EU classification and labeling system.

2.4.1.9   EU Test Guidelines

The legally binding EU standardized testing methods to determine the hazardous properties of chemicals are contained in Annex V of Directive 67/548/EEC (EEC 1967) on the Classification, Packaging and Labelling of Dangerous Substances (ECB 2006). For health effects, the relevant methods are Part B, Methods for the determination of toxicity. These guidelines are addressed in more detail in Chapter 4. The adopted test guidelines, which are listed in Table 2.9, are available for download free of charge at the ECB Web site under the heading "Testing-Methods" (ECB 2006).

The standardized testing methods play a central role in the EU policy on chemicals control and they are referred to in many other pieces of EU legislation (e.g., those related to dangerous preparations, pesticides, cosmetics, and biocides also refer to these methods). The methods are based on those recognized and recommended by competent international bodies, in particular OECD (Section 2.2). When such methods were not available, national standards or scientific consensus methods have been adopted.

2.4.2   Harmonization of Risk Assessment in DG SANCO

Scientific procedures for risk assessment include assessment of risk for human health as well as risk for the environment. A substantial part of the EU risk assessment work was in 1997 delegated to the DG SANCO, in relation to the scandal surrounding BSE (bovine spongiform encephalopathy or "mad cow disease"). Risk assessment work not under DG SANCO includes pharmaceuticals, working environment, and health effects caused by lifestyle factors such as diet, smoking, and alcohol consumption (EU 2006f).

TABLE 2.9
EU Test Guidelines Part B: Methods for the Determination of Toxicity

B.1 bis Acute Oral Toxicity - Fixed Dose Procedure (2004)

B.1 tris Acute Oral Toxicity - Acute Toxic Class Method (2004)

B.2 Acute Toxicity (Inhalation) (1992/93)

B.3 Acute Toxicity (Dermal) (1992)

B.4 Acute Toxicity: Dermal Irritation/Corrosion (2004)

B.5 Acute Toxicity: Eye Irritation/Corrosion (2004)

B.6 Skin Sensitisation (1996)

B.7 Repeated Dose (28 days) Toxicity (Oral) (1996)

B.8 Repeated Dose (28 days) Toxicity (Inhalation) (1992)

B.9 Repeated Dose (28 days) Toxicity (Dermal) (1992)

B.10 Mutagenicity - In Vitro Mammalian Chromosome Aberration Test (2000)

B.11 Mutagenicity - In Vivo Mammalian Bone Marrow Chromosome Aberration Test (2000)

B.12 Mutagenicity Mammalian Erythrocyte Micronucleus Test (2000)

B.13/14 Mutagenicity - Reverse Mutation Test Using Bacteria (2000)

B.15 Gene Mutation - *Saccharomyces cerevisiae* (1988)

B.16 Mitotic Recombination - *Saccharomyces cerevisiae* (1988)

B.17 Mutagenicity - In Vitro Mammalian Cell Gene Mutation Test (2000)

B.18 DNA Damage and Repair - Unscheduled DNA Synthesis - Mammalian Cells In Vitro (1988)

B.19 Sister Chromatid Exchange Assay In Vitro (1988)

B.20 Sex-Linked Recessive Lethal Test in *Drosophila melanogaster* (1988)

B.21 In Vitro Mammalian Cell Transformation Test (1988)

B.22 Rodent Dominant Lethal Test (1988)

B.23 Mammalian Spermatogonial Chromosome Aberration Test (2000)

B.24 Mouse Spot Test (1988)

B.25 Mouse Heritable Translocation (1988)

B.26 Sub-Chronic Oral Toxicity Test: Repeated Dose 90-Day Toxicity Study in Rodents (2001)

B.27 Sub-Chronic Oral Toxicity Test: Repeated Dose 90-Day Toxicity Study in Non-Rodents (2001)

B.28 Sub-Chronic Dermal Toxicity Test: 90-Day Repeated Dermal Dose Study Using Rodent Species (1988)

B.29 Sub-Chronic Inhalation Toxicity Test: 90-Day Repeated Inhalation Dose Study Using Rodent Species (1988)

B.30 Chronic Toxicity Test (1988)

B.31 Teratogenicity Test - Rodent and Non-Rodent (2004)

B.32 Carcinogenicity Test (1988)

B.33 Combined Chronic Toxicity/Carcinogenicity Test (1988)

B.34 One-Generation Reproduction Toxicity Test (1988)

B.35 Two-Generation Reproduction Toxicity Test (2004)

B.36 Toxicokinetics (1988)

B.37 Delayed Neurotoxicity of Organophosphorus Substances Following Acute Exposure (1996)

B.38 Delayed Neurotoxicity of Organophosphorus Substances 28-Day Repeated Dose Study (1996)

B.39 Unscheduled DNA Synthesis (UDS) Test with Mammalian Liver Cells In Vivo (2000)

B.40 Skin Corrosion (In Vitro) (2000)

B.41 Phototoxicity - In Vitro 3T3 NRU Phototoxicity Test (2000)

B.42 Skin Sensitisation: Local Lymph Node Assay (2004)

B.43 Neurotoxicity Study in Rodents (2004)

Source: European Chemicals Bureau. Available at http://ecb.jrc.it/testing-methods/

Note: The date for the latest revision is given in brackets ().

The DG SANCO system of Scientific Committees established in 1997 included a Scientific Steering Committee and eight specific committees. Risk assessment responsibilities, previously carried out by the Steering Committee and the five scientific committees in the field of food and feed safety and animal health and welfare, were transferred to the EFSA in May 2003. DG SANCO retained responsibility for the three remaining nonfood Scientific Committees. Following a review of the work of these Committees and an assessment of the needs of the EU Commission for independent scientific advice, the Commission in 2004 replaced these Committees with three new Scientific Committees, including the Scientific Committee on Consumer Products (SCCP), the Scientific Committee on Health and Environmental Risks (SCHER), and the Scientific Committee on Emerging and Newly Identified Health Risks (SCENIHR).

The present three Scientific Committees under DG SANCO provide the Commission with the scientific advice it needs when preparing policy and proposals relating to consumer safety, public health, and the environment. The Committees also draw the Commission's attention to the new or emerging problems, which may pose an actual or potential threat. An Inter-Committee Coordination Group coordinates the work of the three Scientific Committees, including matters relating to the harmonization of risk assessment. In addition, it deals with questions, which are common to more than one Committee, diverging scientific opinions and exchange of information on the activities of the Committees. The responsibilities of the three Scientific Committees are described below.

2.4.2.1   Scientific Committee on Health and Environmental Risks

The Scientific Committee on Health and Environmental Risks (SCHER) deals with questions relating to examinations of the toxicity and ecotoxicity of chemicals, biochemicals, and biological compounds whose use may have harmful consequences for human health and the environment. In particular, the Committee addresses questions in relation to new and existing chemicals, the restriction and marketing of dangerous substances, biocides, waste, environmental contaminants, plastic, and other materials used for water pipe work (e.g., new organic substances), drinking water, indoor and ambient air quality. The Committee also addresses questions relating to human exposure to mixtures of chemicals, sensitization, and identification of endocrine disrupters (EU 2006f).

2.4.2.2   Scientific Committee on Consumer Products

The Scientific Committee on Consumer Products (SCCP) handles questions concerning the safety of consumer products (nonfood products intended for the consumer). In particular, the Committee addresses questions in relation to the safety and allergenic properties of cosmetic products and ingredients with respect to their impact on consumer health, toys, textiles, clothing, personal care products, domestic products such as detergents, and consumer services such as tattooing (EU 2006f).

2.4.2.3   Scientific Committee on Emerging and Newly Identified Health Risks

The Scientific Committee on Emerging and Newly Identified Health Risks (SCENIHR) deals with questions concerning emerging or newly identified risks, and broad, complex, or multidisciplinary issues requiring a comprehensive assessment of risks to consumer safety or public health, and related issues not covered by other community risk assessment bodies (EU 2006f).

2.4.2.4   EFSA Committees

In the EU until 2003, the European Commission's Scientific Committee for Food (SCF) performed safety evaluations of food additives and contaminants. This task has now been taken over by the EFSA. EFSA's risk assessments and other scientific work are undertaken by its Scientific Committee and nine scientific Panels, each responsible for a different aspect of food and feed safety. The scientific work is also supported by external Scientific Expert Working Groups, each specializing in a specific subject (EFSA 2006).

The following panels exist:

- Panel on food additives, flavorings, processing aids, and materials in contact with food
- Panel on animal health and welfare

- Panel on biological hazards
- Panel on contaminants in the food chain
- Panel on additives and products or substances used in animal feed
- Panel on genetically modified organisms
- Panel on dietetic products, nutrition, and allergies
- Panel on plant health
- Panel on PPPs and their residues

A review of risk assessment in European traditional food has recently been published (Larsen 2006).

2.5 GLOBALLY HARMONIZED SYSTEM OF CLASSIFICATION AND LABELING OF CHEMICALS

The need to develop national programs to ensure the safe use, transport, and disposal of chemicals has grown in recent years and it has been recognized that an internationally harmonized approach to classification and labeling of chemicals would provide the foundation for such programs (UNECE 2006).

The new system, which is called "Globally Harmonized System of Classification and Labeling of Chemicals (GHS)," addresses classification of chemicals by types of hazard and proposes harmonized hazard communication elements, including labels and safety data sheets. It aims at ensuring that information on physical hazards and toxicity from chemicals be available in order to enhance the protection of human health and the environment during the handling, transport, and use of these chemicals. The GHS also provides a basis for harmonization of rules and regulations on chemicals at national, regional, and worldwide level, an important factor also for trade facilitation.

The work about the elaboration of the GHS began with the premise that existing systems should be harmonized in order to develop a single, globally harmonized system to address classification of chemicals, labels, and safety data sheets. The work was coordinated and managed under the auspices of the WHO Interorganization Programme for the Sound Management of Chemicals (IOMC) Coordinating Group for the Harmonization of Chemical Classification Systems (CG/HCCS). The technical focal points for completing the work were (i) ILO for the hazard communication, (ii) OECD for the classification of health and environmental hazards, and (iii) the United Nations Sub-Committee of Experts on the Transport of Dangerous Goods (UNSCETDG) and the ILO for the physical hazards.

The first version of the GHS was adopted in 2002 and The Plan of Implementation of the World Summit on Sustainable Development (WSSD), adopted in Johannesburg in 2002, encouraged countries to implement the GHS as soon as possible with a view to having the system fully operational by 2008. A first revised edition was published in 2005 and is currently the most recent version. The GHS is now ready for worldwide implementation and has already started with pilot countries introducing the system in their national practices in different regions of the world.

More information on GHS can be obtained from the United Nations Economic Commission for Europe Web site (UNECE 2006).

2.6 PRECAUTIONARY PRINCIPLE

A generally accepted definition of the "Precautionary Principle" has never been brought forward. According to a popular definition (Wikipedia 2006) "The precautionary principle is a moral and political principle which states that if an action or policy might cause severe or irreversible harm to the public, in the absence of a scientific consensus that harm would not ensue, the burden of proof falls on those who would advocate taking the action. The precautionary principle is most often applied in the context of the impact of human actions on the environment and human health, as both involve complex systems where the consequences of actions may be unpredictable."

The formal concept evolved out of the German socio-legal tradition in the 1930s, centering on the concept of good household management (O'Riordan and Cameron 1995). In German the concept is *vorsorgeprinzip*, which translates into English as *precaution principle*.

At international level, the precautionary principle was first recognized in the World Charter for Nature, adopted by the UN General Assembly in 1982 (EU 2000). It was subsequently incorporated into various international conventions on the protection of the environment.

In 2000, the European Commission issued a communication on the precautionary principle (EU 2000), in which it adopted a procedure for the application of this concept. The precautionary principle is not defined in the treaty, which prescribes it only once - to protect the environment. But in practice, its scope is much wider, beyond that of environmental policy, and, specifically where preliminary objective scientific evaluation, indicates that there are reasonable grounds for concern that the potentially dangerous effects on the environment, human, animal, or plant health may be inconsistent with the high level of protection chosen for the Community. The precautionary principle is implemented, for example, in the EU food law and also affects, among others, policies relating to consumer protection, trade and research, and technological development.

While a comprehensive definition of the precautionary principle was never formally adopted by the EU, a working definition and implementation strategy for the EU context has been proposed in Fisher et al. (2006): "Where, following an assessment of available scientific information, there are reasonable grounds for concern for the possibility of adverse effects but scientific uncertainty persists, provisional risk management measures based on a broad cost/benefit analysis whereby priority will be given to human health and the environment, necessary to ensure the chosen high level of protection in the Community and proportionate to this level of protection, may be adopted, pending further scientific information for a more comprehensive risk assessment, without having to wait until the reality and seriousness of those adverse effects become fully apparent."

The issue of when and how to use the precautionary principle, both within the EU and internationally, is giving rise to much debate, and to mixed, and sometimes contradictory views. Thus, decision-makers are constantly faced with the dilemma of balancing the freedom and rights of individuals, industry, and organizations with the need to reduce the risk of adverse effects to the environment, human, animal, or plant health. Therefore, finding the correct balance so that the proportionate, nondiscriminatory, transparent, and coherent actions can be taken, requires a structured decision-making process with detailed scientific and other objective information.

REFERENCES

ATSDR. 2006. The United States Agency for Toxic Substances and Disease Registry (ATSDR) website. http://www.atsdr.cdc.gov/

EC. 1994. Commission Regulation (EC) No 1488/94 of 28 June 1994 laying down the principles for the assessment of risks to man and the environment of existing substances in accordance with Council Regulation (EEC) No 793/93. *Off. J. Eur. Communities* L 161, 29.6.1994, pp. 3–11.

EC. 1998. Directive 98/8/EC of the European Parliament and of the Council of 16 February 1998 concerning the placing of biocidal products on the market. *Off. J. Eur. Communities* L 123, 24.4.1998, pp. 1–63.

EC. 1999. Directive 1999/45/EC of the European Parliament and of the Council of 31 May 1999 concerning the approximation of the laws, regulations and administrative provisions of the Member States relating to the classification, packaging and labelling of dangerous preparations. *Off. J. Eur. Communities* L 200, 30.7.1999, pp. 1–68.

EC. 2003. Technical Guidance Document in support of Commission Directive 93/67/EEC on Risk Assessment for new notified substances, Commission Regulation (EC) No 1488/94 on Risk Assessment for existing substances and Directive 98/8/EC of the European Parliament and of the Council concerning the placing of biocidal products on the market. http://ecb.jrc.it/tgd

ECB. 2006. The European Chemicals Bureau (ECB) website. http://ecb.jrc.it/

EEC. 1967. Council Directive 67/548/EEC of 27 June 1967 on the approximation of laws, regulations and administrative provisions relating to the classification, packaging and labelling of dangerous substances. *Off. J. Eur. Communities* L 196, 16.8.1967, p. 1.

EEC. 1976. Council Directive 76/769 of 27 July 1976 on the approximation of the laws, regulations and administrative provisions of the Member States relating to restrictions on the marketing and use of certain dangerous substances and preparations. *Off. J. Eur. Communities* L 262, 27.9.1976, p. 201.

EEC. 1979. Council Directive 79/831/EEC of 18 September 1979 amending for the sixth time Directive 67/548/EEC on the approximation of the laws, regulations and administrative provisions relating to the classification, packaging and labelling of dangerous substances. *Off. J. Eur. Communities* L 259, 15.10.1979, p. 10.

EEC. 1988. Council Directive 88/379/EEC of 7 June 1988 on the approximation of the laws, regulations and administrative provisions relating to the classification, packaging and labelling of dangerous preparations. *Off. J. Eur. Communities* L 187, 16.7.1988, p. 14.

EEC. 1991. Council Directive 91/414/EEC of 15 July 1991 concerning the placing of plant protection products on the market. *Off. J. Eur. Communities* L 230, 19.8.1991, p. 1.

EEC. 1992. Council Directive 92/32/EEC of 30 April 1992 amending for the seventh time Directive 67/548/EEC on the approximation of the laws, regulations and administrative provisions relating to the classification, packaging and labelling of dangerous substances. *Off. J. Eur. Communities* L 154, 5.6.1992, p. 1.

EEC. 1993a. Council Regulation (EEC) No 793/93 of 23 March 1993 on the evaluation and control of the risks of existing substances. *Off. J. Eur. Communities* L 84, 5.4.1993, p. 1.

EEC. 1993b. Commission Directive 93/67/EEC of 20 July 1993, laying down the principles for the assessment of risks to man and the environment of substances notified in accordance with Council Directive 67/548/67.

EFSA. 2006. The European Food Safety Authority (EFSA) website. http://www.efsa.europa.eu

EU. 2000. Communication from the Commission on the Precautionary Principle. Brussels, 2.02.2000, COM (2000) 1. http://ec.europa.eu/dgs/health_consumer/library/pub/pub07_en.pdf

EU. 2001. White Paper on the Strategy for a Future Chemicals Policy. http://ec.europa.eu/consumers/cons_safe/prod_safe/other_EU/chem_policy_en.htm

EU. 2005. The acquis of the European Union under the management of the DG Enterprise and Industry. List of measures (The "Pink Book"). http://ec.europa.eu/dgs/enterprise

EU. 2006a. The European Union website. http://europa.eu/

EU. 2006b. The European Commission website. http://ec.europa.eu/

EU. 2006c. The DG SANCO Consumer Affairs website. http://ec.europa.eu/consumers/

EU. 2006d. The DG Environment website. http://ec.europa.eu/environment/index_en.htm

EU. 2006e. The DG Enterprise website. http://ec.europa.eu/dgs/enterprise

EU. 2006f. The DG SANCO website. http://ec.europa.eu/dgs/health_consumer

EU. 2006g. The Employment, Social Affairs & Equal Opportunities Directorate-General website. http://ec.europa.eu/employment_social

EU. 2006h. The Energy and Transport Directorate-General website. http://ec.europa.eu/energy_transport

EU. 2006i. The DG Environment REACH website. http://ec.europa.eu/environment/chemicals/reach/reach_intro.htm

EU. 2006j. The DG Environment Existing Chemicals website. http://ec.europa.eu/environment/chemicals/exist_subst

FAO. 2006. The United Nations Food and Agriculture Organization (FAO) website. www.fao.org/

Fisher, E., Judith, J., and René von, S., eds. 2006. *Implementing the Precautionary Principle: Perspectives and Prospects*. Cheltenham, UK and Northampton, MA, US: Edward Elgar.

IARC. 2006. The International Agency for Research on Cancer website. http://www.iarc.fr/index.html

IHCP. 2006. The Institute for Health and Consumer Protection website. http://ihcp.jrc.cec.eu.int

INCHEM. 2006. The IPCS INCHEM website. http://www.inchem.org/

IOMC. 2006. The Inter-Organization Programme for the Sound Management of Chemicals (IOMC) website. http://www.who.int/iomc/en/

IPCS. 2006. The International Programme on Chemical Safety (IPCS) website. www.who.int/ipcs/en/

IRIS. 2006. The United States Environmental Protection Agency Integrated Risk Information System (IRIS) website. http://www.epa.gov/iris/

IUCLID. 2006. The International Uniform Chemical Information Database (IUCLID) website. http://ecb.jrc.it/iuclid4/

JECFA. 2006. The Joint FAO/WHO Expert Committee on Food Additives (JECFA) website. http://www.who.int/ipcs/food/jecfa/en/

JMPR. 2006. The Joint FAO/WHO Meetings on Pesticide Residues (JMPR) website. http://www.who.int/foodsafety/chem/jmpr/en/

JRC. 2006. The Joint Research Center website. http://www.jrc.cec.eu.int/default.asp@sidsz%3Dwelcome_first.htm

Larsen, J.C. 2006. Risk assessment of chemicals in European traditional foods. *Food Sci. Technol.* 17:471–481.

NCEA. 2006. The United States Environmental Protection Agency National Center for Environmental Assessment (NCEA) website. http://cfpub.epa.gov/ncea/index.cfm

NTP. 2006. The United States National Toxicology Program (NTP) website. http://ntp.niehs.nih.gov

OECD. 1960. Convention on the Organisation for Economic Co-operation and Development, Paris 14th December 1960. http://www.oecd.org/document/7/0,3343,en_2649_34483_1915847_1_1_1_1,00.html

OECD. 1981. Decision of the Council concerning the Mutual Acceptance of Data in the Assessment of Chemicals. 12 May 1981 - C(81)30/Final amended on 26 November 1997 - C(97)186/FINAL (Annex II). http://webdomino1.oecd.org/horizontal/oecdacts.nsf/linkto/C(81)30

OECD. 1989. Decision-Recommendation of the Council on Compliance with Principles of Good Laboratory Practice. 2 October 1989 - C(89)87/Final amended on 9 March 1995 - C(95)8/Final. http://webdomino1.oecd.org/horizontal/oecdacts.nsf/linkto/C(89)87

OECD. 1997. Decision of the Council concerning the Adherence of non-Member Countries to the Council Acts related to the Mutual Acceptance of Data in the Assessment of Chemicals [C(81)30(Final) and C(89)87 (Final)]. http://webdomino1.oecd.org/horizontal/oecdacts.nsf/linkto/C(97)114

OECD. 2004. *Manual for Investigation of HPV Chemicals.* Paris: OECD. http://www.oecd.org/document/7/0,3343,en_2649_34379_1947463_1_1_1_1,00.html

OECD. 2006a. The OECD website. http://www.oecd.org/

OECD. 2006b. The OECD HPV Database. http://cs3-hq.oecd.org/scripts/hpv/

OECD. 2006c. The OECD Existing Chemicals website. http://www.oecd.org/document/21/0,2340,en_2649_34379_1939669_1_1_1_1,00.html

O'Riordan, T. and Cameron, J. 1995. *Interpreting the Precautionary Principle*, London: Earthscan Publications.

RAF. 2006. The United States Environmental Protection Agency Risk Assessment Forum (RAF) website. http://cfpub.epa.gov/ncea/raf/index.cfm

UN. 2001. The United Nations Stockholm Convention On Persistent Organic Pollutants. http://www.chem.unep.ch/POPS/

UNECE. 2006. The United Nations Economic Commission for Europe website. http://www.unece.org/trans/danger/publi/ghs/ghs_welcome_e.html

UNEP. 2006. The United Nations Environment Programme website. www.unep.org

US-EPA. 2003. Overview: Office of Pollution Prevention and Toxics Programs. Prepared for the U.S. Environmental Protection Agency Office of Pollution Prevention and Toxics. Prepared by Battelle, 505 King Avenue Columbus, Ohio 43201. http://www.chemicalspolicy.org/downloads/TSCA10112-24-03.pdf

US-EPA. 2006a. The United States Environmental Protection Agency website. http://www.epa.gov/

US-EPA. 2006b. The United States Environmental Protection Agency Office of Prevention, Pesticides and Toxic Substances website. http://www.epa.gov/oppts/

US-EPA. 2006c. The United States Environmental Protection Agency Toxic Substances Control Act New Chemicals Program website. www.epa.gov/opptintr/newchems

US-EPA. 2006d. United States Environmental Protection Agency Test guidelines. http://www.epa.gov/opptsfrs/publications/OPPTS_Harmonized/870_Health_Effects_Test_Guidelines/index.html

US-FDA. 2006a. The United States Food and Drug Administration website. http://www.fda.gov/default.htm

US-FDA. 2006b. Redbook 2000. (July 2000; Updated October 2001, November 2003, April 2004, February 2006). http://www.cfsan.fda.gov/~redbook/red-toca.html

WHO. 2006. The World Health Organisation website. www.who.int/en/

Wikipedia. 2006. Wikipedia, the free encyclopedia. http://en.wikipedia.org

3 Data for Hazard Assessment

The data required for the risk assessment in relation to human health can be categorized as data on the identity of the substance, its physico-chemical and toxicological properties, and on exposure. The minimum data set required for a risk assessment depends on the chemical use category (industrial chemical, pesticide, biocide, food additive, food contact material, etc.), the regulation involved, and the goal of the risk assessment. This chapter will focus on the data used in the hazard assessment.

3.1 INTRODUCTION

According to the OECD/IPCS definitions listed in Annex 1 (OECD 2003)

Hazard is "The inherent property of an agent or situation having the potential to cause adverse effects when an organism, system or (sub) population is exposed to that agent."

Hazard assessment is "A process designed to determine the possible adverse effects of an agent or situation to which an organism, system or (sub) population could be exposed. The process includes hazard identification and hazard characterization. The process focuses on the hazard in contrast to risk assessment where exposure assessment is a distinct additional step."

This chapter will describe the various types of data used in the hazard assessment process, including human data, data from laboratory animal studies, data from *in vitro* studies, and nontesting data that can be deducted from the physico-chemical structure of the substance.

Various criteria documents and evaluations from international bodies that are available to the assessor are also mentioned.

The molecular techniques of proteomics, toxicogenomics, and metabolomics are being used increasingly in toxicological research, but have not yet been incorporated into test guidelines for regulatory toxicological studies. These techniques, popularly termed "omics," may serve as adjuncts to conventional toxicological studies, but further research and validation is required before they can be considered for routine use in regulatory toxicological assessments.

Based on the outcome of two OECD workshops (in 2003–2004), strategies concerning the future application of toxicogenomics in regulatory assessment of chemical safety are being developed by some OECD member countries (OECD 2007a).

In 2000, the U.S. National Institute of Environmental Health Sciences (NIEHS) created the National Center for Toxicogenomics (NCT), whose mission is to coordinate a nationwide research effort for the development of a toxicogenomics knowledge base. NCT has five goals: to facilitate the application of gene and protein expression technology; to understand the relationship between environmental exposures and human disease susceptibility; to identify useful biomarkers of disease and exposure to toxic substances; to improve computational methods for understanding the biological consequences of exposure and responses to exposure; and to create a public database of environmental effects of toxic substances in biological systems (NCT 2007).

The U.S. Environmental Protection Agency (US-EPA) published the white paper titled "Potential Implications of Genomics for Regulatory and Risk Assessment Applications at EPA" in 2004 (US-EPA 2004). This paper was issued to present exemplary applications and resultant implications of the use of genomics technologies in US-EPA practice.

The European Center for the Validation of Alternative Methods (ECVAM) uses and evaluates toxicogenomics, among other methods, e.g., in its carcinogenicity program (ECVAM 2007).

The various types of data used in hazard assessment of chemical substances have been extensively addressed in various guidance documents, e.g., in the EU Technical Guidance Document (TGD) on Risk Assessment of New and Existing Chemical Substances and Biocides (EC 2003), the WHO/IPCS Environmental Health Criteria 210 (WHO/IPCS 1999), as well as in the OECD Manual for Investigation of HPV Chemicals (OECD 2004). The US-EPA's Risk Assessment guidelines (Section 2.3.1.8) contain guidance on data sources in relation to the Guideline's toxicological endpoint (cancer, reproductive toxicity, etc.).

In the new EU chemicals regulation (REACH), which entered into force on 1 June 2007, detailed guidance documents on different REACH elements, including data to be used in the hazard assessment of chemical substances, are currently in preparation (spring 2007). These documents will probably be available on the EU DG Environment REACH Web site (EU 2006) when published.

3.2 HUMAN DATA

Human data adequate to serve as the sole basis for a hazard assessment are rare; however, when available, reliable and relevant human data are preferable over animal data.

Human data include information from case reports (e.g., poisonings), clinical examinations, experimental studies in volunteers, experiences from the working environment, epidemiological studies, and meta-analyses.

A major advantage of using human data in the hazard assessment is that man is the relevant species and thus, no uncertainty from extrapolation or interpretation of data obtained from studies in experimental animals arises. In addition, certain types of effects are only observable in humans, including light degrees of mucosal membrane irritation, certain types of nerve damage, and certain types of neuropsychological effects. Less severe degrees of toxic effects, in humans manifested as subjective complaints, are also difficult to detect in animals.

Among the disadvantages is that generally only a small number of subjects have been studied. In addition, the subjects in a volunteer study as well as epidemiological studies in workers are often selected, e.g., young and/or healthy individuals, and the data obtained from such a group of selected subjects have to be extrapolated to all individuals in the heterogeneous population of humans, including susceptible subgroups such as children, elderly people, and sick individuals. In epidemiological studies, exposure data are often uncertain, and exposure is often to a mixture of chemical substances rather than to a single substance.

In relation to a hazard assessment, the relative lack of sensitivity of human data may cause particular difficulty. Therefore, negative human data cannot be used to override the positive findings in animals, unless it has been demonstrated that the mode of action of a certain toxic response observed in animals is not relevant for humans. In such a case, a full justification is required.

3.2.1 CASE REPORTS

Case reports describe a particular effect observed in an individual or a group of individuals in which an exposure to a substance has occurred, often accidentally or in suicidal attempts. Information can be obtained from published case reports or from poison information centers.

Information from case reports is primarily on acute toxic effects and the clinical symptoms can in some cases be rather well described. For some substances, however, effects may be observed only after a latency period and therefore, may not be recognized as being linked to the poisoning incident.

In poisoning cases, the exposure concentration or dose is often unknown and thus, a dose–effect relationship is difficult to evaluate. Therefore, information from poisoning cases generally has a limited use in the hazard assessment. An exception is identification of acute toxic effects. Furthermore, poisoning cases can also indicate whether a substance can become systemically available;

i.e., if systemic effects are observed in a poisoning case, this indicates that the substance has been absorbed following the route of exposure leading to the poisoning.

3.2.2 CLINICAL AND PHYSIOLOGICAL INVESTIGATIONS

For some substances, information on some toxicological endpoints is available from clinical and physiological investigations such as provocation tests for detecting allergy, lung function tests, and analyses of biochemical parameters and biomarkers for exposure or effects.

From such investigations, the exposure concentration or dose is often unknown and thus, a dose–effect relationship is difficult to evaluate. Therefore, information from clinical and physiological investigations generally has a limited use in the hazard assessment; exceptions are identification of skin sensitizers and indications of possible adverse effects to selected organs and tissues depending on the investigations performed.

3.2.3 STUDIES IN VOLUNTEERS

When they are already available, well-conducted human exposure studies in volunteers can be used in the hazard assessment. Studies in human volunteers are experimental studies in which the subjects, often young and healthy persons, are exposed to known, generally low doses or concentrations of a substance for a well-defined, generally short exposure period.

Criteria for a well-designed study include the use of a double-blind study design, inclusion of a matched control group, and an adequate number of subjects to detect an effect. However, the results from human volunteer studies are often limited by a relatively small number of subjects, often young and healthy, and thus, extrapolation of the information to the general population has to be performed. Other limitations include the short duration of exposure precluding an assessment of toxicological effects that are only observed following a certain exposure duration, as well as the low-dose levels resulting in poor ability (power) to detect effects.

As studies in human volunteers are performed under standardized conditions with well-defined exposure concentrations or doses, but only for a short duration, such studies are most relevant for hazard assessment of mild acute effects, mild irritation, sensitization, toxicokinetics, and mode of action(s).

Over the years, scientific research with human subjects has provided valuable information to help characterize and control risks to public health, but its use has also raised particular ethical concerns for the welfare of the human participants in such research as well as scientific issues related to the role of such research in assessing risks. Society has responded to these concerns by defining general standards for conducting human research. As an example, studies carried out for the authorization of a medical product, have to be conducted in line with the World Medical Association's Declaration of Helsinki, which describes the general ethical principles for medical research involving human subjects (World Medical Association 2004). The Helsinki Declaration was first issued by the World Medical Association in 1964 and has been revised several times since then.

In the United States, the National Commission for the Protection of Human Subjects of Biomedical and Behavioral Research issued in 1979 "The Belmont Report: Ethical Principles and Guidelines for the Protection of Human Subjects of Research" (US-EPA 2007a). For many Federal agencies and departments in the United States, the principles of the Belmont Report are implemented through the Federal policy for the protection of human subjects, also known as the "Common Rule." The Common Rule, which was officially announced by 15 Federal departments and agencies, including the US-EPA, on 18 June 1991 (56 FR 28003), applies to all research involving human subjects conducted, supported, or otherwise subject to regulation by any Federal department or agency that has adopted the common rule and has taken appropriate administrative action to make it applicable to such research. The Common Rule contains detailed requirements for establishment of Institutional Review Boards with responsibility for review and approval of research proposals, for information of subjects, for informed consent, for minimizing risks to subjects, for risk–benefit considerations, and for selection of subjects. Special guidance is provided for research involving children, pregnant women, human fetuses and neonates, or prisoners (OHRP 2007).

Human research issues affect all programs in US-EPA. In its Office of Research and Development, US-EPA conducts research with human subjects to provide critical information on environmental risks, exposures, and effects in humans. This is referred to as "first-party research." In both its Office of Research and Development and its program offices (including the Office of Air and Radiation, the Office of Water, the Office of Solid Waste and Emergency Response, and the Office of Prevention, Pesticides and Toxic Substances), US-EPA also supports research with human subjects conducted by others. This is referred to as "second-party research." In all this work US-EPA is committed to full compliance with the common rule. The US-EPA will continue to conduct and support such research, and to consider and rely on its results in US-EPA assessments and decisions.

Much of the scientific information supporting US-EPA's actions is generated by researchers who are not part of or supported by a Federal agency, including a significant portion of the research with human subjects submitted to the US-EPA or retrieved by the US-EPA from published sources. Such research is generally referred to as "third-party research."

Under the Federal Insecticide, Fungicide, and Rodenticide Act (FIFRA), US-EPA is authorized to require pesticide companies to conduct studies with human subjects, e.g., to measure potential exposure to pesticide users or to workers and others who reenter areas treated with pesticides, or to evaluate the effectiveness of pesticide products intended to repel insects and other pests from human skin. In addition, US-EPA sometimes encourages other research with human subjects, including tests of the potential for some pesticides, generally those designed for prolonged contact with human skin, to irritate or sensitize human skin, and tests of the metabolic fate of pesticides in the human body. In addition to these kinds of research, which have been required or encouraged by US-EPA, other kinds of studies involving human subjects intentionally exposed to pesticides have occasionally been submitted to the agency voluntarily.

In February 2005, US-EPA published a notice on "Human Testing; Proposed Plan and Description of Review Process" (US-EPA 2007a). The notice announced US-EPA's plan to establish a comprehensive framework for making decisions about the extent to which it will consider or rely on certain types of research with human participants. Among other actions the plan provided a description of the agency's case-by-case process for reviewing and relying on completed third-party studies that involve intentional dosing of human participants to identify or quantify a toxic endpoint. For further details, the reader is referred to the US-EPA Web site (US-EPA 2007a).

The US-EPA recently significantly strengthened and expanded the protections for human volunteers by prohibiting new research involving intentional exposure of pregnant or nursing women or children, intended for submission to US-EPA under the pesticide laws (US-EPA 2007b). Under the new rules, which were effective from 22 August 2006, all third-party intentional dosing research on pesticides involving pregnant or nursing women and children intended for submission to US-EPA is banned, and US-EPA will neither conduct nor support any intentional dosing studies that involve pregnant or nursing women or children for all substances US-EPA regulates. US-EPA is also extending new ethical protections to adult (nonpregnant, nonnursing) subjects involved in intentional dosing human studies for pesticides. Now such adult volunteers who choose to participate in this research will be protected by US-EPA's new, high ethical standards.

The European Union (EU) and its Member States promote responsible research in order to help society draw the maximum benefit from scientific inquiry and in order to protect the rights of citizens, researchers, and other members of society (EU 2007). All EU-funded research activities must comply with an ethical code according to Article 3 of the Sixth Framework Programme (FP6, 2002–2006) that states: "All the research activities carried out under FP6 must be carried out in compliance with fundamental ethical principles." No further guidance specifically directed toward the use of human data in the risk assessment of chemical substances is provided.

Within the EU, experimental human toxicity studies must not be conducted specifically for the purpose of hazard assessment of biocides according to the EU Biocidal Product Directive (EC 1998), or pesticides according to the EU Plant Protection Product Directive (EC 1991).

According to the Technical Guidance Document (TGD) for risk assessment of new and existing chemical substances (EC 2003), studies conducted with human volunteers are strongly discouraged as they are problematic from an ethical point of view and results from such studies should be used only in justified cases (e.g., tests which were conducted for the authorization of a medical product or when effects in already available human volunteer studies with existing substances have been observed to be more severe than deduced from prior animal testing). However, the potential differences in sensitivity of human studies and studies in animals should be taken into account in the risk assessment, on a case-by-case basis.

3.2.4 OCCUPATIONAL EXPERIENCE

In occupational settings, the concentrations of chemicals are often monitored in the working environment to monitor compliance with occupational exposure limits as required by various national laws. Moreover, medical surveys of workers are often performed including analyses of biomarkers for exposure and/or effects. In addition, workers also generally have the possibility to report signs and symptoms of nuisances related to their working environment. Such data, which in some cases are available in the open literature, are relevant for use in a hazard assessment.

One limitation is that the workforce generally only includes healthy subjects of a certain age and thus, extrapolation of the information to the general population has to be performed. Another limitation is that workers are generally exposed to more than one substance in the working environment, precluding as assessment whether an effect observed is related to a specific substance, or to co-exposure to several of the substances in the working environment.

3.2.5 EPIDEMIOLOGICAL STUDIES

Epidemiology may be defined as the study of the distribution and determinants of health-related states or events (including disease) in human populations, and the application of this study to the control of diseases and other health problems (WHO 2002). The word epidemiology consists of the Greek words *epi* which is among, *demos* which is people, and logos which is doctrine.

Epidemiological studies are nonexperimental trials and involve an investigation of various individuals or groups of subjects as they happen to have been exposed. Endpoints generally measured include mortality, morbidity, medical visits or hospital admissions, and/or clinical signs and symptoms.

Epidemiological studies are relevant in the hazard assessment for long-term effects, i.e., effects occurring following repeated exposure for a longer time. Furthermore, epidemiological studies can reveal effects following a single or a short-term exposure such as subjective symptoms (e.g., irritation, headache). In addition, effects with unknown relation to the duration of exposure may also be detected in epidemiological studies. As an example, for abortion and fetal malformations, it is often not known whether these effects are related to a single exposure at a critical period in the embryogenesis, or to repeated exposure.

Advantages of epidemiological studies include that often a heterogenous population is studied, and the exposure scenarios generally are realistic.

Limitations include difficulties in performing a correct exposure assessment, in some cases even a lack of information on exposure, insufficient sample size (i.e., small number of subjects in the study), selected group of subjects (e.g., the workforce), short length of follow-up, exposure to more than one substance, and potential errors (bias, confounding).

Different methods are used in epidemiology. Epidemiological studies are often divided into descriptive studies and analytic studies.

Descriptive studies may be defined as studies that describe the patterns of disease occurrence by time, place, and person (WHO 2002).

Types of descriptive studies include case reports or case series, and ecological studies.

They examine differences in disease rates among human populations in relation to age, gender, race, and differences in temporal or environmental conditions (EC 2003). Typically these studies can only identify patterns or trends in disease occurrence over time or in different geographical locations but cannot ascertain the causal agent or degree of human exposure.

These studies are not very useful for a hazard assessment but may be useful for identifying priority areas for further research.

Analytic studies may be defined as studies used to test hypotheses concerning the relationship between a suspected risk factor and an outcome, and to measure the magnitude of the association effect, and its statistical significance (WHO 2002).

An analytic study always implies a comparison among two or more groups. Data from analytic studies may be useful for identifying a relationship between human exposure and effects such as biological effect markers, early signs of chronic effects, disease occurrence, or mortality (EC 2003).

Types of analytic studies include case-control studies, cohort studies, and cross-sectional studies.

Case-control studies: A group of individuals with a particular outcome (cases) and a group without that particular outcome (controls) are identified. Information about previous exposures is obtained for cases and controls, and the frequency of exposure is compared for the two groups. In case-control studies, both exposure and disease are normally considered to have occurred prior to enrolment in the study.

Cohort (follow-up) studies: A cohort can be defined as a designated group of people who have had a common experience vis-à-vis exposure, and are then followed up or traced over a period of time. Cohort studies proceed conceptually from exposure to outcome. Study groups are identified by exposure status prior to ascertainment of their disease status, i.e., a group of "exposed" and "unexposed" are identified. Both exposed and unexposed groups are then followed prospectively in an identical manner until they develop the disease under study, until the study ends, or the subjects die or are lost to follow-up. The relative risk (RR) is estimated as the ratio between risk in the exposed group and in the unexposed group. Both cohorts should have similar characteristics except for the exposure under investigation. In some studies, called retrospective cohort studies, exposure and outcome both lie in the past (before enrolment).

Cross-sectional studies (surveys): These examine the relationship between a disease or other health-related characteristic in a population at a given time. The presence or absence (or the level) of a characteristic is examined in each member of the study population or in a representative sample. These studies are used to obtain information not routinely available from surveillance or case series. Cross-sectional studies provide no information on the temporal sequence of cause and effect. In surveys examining the association between an exposure and an outcome, both are measured at the same time and it is often hard to determine whether the exposure preceded the outcome or vice versa. In general, surveys measure the situation at a given moment (prevalence) rather than the occurrence of new events (incidence).

The evaluation of epidemiological data usually requires a more elaborate and in-depth critical assessment of the reliability of the data than for animal data (EC 2003).

Criteria for assessing the adequacy of an epidemiology study include an adequate research design, proper selection and characterization of the exposed and control groups, adequate characterization of exposure, sufficient length of follow-up for the disease as an effect of the exposure to develop, sufficient length of follow-up for disease occurrence, valid ascertainment of effect, proper consideration of potential errors, proper statistical analysis, and a reasonable statistical power to detect an effect. These types of criteria have been described in more detail (Swaen 2006) and can be derived from epidemiology textbooks (e.g., Rothman and Greenland 1998).

Potential errors in epidemiological studies are divided into random error and systematic error (bias) (WHO 2002).

Random error is the divergence, due to chance alone, of an observation on a sample from the true population value, leading to lack of precision in the measurement of an association. There are three major sources of random error: individual/biological variation, sampling error, and measurement error. Random error can be minimized but can never be completely eliminated since only a sample of the population can be studied, individual variation always occurs, and no measurement is perfectly accurate.

Bias: Occurs when there is a tendency to produce results that differ in a systematic manner from the true values. A study with small systematic bias is said to have high accuracy. Bias may lead to over- or underestimation of the strength of an association. The sources of bias in epidemiology are many and over 30 specific types of bias have been identified. The main biases are selection bias, information bias, and bias due to confounding.

Selection bias: Occurs when there is a systematic difference between the characteristics of the people selected for a study, or who agree to participate, and the characteristics of those who are not selected, or who do not agree to participate (e.g., in a study limited to volunteers).

Information bias (observation bias): Occurs when there are quality (accuracy) problems in the collection, recording, coding, or analysis of data among comparison groups. Interviewers might, e.g., interview the cases with more diligence than they interview the control, or a person with a disease may recall previous exposures better than persons who are healthy (this type of bias is called recall bias).

Bias due to confounding: In a study of the association between exposure to a cause (or risk factor or protecting factor) and the occurrence of the disease, confounding can occur when another factor exists in the study population and is associated both with the disease and the initial factor being studied. A problem arises if this second extraneous factor is unequally distributed among the exposure subgroups. Confounding occurs when the effects of two protective or risk factors have not been separated and it is therefore incorrectly concluded that the effect is due to one variable rather than the other. Examples of confounding factors include age, gender, socioeconomics, smoking, and alcohol. Such biases can be controlled for in the analysis if appropriate information has been collected during the study on potential confounding variables and if each factor is properly analyzed and interpreted.

There are no clear-cut criteria to distinguish positive from negative studies. However, the degree of reliability for an epidemiological study should be evaluated using generally accepted causality criteria, such as that of Bradford Hill (1965): (1) strength, (2) consistency, (3) specificity, (4) temporality, (5) biological gradient, (6) plausibility, (7) coherence, (8) experiment, and (9) analogy; these criteria and their contemporary use in epidemiology have recently been reviewed (Lucas and McMichael 2005).

Based on the Bradford Hill criteria, a positive association is indicated if the following criteria are met: (1) no identifiable positive bias, (2) possibility of positive confounding considered, (3) association unlikely to be due to chance alone, (4) association strong, and (5) dose–response relationship.

The strength of the epidemiological evidence for specific health effects depends, among other things, on the type of analyses and on the magnitude and specificity of the response. Confidence in the findings is increased when comparable results are obtained in several independent studies on populations exposed to the same agent under different conditions and using different study designs.

It should be kept in mind that epidemiological studies with negative results cannot prove the absence of an intrinsic hazardous property of a substance but well-documented "negative" studies of good quality may be useful in the risk assessment.

3.2.6 META-ANALYSIS

A meta-analysis is the statistical procedure for combining data from multiple studies. When the effect (or effect size) is consistent from one study to the next, a meta-analysis can be used to characterize this common effect. When the effect varies from one study to the next, a meta-analysis may be used to identify the reason for the variation.

A comprehensive meta-analysis software program has been developed by a team of experts in the United States and the United Kingdom (Borenstein et al. 2005).

A handbook "*Meta Analysis: The Handbook for the Understanding and Practice of Meta-Analysis*" and an accompanying program "META" are available (Leandro 2005).

3.3 DATA FROM STUDIES IN EXPERIMENTAL ANIMALS

3.3.1 ANIMAL TOXICITY STUDIES

For most chemical substances, useful and relevant human data are not available. The risk assessment is therefore most often based on studies in experimental animals. The results of animal studies are used to predict the possible effect in humans, i.e., effects in animals are used to model corresponding effects in humans. Animal data are also used as a supplement to human data, which are equivocal, or to identify the active substances in a mixture to which humans have been exposed.

Advantages of animal studies include standardized test conditions, the possibility to detect far more effects through invasive measurements and postmortem examinations, the possibility to investigate modes and mechanisms of action, and to investigate dose–effect relationships for single chemical substances.

The major disadvantage is that the examined species is different from the species of interest (humans). Cross-species differences in dose response result from a combination of toxicokinetic and toxicodynamic factors. Another problem is that certain types of effects are difficult to disclose in animal studies, including light degrees of mucosal membrane irritation, certain types of nerve damage, and certain types of neuropsychological effects. Less severe degrees of toxic effects, in humans manifested as subjective complaints, are difficult to detect in animals. Furthermore, animal studies are performed in homogenous groups of animals, but the results have to be applied for the protection of all individuals in a heterogeneous population of humans. In consequence of this, interspecies variation must be expected.

Ideally, a full data set should be available for the hazard assessment of a chemical substance, including animal tests to evaluate the toxicokinetics and the following toxicological properties: acute toxicity, irritation, sensitization, toxicity following repeated exposure to the substance, mutagenicity and genotoxicity, carcinogenicity, and effects on fertility and fetal development.

Further, the studies should ideally have been conducted according to agreed principles for testing (e.g., be performed according to internationally accepted guidelines), and the data should be accurate and reliable (e.g., be generated according to GLP).

However, the above-mentioned requirements cannot always be fulfilled. Risk assessment, including hazard assessment, is performed within a political and legal framework, which puts limitations on the availability of information. For example, in the EU, the data requirements for existing substances are not as comprehensive as those for new chemical substances.

3.3.2 TEST GUIDELINES FOR ANIMAL TOXICITY STUDIES

It is possible to conduct animal studies in an infinite number of ways. Although individually designed studies are often scientifically sound, and in many cases serve a particular purpose very well, they pose problems in a regulatory context. Free movement of chemicals between countries is based on the mutual acceptance of the risk evaluation made by each country and this, in turn, relies on the mutual acceptance of the data generated when testing the chemicals. Experience has shown this acceptance to be extremely difficult, if chemicals have been tested by different methods.

Therefore, the OECD has undertaken the development of a set of test guidelines, which are internationally agreed and therefore acceptable in all OECD member countries and nonmember countries accepting OECD standards (Sections 2.2.6 and 3.3.4).

The United States and EU each maintain their own sets of chemical testing guidelines (Sections 2.3.1.9/3.3.5 and 2.4.1.9/3.3.6, respectively). These U.S./EU test guidelines are often identical to the corresponding OECD Test Guideline or similar with minor deviations. It should be noted that some U.S. guidelines do not have an OECD counterpart. Countries outside the United States and EU may have national guidelines; such guidelines are not included in this book.

3.3.3 GOOD LABORATORY PRACTICE

If national regulatory authorities can rely on safety test data developed abroad, duplicative testing can be avoided and costs saved to government and industry. Furthermore, a reduction in the number of laboratory animals used for testing of chemicals is an important goal in relation to animal welfare.

As a tool to make mutual acceptance of risk assessments possible, OECD has developed the concept of Good Laboratory Practice (GLP). The OECD Principles of GLP are an integral part of the 1981 OECD council decision on the Mutual Assessment of Data (MAD) in the Assessment of Chemicals (revised 1997, Section 2.2.2). MAD also harmonizes procedures of GLP compliance monitoring, ensuring that preclinical safety studies are carried out according to the principles of GLP and that countries can have confidence in the quality and rigor of safety tests.

GLP is a managerial concept covering the organizational process and the conditions under which laboratory studies are planned, performed, monitored, recorded, and reported. Its principles must be followed by test facilities carrying out studies to be submitted to national authorities for the purposes of assessment of chemicals and other uses relating to the protection of man and the environment.

The 1989 OECD council decision & Recommendation on Compliance with Good Laboratory Practice requires the establishment of national compliance monitoring program based on laboratory inspections and study audits and recommends the use of the guides for compliance monitoring procedures for good laboratory practice and the guidance for the conduct of laboratory inspections and study audits.

Since 1997 a procedure through which non-OECD countries can adhere to the MAD system has been embodied in an OECD council decision (Council Decision on the Adherence of Non-Member Countries to the Council Acts Related to the Mutual Acceptance of Data in the Assessment of Chemicals C(97)114/FINAL). A series of documents related to specific issues of GLP and compliance monitoring has been published (available on the OECD Web site, OECD 2006).

3.3.4 OECD TEST GUIDELINES

The OECD Guidelines for the testing of chemicals are a collection of the most relevant internationally agreed testing methods used by government, industry, and independent laboratories to characterize potential hazards of new and existing chemical substances and chemical preparations/mixtures.

The OECD Guidelines are available to the public free of charge. The adopted test guidelines are listed in Table 2.3 and the draft test guidelines in Tables 2.4 and 2.5.

3.3.5 US-EPA TEST GUIDELINES

The US-EPA's Office of Prevention, Pesticides and Toxic Substances (OPPTS) has developed a series of test guidelines called the Harmonized Test Guidelines. For health effects, the relevant guidelines are Series 870, Health Effects Test Guidelines.

The harmonized test guidelines have been developed for use in the testing of pesticides and toxic substances, and the development of test data that must be submitted to the US-EPA for review under Federal regulations. The OPPTS guidelines are based on the OECD Guideline methods; however, minor deviations may occur. The guidelines, which are listed in Table 2.7 (adopted test guidelines) and Table 2.8 (draft test guidelines), are available for download at the OPPTS Web site (US-EPA 2006).

3.3.6 EU Test Guidelines

The legally binding EU standardized testing methods to determine the hazardous properties of chemicals are contained in Annex V of Directive 67/548/EEC on the classification, packaging, and labeling of dangerous substances. The testing methods play a central role in the EU policy on chemicals control and they are referred to in many other pieces of EU legislation (e.g., those related to dangerous preparations, pesticides, cosmetics, and biocides also refer to these methods).

The testing methods working area of the EU is closely linked and coordinated with the parallel OECD Test Guidelines program.

Annex V to Directive 67/548/EEC is divided in three parts (A, B, and C), which contain testing methods for chemicals that address all areas of concern. Part B contains methods for the determination of effects on human health. The adopted test guidelines, which are listed in Table 2.9, are available for download free of charge at the ECB Web site (ECB 2006) under the heading "Testing-Methods."

3.3.7 Nonguideline Animal Toxicity Studies

Chemicals are often tested in nonguideline studies in research projects, and when it has been necessary to go into further detail with effects initially identified in guideline studies. Such nonguideline studies may be of high or low quality and relevance, and the evaluation of this requires expert judgment. Nonguideline studies are usually not performed according to GLP principles, and therefore carry no official recognition of the quality and reliability of the data presented.

3.4 IN VITRO METHODS

In 1959, a book titled *The Principles of Humane Experimental Technique* was published, which introduced the concept of the "3Rs" (Russell and Burch 1959). The 3Rs stand for reduction, refinement, and replacement.

The authors, William Russell and Rex Burch (a zoologist and a microbiologist, respectively) recommended the reduction of the number of animals used in experiments to the minimum number required to obtain statistically relevant data, the refinement of procedures to minimize pain and distress in experimental animals and provide for their well-being based on their behavioral needs, and the replacement of whole, living animals with *in vitro* models like tissue and cell cultures when possible.

The Johns Hopkins Center for Alternatives to Animal Testing (CAAT) maintains the alternatives to animal testing Web site (Altweb 2007), where the full text of Russell and Burch's book is freely available for download.

The 3R concept lies behind efforts to improve ethical standards for the use of experimental animals throughout the scientific community, including toxicity testing. A number of *in vitro* methods for genetic toxicology testing have been established as guideline methods for many years, e.g., the bacterial reverse mutation test, more popularly known as the Ames test.

It is imperative that new or alternative methods are not employed for toxicity testing at the cost of reduced relevance and validity of the test data. A thorough validation process of the new test is therefore necessary. Only few new *in vitro* tests have so far been validated to replace *in vivo* testing. Most of these tests are tests for skin corrosion and genotoxicity, but a number of other tests are under way and are undergoing pre-validation. Generally, it is possible to develop *in vitro* tests for the study of single endpoints, while the study of endpoints involving interaction with the entire organism, e.g., hormonal systems, necessitate the use of whole animals.

Isolated cells, tissues, and organs may be grown in culture in a manner where their natural properties (in vivo) are maintained to some extent. Such *in vitro* biological systems have been used for many years in studies of mutagenic and genotoxic properties, and in studies of mechanisms.

In vitro assays may also be used as preliminary studies to *in vivo* testing, e.g., to provide information about endpoints to study in the *in vivo* study.

The term "ex vivo" is often used to describe experiments where either the experiment is performed *in vivo* and then analyzed *in vitro* or where part of the subject is removed, the experiment is performed, and the part is returned. It should be noted that the term *ex vivo* is often used synonymously with the term *in vitro*.

During the last 10 years it has been attempted to develop *in vitro* methods as alternative methods in the study of effects where animal models have previously been necessary. Such effects include skin and eye irritation and specific organ damage. Validation programs have been launched, and some of the above-mentioned methods have been sufficiently validated for use in regulatory risk assessment of chemical substances and may now for certain purposes be used as stand-alone evidence. Results from nonvalidated methods can in some cases be used as supportive evidence to human and animal data.

The United States as well as the EU have formed institutions, which have the responsibility for progress in the development of alternative methods to animal testing. In addition, a number of national and international public and private organizations participate in this work. In the following sections, the U.S. and EU governmental institutions will be presented.

3.4.1 UNITED STATES INTERAGENCY COORDINATING COMMITTEE ON THE VALIDATION OF ALTERNATIVE METHODS

The Interagency Coordinating Committee on the Validation of Alternative Methods (ICCVAM) was established in 1997 by the Director of the U.S. National Institute of Environmental Health Sciences (NIEHS) to implement NIEHS directives in Public Law (P.L.) 103-43. This law directed NIEHS to develop and validate new test methods, and to establish criteria and processes for the validation and regulatory acceptance of toxicological testing methods. P.L. 106–545, the ICCVAM Authorization Act of 2000, established ICCVAM as a permanent committee. The Committee is composed of representatives from 15 Federal regulatory and research agencies including the US-EPA; these agencies generate, use, or provide information from toxicity test methods for risk assessment purposes. The committee coordinates cross-agency issues relating to development, validation, acceptance, and national/international harmonization of toxicological test methods (ICCVAM 2006).

3.4.2 UNITED STATES NATIONAL TOXICOLOGY PROGRAM INTERAGENCY CENTER FOR THE EVALUATION OF ALTERNATIVE TOXICOLOGICAL METHODS

The National Toxicology Program (NTP) Interagency Center for the Evaluation of Alternative Toxicological Methods (NICEATM) was established in 1998 to provide operational support for ICCVAM, and to carry out committee-related activities such as peer reviews and workshops for test methods of interest to Federal agencies. NICEATM and ICCVAM coordinate the scientific review of the validation status of proposed methods and provide recommendations regarding their usefulness to appropriate agencies. NICEATM and ICCVAM seek to promote the validation and regulatory acceptance of toxicological test methods that will enhance the agencies' ability to assess risks and make decisions, and methods that will refine, reduce, and/or replace animal use. The ultimate goal is the validation and regulatory acceptance of test methods that are more predictive of adverse human and ecological effects than currently available methods (ICCVAM 2006).

3.4.3 ICCVAM REVIEW PROCESS

A test method is nominated to ICCVAM for consideration and review. NICEATM, on behalf of ICCVAM, receives proposed test method nominations or submissions and communicates with the submitting organization or individual. Typically, the ICCVAM evaluation process involves an initial assessment by NICEATM of the adequacy and completeness of the proposed test method nomination or submission, and a determination by ICCVAM of the priority that the proposed test method will have for technical evaluation. Once a proposed test method has been accepted for evaluation, ICCVAM assembles an interagency working group of government scientists with scientific and regulatory expertise in the appropriate scientific disciplines to collaborate with NICEATM on the evaluation process. Depending on the validation status of the proposed test method, ICCVAM, in conjunction with NICEATM, develops recommendations and priorities for further efforts. Such efforts might include an expert workshop, an expert panel meeting, a peer review meeting, an expedited peer review process, or a validation study. Following a review process, ICCVAM develops and forwards recommendations on the usefulness and limitations of the proposed test method for regulatory purposes to Federal agencies. Based on their specific statutory mandates, each agency then makes a determination regarding the acceptability of the test method (ICCVAM 2006).

3.4.4 IN VITRO TESTS UNDERGOING VALIDATION BY ICCVAM

ICCVAM has evaluated alternative test methods for acute oral toxicity, genetic toxicity, biologics, immunotoxicity, dermal corrosion and irritation, ocular toxicity, developmental toxicity, pyrogenicity, and endocrine disruptor effects (ICCVAM 2007). As examples, alternative test systems for dermal corrosion and irritation are described in the following text.

ICCVAM has validated three tests for dermal corrosion, including Corrositex®, EPISKIN™/EpiDerm™, and Transcutaneous Electrical Resistance (TER).

Corrositex uses a synthetic membrane-based detection system to determine the UN packing group classification of chemicals, consumer products, or other hazardous materials.

EPISKIN is a three-dimensional human skin model composed of a human collagen (Types III and I) matrix, representing the dermis, covered with a film of Type IV human collagen and stratified differentiated epidermis derived from human keratinocytes.

The EpiDerm (EPI-200) skin model is mechanistically and functionally related to EPISKIN. The assay consists of normal human epidermal keratinocytes, which have been cultured in a chemically defined medium to produce a stratified, highly differentiated, organotypic tissue model of the human epidermis.

The TER assay uses freshly isolated rat skin and is described in Section 4.5.3.2.

3.4.5 EUROPEAN CENTER FOR THE VALIDATION OF ALTERNATIVE METHODS

The EU's European Center for the Validation of Alternative Methods (ECVAM) was established in 1992 (ECVAM 2007). ECVAM is a unit of the Institute for Health and Consumer Protection (Section 2.4.1.1.2).

ECVAM was created in response to a requirement in Directive 86/609/EEC on the protection of animals used for experimental and other scientific purposes, which requires that the EU Commission and the EU Member States should actively support the development, validation, and acceptance of methods, which could reduce, refine, or replace the use of laboratory animals.

The requirement is (Article 7.2): "An experiment shall not be performed if another scientifically satisfactory method of obtaining the result sought, not entailing the use of an animal, is reasonably and practicably available." Further (Article 23): "The Commission and Member States should encourage research into the development and validation of alternative techniques which could provide the same level of information as that obtained in experiments using animals, but which involve fewer animals or which entail less painful procedures, and shall take such other steps as they consider appropriate to encourage research in this field."

The duties of ECVAM are to coordinate the validation of alternative test methods at the EU level, to act as a focal point for the exchange of information on the development of alternative test methods, to set up, maintain, and manage a database on alternative procedures, and to promote dialogue between legislators, industries, biomedical scientists, consumer organizations, and animal welfare groups, with a view to the development, validation, and international recognition of alternative test methods.

ECVAM has its own Scientific Advisory Committee (ESAC) with participation from all Member States, relevant industrial associations, academic toxicology, the animal welfare movement, as well as other Commission services with interest in the alternatives topic area.

3.4.6 IN VITRO TESTS UNDERGOING VALIDATION BY ECVAM

ECVAM has validated alternative test methods for acute oral toxicity, biologics, immunotoxicity, dermal corrosion and irritation, developmental toxicity, and pyrogenicity (ECVAM 2007).

3.4.7 IN VITRO TEST GUIDELINE METHODS

3.4.7.1   OECD *In Vitro* Test Guideline Methods

OECD has, as test guidelines, approved the following *in vitro* tests:

- *In vitro* test for skin absorption potential (TG 428) (Section 4.3.3.2)
- 3T3 NRU PT test for assessment of the phototoxic/photoirritant potential (TG 432)
- Two *in vitro* tests for the assessment of skin corrosion, the Transcutaneous Electrical Resistance [TER] Assay (TG 430) and the human Skin Model Test (TG 431) (Section 4.5.3.2).

3.4.7.2   EU *In Vitro* Test Guideline Methods

The following *in vitro* tests, replacing *in vivo* animal tests, have been endorsed as scientifically valid by ECVAM and have been adopted as EU test methods (EU 2006):

- EU test method B 41, Phototoxicity - *in vitro* 3T3 NRU phototoxicity test
- EU test method B40, Skin corrosion (*in vitro*), which includes two *in vitro* tests for skin corrosion, the rat skin TER assay and a test employing a human skin model (Section 4.5.3.2).

3.5 NONTESTING DATA

3.5.1 PHYSICO-CHEMICAL PROPERTIES

The physico-chemical properties of a substance can be essential elements in the hazard assessment. The physico-chemical properties most relevant in relation to a hazard assessment include:

- Chemical structure
- Molecular weight
- Physical state, i.e., solid, liquid, or vapor at room temperature
- Particle size, for aerosols
- Vapor pressure
- Concentration of saturated vapor
- Water solubility

- LogP$_{octanol/water}$ partition coefficient (logP$_{o/w}$), provides information on the relative solubility of the substance in water and the hydrophobic solvent octanol (used as a surrogate for lipid) and is a measure of lipophilicity
- pK$_a$-value, the pH at which 50% of the substance is in ionized and 50% in nonionized form
- Stability/reactivity

The physico-chemical properties may provide indications about the absorption of the substance for various routes of exposure and may therefore be of importance in the evaluation whether an appropriate administration route has been applied in the available experimental toxicity studies. In order for a substance to be absorbed, it must cross biological membranes. Most substances cross biological membranes by passive diffusion. This process requires a substance to be soluble both in lipid and water. The most useful parameters providing information on the potential for a substance to diffuse across biological membranes are the logP$_{octanol/water}$ and the water solubility.

The prediction of absorption of a substance from its physico-chemical properties is further addressed in Section 4.3.5.3.

The chemical reactivity may have an impact on the toxicokinetics and metabolism of the substance and may therefore be of importance for an evaluation whether the substance itself, or an active metabolite, is likely to reach the target organ(s) and tissue(s).

Data on physico-chemical properties can provide information on the potential corrosivity or irritancy of a substance. For example, substances with a low or high pH (strong acids and bases, respectively) are potential corrosive substances, and substances with a moderate pH may be potential irritants.

3.5.2 Use of Structure–Activity Relationships

A Structure–Activity Relationship (SAR) is the relationship of the molecular structure of a chemical with a physico-chemical property, environmental fate attribute, and/or specific effect on human health or an environmental species. These correlations may be qualitative (simple SAR) or quantitative (QSAR) (OECD 2002).

Section 3.3 in the OECD Manual for Investigation of HPV Chemicals, provides a detailed guidance for the use of SARs in the HPV Chemicals Program (OECD 2002).

When data do not exist for a given toxicological endpoint, or when data are limited, the use of SARs may be considered in the hazard assessment. The potential toxicity of a substance, for which no or limited data are available on a specific toxicological endpoint can, in some cases, be evaluated by read-across from structurally or mechanistically related substances for which experimental data exist. The read-across approach is based on the principle that structurally and/or mechanistically related substances may have similar toxicological properties.

Based on structural similarities between different substances, the toxic potential for a specific endpoint of one substance or a group of substances can be extended (read-across) to a substance, for which there are no or limited data on this endpoint.

Based on mechanistic similarities between different substances, a mechanism of toxicity or mode of action identified for a substance and/or group of substances and causally related to adverse effects in a target organ can be extended (read-across) to a substance for which a similar mechanism or mode of action has been identified, but where no or limited data on a specific endpoint are available. In such cases, the substance under evaluation may reasonably be expected to exhibit the same pattern of toxicity in the target organ(s) and tissue(s).

The concept of grouping, including both read-across and the related "chemical category concept" has been developed under the OECD HPV Chemicals Program (OECD 2004) as an approach to fill data gaps without the need for conduction of tests. A chemical category is a group of chemicals whose physico-chemical and toxicological properties are likely to be similar or follow a regular pattern as a result of structural similarity. In the category approach, not every substance needs to be tested for every endpoint. However, the information finally compiled for the category must prove adequate to support a hazard assessment for the category and its members. That is, the final data set must allow one to assess the untested endpoints, ideally by interpolation between and among the category members.

There are no formal criteria to identify structural alerts for toxicity in general, for a specific endpoint or for read-across to closely related substances. However, for substances where no data for one or more endpoints are available, the assessment at a screening level can be performed using data obtained from closely analogous substance(s) to indicate a potential for toxicity, e.g., if a closely related substance has a potential for inducing a specific toxic effect, the substance for which no data on this specific toxic effect are available may reasonably be expected to exhibit a similar potential for inducing this specific toxic effect. In the case of such knowledge, it should be considered whether these data are adequate for a hazard assessment.

When analog data are used to fill the data gaps for one or more endpoints, the data for the analogs must be compared and discussed in relation to the substance under evaluation in order to shed light on the similarities and differences in the Toxicological Profile of the substance under evaluation and its analog(s). If the available test results show that the substances in a category behave in a similar or predictable manner, then read-across can be used to assess the substance for which no data on a specific endpoint are available.

3.5.3 Quantitative Structure–Activity Relationship

Quantitative Structure–Activity Relationships (QSARs) are estimation methods developed and used in order to predict certain effects or properties of chemical substances, which are primarily based on the structure of the substance. They have been developed on the basis of experimental data on model substances. Quantitative predictions are usually in the form of a regression equation and would thus predict dose–response data as part of a QSAR assessment. QSAR models are available in the open literature for a wide range of endpoints, which are required for a hazard assessment, including several toxicological endpoints.

A (Q)SAR analysis for a substance may give indications for a specific mechanism to occur and identify possible organ or systemic toxicity upon single or repeated exposure. The reliability, applicability and overall scope of (Q)SAR science to identify chemical hazard and assist in risk assessment have been evaluated by various groups and organizations.

The OECD has recently published a guidance document on the validation of (Q)SAR models (OECD 2007b). Anticipating the benefits of adding the in silico technology of (Q)SAR to the well established *in vitro* and *in vivo* test guidelines, experts have been meeting to discuss the barriers to acceptance of (Q)SARs by regulatory agencies. A critical element of regulatory acceptance is the creation of a flexible scientific validation process for (Q)SARs, which allows individual regulatory agencies to establish the reliability of (Q)SAR estimates for specific authorities and regulatory constraints. The report provides a discussion of the "OECD Principles for the Validation, for Regulatory Purposes, of (Q)SAR Models" and provides guidance on how individual regulatory agencies can evaluate specific (Q)SAR models with respect to those principles. The purpose of the report is to provide detailed but nonprescriptive guidance that explains and illustrates the application of the validation principles to different types of (Q)SAR models. The report is needed to provide a harmonized framework for (Q)SAR validation studies to explain and illustrate with examples how the validation principles can be interpreted for different types of (Q)SAR models. While the report provides nonprescriptive guidance on the processes of validation, which address the performance of a wide variety of (Q)SAR models, the report is not intended to establish specific criteria for judging the scientific validity or for regulatory acceptance of individual (Q)SAR models. The report defers explicitly to the appropriate regulatory authorities in the member countries to establish criteria for validity and acceptance.

The OECD has also recently published a report on the use of (Q)SAR in the various member countries (OECD 2007c). The report is part of the OECD effort to develop guidance for regulatory applications of QSAR models, and emphasizes the use of program-specific case studies to highlight the importance of legal and practical constraints and information requirements of individual regulatory programs within OECD member countries in applying (Q)SAR approaches. The report provides a snapshot of the experiences of OECD member countries with respect to the use of (Q)SAR models in chemical assessment. The document provides both current regulatory uses in OECD member countries as well as prospective regulatory applications - especially those within the EU Member States as a result of the new chemicals regulation, REACH. As such, the case studies report should be regarded as a living document, which will be updated periodically as requested by member countries to describe the expanding role of (Q)SAR in regulatory settings.

A review made by ECETOC (European Centre for Ecotoxicology and Toxicology of Chemicals, an industry-financed scientific forum) on the use of (Q)SARs within current regulatory decision-making frameworks in EU, North America, and Japan, and within industry concluded that applicability of currently available (Q)SARs for chronic mammalian toxicity, certainly as a stand-alone approach, was very limited at that time (ECETOC 2003).

The ECB (European Chemicals Bureau, EU's focal point for the assessment procedure on chemicals) has started building a freely accessible inventory of evaluated (Q)SAR models, which help to identify valid (Q)SARs for regulatory purposes. If there are any models relevant for the underlying endpoint these will be included in the ECB inventory (ECB 2007a).

3.6 DATA COLLECTION

The data to be used in a hazard assessment can be collected from publicly available criteria documents and monographs from international bodies, peer reviewed articles and other publications, and international databases. In some cases, unpublished data from, e.g., industry can also be available for the assessor.

3.6.1 CRITERIA DOCUMENTS AND MONOGRAPHS FROM INTERNATIONAL BODIES

Criteria documents and monographs from international bodies relevant for risk assessment of chemical substances are described in the following sections (see also Figure 3.1).

3.6.1.1 Environmental Health Criteria

The Environmental Health Criteria (EHC) monographs (Figure 3.2), which are issued by the International Programme on Chemical Safety (IPCS), are intended to assist national and international authorities in making risk assessments and subsequent risk management decisions. They represent a thorough evaluation of risks and are not, in any sense, recommendations for regulation or standard setting.

A Task Group carries out the peer review. The Task Group members serve as individual scientists, not as representatives of any organization, government, or industry. Their function is to evaluate the accuracy, significance, and relevance of the information in the document and to assess the health and environmental risks from exposure to the chemical. The principles, procedures, and scientific criteria that guide the evaluations are described in the preamble to the EHC monographs.

The EHC monographs provide critical reviews on the effect on human health and the environment of chemicals and of combinations of chemicals and physical and biological agents. As such, they include and review studies that are of direct relevance for the evaluation. However, they do not describe every study carried out. Worldwide data are used and are quoted from original studies, not from abstracts or reviews. Both published and unpublished reports are considered and it is incumbent on the authors to assess all the original articles cited in the references. Preference is

FIGURE 3.1 Examples of criteria documents and monographs from international bodies relevant for risk assessment of chemical substances.

always given to published data. Unpublished data are used only when relevant published data are absent or when they are pivotal to the risk assessment. A detailed policy statement is available that describes the procedures used for unpublished proprietary data so that this information can be used in the evaluation without compromising its confidential nature.

FIGURE 3.2 Examples of Environmental Health Criteria (EHC) documents published by the WHO's International Programme on Chemical Safety (IPCS).

EHC monographs examine: the physical and chemical properties and analytical methods; sources of environmental and industrial exposure and environmental transport; kinetics and metabolism including absorption, distribution, transformation, and elimination; short- and long-term effects on animals, carcinogenicity, mutagenicity, and teratogenicity; and finally, an evaluation of risks for human health and the effects on the environment.

In the evaluation of human health risks, sound human data, whenever available, are preferred to animal data. Animal and *in vitro* studies provide support and are used mainly to supply evidence missing from human studies. It is mandatory that research on human subjects is conducted in full accord with ethical principles, including the provisions of the Helsinki Declaration (World Medical Association 2000).

The EHC monographs are published by the World Health Organization (WHO) and are available at the INCHEM Web site (INCHEM 2007a).

3.6.1.2   Monographs from the International Agency for Research on Cancer

The International Agency for Research on Cancer (IARC) monographs (Figure 3.3) identify environmental factors that can increase the risk of human cancer. These include chemicals, complex

FIGURE 3.3 Examples of an IARC monograph published by the WHO's International Agency for Research on Cancer (IARC).

mixtures, occupational exposures, physical and biological agents, and lifestyle factors. National health agencies use this information as scientific support for their actions to prevent exposure to potential carcinogens.

Interdisciplinary working groups of expert scientists review the published studies and evaluate the weight of the evidence that an agent can increase the risk of cancer. The principles, procedures, and scientific criteria that guide the evaluations are described in the preamble to the IARC monographs.

Each monograph reviews all pertinent epidemiological studies and cancer bioassays in experimental animals. Mechanistic and other relevant data are also reviewed. With regard to epidemiological studies, cancer bioassays, and mechanistic and other relevant data, only reports that have been published or accepted for publication in the openly available scientific literature are reviewed. Data from government agency reports that are publicly available are also considered.

Exposure data and other information on an agent under consideration are also reviewed. In the sections on chemical and physical properties, on analysis, on production and use and on occurrence, published and unpublished sources of information may be considered.

Inclusion of a study does not imply acceptance of the adequacy of the study design or of the analysis and interpretation of the results, and limitations are clearly outlined in square brackets at the end of each study description. The reasons for not giving further consideration to an individual study are also indicated in square brackets.

Since 1971, more than 900 agents have been evaluated, of which approximately 400 have been identified as carcinogenic or potentially carcinogenic to humans.

Summaries and evaluations on the evaluated agents are available at the INCHEM Web site (INCHEM 2007b) whereas the IARC monographs are priced publications, which can be purchased via the IARC Web site (IARC 2007).

3.6.1.3 Monographs from the Joint FAO/WHO Expert Committee on Food Additives and from the Joint FAO/WHO Meeting on Pesticide Residues

Toxicological evaluations of food additives and of contaminants, naturally occurring toxicants and residues of veterinary drugs in food produced by the Joint FAO/WHO Expert Committee on Food Additives (JECFA), and of pesticide residues in food by the Joint FAO/WHO Meeting on Pesticide Residues (JMPR) are used by the Codex Alimentarius Commission and national governments to set international food standards and safe levels for protection of the consumer.

The monographs provide the toxicological information upon which the JECFA or JMPR make their evaluations (Figure 3.4). The monographs are prepared by scientific experts and peer reviewed at the JECFA or JMPR meetings. The monographs are generally based on working papers prepared by temporary advisers. The temporary advisers base their working papers on all the data that have been submitted, and all of these reports are also available to JECFA or JMPR when they make their evaluation. Proprietary unpublished reports on food additives or pesticides are not referenced as these generally are voluntarily submitted to JECFA or JMPR by various producers of the food additives or pesticides under review, and in many cases represent the only data available on those substances.

The JECFA and JMPR monographs are published by the WHO and are available at the INCHEM Web site (INCHEM 2007c,d).

3.6.1.4 WHO Guidelines for Drinking-Water Quality

The WHO published the first and second editions of the Guidelines for Drinking-Water Quality (Figure 3.5), in three volumes, in 1984–1985 and in 1993–1997, respectively, and a third edition of Volume 1 was published in 2004 (Section 9.2.1.2). The publications are available to the public at the WHO Water Sanitation and Health Web site (WHO 2007a).

FIGURE 3.4 Examples of monographs from the Joint FAO/WHO Expert Committee on Food Additives (JECFA) and from the Joint FAO/WHO Meeting on Pesticide Residues (JMPR) published by the WHO.

3.6.1.5 WHO Air Quality Guidelines for Europe

The WHO Regional Office for Europe has published "Air Quality Guidelines for Europe" (Figure 3.5) containing health risk assessments of chemical air contaminants (Section 9.2.1.1). The publication is available at the WHO Regional Office for Europe's Web site (WHO 2007b).

FIGURE 3.5 WHO air quality guidelines for Europe and WHO guidelines for drinking-water quality.

FIGURE 3.6 Examples of Toxicological Profiles from the Agency for Toxic Substances and Disease Registry (ATSDR).

3.6.1.6 Toxicological Profiles from Agency for Toxic Substances and Disease Registry

By Congressional mandate, the Agency for Toxic Substances and Disease Registry (ATSDR) produces monographs, the so-called "Toxicological Profiles" (Figure 3.6), for hazardous substances found at National Priorities List (NPL) sites.

Toxicological Profiles are developed in two stages: (1) drafts that undergo a 90-day public commenting period and (2) finals where ATSDR has considered incorporating all comments into the documents and finalized the Profiles after which the National Technical Information Service (NTIS) distributes them. The Toxicological Profiles are available at the ATSDR Web site (ATSDR 2007).

Each Profile succinctly characterizes the toxicological and adverse health effects information for the hazardous substance described. Each peer-reviewed profile identifies and reviews in detail the key literature that describes a hazardous substance's toxicological properties. The focus of the Profile is on health and toxicological information. Therefore, each profile begins with a Public Health Statement that summarizes in nontechnical language, a substance's relevant properties. A two-page information sheet, the ToxFAQs, is also available; these are also available at the ATSDR Web site (ATSDR 2007).

So far, 289 Toxicological Profiles have been published or are under development as "finals" or "drafts for public comment;" 268 profiles have been published as finals; 118 profiles have been updated. Currently, 14 profiles are being revised based on public comments received. These profiles cover more than 250 substances.

3.6.1.7 EU Risk Assessment Reports on Existing Chemicals

In the EU, 141 of existing high production volume chemical substances have been prioritized for an in-depth risk assessment, see Section 2.4.1.4 for details.

After adoption of the risk assessment, the comprehensive risk assessment report (Figure 3.7) is available at the ECB Web site (ECB 2007b).

FIGURE 3.7 An example of a European Union risk assessment report on existing chemicals.

3.6.1.8 Publications from the International Life Sciences Institute

The International Life Sciences Institute (ILSI) was founded in 1978 as a nonprofit, worldwide foundation that seeks to improve the well-being of the general public through the advancement of science (ILSI 2007a).

Its goal is to further the understanding of scientific issues relating to nutrition, food safety, toxicology, risk assessment, and the environment by bringing together scientists from academia, government, and industry.

ILSI publications include scientific monographs, books, periodicals, and technical catalogs. These are critical references on nutrition and diet, risk assessment, toxicology, pathology, allergy and immunology, and other environmental and health issues. These publications help strengthen scientific and professional expertise, improve occupational practice, enhance and advance scientific research, and supplement academic teaching worldwide. The various publications can be purchased via the ILSI Web site (ILSI 2007b).

3.6.1.9 Publications from the European Centre for Ecotoxicology and Toxicology of Chemicals

ECETOC was established in 1978 as a scientific, nonprofit making, noncommercial association, financed by 51 of the leading industrial companies with interests in the manufacture and use of chemicals (ECETOC 2007a).

FIGURE 3.8 Examples of Technical Reports published by the European Centre For Ecotoxicology and Toxicology of Chemicals (ECETOC).

ECETOC's main objective is to identify, evaluate, and through such knowledge help industry minimize any potentially adverse effects on health and the environment that may arise from the manufacture and use of chemicals. To achieve this mission, ECETOC facilitates the networking of suitably qualified scientists from its member companies and cooperates in a scientific context with intergovernmental agencies, health authorities, and professional institutions.

The output of the ECETOC work program is delivered in published reports and papers, through scientific representation in the activities of international organizations and regulatory groups and through presentations and organization of specialized workshops and fora.

Monographs are comprehensive reviews of generic topics or issues fundamental to the application of sound science in evaluating the hazards and risks of chemicals to human health and the environment.

Technical Reports (Figure 3.8) address specific applications of the science in evaluating the hazards and risks of chemicals to human health and the environment.

JACC Reports (Joint Assessment of Commodity Chemicals) are comprehensive reviews of toxicological and ecotoxicological data on individual chemical substances.

Reports are published following peer review by the Scientific Committee comprising members from various European industrial companies. ECETOC's publications are provided freely to all member companies and to other interested parties, such as the various regulatory authorities, international organizations, and academic groups. For others, the various publications can be purchased via the ECETOC Web site (ECETOC 2007b).

3.6.1.10 BUA Reports from the German Chemical Society

The German Chemical Society (GDCh) is the largest chemical society in continental Europe with members from academy, industry, and other areas (GDCh 2007). The GDCh supports chemistry in teaching, research, and application and promotes the understanding of chemistry in the public.

In 1982, the Advisory Committee on Existing Chemicals of Environmental Relevance (BUA) was established as a committee of the GDCh by an agreement between the German government,

FIGURE 3.9 An example of a BUA Report published by the Advisory Committee on Existing Chemicals of Environmental Relevance (BUA).

scientific community, and chemical industry (BUA 2007). This committee, made up equally of representatives of the chemical industry, scientific community, and agencies, was given the task of drawing up a concept for the systematic evaluation of existing chemicals. The BUA is responsible for an examination of the data sets, search and evaluate the literature, draft a report, and publish the results in BUA Reports (Figure 3.9). For the German government and the responsible assessment agencies, BUA Reports represent the most important decision-making aid in the final assessment of substances and in calling for risk-reducing measures where necessary.

To date, the BUA has published 251 reports on about 335 substances; these can be purchased via the GDCh Web site (BUA 2007). All BUA Reports are also available in English translation in bookstores.

3.6.1.11 American Conference on Governmental Industrial Hygienists

The American Conference on Government Industrial Hygienists (ACGIH) is a member-based organization that advances occupational and environmental health (ACGIH 2007). An example of this includes the annual publication of the Threshold Limit Values (TLVs) and Biological Exposure Indices (BEIs) recommended for chemical substances.

The TLVs and BEIs are health-based values established by committees that review existing published and peer-reviewed literature in various scientific disciplines (e.g., industrial hygiene, toxicology, occupational medicine, and epidemiology). TLVs have been recommended for more than 700 chemical substances and physical agents and there are more than 50 BEIs that cover more than 80 chemical substances. Documentations provide the rationale for the TLVs and BEIs; these can be purchased via the ACGIH Web site (ACGIH 2007).

3.6.1.12 Criteria Documents from the Nordic Expert Group

The Nordic Expert Group for Criteria Documentation of Health Risks from Chemicals (NEG) consisted of scientific experts from the five Nordic countries representing different fields of science, such as toxicology, occupational hygiene, and occupational medicine. The main task was to produce criteria documents (Figure 3.10) to be used by the regulatory authorities of the Nordic countries as the scientific basis for setting Occupational Exposure Limits (OELs) for chemical substances.

The documents, written in English, were published by the Swedish National Institute for Working Life (NIWL, closed down as of 1 July 2007) in the scientific serial *Arbete och Hälsa*. The documents are now available at the Göteborg University's Web site (GU 2007).

FIGURE 3.10 An example of a criteria document published by the Nordic Expert Group for Criteria Documentation of Health Risks from Chemicals (NEG).

3.6.2 DATABASES

Databases relevant for a hazard assessment of chemical substances are described in the following sections.

3.6.2.1 IPCS INCHEM

The IPCS INCHEM was produced through cooperation between the International Programme on Chemical Safety (IPCS) and the Canadian Centre for Occupational Health and Safety (CCOHS), and directly responds to one of the Intergovernmental Forum on Chemical Safety (IFCS) priority actions to consolidate current, internationally peer-reviewed chemical safety-related publications and database records from international bodies, for public access (IPCS INCHEM 2007).

IPCS INCHEM offers quick and easy electronic access to thousands of searchable full-text documents on chemical risks and management of chemicals. IPCS INCHEM contains the following (see also Section 2.1.7):

- Concise International Chemical Assessment Documents (CICADs)
- Environmental Health Criteria (EHC) monographs
- Health and Safety Guides (HSGs)
- International Agency for Research on Cancer (IARC) - Summaries and evaluations
- International Chemical Safety Cards (ICSCs)
- IPCS/CEC Evaluation of Antidotes Series
- Joint Expert Committee on Food Additives (JECFA) - Monographs and evaluations
- Joint Meeting on Pesticide Residues (JMPR) - Monographs and evaluations
- Pesticide Data Sheets (PDSs)
- Poisons Information Monographs (PIMs)
- Screening Information Data Set (SIDS) for High Production Volume Chemicals

3.6.2.2 International Uniform ChemicaL Information Database

International Uniform ChemicaL Information Database (IUCLID) is the basic tool for data collection and evaluation within the EU risk assessment program for existing substances and has also been accepted by the OECD as the data exchange tool under the OECD Existing Chemicals Program (see Section 2.4.1.6 for details).

A nonconfidential IUCLID CD-ROM is available at the ECB Web site (ECB 2007c).

3.6.2.3 US-EPA Integrated Risk Information System

The Integrated Risk Information System (IRIS) is an electronic database containing information on human health effects that may result from exposure to various chemicals in the environment (IRIS 2007). The chemical files in IRIS contain descriptive and quantitative information in the following categories:

- Oral reference doses and inhalation reference concentrations (RfDs and RfCs, respectively) for chronic noncarcinogenic health effects
- Hazard identification, oral slope factors, and oral and inhalation unit risks for carcinogenic effects

3.6.2.4 U.S. National Library of Medicine TOXNET

The Division of Specialized Information Services (SIS), National Library of Medicine (NLM), creates information resources and services in toxicology, environmental health, chemistry, and HIV/AIDS (SIS 2007).

SIS's Toxicology and Environmental Health Information Program (TEHIP) produces TOXNET (TOXNET 2007), a collection of toxicology and environmental health databases that includes, among others, the Hazardous Substances Data Bank (HSDB), a database of potentially hazardous chemicals, TOXLINE (containing references to the world's toxicology literature), and ChemIDplus (a chemical dictionary and structure database).

Some SIS products help to address the toxicology and environmental health information needs of the general public. One such resource is ToxTown, an interactive guide to toxic chemicals and environmental health issues in everyday locations. It is a companion to the extensive information in the TOXNET collection of databases. The Household Products Database is a consumer's guide that provides information on the potential health effects of chemicals contained in more than 5000 common household products used inside and around the home. This database allows consumers, scientists, and health professionals to investigate ingredients in brand-name products.

REFERENCES

ACGIH. 2007. http://www.acgih.org/home.htm

Altweb. 2007. The alternatives to animal testing Web site. http://altweb.jhsph.edu/

ATSDR. 2007. http://www.atsdr.cdc.gov/toxpro2.html

Borenstein, M., Hedges, L., Higgins, J., and Rothstein, H. 2005. Comprehensive Meta-analysis Version 2, Biostat, Englewood NJ. http://www.meta-analysis.com/

Bradford Hill, A. 1965. The environment and diseases: association or causation? *Proc. R. Soc. Med.* 58:295–300.

BUA. 2007. The Advisory Committee on Existing Chemicals of Environmental Relevance (BUA) website http://www.gdch.de/taetigkeiten/bua__e.htm

EC. 1991. Council Directive 91/414/EEC of 15 July 1991 concerning the placing of plant protection products on the market. *Off. J. Eur. Communities* L 230, 19.8.1991, pp. 1–32.

EC. 1998. Directive 98/8/EC of the European Parliament and of the Council of 16 February 1998 concerning the placing of biocidal products on the market. *Off. J. Eur. Communities* L 123, 24.4.1998, pp. 1–63.

EC. 2003. Technical Guidance Document in support of Commission Directive 93/67/EEC on Risk Assessment for new notified substances, Commission Regulation (EC) No 1488/94 on Risk Assessment for existing substances and Directive 98/8/EC of the European Parliament and of the Council concerning the placing of biocidal products on the market. http://ecb.jrc.it/tgd

ECB. 2006. European Chemicals Bureau homepage. http://ecb.jrc.it/

ECB. 2007a. http://ecb.jrc.it/qsar/

ECB. 2007b. http://ecb.jrc.it/existing-chemicals/

ECB. 2007c. http://ecb.jrc.it/iuclid5/

ECETOC. 2003. (*Q*)*SARs: Evaluation of the commercially available software for human health and environmental endpoints with respect to chemical management applications.* ECETOC Technical Report No. 89. Brussels: ECETOC.

ECETOC. 2007a. European Centre for Ecotoxicology and Toxicology of Chemicals website. http://www.ecetoc.org/Content/Default.asp?PageID = 3

ECETOC. 2007b. European Centre for Ecotoxicology and Toxicology of Chemicals (publications) website http://www.ecetoc.org/Content/Default.asp?PageID = 21

ECVAM. 2007. European Center for Validation of Alternative Methods website. http://ecvam.jrc.it/index.htm

EU. 2006. The DG Environment REACH website. http://ec.europa.eu/environment/chemicals/reach

EU. 2007. European Commision, Research: Science and Society website: The role of ethics in EU research. http://ec.europa.eu/research/science-society/page_en.cfm?id = 3172

GDCh. 2007. German Chemical Society website. http://www.gdch.de/gdch__e.htm

GU. 2007. The Göteborg University website. http://www.medicine.gu.se/avdelningar/samhallsmedicin_folkhalsa/amm/aoh/

IARC. 2007. IARC monographs on the evaluation of carcinogenic risks to humans website. http://monographs.iarc.fr/

ICCVAM. 2006. ICCVAM website. http://iccvam.niehs.nih.gov/

ICCVAM. 2007. General information about NICEATM-ICCVAM test method evaluation areas. http://iccvam. niehs.nih.gov/methods/methods.htm

ILSI. 2007a. International Life Sciences Institute (ILSI) website. http://www.ilsi.org/AboutILSI/

ILSI. 2007b. International Life Sciences Institute (ILSI) (publications) website. http://www.ilsi.org/ Publications/

INCHEM. 2007a. IPCS INCHEM Environmental Health Criteria Monographs (EHCs) website. http://www. inchem.org/pages/ehc.html

INCHEM. 2007b. IPCS INCHEM International Agency for Research on Cancer (IARC)—Summaries & evaluations website. http://www.inchem.org/pages/iarc.html

INCHEM. 2007c. IPCS INCHEM JECFA—Monographs & evaluations website. http://www.inchem.org/ pages/jecfa.html

INCHEM. 2007d. IPCS INCHEM JMPR—Monographs & evaluations website. http://www.inchem. org/pages/jmpr.html

IPCS INCHEM. 2007. IPCS INCHEM website. http://www.inchem.org/pages/about.html

IRIS. 2007. U.S. EPA Integrated Risk Information System (IRIS) website. http://www.epa.gov/iris/

Leandro, G. 2005. *Meta-Analysis in Medical Research. The Handbook for the Understanding and Practice of Meta-Analysis.* Oxford: Blackwell Publishing.

Lucas, R. M. and McMichael, A. J. 2005. Association or causation: evaluating links between "environment and disease." *Bull. WHO* 83:792–795. http://www.who.int/bulletin/volumes/83/10/792.pdf

NCT. 2007. The National Center for Toxicogenomics (NCT) website. http://www.niehs.nih.gov/nct/office.htm

OECD. 2002. 3.3 *Guidance for the use of structure–activity relationships (SARs) in the HPV chemicals programme.* In: *Manual for Investigation of HPV Chemicals.* Paris: OECD. http://www.oecd. org/dataoecd/60/24/1947517.pdf

OECD. 2003. *Descriptions of Selected Key Generic Terms Used in Chemical Hazard/Risk Assessment. Joint Project with IPCS on the Harmonisation of Hazard/Risk Assessment Terminology.* OECD Series on Testing and Assessment No. 44. Environment Directorate, Joint Meeting of the Chemicals Committee and the Working Party on Chemicals, Pesticides and Biotechnology. ENV/JM/MONO(2003)15. Paris: OECD.

OECD. 2004. *Manual for Investigation of HPV Chemicals.* Paris: OECD.http://www.oecd.org/document/7/ 0,2340,en_2649_34379_1947463_1_1_1_1,00.html

OECD. 2006. OECD website. http://www.oecd.org

OECD. 2007a. OECD website. OECD activities to explore and evaluate regulatory application of genomic methods; toxicogenomics. http://www.oecd.org/document/29/0,3343,en_2649_34365_34704669_ 1_1_1_1,00.html

OECD. 2007b. *Guidance document on the validation of (quantitative) structure–activity relationships [(Q) SAR] models.* ENV/JM/MONO(2007)2. 15 February 2007. Paris: OECD http://www.oecd.org/ searchResult/0,2665,en_2649_201185_1_1_1_1_1,00.html

OECD. 2007c. *Report on the regulatory uses and applications in OECD member countries of (quantitative) structure activity relationship (Q)SAR models in the assessment of new and existing chemicals.* ENV/JM/MONO(2006)25. 15 February 2007. Paris: OECD http://www.oecd.org/searchResult/ 0,2665,en_2649_201185_1_1_1_1_1,00.html

OHRP. 2007. Office for Human Research Protections (OHRP) homepage. http://www.hhs.gov/ohrp/ about/ohrpfactsheet.htm.

Rothman, K.J. and Greenland, S. 1998. *Modern Epidemiology.* Philadelphia: Lippincott-Raven.

Russell, W. M. S. and Burch, R. L. 1959. *The Principles of Humane Experimental Technique.* London: Methuen. http://altweb.jhsph.edu/publications/humane_exp/foreward.htm

SIS. 2007. Division of Specialized Information Services, National Library of Medicine http://www.nlm.nih. gov/pubs/factsheets/sis.html

Swaen, G. M. H. 2006. A framework for using epidemiological data for risk assessment. *Hum. Exp. Toxicol.* 25:147–155.

TOXNET. 2007. U.S. National Library of Medicine. Databases on toxicology, hazardous chemicals, environ-mental health, and toxic releases website. http://toxnet.nlm.nih.gov/

US-EPA. 2004. U.S. Environmental Protection Agency: Potential implications of genomics for regulatory and risk assessment applications at EPA. http://www.oecd.org/dataoecd/32/61/34706973.pdf

US-EPA. 2006. EPA website. 2006. http://www.epa.gov/

US-EPA. 2007a. Human testing; proposed plan and description of review process. *Federal Register: February 8, 2005*. http://www.epa.gov/EPA-TOX/2005/February/Day-08/t2371.htm

US-EPA. 2007b. Pesticides science and policy website. Expanded protections for subjects in human studies research. http://www.epa.gov/oppfead1/guidance/human-test.htm

WHO. 2002. Introduction to basic epidemiology and principles of statistics for tropical diseases control. WHO/CDS/CPE/SMT/2000.2 Rev.1 Updated July 2002. http://www.who.int/malaria/cmc_upload/0/000/015/866/basicepidemiology_tg-en.pdf

WHO. 2007a. WHO water sanitation and health website. http://www.who.int/water_sanitation_health/dwq/guidelines/en/

WHO. 2007b. WHO regional office for Europe's (air quality and health) website. http://www.euro.who.int/air

WHO/IPCS. 1999. *Principles for the Assessment of Risks to Human Health from Exposure to Chemicals*. Environmental Health Criteria 210. Geneva: WHO. http://www.inchem.org/documents/ehc/ehc/ehc210.htm

World Medical Association. 2004. Declaration of Helsinki: Ethical Principles for Medical Research Involving Human Subjects. http://www.wma.net/e/policy/b3.htm

4 Hazard Assessment

The aim of the hazard assessment of a chemical substance under evaluation is to assess whether exposure to the chemical might result in adverse health effects in humans, based on a critical evaluation of the available data on the inherent toxicological properties of the substance as well as the toxicological mode(s) of action/mechanisms of toxicity.

4.1 INTRODUCTION

According to the OECD/IPCS definitions listed in Annex 1 (OECD 2003a):

Hazard assessment is "A process designed to determine the possible adverse effects of an agent or situation to which an organism, system or (sub) population could be exposed. The process includes hazard identification and hazard characterization. The process focuses on the hazard in contrast to risk assessment where exposure assessment is a distinct additional step."

Hazard assessment is also known as "effect assessment."

Hazard identification is "The identification of the type and nature of adverse effects that an agent has as inherent capacity to cause in an organism, system or (sub) population."

Hazard identification is the first stage in hazard assessment and the first step in the process of risk assessment.

Hazard characterization is "The qualitative and, wherever possible, quantitative description of the inherent properties of an agent or situation having the potential to cause adverse effects. This should, where possible, include a dose–response assessment and its attendant uncertainties."

Hazard characterization, also known as dose–response assessment, is the second stage in hazard assessment, and the second step in the process of risk assessment. At this step, the No-Observed-Adverse-Effect Level (NOAEL) and the Lowest-Observed-Adverse-Effect Level (LOAEL) are derived for the observed effects, where possible and appropriate.

A hazard assessment of a specific chemical substance is carried out based on all the data available on adverse health effects in humans and in experimental animals, i.e., on acute toxicity, irritation/corrosion, sensitization, repeated dose toxicity, mutagenicity/genotoxicity, carcinogenicity, and reproductive toxicity. In addition, the toxicokinetic properties of the substance are also essential in the hazard assessment, i.e., in the interpretation of the toxicological findings and hence in the hazard and risk assessment processes. Furthermore, information on the toxicological mode(s) of action as well as mechanistic data are essential in establishing the relevance to humans of the toxicological effects observed in experimental animals. Also data from *in vitro* tests (as well as nontesting data such as (Q)SAR and physico-chemical properties) are relevant as part of a weight of evidence approach for identifying the inherent toxicological properties of a specific chemical under evaluation.

During both steps of the hazard assessment, hazard identification and hazard characterization, it is important to evaluate the available data with regard to their adequacy and completeness. The evaluation of adequacy shall address the reliability and relevance of the data. These aspects are addressed in detail in Chapter 3.

This chapter will first address (Section 4.2) some general aspects of importance for the hazard assessment: systemic effects versus local effects (Section 4.2.1); adverse effect(s) versus non-adverse effect(s) (Section 4.2.2); dose–response relationships (Section 4.2.3); no-effect levels and lowest-effect levels (Section 4.2.4); the Benchmark Dose concept (Section 4.2.5); toxicological mode(s) of action (Section 4.2.6); and critical effect(s) (Section 4.2.7). Secondly, the assessment of the toxicokinetic properties of a chemical substance and use of such information in the hazard assessment is addressed (Section 4.3). Then the focus is put on the hazard assessment of the various toxicological endpoints including acute toxicity (Section 4.4), irritation and corrosion (Section 4.5), sensitization (Section 4.6), toxicity following repeated exposure to the substance (repeated dose toxicity, Section 4.7), mutagenicity and genotoxicity (Section 4.8), carcinogenicity (Section 4.9), and effects on fertility, and fetal and postnatal development (reproductive toxicity, Section 4.10).

In recent years, concern that chemicals might inadvertently be disrupting the endocrine system of humans and wildlife has increased. The concerns regarding exposure to these "endocrine disruptors" are based on adverse effects observed in certain wildlife, fish, and ecosystems; increased incidences of certain endocrine-related human diseases; and adverse effects observed in laboratory animals exposed to certain chemicals. The main effects reported in both wildlife and humans concern reproductive and sexual development and function; altered immune system, nervous system, and thyroid function; and hormone-related cancers. Endocrine disruption is not considered a toxicological endpoint in its own right, but a functional change or toxicological mode(s) of action that may lead to adverse effects. Endocrine disrupters are addressed further in Section 4.11.

Finally, the concept of hormesis (Section 4.12), threshold of toxicological concern (Section 4.13), and probabilistic methods for effect assessment (Section 4.14) will be briefly addressed.

For each toxicological endpoint as well as for toxicokinetics, relevant definitions are provided, the objectives of investigating the potential for the substance-induced endpoint are summarized, the relevant test guidelines are mentioned, and the principles of the various test methods are briefly described. The specific test guidelines are not included in the reference list to this chapter as an overview of the adopted human toxicity test guidelines with references to the relevant Web sites is included in Chapter 2, the OECD Test Guidelines in Table 2.3, the US-EPA (OPPTS) Harmonized Test Guidelines in Table 2.7, and the EU Test Guidelines (Annex V methods) in Table 2.9.

The hazard assessment of toxicokinetic and toxicological properties of chemical substances have been extensively addressed in various guidance documents, e.g., in the EU Technical Guidance Document (TGD) on Risk Assessment of New and Existing Chemical Substances and Biocides (EC 2003), the WHO/IPCS Environmental Health Criteria 210 (WHO/IPCS 1999a) as well as in the OECD Manual for Investigation of HPV (High Production Volume) Chemicals (OECD 2004a). The US-EPA's Risk Assessment Guidelines (Section 2.3.1.8 and Table 2.6) contain guidance on risk assessment, including hazard assessment, in relation to each guideline's toxicological endpoint. In addition, various criteria documents and evaluations from international bodies have also addressed principles for hazard assessment of chemical substances.

In the new EU chemicals regulation REACH, which entered into force on 1 June 2007, detailed guidance documents on different REACH elements, including hazard assessment of chemical substances, are currently in preparation (spring 2007). These documents will probably be available on the EU DG Environment REACH Web site (EU 2006) when published.

For each toxicological endpoint as well as for toxicokinetics, relevant guidance documents, criteria documents, and evaluations from international bodies are mentioned.

Finally, the use of different types of information (human data, data from studies in experimental animals, *in vitro* test data, and other data such as, e.g., data on physico-chemical properties and (Q)SAR) in the hazard assessment for a specific endpoint is addressed in more detail.

4.2 GENERAL ASPECTS

As mentioned above, the hazard assessment process comprises two steps: hazard identification and hazard characterization.

The purpose of the hazard identification is to evaluate the weight of evidence of adverse effects in humans based on assessment of all available data on toxicity and toxicological mode(s) of action. It is designed to address primarily two questions: (1) whether a substance may pose a health hazard to humans, and (2) under what circumstances an identified hazard may be expressed. The weight of evidence is assessed on the basis of combined strength and coherence drawn from all of the available data. This entails a critical examination of the quality and nature of the results of available epidemiological studies, toxicological studies in experimental animals, *in vitro* studies and (Q)SAR analyses, and information on toxicological mode(s) of action. The latter is particularly important with respect to assessment of relevance to humans. The result of the hazard identification is a scientific judgment as to whether the substance can cause an adverse effect in humans.

The purpose of the hazard characterization is the quantitative assessment of the dose–response relationship, also known as dose–response assessment. Approaches to a quantification of the dose–response relationship vary according to the scope and purpose of the assessment. For most types of toxic effects (e.g., organ-specific, neurological, immunological, non-genotoxic carcinogenicity, reproductive, developmental), it is generally considered that there is a dose or concentration below which adverse effects will not occur, i.e., a threshold exists. For other types of effects (e.g., sensitization, mutagenicity, genotoxicity, genotoxic carcinogenicity) it is assumed that there is some probability of harm at any level of exposure, i.e., no threshold exists. Adverse health effect(s) can thus be considered to be of two types: those considered to have a threshold, known as "threshold effects," and those for which there is considered to be some risk at any exposure level, known as "non-threshold effects." Though it is not possible to demonstrate experimentally the presence or absence of a threshold, differences in the approach to the dose–response assessment of threshold versus non-threshold effects have been adopted widely. The distinction in approaches is based primarily on the premise that simple events such as *in vitro* activation and covalent binding may be linear over many orders of magnitude, i.e., that these events occur even at very low exposure levels. However, a simple pragmatic distinction on this basis is increasingly problematic as it is likely that there is a threshold for a number of genotoxic effects. This is addressed in more detail in Chapter 6.

The concept of thresholds in toxicology and risk assessment has been discussed in an article by Slob (1999). According to Slob, a dose-threshold may be defined in different ways:

- Biological definition: The dose, below which the organism does not suffer from any (adverse) effects from the compound considered.
- Experimental definition: The dose, below which no effects are observed.
- Mathematical definition: The dose, below which the response is zero, and above which it is nonzero.

According to Slob, the definition that toxicologists usually have in mind, probably lies somewhere between the biological and the experimental definition. The toxicologist's definition reflects the observation that a substance does not seem to affect organisms at sufficiently low doses. The observation that organisms are apparently able to handle small amounts of a substance may be explained by biological phenomena such as homeostasis and repair. For example, as long as the damage caused by the substance does not exceed the repair capacity, the organism would not suffer from it, assuming repair has no harmful effect by itself.

According to the OECD/IPCS definitions listed in Annex 1 (OECD 2003a):

Threshold is "Dose or exposure concentration of a substance below that a stated effect is not observed or expected to occur."

For threshold effects, traditionally, a level of exposure below which it is believed that there are no adverse effects estimated, based on an approximation of the threshold termed the No-Observed-(Adverse)-Effect Level (NO(A)EL) and assessment factors; this is addressed in detail in Chapter 5. This estimated level of exposure will in this book be termed "tolerable exposure level." Examples, where this approach is used, include establishment of the Acceptable/Tolerable

Daily Intake (ADI/TDI), or Reference Dose (RfD); these terms are addressed in detail in Chapter 5. As an alternative to the traditional NOAEL approach, the Benchmark Dose (BMD) (a model-derived estimate or its lower confidence limit of a particular incidence level, see Section 4.2.5) for the critical effect has been proposed for use in the quantitative assessment of the dose–response.

At present, there is no clear consensus on an appropriate methodology for the hazard characterization of non-threshold effects. In many countries, cancer risks have traditionally been assessed by mathematical modeling of the dose–response data in the observable range to estimate the risk at much lower human exposure levels, i.e., low dose risk extrapolation (Chapter 6). It should be noted that such low dose risk estimation is uncertain. Owing to this uncertainty, alternative measures of dose–response are increasingly being developed and adopted; for example, there is an increasing reliance on specification of the margin between potency in the experimental range and exposure as the measure of risk for carcinogens (addressed in detail in Chapter 6).

4.2.1 SYSTEMIC EFFECTS VERSUS LOCAL EFFECTS

A substance may exert systemic as well as local effects.

A local effect is an effect that is observed at the site of first contact, caused irrespective of whether a substance is systemically available.

A systemic effect is an effect that is normally observed distantly from the site of first contact, i.e., after the substance has passed through a physiological barrier (mucous membrane of the gastrointestinal tract or of the respiratory tract, or the skin) and becomes systemically available. It should be noted, however, that toxic effects on surface epithelia may reflect indirect effects as a consequence of systemic toxicity or secondary to systemic distribution of the substance or its active metabolite(s).

Commonly, the underlying toxicological mode(s) of action is not clarified by routine toxicity studies. The decision as to whether or not an effect should be considered local or systemic, is based on expert judgment. In addition, the decision as to whether or not a local effect should be considered as a substance-related adverse effect or caused by treatment procedures (e.g., adverse effects in the upper gastrointestinal tract, mediastinum, and lungs following bolus application in oral gavage studies), should also be based on expert judgment.

4.2.2 ADVERSE EFFECTS VERSUS NON-ADVERSE EFFECTS

Central in the hazard assessment of chemical substances is a clear understanding of whether an observed effect constitutes an adverse effect or can be considered as non-adverse.

The distinction between non-adverse effects and adverse effects can seem academic, but is essential in the hazard assessment in relation to, e.g., evaluation of no-effect levels and lowest-effect levels (Section 4.2.4), identification of the critical effect(s) (Section 4.2.7), and to the magnitude of the assessment factor to be used for taking into account the uncertainty due to the nature and severity of effects (Section 5.8).

According to the OECD/IPCS definitions listed in Annex 1 (OECD 2003a):

Adverse effect is "Change in the morphology, physiology, growth, development, reproduction or life span of an organism, system, or (sub) population that results in an impairment of functional capacity, an impairment of the capacity to compensate for additional stress, or an increase in susceptibility to other influences."

In general, there are two types of significant biological responses to external influences. First, there are the normal biological responses, which will occur in response to stress, e.g., sweating in exercise, loss of weight in starvation; these changes often represent normal homeostatic reactions to stimuli. Secondly, there are the abnormal biological responses, which may be caused by exposure to chemicals or other stresses. Either of these types of biological responses could be significantly different from the normal baseline when subjected to statistical analysis.

Thus, the hazard identification, in addition to an identification of the inherent toxicological properties (type of effects), also involves an evaluation of the nature of the observed effects, i.e.: (1) whether an observed effect constitutes an adverse effect and thus results in an impairment of body function(s), and (2) whether an effect is a direct toxic effect exerted by the chemical (biologically relevant) or is due to normal unspecific reactions toward changes in the environment (homeostasis). Examples of effects, which generally are not considered as being adverse, include:

- Fluctuations in enzyme levels or biochemical parameters, which obviously are effects exerted by the chemical, but will generally not result in an impairment of the body function(s); such effects are often interpreted as indicators of adverse effects at higher exposure levels.
- Discoloration of organs and tissues, which obviously is an effect exerted by the chemical, but is not considered as being an adverse effect if other signs of toxicity such as, e.g., biochemical and/or histopathological changes are not observed concurrently. However, it should be noted that, according to the EU classification criteria for dangerous substances and preparations (Annex VI to Directive 67/548 (EC 2001a)), ocular lesions are regarded as being severe if the substance (or preparation) causes irreversible coloration of the eyes.
- Induction of enzymes involved in the metabolism of a test chemical is generally not a direct toxic effect exerted by the chemical, but a reaction toward a xenobiotic entering the body.
- Decreased body weight gain in experimental animals, which can be related to the palatability of the feed and thus is not a direct toxic effect exerted by the chemical.

The distinction between a non-adverse effect and an adverse effect is furthermore evaluated from a more quantitative point of view in terms of statistical significance, i.e., an evaluation whether a biologically relevant effect (adverse from a qualitative point of view) occurs in treated individuals at an incidence, which is statistically significantly different from the incidence of that effect in the control group. From this point of view, the distinction between an effect and an adverse effect relies and depends on the available data and may therefore be encumbered with some uncertainty. Uncertainty in this respect is defined as an absence of information or inadequate information about a specific parameter that could be reduced by undertaking studies to fill in gaps in the knowledge. The following two examples illustrate this aspect:

- Histopathological alterations have been observed in an organ or tissue of experimental animals, but the incidences of this effect in the treated group(s) were not statistically significantly different from that in the control group. The effect might thus be interpreted as being due to biological variation among the individual animals and thus not an adverse effect. However, if the number of animals in the treated groups was increased, then the incidence of the effect in the treated group(s) might possibly achieve statistical significance from that in the control group and thus be interpreted as an adverse effect.
- For certain unusual effects such as, e.g., some types of malformations in the fetus or rare tumors, the number of animals usually included in the various dose groups in test guideline studies is generally too low in order for the incidence of such an effect in treated group(s) to achieve statistical significance from that in the control group.

As can be seen from these two examples, an achievement of statistical significance very much relies on the number of animals included in the various dose groups. If statistical significance is not achieved for a specific biologically relevant endpoint or parameter, the effect should be considered as being adverse if it is plausible that the effect is not related to biological variation and thus occurring by chance. An example of a rare effect, which may be evaluated as biologically significant in spite of lacking statistical significance, is the occurrence of the fetal malformation cleft palate in the rat; this effect is usually regarded as a sign of chemical-induced developmental toxicity, even if only a single pup or a few pups in a study are affected.

The decision as to when a change is biologically significant is usually left to expert judgment. As a starting point, deviations exceeding 5%–10% of the control value, e.g., in body weight, are often considered biologically significant. It should be recognized that such decisions are highly individual and depend on the normal variation in the particular parameter. For example, certain clinical–chemical parameters vary considerably within or among individuals, and changes are usually only considered biologically relevant when they are consistent, or part of a pattern of observations, or very marked.

The European Centre for Ecotoxicology and Toxicology of Chemicals (ECETOC) has published a report entitled "Recognition of, and Differentiation between, Adverse and Non-adverse Effects in Toxicology Studies" (ECETOC 2002a). A review of regulatory and other scientific literature, and of current practices, revealed a lack of consistency in definition and application of a number of frequently used terms including "adverse effect," and no coherent criteria were found that could be used to the recognition of, and differentiation between, adverse and non-adverse effects. The report addresses these issues. First, a standard set of definitions is proposed for key terms frequently used to describe the overall outcome of a toxicity study. Secondly, a structured approach is proposed that will assist in arriving at a consistent study interpretation. There are two main steps in this approach. In the first step, it is decided whether differences from control values are treatment-related effects, or occur by chance. In the second step, only those differences judged to be treatment-related effects are evaluated further, in order to discriminate between those that are adverse and those that are not. For each step, criteria are described that form the bases of consistent judgments.

The report provides the following definition of an adverse effect: "A biochemical, behavioural, morphological or physiological change (in response to a stimulus) that either singly or in combination adversely affects the performance of the whole organism or reduces the organism's ability to respond to an additional environmental challenge. In contrast to adverse effects, non-adverse effects can be defined as those biological effects that do not cause biochemical, behavioural, morphological or physiological changes that affect the general well-being, growth, development or life span of an animal."

Also relevant in relation to the interpretation of the term "adverse effect" is the term "biologically significant effect," which in the report is defined as "A response (to a stimulus) in an organism or other biological system that is considered to have substantial or noteworthy effect (positive or negative) on the well-being of the biological system. The concept is to be distinguished from statistically significant effects or changes, which may or may not be meaningful to the general state of health of the system."

Areas related to the evaluation of the adversity of an effect are reversibility and irreversibility and adaptation to an exposure. Irreversible effects are always of great concern. Reversible effects may also be of great concern depending on the nature of the effect and on the setting in which they occur. It cannot be ruled out that a permanent lesion may have occurred even if the overt effect is transient. Furthermore, when there is a more or less continuous exposure to a substance, the question of reversibility is not relevant because adaptation systems will be counteracted by new insults. In many cases it is not possible to draw any conclusion on whether an effect is reversible or not as such experimental data are rare, and all significant health effects that can impair function, both reversible and irreversible, should therefore be considered in the hazard assessment.

An adaptation, where an organism stabilizes its physiological condition after exposure to a chemical, without any irreversible disruption of a biological system and without exceeding the normal capacities of its response, should be considered as an early, not yet adverse, effect, which later on might lead to an adverse effect and thus needs expert judgment in a case-by-case manner.

4.2.3 DOSE–RESPONSE RELATIONSHIPS

Exposure to a chemical substance may result in a number of different effects varying from mild effects to fatal poisonings depending on the dose or exposure concentration. At low doses (concentrations), only mild effects such as nuisance and irritation may occur while at higher doses (concentrations), serious damage to organs and tissues may be observed, sometimes so severe that the effects result in the death of the individual. In principle, a single study, epidemiological as well as in experimental animals, may thus reveal different types of effects as well as different grades of the severity of a particular effect depending on the various doses (concentrations) administered in the specific study. Consequently, different dose–response relationships can be observed for a specific chemical if the exposure to the chemical results in different types of effects as well as different grades of the severity of a particular effect. Implicit in the hazard characterization of a chemical is therefore an assessment of the dose–response relationship for every type of effect observed in the available studies for a specific chemical.

Other terms often used indiscriminately for the dose–response relationship include concentration–effect relationship and dose–effect relationship. According to the joint OECD/IPCS project (OECD 2003a), which has developed internationally harmonized generic and technical terms used in chemical hazard and risk assessment, the following definitions have been provided although consensus was not achieved:

Dose–response relationship is the "Relationship between the amount of an agent administered to, taken up or absorbed by an organism, system or (sub) population and the change developed in that organism, system or (sub) population in reaction to the agent."

Concentration–effect relationship is the "Relationship between the exposure, expressed in concentration, of a given organism, system or (sub) population to an agent in a specific pattern during a given time and the magnitude of a continuously-graded effect to that organism, system or (sub) population."

Dose–effect relationship is the "Relationship between the total amount of an agent administered to, taken up or absorbed by an organism, system or (sub) population and the magnitude of a continuously-graded effect to that organism, system or (sub) population."

In the TGD (EC 2003), the following definition is provided:

Dose (concentration)–response (effect) relationship is the "Relationship between dose, or level of exposure to a substance, and the incidence and severity of an effect."

As no international consensus has been achieved in the OECD/IPCS project (OECD 2003a) in order to differentiate between "dose (concentration)–response (effect) relationship" and because it in reality is difficult to understand the subtle differences in the different terms as defined in the OECD/IPCS project, the broader and more general definition provided in the TGD (EC 2003) will be used in this book, and will generally be referred to as "dose–response." Consequently, the term "dose" will, in this book, generally mean both dose and exposure concentration unless otherwise stated.

In well-performed test guideline studies, the highest dose level should preferably induce toxicity (but not death or severe suffering) and a descending sequence of dose levels should demonstrate a dose–response with no effect(s) at the lowest dose level. The dose–response, e.g., the number of animals showing a specific effect or an increase in an organ weight, can be presented graphically; this graph is termed the dose–response curve. Generally, the x-axis represents the dose (or the logarithm to the dose) and the y-axis represents the response, e.g., expressed as the incidence of a specific effect in the treated groups as a percentage of the incidence of this effect in the control group. It should be appreciated that toxicity studies are of necessity generally limited to a small number of dose levels although the biological response may represent a continuum of change with changing dose level. For this reason, the derived shape of the dose–response curves for the various effects observed represent only an approximation to a true description of the observed biological effects.

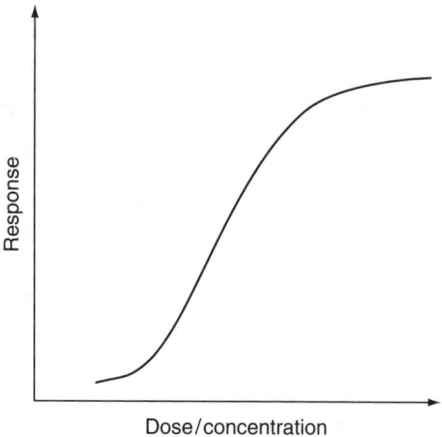

FIGURE 4.1 S-shaped dose–response curve.

For most of the toxic effects that might be exerted by a chemical substance, the dose–response curve is S-shaped as illustrated in Figure 4.1. This means that no response occurs at the lower dose levels, but as the dose level increases the response will become more and more pronounced until a plateau is reached.

Other dose–response relationships may also be seen for certain compounds such as, e.g., essential metals, where symptoms of deficiency may occur if the intake is too low, whereas toxic symptoms may occur if the intake is too high. For such compounds, the dose–response curve is generally U-shaped as illustrated in Figure 4.2. It should be noted that the right part of the U-shaped curve representing the toxic effects in reality is the typical S-shape observed for toxic effects in general.

For "non-threshold effects" (e.g., sensitization, mutagenicity, genotoxicity, genotoxic carcino-genicity), the dose–response curve, in the absence of mechanistic evidence to the contrary, is assumed to be linear.

In relation to hormesis (Section 4.12), the dose–response curve can be an inverted U-shaped curve (see Figure 4.3) or a J-shaped curve (see Figure 4.4).

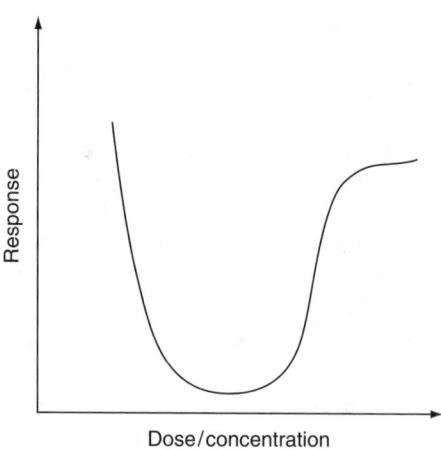

FIGURE 4.2 U-shaped dose–response curve.

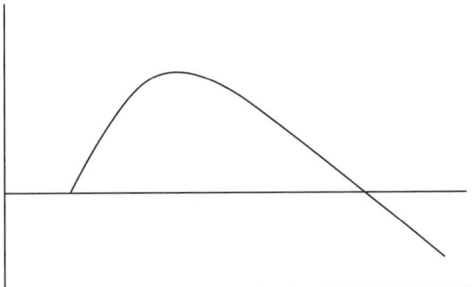

FIGURE 4.3 The "inverted U" curve, characteristic of the hormetic response of low-dose stimulation and high-dose inhibition.

As mentioned above, a single study may reveal different types of effects and in principle, a dose–response curve can be recorded for each type of effect observed in the study as illustrated in Figure 4.5 for three different effects A, B, and C that occur following exposure to a chemical.

One important aspect in the dose–response assessment is the shape of the dose–response curve. For certain effects, the dose should be increased considerably for a response to occur and the dose–response curve will in such cases be rather shallow as illustrated by curve C in Figure 4.5. For other types of effects, even a small increase in the dose results in a marked response and the dose–response curve will in such cases be rather steep as illustrated by curves A and B in Figure 4.5. It should be noted, however, that in general, there is no direct correlation between the shape of the dose–response curve and the severity of the effect, i.e., a steep dose–response curve does not necessarily implicate a severe effect and vice versa.

Another important aspect in the dose–response assessment is the placing of the dose–response curve in the diagram, i.e., the distance from the *y*-axis to the curve. The shorter the distance is to the *y*-axis the lower the dose is for a response to occur. It should be noted, however, that in general, there is no direct correlation between the placing of the dose–response curve and the severity of the effect, i.e., a short distance between the curve and the *y*-axis does not necessarily implicate a severe effect and vice versa.

These aspects can be illustrated by comparing the shape and the placing of the dose–response curves in Figure 4.5 for the various effects A, B, and C:

1. Effects A and C have differently shaped dose–response curves, i.e., curve A is much steeper than curve C; therefore, effect A could be interpreted as being more severe than effect C. Furthermore, effect A is observed at much lower dose levels than effect C, i.e., the

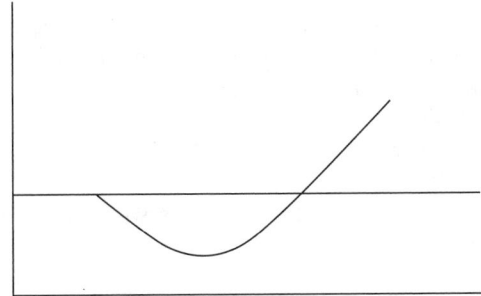

FIGURE 4.4 The "J-shaped" curve, characteristic of the hormetic response of reduction and high-dose enhancement of response.

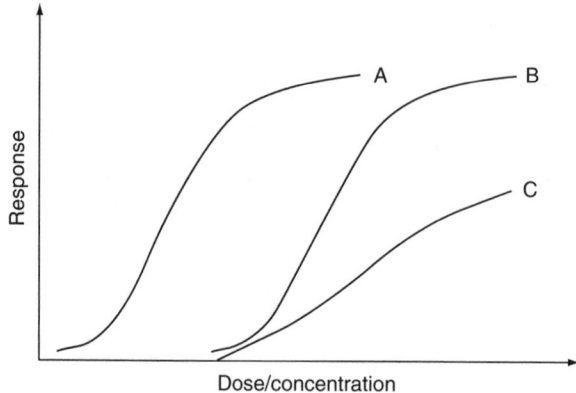

FIGURE 4.5 Dose–response curves for different effects observed in an experimental study.

distance from the y-axis to curve A is much shorter than to curve C; therefore again, effect A could be interpreted as being more severe than effect C.

2. Effects B and C are observed at similar dose levels, i.e., the distance from the y-axis to curves B and C is more or less the same; therefore, effects B and C could be interpreted as being of similar severity. However, effects B and C have differently shaped dose–response curves, i.e., curve B is much steeper than curve C; therefore, effect B could be interpreted as being more severe than effect C.

3. Effects A and B have similarly shaped dose–response curves, i.e., the steepness of the curves for effects A and B is more or less the same; therefore, effects A and B could be interpreted as being of similar severity. However, effect A is observed at much lower dose levels than effect B, i.e., the distance from the y-axis to curve A is much shorter than to curve B; therefore, effect A could be interpreted as being more severe than effect B.

However, in addition to the dose levels at which the various effects are observed as well as to the steepness of the dose–response curve, the type of the various effects observed is also a very important aspect in the dose–response assessment. If effect A is an alteration in an unspecific liver enzyme blood level, effect B is an increase in the relative liver weight, and effect C is the incidence of liver tumors, then both effects B and C obviously are evaluated as being more severe effects than effect A; i.e., an example of an exception from situation 1 and 3 above. Similarly, effect C is evaluated as being more severe than effect B; i.e., an example of an exception from situation 2 above.

4.2.4 No-Observed-Adverse-Effect Level, Lowest-Observed-Adverse-Effect Level

As addressed previously, it is generally agreed that most types of the adverse health effects caused by chemical substances (e.g., organ-specific, neurological, immunological, non-genotoxic carcinogenicity, reproductive, developmental), are not expressed until the substance, or an active metabolite, reaches a threshold dose or concentration at the relevant target.

As a consequence of the distinction between non-adverse effects and adverse effects, a true No-Effect Level (NEL) and No-Adverse-Effect Level (NAEL) as well as a true Lowest-Effect Level (LEL) and Lowest-Adverse-Effect Level (LAEL) exists in theory for a non-adverse effect and adverse effect, respectively. This is illustrated in Figure 4.6 where the NEL is the intersection of the dose–response curve with the x-axis, and the LEL, NAEL, and LAEL are somewhere on the dose–response curve.

Whether or not the threshold is reached, is related to the dose level of the substance to which the individual organism (human or experimental animal) is exposed, i.e., for a given route of exposure, there will be a threshold, which must be attained before the effect is induced. Ideally, in the hazard

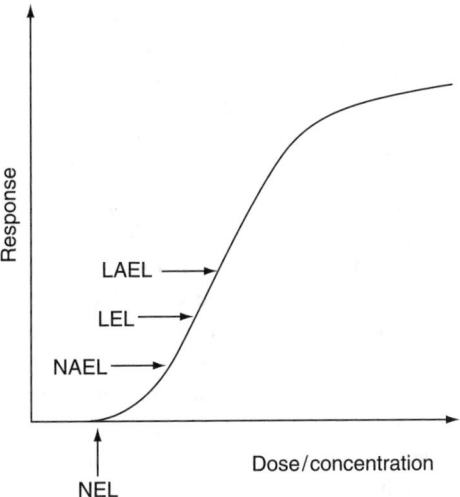

FIGURE 4.6 Dose–response curve illustrating the true no-effect level (NEL) and no-adverse-effect level (NAEL) as well as the true lowest-effect level (LEL) and lowest-adverse-effect level (LAEL).

assessment, a NAEL and LAEL should be identified for each of the various adverse effects exerted by the chemical under evaluation. As mentioned in Section 4.2.3, toxicity studies are of necessity generally limited to a small number of dose levels although the biological response may represent a continuum of change with changing dose level. For this reason, the derived shape of the dose–response curves for the various effects observed represent only an approximation to a true description of the observed biological effects and thus the NAEL and LAEL cannot be identified in practice. Therefore, in practice, the observed threshold for a specific adverse effect, the NOAEL in a toxicity study is a surrogate for the true NAEL and the LOAEL is a surrogate for the true LAEL. The NOAEL and the LOAEL are usually simply one of the dose levels in a toxicity study. The NOAEL will, in a particular study, generally be the highest dose level of the substance used in that study at which no statistically significant adverse effects were observed. The LOAEL will consequently be the next higher dose level, i.e., the NOAEL and LOAEL are operational values derived from a particular toxicity study. This is illustrated in Figure 4.7, where the NAEL and the LAEL are

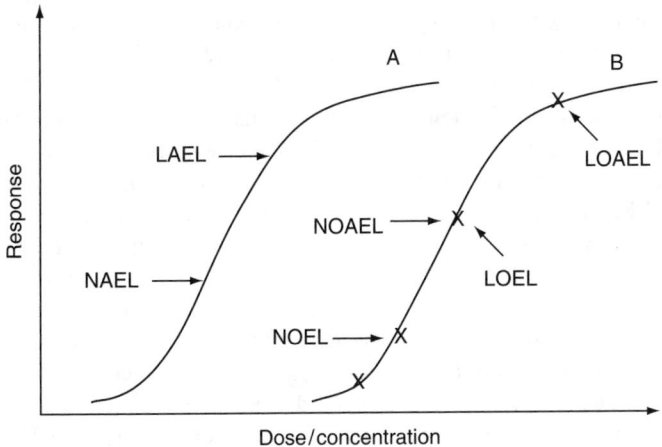

FIGURE 4.7 Dose–response curve illustrating the NAEL and LAEL in relation to the NOAEL and LOAEL.

somewhere on curve A, and the NOAEL is one of the specific dose levels on curve B and the LOAEL the next higher dose level on curve B. For example, if the dose levels used in an oral repeated dose toxicity test were 0, 5, 20, and 100 mg/kg body weight/day and an adverse effect (e.g., histopathological changes in the liver for which the incidences had achieved statistical significance from the incidence in the control group) was observed at 100 mg/kg body weight/day but not at 5 and 20 mg/kg body weight/day, the derived NOAEL would be 20 mg/kg body weight/day and the derived LOAEL would be 100 mg/kg body weight/day.

Similarly, for a non-adverse effect, a No-Observed-Effect Level (NOEL) and a Lowest-Observed-Effect Level (LOEL) are the surrogates for the true NEL and LEL, respectively. The NOEL is one of the specific dose levels on curve B in Figure 4.7 and the LOEL the next higher dose level. It should be realized that, in this case, the LOEL is identical with the NOAEL. It should be noted, however, that from a statistical point of view, it cannot be excluded that the effect at the dose level depicting the NOEL would have been interpreted as being adverse if the number of animals had been increased in the particular study. In the example mentioned above, the derived NOEL would be 5 mg/kg body weight/day if the histopathological changes in the liver were observed at 20 mg/kg body weight per day but the incidence had not achieved statistical significance from the incidence in the control group, and the derived LOEL would be 20 mg/kg body weight/day, i.e., equal to the NOAEL.

According to WHO/IPCS (1994), the following definitions are provided:

No-Observed-Adverse-Effect Level (NOAEL) is "Greatest concentration or amount of a substance, found by experiment or observation, which causes no detectable adverse alteration of morphology, functional capacity, growth, development or life span of the target organism under defined conditions of exposure. Alterations of morphology, functional capacity, growth, development or life span of the target may be detected which are judged not to be adverse."

No-Observed-Effect Level (NOEL) is "Greatest concentration or amount of a substance, found by experiment or observation, that causes no alterations of morphology, functional capacity, growth, development or life span of target organisms distinguishable from those observed in normal (control) organisms of the same species and strain under the same defined conditions of exposure."

Lowest-Observed-Adverse-Effect Level (LOAEL) is "Lowest concentration or amount of a substance, found by experiment or observation, which causes an adverse alteration of morphology, functional capacity, growth, development or life span of the target organism distinguishable from normal (control) organisms of the same species and strain under the same defined conditions of exposure."

The term "Level" is, according to the definitions provided by the WHO/IPCS (1994), applicable for both the dose level and the exposure concentration. However, the terms No-Observed-Adverse-Effect Concentration (NOAEC) and Lowest-Observed-Adverse-Effect Concentration (LOAEC) are often used when the exposure level is expressed as a concentration as, e.g., in inhalation studies. In this book, the terms "NOAEL" and "LOAEL" will be used in general.

It should be noted that the terms "NOAEL/NOEL" and "LOAEL/LOEL" often are used indiscriminately in the scientific literature and with various significances, i.e., without a distinction whether the NOEL or LOEL has been determined for an adverse effect or for a non-adverse effect. Therefore, the scientific literature should always be carefully evaluated in order to interpret the stated NOEL or LOEL. However, in most situations, the NOEL or LOEL could be interpreted as being determined for adverse effects regardless of the term used as the aim of the hazard assessment in general is to assess adverse effects.

The threshold for a specific effect may vary considerably for different exposure routes and for different species because of differences in the toxicokinetics for different species and exposure routes, and possibly also because of differences in the toxicodynamics. The NOAEL and LOAEL derived for a given study will therefore in general depend on the experimental study design, i.e., species, sex, age, strain, and developmental status of animals; number of animals per exposure level; selection of exposure levels; the spacing between the exposure levels; duration of exposure; and sensitivity of methods used to measure the responses. Thus, the sensitivity of a particular study may limit the extent to which it could be possible to derive a reliable NOAEL from that study. The most critical factors in the derivation of the NOAEL and LOAEL are the selected dose levels, the number of animals per dose level and the number of dose levels in a particular study. In case it is impossible to derive a NOAEL from a particular study, i.e., an adverse effect is observed even at the lowest dose level in the study, at least a LOAEL should be derived and used in the hazard assessment.

As mentioned in Section 4.2.3, a single study may reveal different types of effects and in principle, a dose–response curve can be recorded for each type of effect observed in the study as illustrated in Figure 4.5. If a single study reveals different types of effects, it is essential to evaluate all these effects in terms of their NOAELs and LOAELs. The overall NOAEL for a particular study is often, but not necessarily, the lowest NOAEL derived for the various effects.

If there are several valid studies addressing the same effect from which different NOAELs could be derived, the highest reliable NOAEL not exceeding any of the reliable LOAELs should be used in the hazard assessment. If the studies are not quite comparable, i.e., do not examine the same endpoints by equally sensitive methods, expert judgment is used to derive the most relevant NOAEL. When it is not possible to derive a NOAEL, the LOAEL should be used in the hazard assessment.

The NOAEL/LOAEL is generally presented as the amount of a chemical dose per-unit-body-weight-basis, e.g., mg/kg body weight/day. If data are either not available or inadequate to enable derivation of a NOAEL/LOAEL in mg/kg body weight/day (e.g., dietary data on a substance provided in ppm, but no food consumption or body weight data available), standard values for biological parameters such as body weight, feed intake, drinking water intake, and ventilation rate (inhaled volume of air per unit time) could be applied. There are no internationally agreed standard values for these parameters, but various authorities and organizations have suggested standard values for a number of these parameters (Section 7.4). For consistency and transparency between evaluations of NOAELs/LOAELs from various studies, the same set of standard values should be used. Depending on the specific circumstances of the study other standard values may be considered more appropriate in some instances. For transparency, the risk assessor should always indicate which methodology has been used, the use of particular standard values should be clearly stated and justified, and references provided to indicate their origin.

It is recognized that the NOAEL derived by using this traditional approach for dose–response assessment is not very accurate with respect to the degree to which it corresponds with the (unknown) true NAEL. Furthermore, in this traditional approach, only the data obtained at one dose (NOAEL) are used in the hazard assessment rather than the complete dose–response data set. In case sufficient data are available, the shape of the dose–response curve should be taken into account in the hazard assessment. In the case of a steep dose–response curve, the derived NOAEL can be considered as more reliable because the greater the slope, the greater the reduction in response to reduced doses. In the case of a shallow dose–response curve, the uncertainty in the derived NOAEL may be higher and this has to be taken into account in the hazard assessment (see Section 5.7). If a LOAEL has to be used in the hazard assessment, then this value can only be considered reliable in the case of a very steep dose–response curve.

As an alternative to the traditional NOAEL approach, the BMD concept has been proposed for use in the quantitative assessment of the dose–response relationship, see the next section.

4.2.5 THE BENCHMARK DOSE CONCEPT

The concept of the Benchmark Dose (BMD), a benchmark is a point of reference for a measurement, in health risk assessment of chemicals was first mentioned by Crump (1984) as an alternative to the NOAEL and LOAEL for noncancer health effects in the derivation of the ADI/TDI; these terms are addressed in detail in Chapter 5. The BMD approach provides a more quantitative alternative to the dose–response assessment than the NOAEL/LOAEL approach. The goal of the BMD approach is to define a starting point of departure (POD) for the establishment of a tolerable exposure level (e.g., ADI/TDI) that is more independent of the study design. In this respect, the BMD approach is not different from the NOAEL/LOAEL approach; the primary difference between the two approaches is how the starting point is determined.

Since the first description of the BMD approach in health risk assessment of chemicals, the method has been modified and extended by many others. Central in this work was a workshop organized by the International Life Science Institute (ILSI) and reported in Barnes et al. (1995) and a workshop organized by the US-EPA Risk Assessment Forum resulting in a US-EPA report (US-EPA 1995). No consensus was reached at these workshops on which variation and extension of the BMD approach is most appropriate for the use in human health risk assessment.

The BMD or Critical Effect Dose (CED) is defined as the dose that corresponds to a specified, predetermined change in an adverse response in treated individuals compared to the response in untreated individuals. This predetermined level of change in response is termed the Benchmark Response (BMR) or the Critical Effect Size (CES). The BMD is determined by modeling a dose–response curve in the region of the dose–response relationship where biologically observable effects are available, i.e., the BMD is based on a mathematical model being fitted to the experimental data within the observable range for each effect parameter and represents the dose corresponding to the BMR. To take uncertainty of the data into consideration, the dose of interest is the lower confidence limit on the BMD, the BMDL, which refers to the corresponding lower limit of a one-sided 95% confidence interval on the BMD. Using the lower bound accounts for the uncertainty inherent in a given study, and assures (with 95% confidence) that the predefined BMR is not exceeded. The BMDL is suggested as the POD for use in health risk assessments and is defined as the point on a dose–response curve established from experimental data, generally corresponding to a low effect level (1%–10%). These aspects are illustrated in Figure 4.8.

In the BMD approach, a curve is fitted to discrete responses (binary, dichotomous/quantal data, i.e., yes/no) or to continuous mean effect values (a response such as weight that can assume any value in a range). The curve is usually fitted to data using the maximum likelihood approach.

For discrete responses, the BMR is typically chosen at 10% above the control response, the BMD_{10} as an excess risk of 10% is considered to be at or near the limit of sensitivity in most carcinogenicity studies and in some noncarcinogenicity studies as well. If a study has greater than usual sensitivity, then a lower BMR can be used, although the BMD_{10} and $BMDL_{10}$ should always be presented for comparison purposes.

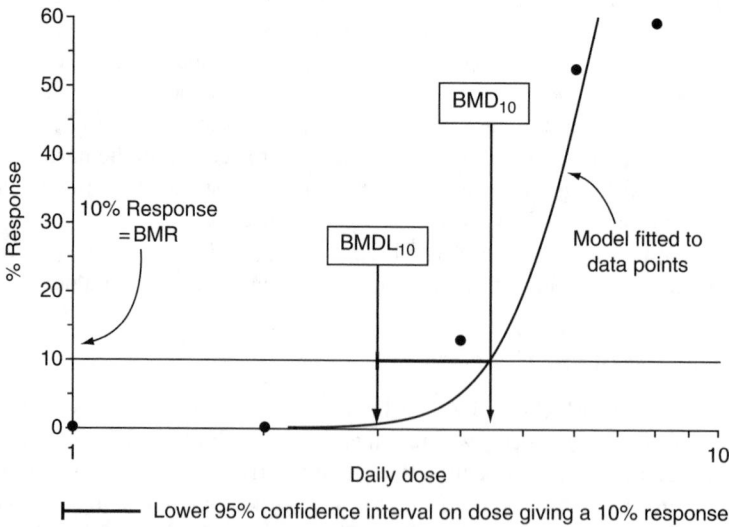

FIGURE 4.8 Benchmark Dose modeling illustrating the Benchmark Dose (BMD), the benchmark response (BMR), and the Benchmark Dose Lower Limit (BMDL).

For continuous data, there are still a number of outstanding issues regarding the benchmark including (Crump 2002): (1) definition of an adverse effect; (2) whether to calculate the BMD from a continuous health outcome, or first convert the continuous response to a binary (yes/no) response; (3) quantitative definition of the BMD, in particular in such a manner that BMD from continuous and binary data are commensurate; (4) selection of a mathematical dose–response model for calculating a BMD; (5) selection of the level of risk to which the BMD corresponds; and (6) selection of a statistical methodology for implementing the calculation.

Several software packages are available for BMD calculations. In a review by Falk Filipsson et al. (2003), the US-EPA software, the Crump software, and the Kalliomaa software have been evaluated. A number of routinely used models are included in the US-EPA BMD software program, which is freely available from the US-EPA Web site (www.epa.gov).

Advantages of the BMD approach over the NOAEL approach are:

- BMD approach makes better use of the experimental data as the BMD is derived using all experimental data and thus reflects the dose–response relationship to a greater degree.
- BMD is independent of predefined dose levels and spacing of dose levels.
- Magnitude of any effect within the observable range can be calculated.
- BMD approach can provide information on the nonlinear region of the dose–response.

A major disadvantage of the BMD approach is that the standard toxicological guideline studies often have too few dose groups and thus there is an uncertainty with respect to the reliability of the approach. For the derivation of reliable dose–response relationships, the classical study design of three dose groups and a vehicle control group is far from ideal, especially if one considers the unfavorable possibility that in a particular experiment, adverse effects may be identified only at the highest dose level. An improved benchmark model fit would be possible by increasing the number of dose groups without changing the total number of animals in the test.

Advantages of the NOAEL approach over the BMD approach are that the NOAEL approach is easy to understand, it is not dependent on a mathematical model being correct, and it can easily be applied on both discrete and continuous data. The major disadvantages are that the NOAEL approach only provides knowledge on the magnitude of risk at the dose levels of the particular study, and that the NOAEL is strongly dependent on study group sizes.

Internationally, the BMD approach is used by the US-EPA to derive health-based limit values (US-EPA 2007a). Within the OECD (OECD 2000) and the European Union (EC 2003), the BMD approach is also mentioned as an alternative to the traditional NOAEL approach in health risk assessment but is not implemented in regulatory toxicology within the European Union.

According to the TGD (EC 2003), the BMD approach needs further development before it can be applied generally in the hazard assessment, particularly in the following areas: (1) optimization of study design; (2) predefinition of internationally accepted CES for each toxicological parameter (sufficiently based on biological, physiological, and toxicological knowledge); and (3) development of specific dose–response analyses for different types of experimental data (continuous, categorical, and quantal). It is also stated that changes in study design to accommodate the BMD approach would generally no longer allow a proper derivation of a NOAEL and thus, in practice, the NOAEL and the benchmark concepts appear to be incompatible. In addition it is noted that the BMD method can be used parallel to derivation of a NOAEL. Especially in cases where a NOAEL cannot be derived for the selected endpoint because only a LOAEL is available, benchmark modeling is considered to be preferable over LOAEL-to-NOAEL extrapolation (Section 5.7) using more or less arbitrary assessment factors. Since generally accepted CES have not yet been established, one may consider postulating a default CES (e.g., 5% over the background level for continuous endpoints). In any case, the chosen value should always be derived in a transparent way using the whole toxicological profile of the substance.

Gephart et al. (2001) have used the BMD methodology to estimate BMDs for 90-day toxico-logical data and several fabricated data sets. According to the authors, there are many variables associated with the BMD that could be set to produce unreasonable BMD estimates; some of these variables and decisions were examined in their study. BMDs were calculated for discrete and continuous endpoints using a variety of different variables (e.g., Maximum Likelihood Estimates (MLEs), Lower Confidence Limits (LCLs), and different risk levels). In addition, the fabricated data sets were manipulated, i.e., dose groups were eliminated, and the BMDs were recalculated. For the 90-day studies, the BMDs were typically within an order of magnitude of the NOAEL for discrete endpoints; the MLEs were typically greater than the NOAEL and the LCLs were typically less than the NOAEL. With the continuous data, the ratios of MLEs and LCLs to the NOAEL were highly variable, and no general trend could be determined. The authors concluded that depending on how the BMD is defined and how the BMD is used in the risk assessment process, BMD estimates may produce Reference Doses that are more or less conservative than the NOAEL approach. The authors recommended that further discussions on how BMDs should be used in the risk assessment process are needed.

4.2.6 TOXICOLOGICAL MODES OF ACTION

As mentioned previously, the assessment of hazard and risk to humans from exposure to chemical substances is generally based on the extrapolation from data obtained in studies with experimental animals. In the absence of comparative data in humans, a basic assumption for toxicological risk assessment is that effects observed in laboratory animals are relevant for humans, i.e., would also be expressed in humans. In assessing the risk to humans, an assessment factor is applied to take account of uncertainties in the differences in sensitivity to the test substance between the species, i.e., to account for interspecies variability (Section 5.3). If data are available from more than one species or strain, the hazard and risk assessment is generally based on the most susceptible of these; except where data strongly indicate that a particular species is more similar to man than the others with respect to toxicokinetics and/or toxicodynamics. Two main aspects of toxicity, toxicokinetics and toxicodynamics, account for the nature and extent of differences between species in their sensitivity to xenobiotics; this is addressed in detail in Chapter 5.

It is generally accepted that the expression of toxicity in a mammalian system is dependent on a sequence of key events taking place, each of which is critical to the manifestation of the toxic endpoint. If at least one of these key events identified in experimental animals does not occur in humans, then it could be concluded that the toxic endpoint would not be observed in humans, and thus this effect is of limited relevance for the prediction of effects in humans.

Therefore, information on the toxicological mode(s) of action as well as mechanistic data are essential in establishing the relevance to humans of the toxicological effects observed in experimental animals. The evaluation of the relevance for humans of data from studies in animals is aided by use of data on the toxicokinetics, including metabolism, of a substance in both humans and the animal species used in the toxicity tests, when they are available, even when they are relatively limited.

Certain effects in laboratory animals are known to be of limited relevance for the prediction of effects in humans. Clear, well-documented evidence for a species-specific effect/response (e.g., light hydrocarbon induced nephropathy in the kidney of male rats, peroxisome proliferation in the liver of rodents) should be used as justification for the conclusion that a particular effect is not expected to occur in humans exposed to the substance.

Positive carcinogenic findings in animals require careful evaluation to determine their relevance to humans. Of key importance is the mechanisms/mode(s) of action of tumor induction. The WHO/IPCS has developed a conceptual framework to provide a structured and transparent approach for the assessment of the overall weight of evidence for a postulated mode of induction for each tumor type observed (Sonich-Mullin et al. 2001). The framework promotes confidence in the conclusions reached by the use of a defined procedure, which mandates clear and consistent documentation of the reasoning used and inconsistencies and uncertainties in the available data. Tumor types and mechanisms/mode(s) of action, which are believed to be of limited relevance for humans, are addressed in Section 4.9.6.

The European Centre for Ecotoxicology and Toxicology of Chemicals (ECETOC) has published a report entitled "Toxicological Modes of Action: Relevance for Human Risk Assessment" (ECETOC 2006). It is stated in this report that there are a growing number of chemicals where application of the traditional default approach to risk assessment has been shown to be too conservative. The aim of the report is to examine, in the context of human risk assessment, the role and use of mode of action and mechanistic data in establishing the relevance of toxicological effects observed in experimental animals. First, to illustrate, with examples, how it can be demonstrated that a mode of action observed in an animal model is not relevant, or cannot be extrapolated to humans. Second, to provide a rational scientific approach by which it might be judged to what extent a mode of action seen in an animal model can be considered not relevant for humans. It is stressed that a clear distinction should be made between the two terms often used in this context, i.e., "mechanism of toxicity" and "mode of action."

The report provides the following descriptions of these two terms: "Identifying a 'mechanism of toxicity' requires rigorous investigation to obtain a comprehensive understanding of the entire sequence of events that result in the toxic effect of interest. In practice, 'mechanisms of toxicity' are rarely established. For the 'mode of action' concept it is sufficient to develop an understanding of key events within the complete sequence of events leading to toxicity. This does not require the amount or depth of data needed to provide a detailed explanation of the complete sequence of events."

The report states "For a specific toxic endpoint, to justify deviating from the default approach in risk assessment, it should be sufficient to identify the key events in the process (mode of action) and establish that these would not occur in humans, or that they would occur only to a much lesser extent. From such knowledge it could be concluded that the toxic outcome would not be observed in humans, or be observed only at much higher and possibly irrelevant exposure levels." A proposal is made for a structured approach for establishing and reviewing the mode of action by which a substance causes a toxicological effect. This approach, which is illustrated with a number of examples, should allow a decision to be made on the appropriate animal endpoint for human hazard identification and risk assessment and guidance is given on testing the hypothesis on which the nonrelevance to humans is based.

4.2.7 CRITICAL EFFECTS

The last step in the hazard assessment is the identification of the critical effect(s).

According to WHO/IPCS (1994), the critical effect(s) is "The adverse effect(s) judged to be most appropriate for determining the tolerable intake."

According to US-EPA's "Glossary of IRIS Terms," the critical effect is "The first adverse effect, or its known precursor, that occurs to the most sensitive species as the dose rate of an agent increases."

These two definitions reflect two sides of the same situation. In this book, the term "critical effect(s)" will be used for the hazard/effect considered as being the essential one(s) for the purpose of the risk characterization, e.g., for the establishment of a health-based guidance value, permissible exposure level, or Reference Dose. It should be noted that the critical effect could be a local as well as a systemic effect. It should also be recognized that the critical effect for the establishment of a tolerable exposure level is not necessarily the most severe effect of the chemical substance. For example, although a substance may cause a serious effect such as liver necrosis, the critical effect for the establishment of, e.g., an occupational exposure limit could be a less serious effect such as respiratory tract irritation, because the irritation occurs at a lower exposure level.

In the first step of the hazard assessment process, all effects observed are evaluated in terms of the type and severity (adverse or non-adverse), the dose–response relationship, and NOAEL/LOAEL (or alternatively BMD) for every single effect in all the available studies if data are sufficient, and the relevance for humans of the effects observed in experimental animals. In this last step of the hazard assessment, all this information is assessed as a whole in order to identify the critical effect(s) and to derive a NOAEL, or LOAEL, for the critical effect(s). It is usual to derive a NOAEL on the basis of effects seen in repeated dose toxicity studies and in reproductive toxicity studies. However, for acute toxicity, irritation, and sensitization it is usually not possible to derive a NOAEL because of the design of the studies used to evaluate these effects. For each toxicological endpoint, these aspects are further addressed in Sections 4.4 through 4.10.

4.3 TOXICOKINETICS

Data on the toxicokinetics of a substance can be very useful in the interpretation of toxicological findings, and may replace the use of some default extrapolation factors used in route-to-route (Section 5.5) or interspecies extrapolations (Section 5.3). In addition, interindividual differences in sensitivity to toxicants may be identified on the basis of toxicokinetic data, thereby making it possible to make the risk assessment more comprehensive by including sensitive subpopulations (Section 5.4). In conjunction with information on the relationship between concentration–dose at the target site and the toxic effect, toxicokinetic information may be an important tool for extrapolation from high to low dose effects.

4.3.1 Definitions

The term "toxicokinetics" is used to describe the time-dependent fate of a substance within the body. This includes absorption, distribution, metabolism, and/or excretion. (EC 2003). The term, often abbreviated "ADME," has essentially the same meaning as pharmacokinetics, but the latter term should be restricted only to pharmaceuticals.

Absorption relates to how, how much, and how fast a substance enters the body.

Distribution relates to the reversible transfer of a substance between various parts of the organism, i.e., body fluids or tissues.

Metabolism (biotransformation) relates to the enzymatic or nonenzymatic transformation of a substance into a structurally different chemical (metabolite).

Excretion relates to the physical loss of the parent substance and/or its metabolite(s). The principal routes of excretion are via the urine, bile (feces), and exhaled air.

Elimination relates to the loss of a substance by the organism, i.e., metabolism and excretion combined.

Disposition relates to the sum of processes following absorption of a substance into the circulatory systems, distribution throughout the body, metabolism, and excretion.

The term "toxicodynamics" means the process of interaction of chemical substances with target sites and the subsequent reactions leading to adverse effects. The toxicodynamic effect is driven by the concentration at the effect site(s) directly or indirectly and may be reversed or modified by several factors such as repair mechanisms for DNA damage and compensatory cell proliferation (EC 2003).

4.3.2 Objectives for Assessing the Toxicokinetics of a Substance

Toxicokinetic studies are designed to obtain species-, dose-, and route-dependent data on the concentration–time course of the parent compound and its metabolites, e.g., in blood, urine, feces, and exhaled air. From these data toxicokinetic parameters can be derived by appropriate techniques. The information, which can be taken from *in vivo/ex vivo* toxicokinetic studies is (EC 2003):

Primary information:

- Concentration–time profile of the substance/metabolites in blood (plasma), tissues, and other biological fluids, such as urine, bile, exhaled air, and the volume of the excreted fluids if appropriate
- Protein binding and binding to erythrocytes (if relevant) (*in vitro/ex vivo* studies)

Derived information:

- Rate and extent of absorption and bioavailability
- Distribution of the substance in the body
- Biotransformation
- Rate and extent of pre-systemic (first pass) and systemic metabolism after oral and inhalation exposure
- Information on the formation of reactive metabolites and possible species differences (e.g., milk, bile, sweat, etc.)
- Half-life and potential for accumulation under repeated or continuous exposure
- Information on enterohepatic circulation

Enterohepatic circulation may pose particular problems for route-to-route extrapolation (Section 5.5) since the fraction of the compound undergoing enterohepatic recirculation after oral administration may be greater than after nonoral administration. This will result in an AUC (area under the plasma/blood concentration versus time curve, representing the total amount of substance reaching the plasma), which reflects both absorption/systemic availability of the compound and the extent of recirculation. As the relative extent of target organ exposure following different routes of exposure is often calculated from the ratio of AUCs, target organ exposure after oral exposure may be overestimated.

It is helpful to have information for the (expected) exposure route(s) in humans (oral, inhalation, dermal) at appropriate dosing level(s). From the plasma/blood concentration time profile and from the excretion over time it can be calculated whether the substance will accumulate when given repeatedly or continuously. However, it is only possible to make this extrapolation for substances that have linear kinetics. Hence, if information on the potential for a substance to accumulate is important for the risk assessment, it will be necessary to gather data from studies with repeated dosing regimes. Conducting toxicokinetic studies in more than one species will enable the presence or absence of interspecies differences to be assessed. In the absence of *in vivo* data some of the toxicokinetic data may be derived from *in vitro* experiments. These include parameters of metabolic steps, such as the rate constants V_{max} (maximal velocity) and K_m (the Michaelis–Menten affinity constant), intrinsic metabolic clearance, as well as skin permeation rate and distribution coefficient. Physiologically based toxicokinetic (PBTK) modeling techniques may be used to simulate the concentration–time profile in blood and at the target site.

4.3.3 Test Guidelines

4.3.3.1 *In Vivo* Tests

The OECD, US-EPA, and EU have adopted *in vivo* test guidelines for the performance of toxicokinetic studies. The various guidelines are shown in Table 4.1 and are described further in the text below.

4.3.3.1.1 OECD
The relevant test guideline for toxicokinetics in general is OECD TG 417 from 1984. This guideline is undergoing revision under the lead of the United States (OECD 2005).

TABLE 4.1

Test Guidelines Adopted for Toxicokinetic Studies

Title

OECD Test Guidelines	Year
417 Toxicokinetics	1984
427 Skin absorption: *In vivo* method	2004
428 Skin absorption: *In vitro* method	2004

US-EPA OPPTS Harmonized Test Guidelines	
870.8500 Toxicokinetic test (Draft)	1996

EU Annex V Test Methods	
B.36 Toxicokinetics	1988

OECD TG 417 gives general guidance, which leaves open the possibility for many different study designs. The radio-labeled or unlabeled test substance is administered by an appropriate route. Depending on the purpose of the study, the substance may be administered in single or repeated doses for defined periods to one or several groups of experimental animals. Subsequently, depending on the type of study, the substance and/or metabolites are determined in body fluids, tissues, and/or excreta.

The OECD TG 427 (Skin Absorption: *In Vivo* Method), is a more specific guideline dealing with dermal absorption only. The test substance, preferably radio-labeled, is applied to the clipped skin of animals at one or more appropriate dose levels. The test preparation is allowed to remain in contact with the skin for a fixed period of time under a suitable cover (nonocclusive, semi-occlusive, or occlusive) to prevent ingestion of the test preparation. At the end of the exposure time the cover is removed and the skin is cleaned with an appropriate cleansing agent, the cover and the cleansing materials are retained for analysis and a fresh cover applied. The animals are housed prior to, during, and after the exposure period in individual metabolism cages and the excreta and expired air over these periods are collected for analysis. The collection of expired air can be omitted when there is sufficient information that little or no volatile radioactive metabolite is formed. Each study will normally involve several groups of animals that will be exposed to the test preparation. One group will be killed at the end of the exposure period. Other groups will be killed at scheduled time intervals thereafter. At the end of the sampling time the remaining animals are killed, blood is collected for analysis, the application site removed for analysis, and the carcass is analyzed for any un-excreted material. The samples are assayed by appropriate means and the degree of percutaneous absorption is estimated.

4.3.3.1.2   US-EPA

The OPPTS Harmonized Test Guidelines, Series 870 Health Effects Test Guidelines, do not contain a guideline method for toxicokinetic testing among the Final Guidelines. The list of public draft guidelines contains a guideline 870.8500 Toxicokinetic Test, which is from 1996. This draft guideline is similar to, but more specific in its demands than the OECD and EU guidelines, e.g., the OPPTS guideline specifically calls for testing in Fischer 344 rats and contains definite time points for sample collection.

4.3.3.1.3   EU

The relevant test guideline is Annex B.36 Toxicokinetics from 1988, which is similar to the OECD TG 417.

4.3.3.2 *In Vitro* Tests

OECD has adopted an *in vitro* test for skin absorption potential (OECD TG 428, Skin Absorption: *In Vitro* Method). According to this guideline, excised skin from human or animal sources can be used. The skin is positioned in a diffusion cell consisting of a donor chamber and a receptor chamber, between the two chambers. The test substance, which may be radio-labeled, is applied to the surface of the skin sample. The chemical remains on the skin for a specified time under specified conditions, before removal by an appropriate cleansing procedure. The fluid in the receptor chamber is sampled at time points throughout the experiment and analyzed for the test chemical and/or metabolites.

4.3.4 GUIDANCE DOCUMENTS

A guidance document for toxicokinetic investigations has been issued by the WHO (WHO/IPCS 1986a). This document provides comprehensive guidance regarding principles of toxicokinetic studies and contains a great deal of information including analytical methods, absorption via the gastrointestinal, pulmonary, dermal, and other routes of exposure, distribution, binding, metabolism, excretion, kinetic models, toxicokinetic methodology in the assessment of human exposure, and assessment of toxicokinetic studies.

A recently published WHO/IPCS document regarding chemical-specific adjustment factors for interspecies differences and human variability (WHO/IPCS 2005) provides guidance for use of toxicokinetic data in dose–response assessment to develop the so-called Compound-Specific Assessment Factors (CSAFs) (Section 5.2.1.12).

In addition, the EU Technical Guidance Document (EC 2003), Chapter 3.5, deals with toxicokinetics.

Specific guidance documents for toxicokinetics have not been published by either the OECD or the US-EPA.

4.3.5 USE OF TOXICOKINETICS IN HAZARD ASSESSMENT

Data on absorption provides important knowledge for the hazard assessment. If a substance is not absorbed or poorly absorbed, it is not surprising that toxicity studies show no or only weak effects. Therefore, in the hazard assessment it should be taken into consideration that a lack of effects in toxicity studies could be caused by nonabsorption of the substance rather than lack of toxicity.

First pass metabolism (metabolism in the intestinal wall or by the liver before the substance reaches the systemic circulation) can also be a cause for nontoxicity, which can be route- and species-dependent. Information about which metabolites, and the relative quantities thereof, are formed, is important. Chemical substances are metabolized to metabolites, which may be toxicologically more or less active than the parent compound, and the relative quantity of the metabolites may differ greatly among animal species. If it is known that toxicity is related to formation of an active metabolite, knowledge about the amount of this particular metabolite in humans and test animals is very important for the hazard assessment. If humans do not form the toxic metabolite, the hazard can be considered limited; and vice versa.

Data on distribution give an indication of whether a particular tissue may be exposed to the substance or not. The extent of chemical distribution into tissues depends on the extent of plasma protein and tissue binding and this may vary among species. This is also the case for passage of chemicals into the brain, which is protected by the blood–brain barrier.

Data on excretion could indicate a possible hazard, e.g., to the breastfeeding infant, if the substance is excreted into breast milk.

The elimination half-life of the substance gives important information about the duration of internal exposure following an episode of exposure. The half-life (half-time, $T_{1/2}$) is the time taken for the concentration of a substance in the blood, tissue/organ, or whole body to decline to half of its original value.

4.3.5.1 Human and Animal Data

The primary endpoint of the toxicokinetic studies is the concentration–time profile of the substance in plasma/blood and other biological fluids as well as in tissues. The excretion rate over time and the amount of metabolites in urine and bile are further possible primary endpoints of kinetic studies, sometimes providing information on the mass balance of the compound. From the primary data, clearance and half-life can be derived by several methods. From the excretion rate over time and from cumulative urinary excretion data and plasma/blood concentration measured during the sampling period, renal clearance can be calculated. The same is the case for the biliary excretion.

If after single dose administration, the blood samples are not collected at time intervals, which allow for a description of the whole plasma concentration time course, including the absorption, distribution, and elimination phase, the information obtained is limited. In particular, data should be available in the first hours after administration to cover the absorption phase. If measurements of the parent compound and its metabolite(s) are made in this period, this will allow assessment of an extensive first pass effect, i.e., when a substance after oral administration is transported via the portal vein to the liver where metabolism takes place before the substance enters the systemic circulation.

When data are available to enable comparison of the plasma concentration time profile after single administration with that after repeated administration, this would enable determination of whether the substance has time dependent kinetics (due to induction of metabolism, inhibition of metabolism, and/or accumulation and saturation of processes involved in distribution, metabolism, and excretion).

Absolute systemic bioavailability (absorbed fraction of the dose or concentration administered) can only be calculated by comparing the so-called Area Under the Plasma Curve (AUC, the area under the curve in a plot of the concentration of a substance in the plasma against time) after oral, inhalation, or dermal administration with the AUC after direct administration into the systemic circulation, e.g., after intravenous administration. In order to obtain a reliable estimate for AUC after single administration, it is necessary to have blood samples for 3–5 half-lives. In case data are not available for a calculation of the AUC, the absorbed fraction can be indicated from data on the amount of the parent compound and its metabolite(s) excreted in the urine, feces, and exhaled air. It should be noted that the amount excreted in the feces stems from both the unabsorbed fraction as well as from the fraction of the substance following biliary excretion.

Distribution, including accumulation of an absorbed substance, will be the same irrespective of the route of administration. However, distribution and accumulation at the site of application (inhalation, oral, dermal) may depend on the route of administration. In such cases, local accumulation may occur and may be responsible for tissue damage. In these cases, systemic toxicokinetics of the substance may be of limited relevance for the risk assessment. It is generally not crucial for risk assessment to determine the precise tissue distribution profile for a substance. In certain special cases, however, specific tissue distribution studies may assist or even be essential for the interpretation of available toxicological data. For example, it may be of interest to know whether the substance will cross the blood–brain barrier, the placenta barrier, or will accumulate in specific tissues.

For repeated dose studies, the relationship between the elimination half-life and the frequency of dosing can provide an indication for the potential of the substance to accumulate. If the dosing interval is much longer than the half-life, then the dose will be effectively eliminated before the next dose is given and each dose can be considered as entirely separate. If, however, the dosing interval is about the half-life or less, then accumulation of the substance occurs until a steady-state concentration is reached, i.e., the rate of uptake into the blood and tissues/organs equals the rate of elimination. The steady-state concentration of a substance is dependent on the administered dose, the absorbed fraction, the half-life, the volume of distribution (a term used to quantify the distribution of a substance throughout the body after oral or parenteral dosing, defined as the volume in which the amount of the substance would need to be uniformly distributed to produce the observed blood concentration), and the dosing interval. As a rule of thumb, the steady-state concentration is generally reached within 4–6 half-lives. Similarly, following cessation of the exposure, it will take 4–6 half-lives until the substance has been eliminated from the body. For example, if the half-life of a substance is 12 h and the dosing interval is 24 h, then the substance will accumulate slightly because, within the dosing interval of two half-lives, 75% of the substance will be excreted whereas 25% will remain in the body. In contrast, for a substance with a half-life of 60 h and a dosing interval of 24 h, the accumulation will be high.

The systemic metabolism is the same irrespective of the route of administration, whereas pre-systemic (e.g., first pass metabolism) and local metabolism at the site of administration differs and this may be relevant from a toxicological perspective. Some substances may undergo first pass metabolism to such an extent that the parent compound does not reach the systemic circulation. Knowledge of the metabolic profile of a substance may also help to consider a mechanistic model of action or at least may allow a mode of action to be ascertained.

The main routes of excretion for a substance as well as for its metabolites are in the urine and bile (feces). For some substances, exhalation is also an important excretion route. In addition, excretion may also take place via biological fluids such as saliva, sweat, and milk. Although the amounts excreted by these routes are relatively small, the presence of a substance in these fluids, particularly breast milk, may be the underlying cause of toxic effects.

In relation to the excretion of a substance, a clear distinction should be made between the excretion of the parent compound and the metabolites because this will help to correlate the presence of the toxicologically active compound (which might be the parent compound and/or the metabolite(s)) with the toxicological effect.

Nonlinearity in toxicokinetics can be assessed by comparing relevant parameters, e.g., AUC, after different dose levels, or after single and repeated exposure. Dose dependency may be indicative of saturation of enzymes involved in the metabolism of the compound. An increase of AUC after repeated exposure as compared to single exposure may be an indication for inhibition of metabolism. A decrease in AUC may be an indication for induction of metabolism.

In conjunction with information on the relationship between concentration/dose at the target site and the toxic effect, toxicokinetic information may be an important tool for extrapolation from high to low dose effects.

Toxicokinetic data can also be used to make informed decisions on testing of chemical substances. In specific circumstances, valid toxicokinetic data may be used to support a decision to omit testing for systemic effects, e.g., in cases where the toxicokinetic data provide sufficient evidence that a substance is not absorbed and therefore not systemically available, i.e., no plasma/blood concentrations were measurable and no parent compound or metabolites could be detected in urine, bile, or exhaled air. For example, *in vivo* testing for mutagenicity, reproductive toxicity, or carcinogenicity may be omitted if toxicokinetic data or other data indicate a lack of systemic availability.

Further refinements of methods require more elaborate models with explicit or implicit assumptions and defaults, e.g., Physiologically Based Pharmacokinetic/Toxicokinetic (PBPK/PBTK) models, which may also provide modeled information on the target concentrations/amounts (Section 4.3.6).

4.3.5.2 *In Vitro* Data

In recent years, several types of *in vitro* approaches have been developed to assess the absorption and metabolic pathways of substances. Except for the "OECD TG 428, Skin Absorption: *In Vitro* Method," none of these test methods have yet been adopted as a test guideline method.

Ex vivo systems derived from animals and from human organs can be used to investigate the *in vitro* metabolism of xenobiotics. Cell lines, which are transfected to express species-specific metabolic enzymes, can also be used to identify the enzymes involved in the metabolism of a specific substance. Blocking the metabolism by an enzyme specific substrate or by antibodies is also helpful for the identification of the enzymes involved in the metabolism of a substance.

The *in vitro* approaches may give qualitative and under special circumstances, also, quantitative information. Information from *in vitro* experiments, in particular data on *in vitro* metabolism, has been used in PBTK models (Section 4.3.6). The use of data derived from *in vitro* test systems should be very carefully considered in the risk assessment until such approaches have been appropriately validated.

4.3.5.3 Other Data

Preliminary predictions of absorption of a substance can be made from its physico-chemical properties if no other information is available. Also elaborate computer programs are available that make predictions about, e.g., dermal penetration or metabolic pathways. However, these systems have often not been extensively validated against appropriate experimental data and it is not always certain if the results obtained in such models reflect the situation *in vivo*. On this basis, modeled data should only be used for risk assessment purposes where it is supported by other kinds of evidence.

The most important characteristics regarding a preliminary prediction of absorption of a substance from its physico-chemical properties are summarized below.

In order for a substance to be absorbed, it must cross biological membranes. Most substances cross by passive diffusion. This process requires a substance to be soluble both in lipid and water. The most useful parameters providing information on the potential for a substance to diffuse across biological membranes are the octanol/water partition coefficient (Log P) value and the water solubility. The Log $P_{o/w}$ value provides information on the relative solubility of the substance in water and the hydrophobic solvent octanol (used as a surrogate for lipid) and is a measure of lipophilicity. Log P values above zero indicate that the substance is more soluble in octanol than water, i.e., is lipophilic, and values below zero (negative values) indicate that the substance is more soluble in water than octanol, i.e., is hydrophilic. In general, moderate Log P values (between 0 and 4) are favorable for absorption. However, a substance with a Log P value around 0 and low water solubility (around 1 mg/l) will also be poorly soluble in lipids and hence not readily absorbed. It is therefore important to consider both the water solubility of a substance and its Log P value when assessing the potential of that substance to be absorbed.

4.3.5.3.1 *Inhalation*
Substances that can be inhaled include gases, vapors, liquid aerosols (both liquid and solid substances in solution), and finely divided powders/dusts (dust aerosols). Such substances may be absorbed from the respiratory tract or, through the action of clearance mechanisms, may be transported out of the respiratory tract and swallowed. This means that absorption from the gastrointestinal tract will contribute to the total body burden of substances that are inhaled.

Physico-chemical factors that determine the extent to which a substance may be absorbed by the inhalation route are presented in Table 4.2.

Most gases and vapors are readily absorbed across the lungs, predominantly in the alveoli by passive diffusion along a concentration gradient. The key determinant of absorption of gases and vapors in the respiratory tract is the solubility in blood. For gases and vapors that readily dissolve into blood, a large proportion of what is inhaled per breath will be absorbed. To be readily soluble in blood, a gas or vapor must be soluble in water and thus increasing water solubility would increase the amount absorbed per breath. However, the gas or vapor must also be sufficiently lipophilic to cross the alveolar and capillary membranes and therefore a moderate Log P value (between 0 and 4) would be favorable for absorption.

The potential for liquid aerosols or finely divided powders to be inhaled will be determined by their particle size. Deposition patterns for dusts will depend not only on the particle size of the dust but also the hygroscopicity, electrostatic properties and shape of the particles, and the respiratory dynamics of the individual. Thus, it is only possible to make very general statements about sites of

TABLE 4.2
Physico-Chemical Factors That Determine the Extent to Which a Substance May Be Absorbed by the Inhalation Route

Vapor pressure	Indicates whether a substance may be available for inhalation as a vapor. As a general guide, highly volatile substances are those with a vapor pressure greater than 25 KPa (or a boiling point below 50°C). Substances with low volatility have a vapor pressure of less than 0.5 KPa (or a boiling point above 150°C).
Particle size	Indicates the presence of inhalable and respirable particles. As a rough guide, particles with aerodynamic diameters below 100 μm have the potential to be inhaled. Particles with aerodynamic diameters of above 1–5 μm have the greatest probability of settling in the nasopharyngeal region whereas particles with aerodynamic diameters below 1–5 μm are most likely to settle in the tracheobronchial or pulmonary regions, i.e., are respirable.
Log $P_{o/w}$	Values above 0 indicate the potential for absorption directly across the respiratory tract epithelium.
Water solubility	Very hydrophilic substances may be retained within the mucus or for low molecular weight substances (MW < 200), could be absorbed through aqueous pores. Very low water solubility (1 mg/l or less) and small particle size (below 1 μm) indicates a potential for accumulation in the lung tissue.

Source: Modified from EC, 2003. Technical guidance document. Available at http://ecb.jrc.it/tgd

deposition for inhaled dusts. Therefore, any powder that contains particles with aerodynamic diameters below 100 μm is potentially of concern.

Once a liquid droplet or dust particle has deposited in the airways, it can be absorbed across the respiratory tract epithelium, cleared from the lungs via the mucociliary mechanism or lymphatic system, or retained within the lungs. Generally, liquids, solids in solution, and water-soluble dusts would readily diffuse/dissolve into the mucus lining the respiratory tract. Lipophilic substances (Log P > 0) would then have the potential to be absorbed directly across the respiratory tract epithelium. For poorly water-soluble dusts, the rate at which the particles dissolve into the mucus will limit the amount that could be absorbed directly. Poorly water-soluble dusts depositing in the nasopharyngeal region could be coughed or sneezed out of the body or swallowed. Poorly water-soluble dusts depositing in the tracheobronchial region would mainly be cleared from the lungs by the mucociliary mechanism and swallowed; however, a small amount may be phagocytozed by macrophages and transported to the blood via the lymphatic system. Poorly water-soluble dusts depositing in the alveolar region would mainly be engulfed by alveolar macrophages. The macrophages will then either translocate particles to the ciliated airways or carry particles into the pulmonary interstitium and lymphoid tissues. Particles may also migrate directly to the pulmonary interstitium where clearance depends on the rate at which the particle dissolves. Those particles most likely to be retained are those that are poorly soluble in both water and lipids.

Substances, which can be inhaled, are sparingly soluble in water and fat, and of low systemic toxicity may cause adverse effects in the lung (irreversible impairment of lung clearance, lung fibrosis, and lung tumor formation), which can be explained by "overload phenomena" (EC 2003).

4.3.5.3.2 Oral Administration
Absorption can occur along the entire length of the gastrointestinal tract. Since most substances are absorbed by passive diffusion, some general physico-chemical characteristics can be identified which favor absorption. These are listed in Table 4.3.

However, substances with physico-chemical characteristics that are not favorable for absorption could still reach the systemic circulation because specific mechanisms exist to enable, e.g., dietary fats and electrolytes to be absorbed. Occasionally a substance may be sufficiently similar to a nutrient substance to compete with that nutrient for a carrier mediated or active transport

TABLE 4.3

Physico-Chemical Factors That Determine the Extent to Which a Substance May Be Absorbed by the Oral Route

Molecular weight	Generally the smaller the molecule the more easily it will be taken up. Molecular weights below 500 are favorable for absorption; molecular weights in the 1000s do not favor absorption.
Particle size	Generally solids have to dissolve before they can be absorbed. It is possible for small amounts of particles in the nanometer size range to be taken up by pinocytosis. The absorption of very large particles, several hundreds of micrometers in diameter, that are administered dry (e.g., in the diet) or in a suspension may be reduced because of the time taken for the particle to dissolve. This would be particularly relevant for poorly water-soluble substances.
Log $P_{o/w}$	Moderate values (0–4) are favorable for absorption by passive diffusion. Any lipophilic compound may be taken up by micellular solubilization but this mechanism may be of particular importance for highly lipophilic compounds (Log $P > 4$), particularly those that are poorly soluble in water (1 mg/l or less) that would otherwise be poorly absorbed.
Water solubility	Water-soluble substances will readily dissolve into the gastrointestinal fluids; however, absorption of very hydrophilic substances by passive diffusion may be limited by the rate at which the substance partitions out of the gastrointestinal fluid. If the molecular weight is low (less than 200), the substance may pass through aqueous pores or be carried through the epithelial barrier by the bulk passage of water.
Dosing vehicle	If the substance has been dosed using a vehicle, the water solubility of the vehicle and the vehicle/water partition coefficient of the substance may affect the rate of uptake. Compounds delivered in aqueous media are likely to be absorbed more rapidly than those delivered in oils, and compounds delivered in oils that can be emulsified and digested, e.g., corn oil or arachis oil are likely to be absorbed to a greater degree than those delivered in nondigestible mineral oil (liquid petrolatum).

Source: Modified from EC, 2003. Technical guidance document. Available at http://ecb.jrc.it/tgd

---

mechanism; however, it is rare that an exogenous compound will be absorbed in this manner and it is not generally possible to predict, which substances could be absorbed by such a mechanism. It should also be noted that substances could undergo chemical changes in the gastrointestinal fluids as a result of metabolism by gastrointestinal flora, by enzymes released into the gastrointestinal tract, or by simple hydrolysis. These changes will alter the physico-chemical characteristics of the substance and hence predictions based upon the physico-chemical characteristics of the parent substance may no longer apply.

One consideration that could influence the absorption of ionic substances, i.e., acids and bases, is the varying pH of the gastrointestinal tract. It is generally thought that ionized substances do not readily diffuse across biological membranes. When assessing the potential for an acid or base to be absorbed, knowledge of its pKa value (the pH at which 50% of the substance is in ionized and 50% in nonionized form) is advantageous. Absorption of acids is favored at pHs below their pKa whereas absorption of bases is favored at pHs above their pKa. A substance that is ionized at a pH of around 5–6, i.e., that of the small intestine would be anticipated to be poorly absorbed. Substances that contain groups with oxygen, sulfur, or nitrogen atoms, e.g., thiol (SH), sulfonate (SO_3H), hydroxyl (OH), carboxyl (COOH), or amine (NH_2) groups are all potentially ionizable.

Active transport mechanisms exist to remove exogenous substances from gastrointestinal epithelial cells (efflux mechanisms) thereby limiting entry into the systemic circulation. It is not possible to identify, which substances could be removed by efflux mechanisms from physico-chemical data.

4.3.5.3.3   Dermal Administration

Substances that can potentially be taken up across the skin include gases and vapors, liquids and particulates. Liquids and substances in solution are taken up more readily than dry particulates. Dry particulates will have to dissolve into the surface moisture of the skin before uptake can begin. Absorption of volatile liquids across the skin may be limited by the rate at which the liquid evaporates off the skin surface. As a result of binding to skin components, the uptake of chemicals with the following groups can be slowed: certain metal ions, acrylates, quaternary ammonium ions, heterocyclic ammonium ions, and sulfonium salts.

Physico-chemical factors that determine the extent to which a substance may be absorbed by the dermal route are presented in Table 4.4.

4.3.5.3.4   Distribution and Accumulation

It is sometimes possible to get an indication of how widely the parent compound may distribute in the body from the available physico-chemical data. The sites to which the parent compound distributes (pattern of distribution) once it has entered the systemic circulation are likely to be similar for all routes of administration. In general, substances and their metabolites that readily diffuse across membranes will distribute throughout the body and may be able to cross the blood–brain and blood–testes barriers, although the concentrations within the brain or testes may be lower than that in the plasma. The rate at which highly water-soluble molecules distribute may be limited by the rate at which they cross cell membranes and access of such substances to the central nervous system (CNS) or testes is likely to be restricted (though not entirely prevented) by the blood–brain and blood–testes barriers.

It is also important to consider the potential for a substance to accumulate or to be retained within the body. Lipophilic substances have the potential to accumulate within the body if the

---

TABLE 4.4

Physico-Chemical Factors That Determine the Extent to Which a Substance May Be Absorbed by the Dermal Route

Vapor pressure	The rate at which gases and vapors partition from the air into the stratum corneum will be offset by the rate at which evaporation occurs. Therefore, although a substance may readily partition into the stratum corneum, it may be too volatile to penetrate further. This can be the case for substances with vapor pressures above 100–10,000 Pa (ca. 0.76–76 mm Hg) at 25°C. Vapors of substances with vapor pressures below 100 Pa are likely to be well absorbed and the amount absorbed dermally may be more than 10% of the amount that would be absorbed by inhalation.
Molecular weight	Less than 100 favors dermal uptake. Above 500 the molecule may be too large.
Log $P_{o/w}$	For substances with values below 0, poor lipophilicity will limit penetration into the stratum corneum and hence dermal absorption. Values below -1 suggest that a substance is not likely to be sufficiently lipophilic to cross the stratum corneum; therefore dermal absorption is likely to be low. Log P values between 1 and 4 favor dermal absorption (values between 2 and 3 are optimal) particularly if water solubility is high. Above 4, the rate of penetration may be limited by the rate of transfer between the stratum corneum and the epidermis, but uptake into the stratum corneum will be high. Above 6, the rate of transfer between the stratum corneum and the epidermis will be slow and will limit absorption across the skin. Uptake into the stratum corneum itself may be slow.
Water solubility	The substance must be sufficiently soluble in water to partition from the stratum corneum into the epidermis. Therefore, if the water solubility is below 1 mg/l, dermal uptake is likely to be low. Between 1 and 100 mg/l absorption is anticipated to be low to moderate and between 100 and 10,000 mg/l moderate to high. However, if water solubility is above 10,000 mg/l and the Log P value below 0, the substance may be too hydrophilic to cross the lipid rich environment of the stratum corneum and dermal uptake for these substances will be low.

Source: Modified from EC, 2003. Technical guidance document. Available at http://ecb.jrc.it/tgd

dosing interval is shorter than four times the whole body half-life (see Section 4.3.5.1). Although there is no direct correlation between the lipophilicity of a substance and its biological half-life, substances with high log P values tend to have longer half-lives. On this basis, there is a potential for highly lipophilic substances (Log P > 4) to accumulate in individuals that are frequently exposed to that substance. Once exposure stops the concentration within the body will decline at a rate determined by the half-life of the substance.

Other substances that can accumulate within the body include poorly soluble particulates that are deposited in the alveolar region of the lungs, substances that bind irreversibly to endogenous proteins, and certain metals and ions that interact with the crystal matrix of bone. The properties of these substances are such that the body cannot readily remove them; hence they gradually build up with successive exposures and the body burden can be maintained for long periods of time.

Protein binding can limit the amount of a substance available for distribution; however, it will generally not be possible to determine from the available data, which substances will bind to proteins and how avidly they will bind.

Physico-chemical factors that determine the potential distribution and accumulation of a substance are presented in Table 4.5.

4.3.5.3.5 Metabolism

It is very difficult to predict what metabolic changes a substance may undergo on the basis of physico-chemical data alone. Although it is possible to look at the structure of a molecule and identify potential metabolites, it is by no means certain that these reactions will occur *in vivo*. The molecule may have the wrong three-dimensional shape or may not reach the necessary site for a particular reaction to take place. It is even more difficult to predict the extent to which a substance will be metabolized along different pathways. Therefore, although predictive models have been developed, at present such models are not able to mimic the complexities of the *in vivo* situation.

4.3.5.3.6 Excretion

There are a limited number of conclusions that can be drawn from physico-chemical data about the excretion of a substance from the body. Depending on the metabolic changes that may have occurred, the compound that is finally excreted may have few or none of the physico-chemical characteristics of the parent compound. Also, depending on whether the substance is conjugated, the molecular weight of the final product may be smaller or greater than that of the parent compound.

For volatile substances and metabolites, exhaled air is an important route of excretion. Substances that are excreted in the urine tend to be water-soluble and of low molecular weight (below 300 in the rat). Substances that are excreted in the bile tend to have higher molecular weights. In the

---

TABLE 4.5

Physico-Chemical Factors That Determine the Potential Distribution and Accumulation of a Substance

Molecular weight	In general, the smaller the molecule, the wider the distribution.
Log $P_{o/w}$	If the molecule is lipophilic (Log P > 0), it is likely to distribute into cells and the intracellular concentration may be higher than the extracellular concentration particularly in fatty tissues. Generally, substances with high values have long biological half-lives. On this basis, daily exposure to a substance with a value of 4 or higher could result in a build up of that substance within the body. Substances with values of 3 or less would be unlikely to accumulate with the repeated intermittent exposure patterns but may accumulate if exposures are continuous.
Water solubility	Small water-soluble molecules and ions will diffuse through aqueous channels and pores. The rate at which very hydrophilic molecules diffuse across membranes could limit their distribution.

Source: Modified from EC, 2003. Technical guidance document. Available at http://ecb.jrc.it/tgd

rat, it has been found that substances with molecular weights below around 300 tend not to be excreted into the bile; however, it is not clear if a similar cutoff exists for humans and if so, where this cutoff lies.

4.3.6 PBPK/PBTK Models

Traditional, default approaches for toxicological risk assessment are not based on specific understanding of modes of action and tissue dose metrics (e.g., tissue concentrations, body burdens, AUCs). In recent years, PBPK/PBTK modeling has found frequent application in risk assessments where PBPK models serve as important adjuncts to studies on modes of action of xenobiotics.

Andersen (2003) gives a comprehensive review outlining the history of PBPK modeling, emphasizes more recent applications of PBPK models in health risk assessment, and discusses the risk assessment perspective provided by modern uses of these modeling approaches.

Pharmacokinetic models are used as tools to extrapolate from the results obtained in studies with experimental animals to predict effects in human populations that generally are exposed at lower environmental exposure levels compared to the generally higher exposure levels used in animal experiments. In such models, target tissue doses in different animal species under a variety of exposure conditions are predicted, using computer simulation.

The newer physiologically based pharmacokinetic (PBPK) models take nonlinearity of physiological processes such as chemical metabolism and excretion into consideration. At the high dose levels used in animal experiments, these mechanisms become saturated with the result that the tissues may be exposed to a different composition of pure compound and metabolites than at the low dose levels encountered in real-life human exposure.

Extensive preparatory work is necessary before PBPK modeling is possible, just like it is for default approaches to risk assessment. The process for application of a PBPK or PT model (PT: physiological toxicokinetics, used in Europe/Canada for PBPK) generally includes the identification of toxic effects in animals and humans, and the identification of critical effects for risk assessment; the organization of data for modes of action, metabolism, chemistry of compounds and metabolites, and similar information for related compounds; a description of the potential mode of action involved in the critical effect; proposals for relationships between response and tissue dose, specifying the tissue dose metric associated with toxicity; the development of an appropriate PBPK/PT model to estimate the tissue dose metric for various routes of administration, at various doses, in test species and in humans; the estimation of tissue dose metrics during exposures that produce toxicity; and the estimation of risks in humans based on the tissue doses during human exposures, assuming a similar dose–response relationship in humans and rats based on the tissue dose metrics.

The use of PBPK models will undoubtedly increase. However, a standard for the development of PBPK models is needed, and reliability of tissue dose estimates must be ensured.

4.4 ACUTE TOXICITY

Acute toxicity refers to the adverse effects, which occur within a given, usually short time, following a single, usually high exposure to a substance. In older literature, acute toxicity is sometimes used synonymously with lethal effect or "LD_{50}," which was the only endpoint in older acute toxicity tests. Nowadays, acute toxicity studies are designed to reveal more subtle effects.

4.4.1 Definitions

A general definition of the term "acute toxicity" is: The adverse effects occurring within a given time, following a single exposure to a substance. The term usually excludes local irritant or corrosive effects arising from a single application of a substance to the skin or eye (Section 4.5) (EC 2003).

According to the definitions provided in the OECD test guidelines (TG 420 and 423), acute oral toxicity refers to those adverse effects that occur following oral administration of a single dose of a substance or multiple doses given within 24 h.

Acute dermal toxicity is the adverse effects occurring within a short time of dermal application of a single dose of a test substance. The duration of exposure in the OECD TG 402 is 24 h, at the end of which residual test substance should be removed.

Acute inhalation toxicity is the total of adverse effects caused by a substance following a single, uninterrupted exposure by inhalation over a short period of time to an airborne substance. For testing, a fixed duration of exposure of 4 h is generally recommended in the OECD TG 403.

In all OECD TG studies for acute toxicity, the minimum observation period following exposure is 14 days.

In a classical acute toxicity study, the LD/LC_{50} value is the dose or concentration, which causes a 50% mortality.

4.4.2 Objectives for Assessing the Acute Toxicity of a Substance

Generally the objectives of investigating the acute toxicity are to find out (EC 2003):

- Whether single exposures of humans to the substance of interest could be associated with adverse effects on health
- In studies in animals, the lethal potency of the substance based on the LD_{50}, the LC_{50}, the discriminating dose, and/or the acute toxic class
- What toxic effects are induced following a single exposure to a substance, their time of onset, duration, and severity (all to be related to dose)
- When possible, the slope of the dose–response curve
- When possible, whether there are marked sex differences in response
- To obtain information necessary for the classification and labeling of the substance for acute toxicity

In relation to the second point above, it should be noted, that there is a general objective to move away from the induction of lethality in animal tests for animal welfare reasons.

4.4.3 Test Guidelines

4.4.3.1 *In Vivo* Tests

Test guidelines for acute toxicity testing have been adopted by the OECD, US-EPA, and EU, see Table 4.6.

There are six OECD test guidelines concerning acute toxicity, one of which (TG 401) has now been deleted because of concerns for animal welfare and the number of animals used. Although this test guideline is now deleted, the toxicologist needs to be familiar with the method, as chemicals have been tested according to TG 401 up to the point of deletion.

TG 402 and 403 are similar to the deleted TG 401, except for the route of administration, and are therefore expected to be deleted when new test guidelines for the dermal and inhalation routes become adopted.

There are three US-EPA OPPTS Test Guidelines for acute toxicity testing. OPPTS 870.1100 is similar to OECD TG 425 (Acute Oral Toxicity: Up-and-Down Procedure). OPPTS 870.1200 and OPPTS 870.1300 are similar to OECD TG 402 and 403, respectively.

The EU test guidelines for acute toxicity testing, the Annex V methods, are similar to the corresponding OECD test guidelines.

The following draft guidelines are being developed by the OECD: TG 433 (Acute Inhalation Toxicity-Fixed Concentration Procedure), TG 434 (Acute Dermal Toxicity-Fixed Dose Procedure),

TABLE 4.6

Test Guidelines Adopted for Acute Toxicity Testing

Title

OECD Test Guidelines	Year
401 Acute oral toxicity - Deleted 2002	1981
402 Acute dermal toxicity	1987
403 Acute inhalation toxicity	1981
420 Acute oral toxicity - Fixed dose method	2001
423 Acute oral toxicity - Acute toxic class method	2001
425 Acute oral toxicity: Up-and-down procedure	2001
418 Delayed neurotoxicity of organophosphorus substances following acute exposure	1995
US-EPA OPPTS Harmonized Test Guidelines	
870.1100 Acute oral toxicity	1998
870.1200 Acute dermal toxicity	1998
870.1300 Acute inhalation toxicity	1998
870.6100 Acute and 28-day delayed neurotoxicity of organophosphorus substances	1998
EU Annex V Test Methods	
B.1bis Acute oral toxicity - Fixed dose procedure	2004
B.1tris Acute oral toxicity - Acute toxic class method	2004
B.2 Acute toxicity (Inhalation)	1992/1993
B.3 Acute toxicity (Dermal)	1992
B.37 Delayed neurotoxicity of organophosphorus substances following acute exposure	1996

and TG 436 (Acute Inhalation Toxicity - Acute Toxic Class (ATC) Method); the latest draft versions are from 2004.

The principle of the acute toxicity studies is that the test substance is administered in graduated doses or concentrations to several groups of experimental animals, one dose being used per group. The doses chosen may be based on the results of a range finding test. Subsequently observations of effects and deaths are made. Animals, which die during the test are necropsied, and at the conclusion of the test the surviving animals are sacrificed and necropsied. Common observations in acute toxicity studies include CNS effects (consciousness, respiration, cardiac function), effects on the autonomic nervous system (sympathic and parasympathic), effects on neuromuscular function, effects on sensory function, and effects on the liver and kidney. Histopathology is rarely performed in acute toxicity studies. Expected findings are unspecific signs of cytotoxicity, cellular degeneration, and necrosis. Histopathology may, however, provide information about target organs in the body.

All three OECD test guidelines on acute oral toxicity testing involve the administration of a single bolus dose of a test substance to fasted healthy young adult rodents by oral gavage, observation for up to 14 days after dosing, recording of body weight, and the necropsy of all animals. The primary endpoint for the Acute Toxic Class Method and the Up-and-Down Procedure is mortality, but for the Fixed Dose Method it is the observation of clear signs of toxicity. The Fixed Dose Method and the Acute Toxic Class Method provide a range estimate of the LD_{50}-value; the ranges are defined by cutoff values of the applied classification system and not as a calculated lower and upper level. In the Fixed Dose Method, which is not intended to involve mortality, the LD_{50}-value range is inferred from the fixed dose, which produces evident toxicity. The Up-and-Down Procedure provides a point-estimate of the LD_{50}-value with confidence intervals (OECD 2001a).

In addition to these standard acute toxicity studies, the OECD, US-EPA, and EU have developed a specific test guideline for delayed neurotoxicity of organophosphorus substances in the domestic laying hen following acute exposure, see Table 4.6.

The test substance is administered orally in a single dose to domestic hens, which have been protected from acute cholinergic effects, when appropriate. The animals are observed for 21 days for behavioral abnormalities, ataxia, and paralysis. Biochemical measurements, in particular NTE (Neuropathy Target Esterase), are undertaken on hens randomly selected from each group (normally 24 and 48 h after dosing). Twenty-one days after exposure, the remainder of the hens is sacrificed and histopathological examination of selected neural tissues is undertaken.

4.4.3.2   *In Vitro* Tests

The OECD, US-EPA, and EU have no adopted test guidelines for *in vitro* testing of acute toxicity.

4.4.4   GUIDANCE DOCUMENTS

4.4.4.1   WHO

The WHO/FAO Joint Meeting of Experts on Pesticide Residues (JMPR) has given recommendations on interpretation of cholinesterase inhibition (FAO 1998, 1999), see Section 4.7.7.3.1.

4.4.4.2   OECD

A Guidance Document on Acute Oral Toxicity Testing has been published (OECD 2001a).

The guidance document offers a comparison of TG 420 (Fixed Dose Method), TG 423 (Acute Toxic Class Method), and TG 425 (Up-and-Down Procedure). The purpose of this Guidance Document is to provide information to assist with the choice of the most appropriate Guideline to enable particular data requirements to be met while reducing the number of animals used and animal suffering. The Guidance Document also contains additional information on the conduct and interpretation of test guidelines 420, 423, and 425.

A Guidance Document on Acute Inhalation Toxicity Testing is being developed and presently exists as a draft (OECD 2004b). The document recommends the Acute Toxic Class (ATC) Method with a group size of three animals per sex, if the objective of the test is solely related to hazard classification. Limits for particle-size distribution of aerosolized test substances are suggested. The preferred mode of exposure is the nose-only, head-only, or head/nose-only exposure technique, because this mode of exposure minimizes exposure or uptake by noninhalation routes.

OECD has published a document on a "Harmonised integrated classification system for human health and environmental hazards of chemical substances and mixtures" (OECD 2001b). Chapter 2.1 addresses a harmonized system for the classification of chemicals which cause acute toxicity, and Chapter 2.8 addresses the chemicals which cause specific target organ oriented systemic toxicity following a single exposure.

4.4.4.3   US-EPA

The series of Risk Assessment Guidelines does not include a guideline for acute toxicity risk assessment.

4.4.4.4   EU

The TGD (EC 2003), Chapter 3.6, addresses acute toxicity, provides guidance on data requirements, evaluation of data, and dose–response assessment for acute toxicity.

4.4.5   Use of Acute Toxicity Data in Hazard Assessment

Acute toxic effects are considered as being "threshold effects," i.e., effects for which there are expected to be a threshold of substance concentration below which the effects will not be manifested. For the hazard and risk assessment, it is important to identify those dose levels at which signs of acute toxicity are observed, and the dose level at which acute toxicity is not observed, i.e., to derive a NOAEL for acute toxicity. However, it should be noted that a NOAEL is usually not derived in the classic acute toxicity studies, partly because of the limitations in study design.

4.4.5.1   Human Data

First of all, the very existence of human acute toxicity data points to a hazard. Human acute toxicity data often provide more qualitative information than animal data in the form of details about signs and symptoms and targets in the body, and follow-up data on duration of symptoms and reversibility. The data usually derive from case stories of accidents and poisonings, and exposure data may or may not be available and reliable, see Section 3.2.1.

It may sometimes be possible to derive reliable NOAEL values for specific subpopulations from well-documented human data.

4.4.5.2   Animal Data

Animal acute toxicity data have traditionally been used almost exclusively for classification purposes, and the classic study design for acute toxicity studies has only provided information about lethality. However, with the development of the newer designs for acute toxicity studies, more emphasis has been put on observation of animals and description of signs and symptoms of acute toxicity.

As mentioned above, a NOAEL is usually not derived in acute toxicity studies. It is more usual that the only numerical value derived is the LD_{50} or LC_{50} value. The LD_{50} or LC_{50} values (or the discriminating dose if the Fixed Dose Procedure was used or the result of the Acute Toxic Class Method) give an indication of the relative lethal potency of a substance. The slope of the dose–response curve is a particularly useful parameter as it indicates the extent to which reduction of exposure will reduce the lethality: the steeper the slope, the greater the reduction in response for a particular finite reduction in exposure.

Where information is available on toxic signs of acute toxicity and the dose levels at which these signs occur, then this is useful information that can aid in the hazard and risk assessment for acute toxicity. Equally, dose levels leading to no acute effects can provide useful information.

The delayed neurotoxicity study of organophosphorus substances provides information on the delayed neurotoxicity arising from a single exposure. The study is used in the assessment and evaluation of the neurotoxic effects of organophosphorus substances (Section 4.7.7.3.1).

4.4.5.3   *In Vitro* Data

Currently, there are no validated and regulatory accepted *in vitro* methods for assessing acute toxicity.

4.4.5.4   Other Data

"Structural alerts" may provide useful information for assessment of the acute toxicity of certain substances. One example is that for highly water-soluble salts of substances with well-characterized toxic properties, the systemic toxicity can be expected to be similar. Another example is the group of organophosphorus insecticides for which the acute toxicity is due to their anticholinesterase action.

Use of structure–activity relationships ((Q)SAR) may be valuable in the evaluation of the acute toxicity of a substance for which no data are available (Sections 3.5.2 and 3.5.3).

4.5 IRRITATION AND CORROSION

A substance may cause changes at the site of first contact (skin, eye, mucous membrane in the respiratory and the gastrointestinal tract), irrespective of whether it can become systemically available. These changes are considered local effects. Local effects may occur as a result of a single exposure or repeated exposure. This section deals with local effects caused by a single ocular, dermal, or inhalation exposure. Local effects following repeated exposure are addressed in Section 4.7.

4.5.1 DEFINITIONS

Substances causing local effects following a single exposure can be further distinguished as irritant or corrosive substances, depending on the severity and (ir)reversibility of the effects observed.

A general definition of an "irritant substance" is: A noncorrosive substance, which, through immediate contact with the tissue under consideration, may cause inflammation.

A general definition of a "corrosive substance" is: A substance, which may destroy living tissues with which it comes into contact.

4.5.1.1 OECD

In the OECD test guideline for acute dermal irritation/corrosion (OECD TG 404), the following definitions are provided:

"Dermal irritation is the production of reversible damage of the skin following the application of a test substance for up to four hours."

"Dermal corrosion is the production of irreversible damage of the skin; namely, visible necrosis through the epidermis and into the dermis, following the application of a test substance for up to four hours. Corrosive reactions are typified by ulcers, bleeding, bloody scabs, and, by the end of observation at 14 days, by discoloration due to blanching of the skin, complete areas of alopecia, and scars. Histopathology should be considered to evaluate questionable lesions."

In the OECD test guideline for acute eye irritation/corrosion (OECD TG 405), the following definitions are provided:

"Eye irritation is the production of changes in the eye following the application of a test substance to the anterior surface of the eye, which are fully reversible within 21 days of application."

"Eye corrosion is the production of tissue damage in the eye, or serious physical decay of vision, following application of a test substance to the anterior surface of the eye, which is not fully reversible within 21 days of application."

4.5.1.2 US-EPA

In the US-EPA test guideline for acute dermal irritation (OPPTS 870.2500), the following definitions are provided:

"Dermal corrosion is the production of irreversible tissue damage in the skin following the application of the test substance."

"Dermal irritation is the production of reversible inflammatory changes in the skin following the application of a test substance."

In the US-EPA test guideline for acute eye irritation (OPPTS 870.2400), the following definitions are provided:

"Eye corrosion is the production of irreversible tissue damage in the eye following application of a test substance to the anterior surface of the eye."

"Eye irritation is the production of reversible changes in the eye following the application of a test substance to the anterior surface of the eye."

4.5.1.3 EU

In the EU test guidelines for acute dermal irritation/corrosion (Annex V B.4) and for acute eye irritation/corrosion (Annex V B.5), the definitions for dermal and eye irritation/corrosion are identical to those provided in the respective OECD test guidelines (Section 4.5.1.1).

4.5.2 OBJECTIVES FOR ASSESSING IRRITATION AND CORROSION OF A SUBSTANCE

The general objectives for assessing the potential of a substance to induce irritation or corrosion are to evaluate:

- Whether information from physico-chemical data, from nontesting methods, from *in vitro* studies, from animal studies, or from human experience provides evidence that the substance is, or is likely to be, corrosive.
- Whether information from physico-chemical data, from nontesting methods, from *in vitro* studies, from animal studies, or from human experience provides evidence of significant skin, eye, or respiratory irritation.
- Time of onset and the extent and severity of the responses and information on reversibility.

4.5.3 TEST GUIDELINES

4.5.3.1 *In Vivo* Tests

The OECD, US-EPA, and EU have developed specific test guidelines for *in vivo* testing of skin and eye irritation/corrosion, see Table 4.7. The EU Annex V methods are equivalent to the corresponding OECD test guidelines, and the US-EPA OPPTS test guidelines are, in principle, similar to the corresponding OECD test guidelines. There are no test guideline methods for respiratory irritation.

The principle of the acute dermal irritation/corrosion tests is that the substance to be tested is applied in a single dose to the skin of an experimental animal; untreated skin areas of the test animal serve as control (except when severe irritation/corrosion is suspected and a stepwise procedure is used). The degree of irritation/corrosion is read and scored at specified intervals and is further

TABLE 4.7
Test Guidelines Adopted for Irritation/Corrosion Testing

Title

OECD Test Guidelines	Year
404 Acute dermal irritation/corrosion	2002
405 Acute eye irritation/corrosion	2002
430 *In vitro* skin corrosion: Transcutaneous electrical resistance test (TER)	2004
431 *In vitro* skin corrosion: Human skin model test	2004
435 *In vitro* membrane barrier test method for skin corrosion (Draft)	2004 (Draft)
US-EPA OPPTS Harmonized Test Guidelines	
870.2400 Acute eye irritation	1998
870.2500 Acute dermal irritation	1998
EU Annex V Test Methods	
B.4 Acute toxicity: Dermal irritation/corrosion	2004
B.5 Acute toxicity: Eye irritation/corrosion	2004
B.40 Skin corrosion (*in vitro*)	2000

TABLE 4.8

Draize Scoring System for Evaluation of Skin Irritating and Corrosive Properties

Erythema and Eschar Formation	Value
No erythema	0
Very slight erythema (barely perceptible)	1
Well-defined erythema	2
Moderate to severe erythema	3
Severe erythema (beet redness) to slight eschar formation (injuries in depth)	4
Maximum possible	4
Edema Formation	
No edema	0
Very slight edema (barely perceptible)	1
Slight edema (edges of area well defined by definite raising)	2
Moderate edema (raised approximately 1 mm)	3
Severe edema (raised more than 1 mm and extending beyond area of exposure)	4
Maximum possible	4

described in order to provide a complete evaluation of the effects. The duration of the study should be sufficient to permit a full evaluation of the reversibility or irreversibility of the effects observed. The period of observation should not exceed 14 days.

The principle of the acute eye irritation/corrosion tests is that the substance to be tested is applied in a single dose to one of the eyes of the experimental animal; the untreated eye serves as the control. The degree of eye irritation/corrosion is evaluated by scoring lesions of conjunctiva, cornea, and iris, at specific intervals. Other effects in the eye as well as adverse systemic effects are also described to provide a complete evaluation of the effects. The duration of the study should be sufficient to permit a full evaluation of the reversibility or irreversibility of the effects. The period of observation should be at least 72 h, but should not exceed 21 days.

The test guideline methods use the scoring system developed by Draize (1944), see Tables 4.8 and 4.9. The EU criteria for classification are based on the mean tissue scores obtained over the first 24–72 h period after exposure and on the reversibility or irreversibility of the effects observed.

There are no internationally adopted test guideline methods for respiratory tract irritation.

4.5.3.2 *In Vitro* Tests

The OECD and EU have developed specific test guidelines for *in vitro/ex vivo* testing of skin corrosion, the "Transcutaneous Electrical Tesistance (TER) Test" and the "Human Skin Model Test" (see Table 4.7). In fact, the EU Annex V, B.40 test guideline includes the two adopted OECD *in vitro* tests for skin corrosion (TG 430 and 431). OECD has also developed a third test guideline for skin corrosion, the "*In Vitro* Membrane Barrier Test;" this test guideline has not yet been adopted.

There are no adopted test guideline methods for *in vitro/ex vivo* testing of eye corrosion or skin and eye irritation.

The principle of the TER test is that the test material is applied for up to 24 h to the epidermal surfaces of skin discs taken from the pelts of humanely killed young rats. Corrosive materials are identified by their ability to produce a loss of normal stratum corneum integrity and barrier function, which is measured as a reduction in the inherent TER below a threshold level (5 kΩ). Generally, materials that are noncorrosive in animals, but are irritating or nonirritating, do not reduce the TER below the threshold level. A dye-binding step can be incorporated into the test procedure for

TABLE 4.9
Draize Scoring System for Evaluation of Eye Irritating and Corrosive Properties

Cornea	Value
Opacity: Degree of density (readings should be taken from most dense area); the area of corneal opacity should be noted	
No ulceration or opacity	0
Scattered or diffuse areas of opacity (other than slight dulling of normal lustre), details of iris clearly visible	1
Easily discernible translucent area, details of iris slightly obscured	2
Nacrous area, no details of iris visible, size of pupil barely discernible	3
Opaque cornea, iris not discernible through the opacity	4
Maximum possible	4

Iris	
Normal	0
Markedly deepened rugae, congestion, swelling, moderate circumcorneal hyperemia, or injection, iris reactive to light (a sluggish reaction is considered to be an effect)	1
Hemorrhage, gross destruction, or no reaction to light	2
Maximum possible	2

Conjunctivae	
Redness (refers to palpebral and bulbar conjunctivae; excluding cornea and iris)	
Normal	0
Some blood vessels hyperemic (injected)	1
Diffuse, crimson color, individual vessels not easily discernible	2
Diffuse beefy red	3
Maximum possible	3
Chemosis - Swelling (refers to lids and/or nictating membranes)	
Normal	0
Some swelling above normal	1
Obvious swelling, with partial eversion of lids	2
Swelling, with lids about half closed	3
Swelling, with lids more than half closed	4
Maximum possible	4

confirmation testing of positive results in the TER including values around $5\,k\Omega$. The dye-binding step determines if the increase in ionic permeability is due to physical destruction of the stratum corneum. The TER method utilizing rat skin has shown to be predictive of *in vivo* corrosion in the rabbit assessed under OECD TG 404.

The principle of the human skin model test is that the test material is applied topically for up to 4 h to a three-dimensional human skin model, comprising at least a reconstructed epidermis with a functional stratum corneum (outermost layer of the skin). The human skin models can come from various sources, but they must meet certain criteria. Corrosive materials are identified by their ability to produce a decrease in cell viability (as determined, e.g., by using a dye reduction assay) below defined threshold levels at specified exposure periods. The principle of the test is in accordance with the hypothesis that corrosive chemicals are able to penetrate the stratum corneum (by diffusion or erosion) and are sufficiently cytotoxic to cause cell death in the underlying cell layers.

The membrane barrier test is composed of two components, a synthetic macromolecular bio-barrier and a chemical detecting system (CDS). The basis of the test method is that it detects membrane barrier damage caused by corrosive test substances after the application of the test substance to the surface of the artificial membrane barrier presumably by the same mechanism(s) of corrosion that operate on living skin. Penetration of the membrane barrier (or breakthrough) might be measured by a number of procedures, including a change in the color of a pH indicator dye or in some other property of the indicator solution below the barrier.

4.5.4 GUIDANCE DOCUMENTS

OECD has published a document on a "Harmonised Integrated Classification System for Human Health and Environmental Hazards of Chemical Substances and Mixtures" (OECD 2001b). Chapters 2.2 and 2.3 address a harmonized system for the classification of chemicals, which cause skin and eye irritation/corrosion, respectively.

The TGD (EC 2003), Chapter 3.7, addresses irritation and corrosion and provides guidance on data requirements, evaluation of data, and dose–response assessment.

Guidance on how industry interprets eye irritation data in the light of EU classification and labeling has been summarized in a publication by ECETOC (1997). In a more recent ECETOC publication, examples on how industry uses existing human data in hazard classification for irritancy are provided (ECETOC 2002b).

A specific guidance document for irritation/corrosion has not been published by WHO/IPCS. The series of US-EPA Risk Assessment Guidelines does not include a guideline regarding risk assessment of irritation/corrosion.

4.5.5 USE OF INFORMATION ON IRRITATION AND CORROSION IN HAZARD ASSESSMENT

Usually it is possible unequivocally to identify a substance as being corrosive, whatever type of study provides the information.

Skin, eye, and respiratory tract irritation are considered as being "threshold effects," i.e., effects for which there are expected to be a threshold of substance concentration below which the effects will not be manifested. However, the classic skin and eye irritation studies are conducted using a single amount of the undiluted substance and therefore it is not possible to derive a NOAEL and/or LOAEL on the basis of such studies.

4.5.5.1 Human Data

Well-documented human data can often provide very useful information on skin and/or respiratory tract irritation, sometimes for a range of exposure levels. Often, the only useful information on respiratory tract irritation, which can be a threshold effect in the workplace, is obtained from human experience. It may in some cases be possible to derive a reliable NOAEL and/or LOAEL from human studies; however, usually there is only the information that a substance is irritating or, often by inference only, that it is not. The usefulness of human data on irritation will depend on the extent to which the effect, and its magnitude, can be reliably attributed to the substance under evaluation. Furthermore, there may be a significant level of uncertainty in human data on irritant effects because of poor reporting, lack of specific information on exposure, subjective or anecdotal reporting of effects, small numbers of subjects, etc.

Experience has shown that it is difficult to obtain useful data on substance-induced eye irritation (EC 2003).

4.5.5.2 Animal Data

Data from studies in animals according to test guideline methods, particularly if conducted in accordance with principles of GLP, will usually give very good information in order to identify whether a substance would be considered to be, or not to be, corrosive or irritant to the skin or eye in the test species. In general, it is assumed that substances, which are irritant in test guideline studies in animals will be skin and/or eye irritants in humans, and those which are not irritant in test guideline studies will not be irritant in humans.

There may be a number of skin or eye irritation studies available for a specific substance, none of which have been performed fully equivalent to a test guideline. If the results from such a batch of studies are consistent, they may, together, provide sufficient information on the skin and/or eye irritation potential of the substance. If the results from such a batch of studies are not consistent, it will be necessary to decide which of the studies are the most reliable ones.

Attention should be given to the occurrence of persisting irritating effects. Effects such as erythema, edema, fissuring, scaling, desquamation, hyperplasia, and opacity, which do not reverse within the test period may indicate that a substance will cause persistent damage to the human skin and eye.

Data from studies other than skin or eye irritation studies, e.g., other toxicological studies on the substance in which local responses of skin, eye, and/or respiratory system were reported, may provide useful information. However, they may not be well reported in relation to, e.g., the basic requirements for information on skin and eye irritation.

As mentioned previously, there are no test guideline methods for respiratory irritation. Good data, often clearly related to exposure levels, can be obtained on respiratory and mucous membrane irritation, from well-designed and well-reported inhalation studies in animals. Also the Alarie test (Alarie 1973, 1981), an experimental animal test assessing the concentration that results in a 50% reduction of the breathing frequency, may provide useful information on sensory irritation of the upper respiratory tract and the results may be used for hazard identification.

Within the framework of risk assessment of existing substances in the EU, the Alarie test has been considered as being inappropriate to evaluate respiratory tract irritation (Bos et al. 2002). Alarie has discussed this at his Web site (Alarie 2007).

As mentioned above, a NOAEL can usually not be derived from the classic test guideline methods for skin and eye irritation. Based on information from acute and/or repeated dose toxicity studies using inhalation, it may be possible to derive a NOAEL and/or LOAEL for respiratory tract irritation. In such studies, the slope of the dose–response curve is a particularly useful parameter as it indicates the extent to which reduction of exposure will reduce the irritative response: the steeper the slope, the greater the reduction in response for a particular finite reduction in exposure.

4.5.5.3 *In Vitro* Data

There are adopted *in vitro* test methods (Section 4.5.3.2) under which a substance can be identified as corrosive. A negative result in these tests should, however, be supported by a weight of evidence assessment using other data. It should be noted that these tests do not provide information on skin irritation.

There is a wide range of *in vitro* test methods, which give information on the potential irritancy of a substance; however, none of these test methods have yet been adopted as a guideline test method. Some of the methods are designed specifically to address a particular type of irritation while others are more general. If there are clear indications from such studies that a substance is likely to be irritating, this may be sufficient for hazard identification purposes. However, the uncertainty in the state of the art for identification of a substance as being nonirritant from testing *in vitro* is too high for a definitive use in hazard assessment for the time being.

4.5.5.4 Other Data

Physico-chemical data can be used to identify a substance as being corrosive, but not as being nonirritant. Substances exhibiting strong acidity ($pH \leq 2$) or alkalinity ($pH \geq 11.5$) in solution are predicted to be corrosive (EU 2001). However, no conclusion can be made regarding corrosion when the pH has an intermediate value ($2 < pH < 11.5$).

Physico-chemical data may also indicate that a substance has defatting properties. Defatting of exposed skin may cause irritation.

Use of (Q)SAR may be valuable in the evaluation of the irritant and corrosive properties of a substance for which no data are available, see Sections 3.5.2 and 3.5.3.

4.6 SENSITIZATION

In the context of allergies, sensitization is the process by which a person becomes, over time, increasingly allergic to a substance (sensitizer) through repeated exposure to that substance. Allergies are inappropriate or exaggerated reactions of the immune system to substances that, in the majority of people, cause no symptoms. Symptoms of the allergic diseases may be caused by exposure of the skin, the respiratory tract, or of the stomach and intestines to a protein or a chemical substance.

A number of diseases are recognized as being, or presumed to be, allergic in nature. These include asthma, rhinitis, conjunctivitis, allergic contact dermatitis, urticaria (a condition in which red or pale, itchy, and swollen areas appear on the skin, often called "hives"), and food allergies. In this section, the endpoints discussed are those traditionally associated with occupational and consumer exposure. Photosensitization is potentially important but its mechanism of action is poorly understood, so it has been considered but not discussed in detail.

Test Guideline methods have so far only been developed for skin sensitization, while standard methods for sensitization via the respiratory and oral route have not yet been established. Therefore, the emphasis in this section is on skin sensitization. This does not mean that sensitization via the other routes is not important. Certain substances are known sensitizers via inhalation, e.g., the isocyanates are inducers of asthma in humans.

4.6.1 Definitions

In the OECD test guideline for skin sensitization (OECD TG 406), the following definition is given: "Skin sensitization (allergic contact dermatitis) is an immunologically mediated cutaneous reaction to a substance. In the human, the responses may be characterized by pruritis, erythema, edema, papules, vesicles, bullae, or a combination of these. In other species, the reactions may differ and only erythema and edema may be seen."

This definition is also used by the US-EPA and by the EU in their respective guidelines for skin sensitization (OPPTS 870.2600 Skin Sensitization and B.6. Skin Sensitization).

4.6.2 Objectives for Assessing Sensitization of a Substance

The general objectives for assessing the sensitization potential of a substance are to examine whether there are indications from human experience of skin allergy or respiratory hypersensitivity following exposure to the substance, and whether the substance has a skin sensitization potential based on tests in animals.

The likelihood that a substance will induce skin sensitization or respiratory hypersensitivity in humans who are exposed to this substance is determined by several factors including the route, duration and magnitude of exposure, and the potency of the substance (EC 2003).

4.6.3 Test Guidelines

4.6.3.1 *In Vivo* Tests for Skin Sensitization

Specific test guidelines for skin sensitization have been developed by the OECD, US-EPA, and EU, see Table 4.10.

There are currently three OECD test guideline methods for test of skin sensitization in animals. These include the Guinea-Pig Maximization Test (GPMT), the Buehler test, and the murine Local Lymph Node Assay (LLNA).

TABLE 4.10

Test Guidelines Adopted for Sensitization Studies

Title

OECD Test Guidelines	Year
406 Skin sensitisation (GPMT and Buehler test)	1992
429 Skin sensitisation: Local lymph node assay	2002
US-EPA OPPTS Harmonized Test Guidelines	
870.2600 Skin sensitization	1998
EU Annex V Test Methods	
B.6 Skin sensitisation	1996
B.42 Skin sensitisation: Local lymph node assay	2004

According to the US-EPA test guideline for skin sensitization testing (870.2600 Skin Sensitization), any of the following test methods is considered to be acceptable: Buehler test, GPMT, other tests including the "Open epicutaneous test," the "Maurer optimization test," the "Split adjuvant technique," the "Freund's complete adjuvant test," and the "Draize sensitization test." The Mouse Ear Swelling Test (MEST) or the Local (auricular) Lymph Node Assay (LLNA) in the mouse may be used as screening tests to detect moderate to strong sensitizers. If a positive result is seen in either assay, the test substance may be designated a potential sensitizer, and it may not be necessary to conduct a further test in guinea pigs. If the LLNA or MEST does not indicate sensitization, the test substance should not be designated a non-sensitizer without confirmation in an accepted test using guinea pigs.

The EU test guidelines for sensitization (B.6 Skin Sensitisation and B.42 Skin Sensitisation: Local Lymph Node Assay) are in principle similar to the corresponding OECD test guideline methods.

The relative ability of a chemical to induce sensitization is an intrinsic property of the chemical, and is determined by the amount of chemical per unit area required for the acquisition of skin sensitization in a previously naïve individual. The tests for skin sensitization include two phases: induction and challenge.

During induction, it is attempted to induce a hypersensitive state in the test animals by exposing them to a high concentration of the test substance. This requires a certain severity of provocation, and a certain period of time (one to several weeks).

Following the induction, it is tested whether a hypersensitive state now exists in the test animals; this test is the challenge. For challenge, a low concentration of the test substance is used.

The GPMT is an adjuvant type test in which the allergic state (sensitization) is potentiated by the use of Freund's Complete Adjuvant (FCA). Adjuvants are used to boost the immune system, and to ensure that the severity of provocation in the test is maximal. However, FCA may cause inflammation, induration, pain, and necrosis at the injection site, and its use is therefore ethically controversial. The Buehler test is a nonadjuvant type test involving for the induction phase topical application rather than the intradermal injections used in the GPMT. Both the GPMT and the Buehler test have demonstrated the ability to detect chemicals with a moderate to strong sensitization potential as well as those with a relatively weak sensitization potential. These guinea pig methods provide information on skin responses, which are evaluated for each animal after several applications of the substance, and on the percentage of animals sensitized.

The murine LLNA has been validated internationally and has been shown to have clear animal welfare and scientific advantages compared with the guinea pig tests. The OECD, in the initial considerations in TG 429, recommends that the LLNA can be used as a stand-alone test as an addition to the existing guinea-pig test methods.

The basic principle underlying the LLNA is that sensitizers induce a primary proliferation of lymphocytes in the lymph node draining the site of chemical application. This proliferation is proportional to the dose applied and to the potency of the allergen, and provides a measurement of sensitization. The LLNA assesses this proliferation as a dose–response in which the proliferation in test groups is compared to that in vehicle treated controls. The ratio of the proliferation in treated groups to that in vehicular controls, termed the "Stimulation Index", is determined, and must be at least three before a test substance can be further evaluated as a potential skin sensitizer. The test substance, plus vehicle and positive control, is applied for three consecutive days to the ears of test mice, on days 4 and 5 the animals are left alone, and on the 6th day they are prepared for the proliferation assay, sacrificed, and the measurements are done.

A very recent WHO/IPCS international workshop on skin sensitization in chemical risk assessment (WHO/IPCS 2007) concluded:

The Local Lymph Node Assay (LLNA) is the preferred test method for assessing the skin sensitization ability of chemicals in view of animal welfare considerations. It has been validated for the purpose of hazard identification. However, presently there is still a need for guinea-pig tests. Guinea-pig tests may still have a place for the testing of aqueous solutions, extracts, fabrics, mixtures, and preparations. When conducting guinea-pig assays, the Buehler assay is preferred over the GPMT from an animal welfare point of view. However, the GPMT is generally considered to be more sensitive than the Buehler Assay, for which reason some regulatory authorities prefer the GPMT.

The Workshop made the following recommendations:

Methodology to assess skin penetration, deposition, and metabolism needs to be further advanced. The LLNA needs to be further developed with a view to testing of aqueous solutions, preparations, and complex mixtures. The effects of irritant activity in the LLNA should be further explored. It is recommended that nonradioactive active forms of the LLNA, or LLNA-type assays that use reduced amounts of radioactivity, get more attention.

4.6.3.2 *In Vivo* Tests for Respiratory or Oral Sensitization

As mentioned in the introduction, standard test methods for respiratory or oral sensitization have not been developed.

In a review, Arts and Kuper (2007) have summarized the animal test methods, which have been used to detect immune-mediated respiratory disease. The tests for respiratory sensitization include dermal as well as inhalatory or topical exposure of mice, rats, or guinea pigs for induction and challenge, and may measure various endpoints to evaluate respiratory sensitization. The review concludes that standardized and validated dose–response test methods are urgently required in order to allow identification of respiratory allergens and to make it possible to recommend safe exposure levels for consumers and workers.

With respect to oral sensitization, attempts have been made to develop animal models of food allergy, which so far has proven to be complicated. One crucial point is the route of exposure; experience indicates that it may not be possible to develop an animal model, which mimics the human sensitization via the oral route. Exposure via the diet or in drinking water appears in rodents to be more likely to cause immunological hyporesponsiveness (i.e., tolerance) than sensitization, and therefore it may be necessary to use parenteral induction in animal testing (Dearman and Kimber 2007).

4.6.3.3 *In Vitro* Tests

The OECD, US-EPA, and EU have no adopted test guidelines for *in vitro* testing of sensitization.

4.6.4 GUIDANCE DOCUMENTS

4.6.4.1 WHO

A comprehensive WHO/IPCS criteria monograph on allergic hypersensitization has been published (WHO/IPCS 1999b). The monograph addresses the immune system, hypersensitivity, and auto-immunity, factors influencing allergenicity, clinical aspects of the most important allergic diseases, epidemiology of asthma and allergic disease, hazard identification, and risk assessment.

Very recently, a WHO/IPCS international workshop on skin sensitization in chemical risk assessment has taken place. The general conclusions and recommendations of the workshop have been published (WHO/IPCS 2007). These conclusions and recommendations are cited in Section 4.6.5.

4.6.4.2 OECD

An OECD guidance document on sensitization has not been published.

OECD has prepared a Detailed Review Document on classification systems for sensitizing substances in OECD Member countries as a step toward consensus on a harmonized classification system concerning criteria and classification systems for sensitizing substances in OECD countries (OECD 1999a).

OECD has also published a document on a "Harmonised integrated classification system for human health and environmental hazards of chemical substances and mixtures" (OECD 2001b). Chapter 2.4 addresses a harmonized system for the classification of chemicals, which cause respiratory or skin sensitization.

4.6.4.3 US-EPA

The series of Risk Assessment Guidelines does not include a guideline regarding risk assessment of sensitization.

4.6.4.4 EU

The TGD (EC 2003), Chapter 3.8, on sensitization gives definitions of skin and respiratory sensitization, and provides advice on the data to be used in the effects assessment, evaluation on the available data, and assessment of the dose–response relationship to be used in the EU-specific risk assessments.

The ECETOC Monograph 29 (ECETOC 2000) addresses the three OECD test guideline methods for test of skin sensitization in animals, the GPMT, the Buehler test, and the murine LLNA, further.

In another ECETOC publication, examples on how industry uses existing human data in hazard classification for sensitization are provided (ECETOC 2002b).

4.6.5 USE OF INFORMATION ON SENSITIZATION IN HAZARD ASSESSMENT

The issue of adverse versus non-adverse effects in the context of allergy has been discussed in an article by Kimber and Dearman (2002). The authors concluded that it has proven difficult to define a clear distinction between adverse and non-adverse changes in the immune system associated with chemical exposure. Furthermore, with regard to allergy, exposure to many known contact allergens, at levels below the threshold necessary for sensitization, can be well tolerated without the mani-festation of allergic disease. However, it is possible, probably with some potent chemical allergens, that subclinical immune priming may occur wherein clinically apparent sensitization is more readily achieved following subsequent exposure.

There is evidence that dose–response relationships exist for both skin sensitization and respiratory hypersensitivity, although these are frequently less well defined in the case of respiratory hypersensitivity (EC 2003). The dose of a substance required to induce sensitization in a previously naïve subject or animal is usually greater than that required to elicit a reaction in a previously sensitized individual; therefore, the dose–response relationship for these two phases will differ. Elicitation responses depend on several factors, among which are potency of the allergen and exposure conditions. Appropriate dose–response data can provide important information on the potency of the substance under evaluation. For sensitizers it is considered prudent to assume that a threshold cannot be identified, i.e., it is not possible to identify an elicitation dose or concentration of a sensitizing substance below which adverse effects are unlikely to occur in people already sensitized to a substance (EC 2003).

The WHO/IPCS international workshop on skin sensitization in chemical risk assessment (WHO/IPCS 2007) recommended:

"There is a need for a standardized system of classifying and determining limits according to potency. The use of the LLNA for potency categorization of induction of skin sensitization needs to be validated. An abbreviated test validation approach may be appropriate to assess the validity of potency assessment based on the LLNA and its appropriateness for predicting sensitizing induction potency in humans. It is recommended to derive dose–response curves from patch testing and/or open testing in individuals diagnosed with contact allergy, and thereby establish a threshold, which can be used to derive a point of departure/risk assessment. Existing human data on variability in individual thresholds should be evaluated to derive adjustment factors for risk assessment. It is recommended that further studies are carried out regarding ranking of chemicals according to their potency to elicit allergic responses in individuals diagnosed with contact allergy. Comparison of information on responses after occluded versus non-occluded exposures, and single versus repeated exposures, should be done to inform adjustment factors for risk assessment that may account for specific exposure conditions. Approaches to evaluate respiratory sensitization induction potency need to be developed."

It should be noted that, although respiratory allergens tested so far were positive in current tests evaluating the skin sensitization potential, skin sensitization potency data available from current test methods do not predict respiratory sensitization potency in general.

4.6.5.1 Human Data

Well-conducted human studies can provide very valuable information on skin sensitization.

However, in some instances (due to lack of information on exposure, a small number of subjects, concomitant exposure to other substances, local or regional differences in patient referral, etc.) there may be a significant level of uncertainty associated with human data. Moreover, diagnostic tests are carried out to see if an individual is sensitized to a specific agent, and not to determine whether the agent can cause sensitization.

Although human studies may provide some information on respiratory and oral hypersensitivity, the data are frequently limited and subject to the same constraints as human skin sensitization data. However, as no standard animal test exists, human data may be the only source of information.

It is generally difficult to obtain dose–response information from existing human data as exposure measurements may not have been taken at the same time as the disease was evaluated, adding to the difficulty of determining a dose–response.

The WHO/IPCS international workshop on skin sensitization in chemical risk assessment (WHO/IPCS 2007) concluded:

"Any test of skin sensitizing capability that includes dose–response assessment can be used to assess potency. Even though potency cannot be directly derived from human elicitation data, a low elicitation threshold is suggestive of a high potency. Where possible, attempts should be made to use clinical data for quantitative risk assessment."

4.6.5.2 Animal Data

Reliable data can be generated on skin sensitization from well-designed and well-conducted studies in animals. However, guinea-pig tests in particular may be difficult to interpret when irritancy or skin staining occurs as the result of challenge. The use of adjuvant in the GPMT may lower the threshold for irritation and so lead to false positive reactions, which can therefore complicate interpretation (running a pretest with FCA-treated animals can provide helpful information). In international trials, the LLNA has been shown to be reliable, but like the guinea-pig tests is dependent on the vehicle used, and it can occasionally give false positive results with irritants. Where tests (guinea pig/mouse) rely on topical exposure rather than intradermal injection, false negatives may occur where the substance fails to be absorbed into the skin as, e.g., with some metal salts. Therefore, careful consideration should be given to the vehicle used and the type of test performed. In some circumstances inconsistent results from similar guinea-pig studies, or between guinea-pig and LLNA studies, might increase the uncertainty of making a correct interpretation. Note that, in some instances sensitization may be due to impurities rather than the test material itself.

The classic test guideline methods for skin sensitization as carried out in the guinea pig (GPMT/Buehler) employ only a single (maximized) concentration of the substance during the induction phase and therefore it is not possible to derive a NOAEL and/or LOAEL for induction on the basis of such studies. Dose–response data can, however, be generated using specially designed guinea-pig test methods (the open epicutaneous test being the most appropriate) or from local LLNA. Appropriate dose–response data can provide important information on the potency (in the induction phase of sensitization) of the substance under evaluation. The LLNA is carried out using multiple concentrations and information on a dose–response relationship; information from such a study may be useful in the hazard characterization of sensitization. Neither the GPMT/Buehler nor the standard LLNA is specifically designed to evaluate the skin sensitizing potency of a substance as they have been designed to identify the skin sensitization potential for classification purposes. However, the relative potency of substances may be indicated by the percentage of positive animals in the guinea-pig studies in relation to the concentrations tested. Likewise, in the LLNA, the EC3 value (the dose estimated to cause a threefold increase in local lymph node proliferative activity) can be used as a measure of relative potency (ECETOC 2000). The dose–response data generated by the LLNA makes this test more informative than guinea-pig assays for the assessment of skin sensitizing potency. It should be remembered that it is considered prudent to assume that a threshold cannot be identified for elicitation, i.e., it is not possible to identify an elicitation dose or concentration of a sensitizing substance below which adverse effects are unlikely to occur in people already sensitized to a substance.

The WHO/IPCS international workshop on skin sensitization in chemical risk assessment (WHO/IPCS 2007) concluded:

"Any test of skin sensitizing capability that includes dose–response assessment can be used to assess potency. Currently the LLNA is the most appropriate assay for single chemical substances, as it is the only test for which guidelines indicate to include dose–response assessment. Guinea-pig data may also be used to categorize a chemical according to its skin sensitizing potency. It is acknowledged that categorization of skin sensitizing potency is associated with a degree of uncertainty. Neither the approach using the LLNA, nor the approach using guinea-pig data have been validated for the purpose of assessment of potency."

4.6.5.3 *In Vitro* Data

Currently, there are no validated and regulatory accepted *in vitro* methods for assessing sensitization.

The WHO/IPCS international workshop on skin sensitization in chemical risk assessment (WHO/IPCS 2007) concluded: "No *in vitro* assay systems for identification of skin sensitizing capacity have been validated to date. Some of the available systems may be useful in a weight of evidence approach or as a preliminary screen."

4.6.5.4 Other Data

Use of (Q)SAR may be valuable in the evaluation of the sensitization of a substance for which no data are available, see Sections 3.5.2 and 3.5.3.

The WHO/IPCS international workshop on skin sensitization in chemical risk assessment (WHO/IPCS 2007) concluded: "(Q)SARs and expert systems for identification of sensitizing capacity have not been validated to date, but may be used as part of a weight of evidence approach for identifying the sensitizing capacity of chemicals. There are certain local (Q)SARs that can be used for a small range of chemicals. However, these are currently insufficient to cover the full range of chemicals. The Workshop recommended that QSAR models need to be further developed, and the applicability domain of each model needs to be established."

Certain well-known groups of chemicals such as isocyanates and acid anhydrides are currently considered to cause respiratory hypersensitivity unless proved otherwise (EC 2003).

4.7 REPEATED DOSE TOXICITY

Repeated dose toxicity studies provide information on possible adverse general toxicological effects likely to arise from repeated exposure of target organs, and on dose–response relationships. Furthermore, these studies may provide information on, e.g., reproductive toxicity and carcinogenicity, even though they are not specifically designed to investigate these endpoints.

Repeated dose toxicity studies differ with respect to duration. In principle, any duration is possible, but for the sake of harmonization it has become necessary to limit the study durations to a number of standard durations in the test guideline studies.

The repeated dose toxicity studies also differ with respect to the doses, group sizes, and type and number of parameters studied. Test guideline studies of longer duration generally use lower doses, have larger group sizes, and include more parameters. It is generally assumed that studies with a longer duration are more likely to reveal more effects, and that effects will be revealed at lower doses, i.e., that the longer-term studies are more sensitive. In a study of short duration, there may not be sufficient time for histopathological changes to develop characteristically, and therefore such findings may be rather unspecific.

In the EU, the amount of information on repeated dose toxicity for a chemical substance under evaluation depends on the minimum data requirements laid down in the relevant legislation for new and existing chemical substances, biocides, and pesticides, respectively. No data on repeated dose toxicity are required for chemicals produced at tonnage levels less than 10 tons per annum (tpa). At higher production volumes, the standard data requirements are, in general, proportional to the tonnage level. At 10 tpa or more, a 28-day study is usually required and at 100 tpa or more, 28-/90-day studies are usually required. At 1000 tpa or more, a long-term repeated toxicity test (≥ 12 months) is usually required in case severe toxic effects were observed in the 28-/90-day studies for which the available evidence is inadequate for a toxicological evaluation and risk characterization, or the substance may have a dangerous property that cannot be detected in a 90-day study.

4.7.1 DEFINITIONS

The term "repeated dose toxicity" comprises the adverse general (i.e., excluding reproductive, genotoxic, or carcinogenic effects) toxicological effects occurring as a result of repeated daily dosing with, or exposure to, a substance for a part of the expected life span (subacute or subchronic exposure) or for the major part of the life span, in case of chronic exposure (EC 2003).

The term "general toxicological effects" (in this book often referred to as "general toxicity") includes effects on, e.g., body weight and/or body weight gain, absolute and/or relative organ and tissue weights, alterations in clinical chemistry, urinalysis and/or hematological parameters, functional disturbances in organs and tissues in general, and non-neoplastic pathological alterations in organs and tissues as examined macroscopically and microscopically.

The terms "subacute," "subchronic," and "chronic" are often used in the context of repeated dose toxicity. These terms are not used consistently; however, there is consistency in the use of the terms subchronic and chronic by OECD, US-EPA, and EU, see below.

The term subacute is not an OECD term. Presumably it relates to a study, which is shorter than a subchronic study, i.e., in OECD terms, it would be a study shorter than 90 days.

4.7.1.1   OECD

In the test guidelines for 90-day dermal (OECD TG 411) and inhalation (OECD TG 413) toxicity studies, the following definition in relation to the term subchronic is provided: "Subchronic dermal/inhalation toxicity is the adverse effects, which follow repeated daily dermal application/inhalation of a chemical for part (not exceeding 10%) of a life span."

In the test guideline for chronic toxicity (OECD TG 452), it is stated that the duration of the exposure period should be at least 12 months implicating that chronic toxicity is associated with the adverse effects, which follow repeated daily exposure to a chemical for the major part of a life span.

4.7.1.2   US-EPA

In the test guidelines for 90-day toxicity studies (OPPTS 870.3100, 870.3150, 870.3250, 870.3465), the following definition in relation to the term 'subchronic' is provided: "Subchronic . . . toxicity is the adverse effects occurring as a result of the repeated daily exposure of experimental animals to a chemical by the . . . route for a part (approximately 10 percent) of the test animal's life span."

In the test guidelines for chronic toxicity (OPPTS 870.4100, 870.4300), the following definition is provided: "Chronic toxicity is the adverse effects occurring as a result of the repeated daily exposure of experimental animals to a chemical by the oral, dermal, or inhalation routes of exposure." The guidelines request an exposure period of at least 12 months. The preferred species is the rat.

4.7.1.3   EU

In the general introduction to the EU test guidelines for toxicity (Part B), the following definition is provided: "Repeated dose/sub-chronic toxicity comprises the adverse effects occurring in experimental animals as a result of repeated daily dosing with, or exposure to, a chemical for a short part of their expected life-span."

The 28-day studies are termed "repeated dose (28 days)" while the 90-day studies are termed "subchronic." The chronic toxicity test (B.30), requests at least 12 months' duration. The preferred species is the rat.

4.7.2   OBJECTIVES FOR ASSESSING THE REPEATED DOSE TOXICITY OF A SUBSTANCE

The general objectives for assessing the potential of a substance to induce repeated dose toxicity are to evaluate:

- Whether exposure of humans to a substance has been associated with adverse toxicological effects occurring as a result of repeated daily exposure for a part of the expected lifetime or for the major part of the lifetime.

- Whether administration of a substance to experimental animals causes adverse toxico-logical effects as a result of repeated daily exposure for a part of the expected life span or for the major part of the life span; effects that are predictive of possible adverse human health effects.
- Whether individuals and/or subgroups in the population are more susceptible to general toxicity.
- Target organs, potential cumulative effects, and the reversibility of the adverse toxico-logical effects.
- Dose–response relationship and threshold for any of the adverse toxicological effects observed in the repeated dose toxicity studies.

4.7.3 Test Guidelines

4.7.3.1 *In Vivo* Tests

The OECD, US-EPA, and EU have developed specific test guidelines for *in vivo* testing of repeated dose toxicity, see Table 4.11. The EU Annex V methods are generally equivalent or very similar to the corresponding OECD test guidelines. Most of the US-EPA OPPTS test guidelines are, in principle, comparable to the corresponding OECD test guidelines.

The principle of the repeated dose toxicity studies is that the test substance is administered daily in graduated doses or concentrations to several groups of experimental animals, one dose level per group for a period of 28 days, 90 days, or at least 12 months. During the period of administration, the animals are observed each day for signs of toxicity. Animals, which die during the test are necropsied and at the conclusion of the test surviving animals are sacrificed and necropsied. Organs and tissues investigated in repeated dose toxicity studies include vital organs such as heart, brain, liver, kidneys, pancreas, spleen, immune system, lungs, etc. Effects examined may include changes in morphology, physiology, growth or life span, and/or behavior, which result in impairment of functional capacity or impairment of capacity to compensate for additional stress, or increase in the susceptibility to the harmful effects of other environmental influences.

Table 4.12 summarizes the various OECD test guideline studies for repeated dose toxicity in more detail, including the parameters examined in each test, in order to provide a brief overview of the similarities and differences between the various repeated dose toxicity studies.

For the 28-/90-day studies, separate guidelines are available for studies using oral administration, dermal application, or inhalation. The principle of these study protocols is identical although the updated protocols for oral administration includes additional parameters compared to those for dermal and inhalation administration, see Table 4.12. The 28-day inhalation test guideline is currently undergoing revision, see Table 2.4.

The combined repeated dose toxicity study with the reproduction/developmental toxicity screening test (OECD TG 422) is, for the repeated dose toxicity part, concordant with the standard 28-day oral toxicity study (OECD TG 407) except for use of pregnant females and longer exposure duration (about 6 weeks for males and approximately 54 days for females) in the combined study compared to the standard 28-day study.

The test guideline studies for repeated dose toxicity are comprehensive and basically include all target organs. In addition, for certain targets, more specific test guidelines have been developed. These targets include the immune system and the nervous system.

The standard repeated dose toxicity guideline studies include a number of parameters relevant for the evaluation of a substance's immunotoxic potential. While some information on potential immunotoxic effects may be obtained from the evaluation of hematology, lymphoid organ weights, and histopathology in these studies, there are data which demonstrate that these endpoints alone are not sufficient to predict immunotoxicity. In addition to these standard studies, the US-EPA has developed a specific test guideline for immunotoxicity testing in rodents (OPPTS 870.7800). This

TABLE 4.11
Test Guidelines Adopted for Repeated Dose Toxicity Testing

Title

OECD Test Guidelines	Year
407 Repeated dose 28-day oral toxicity study in rodents	1995
410 Repeated dose dermal toxicity: 21/28-Day study	1981
412 Repeated dose inhalation toxicity: 28-Day or 14-day study	1981
408 Repeated dose 90-day oral toxicity study in rodents	1998
409 Repeated dose 90-day oral toxicity study in non-rodents	1998
411 Subchronic dermal toxicity: 90-Day study	1981
413 Subchronic inhalation toxicity: 90-Day study	1981
452 Chronic toxicity studies	1981
453 Combined chronic toxicity/carcinogenicity studies	1981
422 Combined repeated dose toxicity study with the reproduction/developmental toxicity screening test	1996
424 Neurotoxicity study in rodents	1997
419 Delayed neurotoxicity of organophosphorus substances: 28-Day repeated dose study	1995

US-EPA OPPTS Harmonized Test Guidelines	
870.3050 Repeated dose 28-day oral toxicity study in rodents	2000
870.3200 21/28-Day dermal toxicity	1998
870.3100 90-Day oral toxicity in rodents	1998
870.3150 90-Day oral toxicity in non-rodent	1998
870.3250 90-Day dermal toxicity	1998
870.3465 90-Day inhalation toxicity	1998
870.4100 Chronic toxicity	1998
870.4300 Combined chronic toxicity/carcinogenicity	1998
870.8355 Combined chronic toxicity/carcinogenicity testing of respirable fibrous particles	2001
870.3650 Combined repeated dose toxicity study with the reproduction/developmental toxicity screening test	2000
870.6200 Neurotoxicity screening battery	1998
870.6500 Schedule-controlled operant behavior	1998
870.6850 Peripheral nerve function	1998
870.6855 Neurophysiology sensory evoked potentials	1998
870.6100 Acute and 28-day delayed neurotoxicity of organophosphorus substances	1998
870.7800 Immunotoxicity	1998

EU Annex V Test Methods	
B.7 Repeated dose (28 days) toxicity (Oral)	1996
B.8 Repeated dose (28 days) toxicity (Inhalation)	1992
B.9 Repeated dose (28 days) toxicity (Dermal)	1992
B.26 Sub-chronic oral toxicity test. Repeated dose 90-day toxicity study in rodents	2001
B.27 Sub-chronic oral toxicity test: Repeated dose 90-day toxicity study in non-rodents	2001
B.28 Sub-chronic dermal toxicity test: 90-Day repeated dermal dose study using rodent species	1988
B.29 Sub-chronic inhalation toxicity test: 90-Day repeated inhalation dose study using rodent species	1988
B.30 Chronic toxicity test	1988
B.33 Combined chronic toxicity/carcinogenicity test	1988
B.43 Neurotoxicity study in rodents	2004
B.38 Delayed neurotoxicity of organophosphorus substances 28-day repeated dose study	1996

guideline only addresses potential immune suppression. In order to obtain data on the functional responsiveness of major components of the immune system to a T-cell-dependent antigen (sheep red blood cells (SRBC)), rats and/or mice are exposed to the test and control substances for at least 28 days. The animals are immunized by injection of SRBCs approximately 4 days (depending on

TABLE 4.12

Overview of *In Vivo* Repeated Dose Toxicity OECD Test Guideline Studies

Test	Design	Endpoints
OECD TG 407 Repeated dose 28-day oral toxicity study in rodents	Exposure for 28 days At least three dose levels plus control At least five males and females per group Preferred rodent species: Rat	Clinical observations Functional observations (4th exposure week - sensory reactivity to stimuli of different types, grip strength, motor activity) Body weight and food/water consumption Hematology (haematocrit, hemoglobin, erythrocyte count, total and differential leucocyte count, platelet count, blood clotting time/potential) Clinical biochemistry Urinalysis Gross necropsy (full, detailed, all animals) Organ weights (all animals - liver, kidneys, adrenals, testes, epididymides, thymus, spleen, brain, heart) Histopathology (full, at least control and high-dose groups - all gross lesions, brain, spinal cord, stomach, small and large intestines, liver, kidneys, adrenals, spleen, heart, thymus, thyroid, trachea and lungs, gonads, accessory sex organs, urinary bladder, lymph nodes, peripheral nerve, a section of bone marrow)
OECD TG 410 Repeated dose dermal toxicity: 21/28-Day study	Exposure for 21/28 days At least three dose levels plus control At least five males and females per group Rat, rabbit, or guinea pig	Clinical observations Body weight and food/water consumption Hematology (haematocrit, hemoglobin, erythrocyte count, total and differential leucocyte count, clotting potential) Clinical biochemistry Urinalysis Gross necropsy (full, detailed, all animals) Organ weights (all animals - liver, kidneys, adrenals, testes) Histopathology (full, at least control and high-dose groups - all gross lesions, normal and treated skin, liver, kidney)
OECD TG 412 Repeated dose inhalation toxicity: 28-Day or 14-day study	Exposure for 28 or 14 days At least three concentrations plus control At least five males and females per group Rodents: Preferred species: rat	Clinical observations Body weight and food/water consumption Hematology (haematocrit, hemoglobin, erythrocyte count, total and differential leucocyte count, clotting potential) Clinical biochemistry Urinalysis Gross necropsy (full, detailed, all animals) Organ weights (all animals - liver, kidneys, adrenals, testes) Histopathology (full, at least control and high-dose groups - all gross lesions, lungs, liver, kidney, spleen, adrenals, heart)
OECD TG 408 Repeated dose 90-day oral toxicity study in rodents	Exposure for 90 days At least three dose levels plus control At least 10 males and females per group Preferred rodent species: Rat	Clinical observations Ophthalmological examination Functional observations (toward end of exposure period - sensory reactivity to stimuli of different types, grip strength, motor activity)

TABLE 4.12 (continued)

Overview of *In Vivo* Repeated Dose Toxicity OECD Test Guideline Studies

Test	Design	Endpoints
		Body weight and food/water consumption
		Hematology (haematocrit, hemoglobin, erythrocyte count, total and differential leucocyte count, platelet count, blood clotting time/potential)
		Clinical biochemistry
		Urinalysis
		Gross necropsy (full, detailed, all animals)
		Organ weights (all animals - liver, kidneys, adrenals, testes, epididymides, uterus, ovaries, thymus, spleen, brain, heart)
		Histopathology (full, at least control and high-dose groups - all gross lesions, brain, spinal cord, pituitary, thyroid, parathyroid, thymus, oesophagus, salivary glands, stomach, small and large intestines, liver, pancreas, kidneys, adrenals, spleen, heart, trachea and lungs, aorta, gonads, uterus, accessory sex organs, female mammary gland, prostate, urinary bladder, gall bladder (mouse), lymph nodes, peripheral nerve, a section of bone marrow, and skin/eyes on indication)
OECD TG 409 Repeated dose 90-day oral toxicity study in non-rodents	Exposure for 90 days At least three dose levels plus control At least four males and females per group Preferred species: Dog	Clinical observations Ophthalmological examination Body weight and food/water consumption Hematology (as in TG 408) Clinical biochemistry Urinalysis Gross necropsy (full, detailed, all animals) Organ weights (as in TG 408 - additional: gall bladder, thyroid, parathyroid) Histopathology (as in TG 408 - additional: gall bladder, eyes)
OECD TG 411 Subchronic dermal toxicity: 90-Day study	Exposure for 90 days At least three dose levels plus control At least 10 males and females per group Rat, rabbit, or guinea pig	Clinical observations Ophthalmological examination Body weight and food/water consumption Hematology (haematocrit, hemoglobin, erythrocyte count, total and differential leucocyte count, clotting potential) Clinical biochemistry Urinalysis Gross necropsy (full, detailed, all animals) Organ weights (all animals - liver, kidneys, adrenals, testes) Histopathology (full, at least control and high-dose groups - all gross lesions, normal and treated skin, and essentially the same organs and tissues as in TG 408)
OECD TG 413 Subchronic inhalation toxicity: 90-Day study	Exposure for 90 days At least three concentrations plus control At least 10 males and females per group Rodents: Preferred species—Rat	Clinical observations Ophthalmological examination Body weight and food/water consumption Hematology (haematocrit, hemoglobin, erythrocyte count, total and differential leucocyte count, clotting potential)

(continued)

TABLE 4.12 (continued)

Overview of *In Vivo* Repeated Dose Toxicity OECD Test Guideline Studies

Test	Design	Endpoints
		Clinical biochemistry
		Urinalysis
		Gross necropsy (full, detailed, all animals)
		Organ weights (all animals - liver, kidneys, adrenals, testes)
		Histopathology (full, at least control and high-dose groups - all gross lesions, respiratory tract, and essentially the same organs and tissues as in TG 408)
OECD TG 452 Chronic toxicity studies	Exposure for at least 12 months At least three dose levels plus control Rodents: At least 20 males and females per group Non-rodents: At least four males and females per group Preferred rodent species: Rat Preferred non-rodent species: Dog	Clinical observations, including neurological changes Ophthalmological examination Body weight and food/water consumption Hematology (haematocrit, hemoglobin, erythrocyte count, total leucocyte count, platelet count, clotting potential) Clinical biochemistry Urinalysis Gross necropsy (full, detailed, all animals) Organ weights (all animals - brain, liver, kidneys, adrenals, gonads, thyroid/parathyroid (non-rodents only)) Histopathology (full, at least control and high-dose groups - all grossly visible tumours and other lesions, as well as essentially the same organs and tissues as in the 90-day studies (TG 408/409))
OECD TG 453 Combined chronic toxicity/ carcinogenicity studies	Exposure for at least 12 months (satellite groups) or majority of normal life span (carcinogenicity part) At least three dose levels plus control At least 50 males and females per group Satellite group: At least 20 males and females per group Preferred species: Rat	Essentially as in TG 452
OECD TG 422 Combined repeated dose toxicity study with the reproduction/ developmental toxicity screening test	Exposure for a minimum of 4 weeks (males) or from 2 weeks prior to mating until at least postnatal day 4 (females - at least 6 weeks of exposure) At least three dose levels plus control At least 10 males and females per group	Clinical observations as in TG 407 Functional observations as in TG 407 Body weight and food/water consumption Hematology as in TG 407 Clinical biochemistry Urinalysis Gross necropsy (full, detailed, all adult animals) Organ weights (testes and epididymides - all males; liver, kidneys, adrenals, thymus, spleen, brain, heart - in five animals of each sex per group, i.e., as in TG 407) Histopathology (ovaries, testes, epididymides, accessory sex organs, all gross lesions - all animals in at least control and high-dose groups; brain, spinal cord, stomach, small and large intestines, liver, kidneys, adrenals, spleen, heart, thymus, thyroid, trachea and lungs, urinary bladder, lymph nodes, peripheral nerve, a section of bone marrow - in five animals of each sex in at least control and high-dose groups, i.e., as in TG 407)

TABLE 4.12 (continued)
Overview of *In Vivo* Repeated Dose Toxicity OECD Test Guideline Studies

Test	Design	Endpoints
US-EPA OPPTS 870-7800 Immunotoxicity	Exposure for at least 28 days At least three dose levels plus control At least eight animals per group Preferred species: Rats and/or mice Generally oral route of administration	Detailed clinical observations Body weight and food/water consumption Gross necropsy (moribund animals, animals that die during the study) Organ weights (spleen, thymus, all animals) Immunotoxicity tests: (1) functional tests (either a splenic plaque-forming cell (PFC) assay or an Enzyme-Linked Immunosorbent Assay (ELISA) to determine the response to antigen administration); (2) enumeration of splenic or peripheral blood total B cells, total T cells, and T-cell subpopulations
OECD TG 424 Neurotoxicity study in rodents	Exposure for at least 28 days Dose levels: Not specified At least 10 males and females per group Preferred rodent species: Rat Generally oral route of administration	Detailed clinical observations Functional observations (sensory reactivity to stimuli of different types, grip strength, motor activity, more specialized tests on indication) Ophthalmological examination Body weight and food/water consumption Hematology (haematocrit, hemoglobin, erythrocyte count, total and differential leucocyte count, platelet count, blood clotting time/potential) Clinical biochemistry Histopathology: at least five animals/sex/ group) for neuropathological examinations (brain, spinal cord, and peripheral nerves); remaining animals to be used either for specific neurobehavioral, neuropathological, neurochemical, or electrophysiological procedures that may supplement the histopathology or alternatively, for routine pathological evaluations according to the guidelines for standard repeated dose toxicity studies
OECD TG 419 Deleayed neurotoxicity of organophosphorus substances: 28-Day repeated dose study	Exposure for 28 days At least 3 dose levels plus control At least 12 birds per group Species: Domestic laying hen	Detailed clinical observations Body weight and food/water consumption Clinical biochemistry (NTE activity, acetylcholinesterase activity) Gross necropsy (all animals) Histopathology (neural tissue)

the strain of animal) prior to the end of the exposure. At the end of the exposure period, either the plaque forming cell (PFC) assay or an enzyme linked immunosorbent assay (ELISA) are performed to determine the effects of the test substance on the splenic anti-SRBC (IgM) response or serum anti-SRBC IgM levels, respectively. Further assays are performed depending on whether the test substance produces significant suppression of the anti-SRBC response or not. It is stated in the test guideline that the tests included in the guideline do not represent a comprehensive assessment of the immune function.

The standard repeated dose toxicity guideline studies include a number of parameters relevant for the evaluation of a substance's neurotoxic potential. In addition to these standard studies, the OECD, US-EPA, and EU have developed specific test guidelines for neurotoxicity testing in rodents as well as for delayed neurotoxicity of organophosphorus substances in the domestic laying hen.

In the neurotoxicity study (OECD TG 424/EU Annex B.43), the animals are tested to allow the detection or the characterization of behavioral and/or neurological abnormalities. A range of behaviors that could be affected by neurotoxicants is assessed during each observation period. At the end of the test, a subset of animals of each sex from each group are perfused in situ and sections of the brain, spinal cord, and peripheral nerves are prepared and examined.

The US-EPA has adopted four different test guidelines for neurotoxicity:

The Neurotoxicity Screening Battery test guideline (OPPTS 870.6200) consists of a functional observational battery, motor activity, and neuropathology. The test battery is not intended to provide a complete evaluation of neurotoxicity, and additional functional and morphological evaluation may be necessary to assess completely the neurotoxic potential of a chemical.

The Schedule-Controlled Operant Behavior test guideline (OPPTS 870.6500) defines procedures for conducting studies of schedule-controlled operant behavior, one way of evaluating the rate and pattern of a class of learned behavior. The purpose of the guideline is to evaluate the effects of acute and repeated exposures on the rate and pattern of responding under schedules of reinforcement. Additional tests may be necessary to completely assess the effects of any substance on learning, memory, or behavioral performance.

The Peripheral Nerve Function test guideline (OPPTS 870.6850) defines procedures for evaluating certain aspects of the neurophysiological functioning of peripheral nerves. The purpose of the guideline is to evaluate the effects of exposures on the velocity and amplitude of conduction of peripheral nerves. Additional tests may be necessary to completely assess the neurophysiological effects of any substance.

The Neurophysiology Sensory Evoked Potentials test guideline (OPPTS 870.6855) is designed to detect and characterize changes in the sensory aspects of nervous system function that result from exposure to chemical substances. The techniques involve neurophysiological measurements from adult animals and are sensitive to changes in the function of auditory, somatosensory (body sensation), and visual sensory systems.

In the delayed neurotoxicity study (OECD TG 419/OPPTS 870.6100/EU Annex B.38), the animals are observed at least daily for behavioral abnormalities, ataxia, and paralysis, until 14 days after the last dose. Biochemical measurements, in particular neuropathy target esterase (NTE), are undertaken on hens randomly selected from each group (normally 24 and 48 h) after the last dose. Two weeks after the last dose, the remainder of the hens are killed and histopathological examination of selected neural tissues is undertaken.

4.7.3.2 *In Vitro* Tests

The OECD, US-EPA, and EU have no adopted test guidelines for *in vitro* testing of repeated dose toxicity.

4.7.4 GUIDANCE DOCUMENTS

4.7.4.1 WHO

WHO/IPCS have published Environmental Health Criteria Monographs on principles and methods for the assessment of toxicity associated with exposure to chemicals in relation to the following specific target organs and tissues: immunotoxicity (WHO/IPCS 1996) (Section 4.7.6.3), nephrotoxicity (WHO/IPCS 1991), and neurotoxicity (WHO/IPCS 1986b, 2001) (Section 4.7.7.3).

The WHO/IPCS criteria monograph on nephrotoxicity (WHO/IPCS 1991) addresses nephrotoxicity in general, kidney structure and function, the mechanistic basis of chemically induced renal injury, chemicals that have the potential to cause nephrotoxicity, renal cancer, assessment of nephrotoxicity, and detection of nephrotoxicity in humans.

The WHO/FAO Joint Meeting of Experts on Pesticide Residues (JMPR) has given recommendations on interpretation of cholinesterase inhibition (FAO 1998, 1999), see Section 4.7.7.3.

4.7.4.2 OECD

OECD has published "Guidance Notes for Analysis and Evaluation of Repeat-dose Toxicity Studies" (OECD 2000). The initial aim of the project was to develop harmonized guidance on the conduction of independent evaluations and the writing of reviews of subchronic oral toxicity tests. Although the guidance notes have been prepared for subchronic oral repeated dose toxicity, it is stated that they are also applicable to other routes of administration as well as to repeated dose toxicity studies other than just those classified as subchronic (90-day), e.g., to chronic toxicity, carcinogenicity, reproductive toxicity, neurotoxicity, and immunotoxicity studies. Furthermore, while the guidance notes have been prepared for the purpose of assisting in the interpretation and transparent reporting of toxicological data on pesticides, they could also be used as guidance on evaluating studies in other programs. If used as such, the guidance notes would enable repeated dose toxicity studies for different groups of chemicals (e.g., pesticides, biocides, and industrial chemicals) to be assessed in the same way.

The objective of the guidance notes is to outline core concepts in order to avoid the need to make reference to large numbers of textbooks, but to refer the reader to other useful sources when more detailed and specific information is required. They are intended to complement OECD Test Guidelines and other publications by the OECD, including the Guidance for Industry Data Submissions (Dossier Guidance) and Guidance for Country Data Review Reports (Monograph Guidance) on Plant Protection Products and their Active Substances. However, whereas the latter publication provides guidance on the format and presentation of entire review reports (monographs), the Guidance Notes place emphasis on data interpretation, scientific judgment, and report writing in the context of regulatory toxicology evaluations.

More recently, OECD has also published "Guidance Notes for Analysis and Evaluation of Chronic Toxicity and Carcinogenicity Studies" (OECD 2002a). The aim of this OECD project was to develop harmonized guidance on conducting independent evaluations of, and writing reviews of, chronic oral toxicity and carcinogenicity tests. These guidance notes contain much material identical to or derived from the above-mentioned "Guidance Notes for Analysis and Evaluation of Repeat-dose Toxicity Studies," because many concepts apply to both types of studies. The objective of these Guidance Notes is identical to that for the "Guidance Notes for Analysis and Evaluation of Repeat-dose Toxicity Studies" (see above).

A "Guidance Document for Neurotoxicity Testing" is also available (OECD 2004c), see Section 4.7.7.3.

OECD has also published a document on a "Harmonised integrated classification system for human health and environmental hazards of chemical substances and mixtures" (OECD 2001b). Chapter 2.9 addresses a harmonized system for the classification of chemicals, which cause specific target organ oriented systemic toxicity following repeated exposure.

4.7.4.3 US-EPA

The series of Risk Assessment Guidelines includes a guideline for neurotoxicity risk assessment (US-EPA 1998), see Section 4.7.7.3.

4.7.4.4 EU

The TGD (EC 2003), Chapter 3.9, addresses repeated dose toxicity and provides guidance on data requirements, evaluation of data, and dose–response assessment. Included is a section on specific system/organ toxicity dealing with guidance on investigation of neurotoxicity (Section 4.7.7.3) and immunotoxicity (Section 4.7.6.3), effects on the endocrine system, and lung overload phenomena and pulmonary fibrosis.

4.7.5   Use of Information on Repeated Dose Toxicity in Hazard Assessment

The adverse general (i.e., excluding reproductive, genotoxic, or carcinogenic effects) toxicological effects occurring as a result of repeated daily dosing with, or exposure to, a substance are considered as being "threshold effects," i.e., effects for which there are expected to be a threshold of substance concentration below which the effects will not be manifested. For the hazard and risk assessment, it is important to identify those dose levels at which adverse effects are observed, and the dose level at which adverse effects are not observed, i.e., to derive a NOAEL for repeated dose toxicity. Crucial in the derivation of the NOAEL and/or LOAEL, is the definition of "adverse effects" (Section 4.2.2). In the derivation of the NOAEL and/or LOAEL, a number of factors need to be considered; these issues are addressed in detail in Sections 4.2.3 and 4.2.4. An alternative approach to the derivation of the NOAEL, the BMD concept, and the limitations of its applicability are addressed in Section 4.2.5.

If local effects are clearly identified after repeated dosing, a NOAEL and/or LOAEL should be derived for these effects in addition to NOAEL and/or LOAEL derived for systemic effects. Supportive evidence for the occurrence or absence of local effects after repeated dermal and inhalation exposure may be available from the total toxicity profile of the substance. It should be noted that lack of evidence for local effects in any type of study (i.e., skin or eye irritation, sensitization, repeated dose toxicity study by routes other than the route of interest) does not exclude the possible occurrence of local effects upon repeated respiratory or dermal exposure (EC 2003).

Substances, which are skin or eye irritating or corrosive after single exposure (Section 4.5) should be suspected of inducing local effects upon repeated respiratory exposure to low concentrations. In contrast, local effects reported from skin sensitization studies as well as dermal repeated dose toxicity studies are not predictive of local effects on the respiratory tract. In addition, observations from irritation and/or sensitization studies as well as repeated dose inhalation toxicity studies are not predictive of local effects on the skin upon repeated dermal exposure (EC 2003).

4.7.5.1   Human Data

When reliable and relevant human data are available, they can be highly useful for hazard assessment and even preferable over animal data. However, the relative lack of sensitivity of human data may cause particular difficulty. Therefore, negative human data cannot be used to override positive findings in animals, unless it has been demonstrated that the mode of action of a certain toxic response observed in animals is not relevant for humans. In such a case, a full justification is required.

In addition, human data adequate to serve as the sole basis for the dose–response assessment are rare. In many human studies, the circumstances of exposure and the exposure levels themselves are not well known, mixed exposure may have occurred, the incidence of effects is low, the number of exposed individuals is small, and the latency period between exposure and disease may be long.

It is emphasized that testing with human volunteers is strongly discouraged, but when there are good quality data already available they can be used in the overall weight of evidence.

4.7.5.2   Animal Data

4.7.5.2.1   General Aspects
The number of repeated dose toxicity studies available for a substance under evaluation is likely to be variable, ranging from none to the 28-day repeated dose toxicity guideline study, to a series of guideline studies for some substances, including subchronic and/or chronic studies. There may also

be studies employing different species and routes of exposure. In addition, special toxicity studies investigating further the nature, mechanism, and/or dose relationship of a critical effect in a target organ or tissue may also have been performed for some substances. While data available from repeated dose toxicity studies not performed according to conventional guidelines and/or GLP may still provide information of relevance for hazard identification, data from such studies need to be carefully evaluated.

In circumstances where repeated dose toxicity studies have not been carried out according to conventional guidelines and/or GLP, data from such studies could be considered to be equivalent to data generated by corresponding test guideline methods if the following conditions are met:

- Adequate for the purpose of hazard assessment
- Adequate and reliable coverage of the key parameters foreseen to be investigated in the corresponding test methods
- Exposure duration comparable to or longer than the corresponding test methods if exposure duration is a relevant parameter
- Adequate and reliable documentation of the study is provided

In all other situations, nonguideline studies cannot stand alone for a hazard assessment of a substance and thus cannot serve as the sole basis for an assessment of repeated dose toxicity, i.e., cannot be used to identify a substance as being of no concern in relation to repeated dose toxicity.

The following general guidance is provided for the evaluation of repeated dose toxicity data in relation to hazard assessment (EC 2003):

- Studies conducted in a species in which the toxicokinetics and toxicodynamics of the substance are most similar to those in humans have greatest weight. In the absence of a species that is clearly the most relevant one, studies on the most sensitive animal species should be selected as the significant ones.
- Studies using an appropriate route, duration, and frequency of exposure in relation to the expected route(s), and frequency and duration of human exposure have greater weight.
- Studies enabling the identification of a NOAEL have greater weight.
- Reliable and sufficiently detailed studies of longer duration should generally have greater weight, e.g., for hazard assessment, a 90-day repeated dose toxicity study should be given greater weight than a 28-day repeated dose toxicity study in the determination of the most relevant NOAEL.
- If sufficient evidence is available to identify the critical effect(s) (with regard to the dose–response relationship(s) and to the relevance for humans) and the target organ(s) and/or tissue(s), greater weight should be given to specific studies investigating this effect in the identification of the NOAEL. The critical effect can be a local as well as a systemic effect.

4.7.5.2.2 Information To Be Obtained from Repeated Dose Toxicity Studies

The information that can be obtained from the various types of repeated dose toxicity studies is briefly summarized below.

The repeated dose 28-day toxicity studies as well as the combined repeated dose toxicity study with the reproduction/developmental toxicity screening test provide information on the possible health hazards likely to arise from a relatively limited period of the animal's life span. The updated oral protocol places more emphasis on neurological effects as a specific endpoint and should identify chemicals with a neurotoxic potential, which may warrant further in-depth investigation of this aspect (Section 4.7.7). In addition, the study may give an indication of immunological effects (Section 4.7.6) and of reproductive organ toxicity (Section 4.10.3). It should be noted that potential effects in certain target organs (e.g., the thyroid) following repeated exposure may not be observed within the span of the 28-day study.

Technical and methodological differences exist between the standard 28-day repeated dose toxicity study and the combined study; therefore, the use of the combined study for assessment of repeated dose toxicity should be evaluated on a case-by-case basis in order to assess the adequacy of the information provided. In the combined study, the dosing period is longer than in a standard 28-day study and thus more information on repeated dose toxicity could be expected from the combined study. However, in the combined study, repeated dose toxicity is assessed in the pregnant population, and it is generally assumed that there are differences in sensitivity between pregnant and nonpregnant animals. Thus, interpretation of the results from the combined study may be complicated due to differences in sensitivity between pregnant and nonpregnant animals, and an assessment of the general toxicity may be more difficult especially when serum and histopathological parameters are not evaluated at the same time in the study. Consequently, where the combined study is used for the assessment of repeated dose toxicity, the use of data obtained from such a study should be clearly indicated.

The repeated dose 90-day toxicity studies provide information on the possible health hazards likely to arise from repeated exposure over a prolonged period of the animal's life span covering postweaning maturation and growth well into adulthood. The 90-day study will provide information on the major toxic effects, indicate target organs and the possibility of accumulation, and can provide an estimate of a NOAEL. The updated oral protocol (OECD TG 408 from 1998) places additional emphasis on neurological, immunological, and reproductive effects and should allow for the identification of substances with the potential to cause neurotoxic, immunological, or reproductive organ effects, which may warrant further in-depth investigation (see Sections 4.7.6, 4.7.7, and 4.10.3).

The chronic toxicity studies provide information on the possible health hazards likely to arise from repeated exposure over a prolonged period of time covering the major part of the animal's life span. The chronic study provides information on the majority of chronic toxic effects, indicates target organs and the possibility of accumulation, determines dose–response relationships, and can form the basis for an estimate of a NOAEL. Ideally, the chronic studies should allow for the detection of general toxicity including neurological, physiological, biochemical, and hematological effects, and exposure-related morphological (pathology) effects. However, nonspecific life shortening effects, which require a long latent period or are cumulative, may possibly not become manifest in this study type.

The immunotoxicity study will provide information on suppression of the immune system, which might occur as a result of repeated exposure to a test chemical. The protocol for the US-EPA OPPTS test guideline is intended to be used along with data from routine toxicity testing, to provide more accurate information on risk to the immune system. See also Section 4.7.6.

The neurotoxicity studies will provide information on major neurobehavioral and neuropathological effects in adult rodents. The protocol for the OECD/EU test guideline studies has been developed so that it can be tailored to meet particular needs to confirm the specific histopathological and behavioral neurotoxicity of a substance as well as to provide a characterization and quantification of the neurotoxic responses, and can thus form the basis for an estimate of a NOAEL for neurotoxicity. See also Section 4.7.7.

The delayed neurotoxicity studies of organophosphorus substances provide information on the delayed neurotoxicity arising from repeated exposure over a relatively limited period of the animal's life span. The study is used in the assessment and evaluation of the neurotoxic effects of organophosphorus substances and will provide information on dose–response relationship and can provide an estimate of a NOAEL for delayed neurotoxicity. See also Section 4.7.7.

4.7.5.2.3 Information To Be Obtained from Other Studies Involving Repeated Exposure

Although not aiming directly at investigating repeated dose toxicity per se, other available *in vivo* test guideline studies involving repeated exposure of experimental animals may provide useful information on repeated dose toxicity.

The one- and two-generation studies (Section 4.10.3) may provide information on the general toxicological effects arising from repeated exposure over a prolonged period of the animal's life span (about 90 days for parental animals) as clinical signs of toxicity, body weight, selected organ weights, and gross and microscopic changes of selected organs are recorded.

However, due to technical and methodological differences between a standard repeated dose toxicity study and the generation studies, the use of the generation studies should be evaluated on a case-by-case basis to assess the adequacy of the information provided. One point is that the major focus in the generation studies is to investigate the reproductive toxicity and therefore these studies do not include an investigation of all the general toxic effects in order to characterize the repeated dose toxicity potential of a substance as investigated in standard repeated dose toxicity studies. Particularly, it is worth noting that in the generation studies, the possible general toxicity is observed in the pregnant population, and it is generally assumed that there are differences in sensitivity between pregnant and nonpregnant animals. It may thus be more complicated to interpret the results with respect to general toxicity. Consequently, where a generation study is used for the assessment of repeated dose toxicity, the use of data obtained from such a study as well as the adequacy of the results for an assessment of repeated dose toxicity should be clearly indicated.

Other types of reproductive toxicity studies, e.g., the prenatal developmental toxicity study, the reproduction/developmental toxicity screening study, and the developmental neurotoxicity study (Section 4.10.3) may give some indications of general toxicological effects arising from repeated exposure over a relatively limited period of the animal's life span as clinical signs of toxicity and body weight are recorded.

However, the possible general toxicity is only observed in the pregnant population and is therefore subject to the same reservations as mentioned above for the generation studies. Such studies cannot stand alone for an assessment of the repeated dose toxicity potential of a substance and thus cannot serve as the sole basis for the assessment of the repeated dose toxicity.

The carcinogenicity study (Section 4.9.3) will, in addition to information on neoplastic lesions, also provide information on the general toxicological effects arising from repeated exposure over a major portion of the animal's life span as clinical signs of toxicity, body weight, and gross and microscopic changes of organs and tissues are recorded. More information on repeated dose toxicity could be expected from the carcinogenicity study compared with the standard 28-/90-day studies due to the longer exposure duration. However, as the major focus in the carcinogenicity study is to investigate the development of neoplastic lesions, this study does not include an investigation of as many of the general toxic effects in order to characterize the repeated dose toxicity potential of a substance as investigated in standard repeated dose toxicity studies (e.g., organ weights, hematology, and clinical chemistry). Therefore, such a study cannot stand alone for a hazard and risk assessment of the repeated dose toxicity potential of a substance and thus cannot serve as the sole basis for the assessment of repeated dose toxicity.

Although contributing limited information to the overall assessment of the repeated dose toxicity potential of a substance, studies such as acute toxicity and irritation studies as well as *in vivo* genotoxicity studies may provide some useful information on repeated dose toxicity.

Acute toxicity studies (Section 4.4.3) can give some indications of general toxicity based on the observations for clinical signs of toxicity, recording of body weight, and gross and microscopic examination of organs and tissues.

Results from standard *in vivo* studies on skin/eye irritation and corrosion (Section 4.5.3) can indicate whether local effects (damage) might be expected to occur in the mucous membrane of the gastrointestinal tract following repeated oral exposure, in the respiratory tract following repeated inhalation exposure, or to the skin following repeated dermal contact to a substance as a liquid and/or in the vapor phase. As a rule of thumb, substances which are skin or eye irritating or corrosive in a conventional guideline study should generally be suspected to have a potential of inducing local effects in the respiratory tract upon repeated inhalation exposure, or to the gastrointestinal tract following repeated oral exposure. However, it should be noted that lack of evidence for local effects in studies on skin/eye irritation and corrosion does not exclude the possible occurrence of local effects upon repeated respiratory, oral, or dermal exposure. Studies on skin/eye irritation and corrosion can also give some indications of systemic effects based on the observations for clinical signs of toxicity and recording of body weight.

In vivo genotoxicity studies (Section 4.8.3) can give some indications of general toxicity based on the observations for clinical signs of toxicity.

Toxicokinetic studies (Section 4.3.3), especially a repeated dose toxicokinetic study, can give some indications of general toxicity based on the observations for clinical signs of toxicity. They may also be helpful in the evaluation and interpretation of repeated dose toxicity data, e.g., in relation to accumulation of a substance or its metabolites in certain tissues or organs as well as in relation to mechanistic aspects of repeated dose toxicity and species differences.

4.7.5.3 *In Vitro* Data

Currently, there are no validated and regulatory accepted *in vitro* methods for assessing repeated dose toxicity. Numerous *in vitro* systems have been developed over the last decades and have been discussed and summarized in recent ECVAM reports on repeated dose toxicity testing (Worth and Balls 2002, Prieto et al. 2005, Prieto et al. 2006). Human *in vitro* data, particularly on kinetics and metabolism, and *in vitro* test data from well-characterized target organ and target system models on, e.g., mode of action(s)/mechanism(s) of toxicity may be useful in the interpretation of observed repeated dose toxicity.

4.7.5.4 Other Data

The chemical structure of a substance may contain structural alerts for certain endpoints, based on clear evidence for such endpoints of structural analogues.

Use of (Q)SAR may be valuable in the evaluation of the repeated dose toxicity of a substance for which no data are available, see Sections 3.5.2 and 3.5.3.

4.7.6 IMMUNOTOXICITY

The immune system is critical in maintaining health. Immunotoxic responses may occur when the immune system is the target of an insult, e.g., a chemical substance. This in turn can result in either immunosuppression and a subsequent decreased resistance to infectious diseases as well as some types of cancer, or immune dysregulation, which exacerbates allergy or autoimmunity. Alternatively, toxicity may arise when the immune system responds to an antigenic specificity of, e.g., a chemical substance as part of a specific immune response (i.e., allergy or autoimmunity). Allergy (sensitization) is dealt with as a separate toxicity endpoint (Section 4.6). In autoimmune diseases, the healthy tissue is attacked by the immune system as if it was a foreign compound. Both suppression and enhancement of immune function are therefore potentially harmful events, and chemical substances may exert immunotoxic effects of either type.

4.7.6.1 Definitions

The US-EPA OPPTS test guideline on immunotoxicity (OPPTS 870.7800) provides the following definition: "Immunotoxicity refers to the ability of a test substance to suppress immune responses that could enhance the risk of infectious or neoplastic disease, or to induce inappropriate stimulation of the immune system, thus contributing to allergic or autoimmune disease."

According to the TGD (EC 2003), immunotoxicity is "the ability of a substance to adversely affect the immune system and the immune response of affected individuals is altered."

4.7.6.2 Test Guidelines

The adopted test guideline methods for repeated dose toxicity testing include a number of parameters relevant for the evaluation of a substance's immunotoxic potential. Furthermore, a specific test guideline for immunotoxicity has also been adopted by the US-EPA. These test methods are addressed in Section 4.7.3.1 and Tables 4.11 and 4.12.

The OECD protocols of the oral 28-/90-day studies (see Table 4.12) in their most recent versions now include the measurement of thymus and spleen weights and histopathological examination of certain lymphoid tissues in addition to the total and differential white blood cell counts and spleen histopathology required in the previous test guideline methods. Primarily the test guidelines are intended as a screening for immunotoxicity and depending on the results, further testing may be needed.

Currently there are few methods for specific investigation of immunotoxic effects, which are regarded as sufficiently validated for routine use (EC 2003). The plaque forming assay or the equivalent using the ELISA method (Enzyme-linked Immunosorbent Assay) are recommended to identify altered T-cell-dependent humoral responses. Of particular value for hazard assessment are the so-called host resistance models, in which the clinical relevance of immunotoxicity can be evaluated. Other methods may also be of value to provide information on the mode of immunotoxic action, e.g., mitogen stimulation tests and leucocyte phenotyping. However, further work is needed on standardization and validation of these test methods.

Autoimmune diseases are another important area of substance-induced immunotoxicity. At present there are no specific assays to assess substances for their potential to induce autoimmune reactions (EC 2003).

4.7.6.3 Guidance Documents

WHO/IPCS have published an Environmental Health Criteria Monograph specifically dealing with immunotoxicity "Principles and Methods for Assessing Direct Immunotoxicity Associated with Exposure to Chemicals" (WHO/IPCS 1996). This monograph is concerned mainly with one aspect of immunotoxicology: the direct or indirect effect of xenobiotic compounds (or their biotransformation products) on the immune system. This effect is usually immunosuppression, or the induction of a state of deficiency or unresponsiveness. Allergy and autoimmunity are dealt with in another monograph (WHO/IPCS 1999b), see Section 4.6.4.1.

The TGD (EC 2003) contains a section on specific system/organ toxicity, which includes a section concerning guidance on investigation of immunotoxicity and EU-specific risk assessment.

Specific guidance documents on immunotoxicity have not been published by either OECD or US-EPA.

4.7.6.4 Use of Information from Repeated Dose Toxicity Studies in the Hazard Assessment of Immunotoxicity

The basis of the recommended approach to a hazard assessment of the potential immunotoxicity of a substance is that many immunotoxic substances can be identified via the standard tests for systemic toxicity. Special studies to characterize effects of concern for immunotoxicity are used only when necessary for adequate hazard assessment. The nature of special studies, and when they should be conducted, need to be decided on a case-by-case basis. A tiered approach to the identification of immunotoxic hazard in routine toxicology is described in WHO/IPCS (1996).

The thymus, spleen, and lymphoid tissues all have immunological function and changes to them can be indicative of adverse effects on the immune system. Indications of immunotoxicity from standard repeated dose toxicity studies include one or more of the following signs:

- Morphological changes of lymphoid organs and tissues including bone marrow (e.g., altered cellularity/size of major compartments)

- Weight changes of lymphoid organs
- Changes in hematology parameters (e.g., white blood cell number, differential cell counts of lymphocytic, monocytic and granulocytic cells)
- Changes in clinical chemistry parameters (e.g., serum protein levels, immunoglobulin concentrations if determined)

The issue of adverse versus non-adverse effects in the context of immunotoxicity has been discussed in an article by Kimber and Dearman (2002). The authors concluded that it has proven difficult to define a clear distinction between adverse and non-adverse changes in the immune system associated with chemical exposure. Furthermore, the pivotal issue is defining whether any degree of chemical related change in the immune function can be tolerated if it is revealed to be of statistical significance with respect to concurrent and/or historical control values. It would appear reasonable to suppose that a modest change in, for instance, the vigor of antibody responses or total T-cell numbers, would normally have little impact on the integrity of host resistance. Whereas this is probably true for the healthy and well-nourished adult, it is difficult to have the same level of confidence about potentially at-risk populations.

Immunotoxicity is of particular concern for substances that induce toxicity on the immune system at dose levels below those which induce toxicity at other target sites (WHO/IPCS 1996).

It should be noted that the observation of the immunological changes listed above may not necessarily reflect a primary immunotoxic effect but may be secondary to other effects, e.g., a secondary response to stress resulting from effects on other organ systems. Therefore, it must be recognized that in principle all chemical substances may be able to influence parameters of the immune system if administered at sufficiently high dosages; however, an immunotoxic effect should not be disregarded until a thorough investigation has been performed (EC 2003).

4.7.7 Neurotoxicity

There is worldwide concern about the potential neurotoxic effects of chemical substances but only a small fraction of chemicals have been adequately evaluated for neurotoxicity (WHO/IPCS 2001a). Of particular concern is the possibility of a relationship between exposures to low levels of environmental chemicals and effects on neurobehavioral development in children and neurodegenerative diseases in the elderly.

The complexity of the nervous system results in multiple potential target sites and adverse sequelae. No other organ system has the wide variety of specialized cell functions seen in the nervous system. Different expressions of neurotoxicity are generally based on the different susceptibilities of the various subpopulations of cells that make up the nervous system. The status and role of the blood–brain barrier in the central nervous system (CNS) and similar structures in the peripheral nervous system in modulating the access of some chemicals to the nervous system are also unique considerations in assessing neurotoxicity. Moreover, certain specialized cells outside the barrier have important integrative neuro-immuno-endocrine functions that orchestrate numerous physiological, metabolic, and endocrine processes. These integrative functions are fundamental for cognition and higher-order neural functions, but knowledge on how they can be disrupted by chemical exposures is limited. In contrast to other tissues, the ability of nerve cells to replace or regenerate is severely constrained and is a limiting factor in achieving full recovery from neurotoxicity under conditions where cell death has occurred (WHO/IPCS 2001a).

The biological basis for identification of certain susceptible populations, including the young, the aged, and people with genetic predispositions to certain forms of toxicity is an important consideration in the risk assessment process for neurotoxicity. Many of the factors that convey susceptibility to neurotoxicity will not differ from those that need to be considered in risk assessments of toxicity to other target organs, because they involve metabolic processes that are common to many organ systems. However, the complexity and critically timed events of the long postnatal CNS development process may make the developing nervous system differentially susceptible to certain exposures. Also, the aging process results in a reduction of plasticity and diminished compensatory capacity of the nervous system, making it potentially more susceptible to neurotoxic insults (WHO/IPCS 2001a).

4.7.7.1 Definitions

Neurotoxicity can be defined as "any adverse effect on the structure or function of the nervous system related to exposure to a chemical substance" (US-EPA 1998, OECD 2004c). According to the TGD (EC 2003), neurotoxicity can be defined as "the induction by a chemical of adverse effects in the central or peripheral nervous system, or in sense organs" and a substance is "neurotoxic" if it "induces a reproducible lesion in the nervous system or a reproducible pattern of neural dysfunction."

WHO provides the following additional definitions related to neurotoxicity: structural neurotoxic effects are defined as "neuroanatomical changes occurring at any level of nervous system organization;" functional changes are defined as "neurochemical, neurophysiological or behavioral effects;" and functional neurotoxic effects include "adverse changes in somatic/autonomic, sensory, motor and cognitive function" (WHO/IPCS 2001a).

4.7.7.2 Test Guidelines

The adopted test guideline methods for repeated dose toxicity testing include a number of parameters relevant for the evaluation of a substance's neurotoxic potential. Furthermore, specific test guidelines for neurotoxicity testing have also been adopted. These test methods are addressed in Section 4.7.3.1 and Tables 4.11 and 4.12.

The 1998 OECD test guidelines for the oral 28-/90-day studies (see Table 4.12) examine a number of simple nervous system endpoints, e.g., clinical observations of motor and autonomous nervous system activity, and histopathology of nerve tissue. It should be recognized that the standard 28-/90-day tests measure only some aspects of nervous system structure and function, while other aspects, e.g., learning and memory and sensory function is not or only superficially tested. Primarily the standard 28-/90-day tests are intended as a screening for neurotoxicity and depending on the results, further testing may be needed.

In addition to the test guidelines mentioned above, the following test guideline methods are also directed toward neurotoxicity testing:

- Delayed Neurotoxicity of Organophosphorus Substances Following Acute Exposure (OECD TG 418/US-EPA OPPTS 870.6100/EU Annex V method B.37) (Section 4.4.3.1)
- Developmental Neurotoxicity Study (US-EPA OPPTS 870.6300/OECD TG 426 Draft of September 2003) (Section 4.10.3.1)

4.7.7.3 Guidance Documents

4.7.7.3.1 WHO
WHO/IPCS have published two Environmental Health Criteria Monographs specifically dealing with neurotoxicity. In 1986, WHO/IPCS issued its monograph on "Principles and Methods for the Assessment of Neurotoxicity Associated with Exposure to Chemicals" (WHO/IPCS 1986b), which in 2001 was updated and expanded with the monograph on "Neurotoxicity Risk Assessment for Human Health: Principles and Approaches" (WHO/IPCS 2001a). The most recent WHO/IPCS monograph addresses the major scientific principles underlying hazard identification, testing methods, and risk assessment strategies in assessing human neurotoxicity. It provides an overview of the current state of neurotoxicity risk assessment for public health officials, research and regulatory scientists, and risk managers on the use and interpretation of neurotoxicity data from

human and animal studies, and it discusses emerging methodological approaches to studying neurotoxicity. It does not provide practical advice or specific guidance for the conduct of specific tests and studies. This comprehensive document provides extensive information and is highly recommended reading for anyone involved in risk assessment of neurotoxicants.

The WHO/FAO Joint Meeting of Experts on Pesticide Residues (JMPR) has given recommendations on interpretation of cholinesterase inhibition (FAO 1998, 1999). According to the recommendations, the inhibition of brain acetylcholinesterase activity and clinical signs are considered to be the primary endpoints of concern in toxicological studies on compounds that inhibit acetylcholinesterases. Inhibition of erythrocyte acetylcholinesterase is also considered to be an adverse effect, insofar as it is used as a surrogate for brain and peripheral nerve acetylcholinesterase inhibition, when data on the brain enzyme are not available. Plasma acetylcholinesterase inhibition is considered not relevant. Statistically significant inhibition of brain and erythrocyte acetylcholinesterase by 20% or more represents a clear toxicological effect. Statistically significant inhibition of less than 20% or statistically insignificant inhibition above 20% indicate that a more detailed analysis of the data should be undertaken.

4.7.7.3.2 OECD
The OECD "Guidance Document for Neurotoxicity Testing" (OECD 2004c) is intended to provide guidance on strategies and methods for testing of chemicals for potential neurotoxicity. The primary objective of the guidance is to ensure that necessary and sufficient data are obtained to enable adequate evaluation of the risk of neurotoxicity arising from exposure to a chemical. The Guidance Document does not specifically address developmental neurotoxicity testing; this issue is covered by another OECD Guidance Document (Section 4.10.4.2).

4.7.7.3.3 US-EPA
The series of Risk Assessment Guidelines includes a guideline for neurotoxicity risk assessment (US-EPA 1998). This Guideline sets forth principles and procedures to guide US-EPA scientists in evaluating environmental contaminants that may pose neurotoxic risks, and inform US-EPA decision-makers and the public about these procedures. The Guideline includes a discussion of general definitions and issues, an overview of test methods, and the interpretation of data within the U.S. framework for risk assessment.

4.7.7.3.4 EU
The TGD (EC 2003) contains a section on specific system/organ toxicity, which includes a section concerning guidance on investigation of neurotoxicity and EU-specific risk assessment. For the evaluation of organophosphate pesticides, it has been agreed to use the WHO/FAO JMPR recommendations (FAO 1998, 1999) (Section 4.7.7.3.1) as the applicability of these recommendations also can be extended to biocides and new/existing substances.

4.7.7.4 Use of Information from Repeated Dose Toxicity Studies in the Hazard Assessment of Neurotoxicity

The detection of neurotoxicity in human studies provides the most direct means of assessing health risk, but is often complicated by confounding factors and inadequate data. Exposure levels in humans are difficult to establish, and the neurological status of populations is extremely heterogeneous. Nevertheless, there has been significant progress in the last decade in developing validated methods for detecting neurotoxicity in humans. Standardized neuropsychological tests, validated computer-assisted test batteries, neurophysiological and biochemical tests, and refined imaging techniques have been improved and become well established. These methods can be used to assess a variety of human neurotoxic endpoints and have provided useful data for the purpose of neurotoxicity risk assessment (WHO/IPCS 2001a).

For most neurotoxicological assessments, it is still necessary to rely on information derived from experimental animal models. It is generally assumed that a substance that produces detectable adverse neurotoxic effects in experimental animal studies will pose a potential hazard to humans. This assumption is based on the comparisons of data for known human neurotoxicants, which indicate that experimental animal data are frequently predictive of a neurotoxic effect in humans. However, there are also notable differences between animals and humans in sensitivity to some neurotoxicants. Although most clinical neurotoxicity signs can be reproduced in animal models using rodents, this is not always the case. Therefore, it may be difficult to determine, which will be the most appropriate species in terms of predicting the specific types of effects seen in humans. The fact that every species may not react in the same way may be due to species-specific differences in maturation of the nervous system, differences in timing of exposure, or biochemical and pharmacokinetic factors. There are also basic structural differences (e.g., pigmentation of substantia nigra) that may underlie species differences.

Neurotoxicity may be indicated by the following signs:

- Morphological (structural) changes in the central or peripheral nervous system or in special sense organs
- Neurophysiological changes, e.g., electroencephalographic changes
- Behavioral (functional) change
- Neurochemical changes, e.g., neurotransmitter levels

The basis of the recommended approach to a hazard assessment of the potential neurotoxicity of a substance is that a number of neurotoxic substances can be identified via the standard tests for systemic toxicity. Special studies to characterize effects of concern for neurotoxicity are used only when necessary for adequate hazard assessment. The nature of special studies, and when they should be conducted, need to be decided on a case-by-case basis.

Signs of neurotoxicity in standard acute or repeated dose toxicity tests may be secondary to other systemic toxicity or to discomfort from physical effects such as a distended or blocked gastrointestinal tract. In acute toxicity studies where high doses are administered, clinical signs are often observed, which are suggestive of effects on the nervous system, e.g., observations of lethargy, postural or behavioral changes, and a distinction should be made between specific and nonspecific signs of neurotoxicity.

The type, severity, number, and reversibility of the effect should be considered. Generally a pattern of related effects is more persuasive evidence of neurotoxicity than one or a few unrelated effects.

It is important to ascertain whether the nervous system is the primary target organ. The reversibility of neurotoxic effects should also be considered. Reversible effects may be of high concern depending on the severity and nature of effect. In this context it should be kept in mind that effects observed in experimental animals that appear harmless might be of high concern in humans depending on the setting in which they occur (e.g., sleepiness in itself may not be harmful, but in relation to operation of machinery it is an effect of high concern). Furthermore, the possibility that a permanent lesion has occurred cannot be excluded even if the overt effect is transient. The nervous system possesses reserve capacity, which may compensate for the damage, but the resulting reduction in the reserve capacity should be regarded as an adverse effect. Compensation may be suspected if a neurotoxic effect slowly resolves during the life span. Irreversible neurotoxic effects are of high concern and usually involve structural changes, though, at least in humans, lasting functional effects (e.g., depression and involuntary motor tremor) are suspected to occur as a result of neurotoxicant exposure, apparently without morphological abnormalities.

4.7.8 NASAL TOXICITY

The nose is usually the first site of contact in the respiratory tract for many airborne chemicals of environmental and occupational concern. Examples of human nasal effects include loss of olfactory function (e.g., anosmia and hyposmia), atrophy of the nasal mucosa, mucosal ulcers, perforated nasal septum, or sinonasal cancer related to exposure to certain metal dusts and vapors (Sunderman 2001).

Nasal effects are commonly observed in toxicity studies using inhalation exposure. The uptake and deposition, and therefore also toxicity, of inhaled gases and particles in the nasal passages depends on nasal anatomy, physiology, and respiratory airflow patterns. These factors vary among rodents, primates, and humans. Rats are obligate nose-breathers, and the nasal tract of the rat is more complex than in man, with larger nasal turbinates, giving a larger surface area and a different flow of air in the nasal cavity. This is often used as an argument for the claim that the rat is more sensitive than humans to airborne toxicants due to enhanced deposition of inhaled gases in the nasal tract. Although this may sometimes be true, it should not be automatically assumed to be the case without supportive evidence such as airflow analysis data, or sufficient human data indicating a lower sensitivity. Nasal deposition in the rat has been studied experimentally using acrylic molds made from postmortem casts and *in vivo* techniques (Kelly and Asgharian 2003); computer fluid dynamics models of airflow and of gas and particle transport have been used to explore the possible consequences for toxicity (e.g., Connolly et al. 2004).

Differences among species in distribution patterns of histological changes may be caused by species variations in the distribution in the nasal epithelium of chemical-metabolizing enzymes. For example, in rats exposed to methyl methacrylate, nasal lesions were shown to be caused by the carboxylesterase mediated metabolism of methyl methacrylate to methacrylic acid, an irritant and corrosive metabolite. The distribution of these enzymes in the nasal tissues of man, rat, and hamster indicated a lower rate of metabolism in man compared to rat and hamster, suggesting a lower sensitivity to methacrylate in humans (Mainwaring et al. 2001).

The normal gene expression in the rat nasal epithelium has been studied by using cDNA array technology (Hester et al. 2001).

4.8 MUTAGENICITY

A mutation is a permanent change in the genetic material in a cell. A mutation may occur spontaneously or be induced as a result of exposure to ionizing or ultraviolet radiation, or to chemical substances.

In principle, human exposure to substances that are mutagens can be expected to result in increased frequencies of mutations above background.

Mutations in somatic cells may be lethal or may be transferred to daughter cells with deleterious consequences for the affected organism, e.g., when they occur in protooncogenes, tumor suppressor genes, and/or repair genes, ranging from trivial to detrimental or lethal.

Substances that are mutagenic in somatic cells may produce heritable effects if they, or their active metabolites, reach the genetic material of germ cells. Heritable damage to the offspring, and possibly to subsequent generations, of parents exposed to mutagens may follow if mutations are induced in parental germ cells. To date, all known germ cell mutagens are also mutagenic in somatic cells *in vivo* (EC 2003).

4.8.1 DEFINITIONS

The chemical and structural complexity of the chromosomal DNA and associated proteins of mammalian cells, and the multiplicity of ways in which changes to the genetic material can be effected make it difficult to give precise, discrete definitions (EC 2003).

Both the terms "mutagenicity" and "genotoxicity" are used in this section.

4.8.1.1 OECD

In the OECD document "Harmonised integrated classification system for human health and environmental hazards of chemical substances and mixtures" (OECD 2001b), the following definitions are provided.

A mutation is defined as a permanent change in the amount or structure of the genetic material in a cell.

The term "mutation" applies both for heritable genetic changes that may be manifested at the phenotypic level, and for the underlying DNA modifications when known (including, e.g., specific base pair changes and chromosomal translocations). The terms "mutagenic" and "mutagen" are used for agents giving rise to an increased occurrence of mutations in populations of cells and/or organisms.

The more general terms "genotoxic" and "genotoxicity" apply to agents or processes, which alter the structure, information content, or segregation of DNA, including those which cause DNA damage by interfering with normal replication processes, or which in a nonphysiological manner (temporarily) alter its replication. Genotoxicity test results are usually taken as indicators for mutagenic effects.

In the various OECD test guidelines for mutagenicity/genotoxicity testing, the definitions relevant for a specific test are provided in the Annex.

4.8.1.2 US-EPA

In the US-EPA's Glossary of IRIS (Integrated Risk Information System) Terms (US-EPA 2007b), the following definition is provided:

"A mutagen is a substance that can induce an alteration in the structure of DNA."

4.8.1.3 EU

In the TGD (EC 2003), the following definitions are provided:

The term " 'mutagenicity' refers to the induction of permanent transmissible changes in the amount or structure of the genetic material of cells or organisms. These changes may involve a single gene or gene segment, a block of genes or whole chromosomes. Effects on whole chromosomes may be structural and/or numerical."

The term " 'genotoxicity' is a broader term and refers to potentially harmful effects on genetic material, which are not necessarily associated with mutagenicity. Thus, tests for genotoxicity include tests, which provide an indication of induced damage to DNA (but not direct evidence of mutation) via effects such as unscheduled DNA synthesis (UDS), sister chromatid exchange (SCE), DNA strand breaks, DNA adduct formation or mitotic recombination, as well as tests for mutagenicity."

4.8.2 Objectives for Assessing the Mutagenicity of a Substance

There is considerable evidence of a positive correlation between the mutagenicity of a substance *in vivo* and its carcinogenicity in long-term studies with animals. In the risk assessment of a substance, it is therefore necessary to address the potential mutagenicity. It can be expected that some of the available data will have been derived from tests conducted to investigate harmful effects on the genetic material in general, i.e., the genotoxicity of a substance.

The aims of testing for genotoxicity are, therefore, to assess the potential of a substance to be a genotoxic carcinogen, or to cause heritable damage in humans, which can be manifested as impaired male and/or female fertility, or adverse effects on fetal or postnatal development.

4.8.3 Test Guidelines

4.8.3.1 *In Vivo* Tests

The OECD, US-EPA, and EU have developed a number of specific test guidelines for *in vivo* testing of mutagenicity/genotoxicity, see Table 4.13. The EU Annex V methods are generally replicates of or very similar to the corresponding OECD test guidelines. The US-EPA OPPTS test guidelines are, in principle, comparable to the corresponding OECD test guidelines.

The principle of the various *in vivo* mutagenicity/genotoxicity tests is that the test substance is administered, usually to rodents, by gavage or intraperitoneal injection as a single dose or a few doses. The number of dose levels differs for the various tests. The animals are observed for clinical signs of toxicity and the relevant endpoints for an evaluation of the mutagenicity/genotoxicity in the specific test are investigated. Table 4.14 summarizes the various *in vivo* mutagenicity/genotoxicity test guideline studies in more detail.

4.8.3.2 *In Vitro* Tests

The OECD, US-EPA, and EU have developed a number of specific test guidelines for *in vitro* testing of mutagenicity/genotoxicity, see Table 4.15. The EU Annex V methods are generally

---

TABLE 4.13
Test Guidelines Adopted for *In Vivo* Testing of Mutagenicity/Genotoxicity

Title

OECD Test Guidelines	Year
474 Mammalian erythrocyte micronucleus test	1997
475 Mammalian bone marrow chromosomal aberration test	1997
477 Genetic toxicology: Sex-linked recessive lethal test in *Drosophila melanogaster*	1984
478 Genetic toxicology: Rodent dominant lethal test	1984
483 Mammalian spermatogonial chromosome aberration test	1997
484 Genetic toxicology: Mouse spot test	1986
485 Genetic toxicology: Mouse heritable translocation assay	1986
486 Unscheduled DNA synthesis (UDS) test with mammalian liver cells *in vivo*	1997

US-EPA OPPTS Harmonized Test Guidelines	
870.5195 Mouse biochemical specific locus test	1998
870.5200 Mouse visible specific locus test	1998
870.5275 Sex-linked recessive lethal test in *Drosophila melanogaster*	1998
870.5380 Mammalian spermatogonial chromosomal aberration test	1998
870.5385 Mammalian bone marrow chromosomal aberration test	1998
870.5395 Mammalian erythrocyte micronucleus test	1998
870.5450 Rodent dominant lethal assay	1998
870.5460 Rodent heritable translocation assays	1998
870.5915 *In vivo* sister chromatid exchange assay	1998

EU Annex V Test Methods	
B.11 Mutagenicity - *In vivo* mammalian bone-marrow chromosome aberration test	2000
B.12 Mutagenicity mammalian erythrocyte micronucleus test	2000
B.20 Sex-linked recessive lethal test in *Drosophila melanogaster*	1988
B.22 Rodent dominant lethal test	1988
B.23 Mammalian spermatogonial chromosome aberration test	2000
B.24 Mouse spot test	1988
B.25 Mouse heritable translocation	1988
B.39 Unscheduled DNA synthesis (UDS) test with mammalian liver cells *in vivo*	2000

TABLE 4.14
Overview of *In Vivo* Mutagenicity/Genotoxicity Test Guideline Studies

Test	Design	Endpoints
The mammalian *in vivo* erythrocyte micronucleus test OECD TG 474 US-EPA OPPTS 870.5395 EU Annex V B.12	Animals, usually rodents, are exposed to the test substance by an appropriate route, usually by gavage or by intraperitoneal injection. Each treated and control group must include at least five animals per sex. Positive controls should produce micronuclei *in vivo* at exposure levels expected to give a detectable increase over the background. No standard treatment schedule (i.e., 1, 2, or more treatments at 24-h intervals) has been recommended. Three dose levels are generally used; these should cover a range from the maximum to little or no toxicity. The erythrocytes are sampled from the bone marrow and/or peripheral blood of the animals. If bone marrow is used, the animals are sacrificed at appropriate times after treatment, the bone marrow extracted, and preparations made and stained. When peripheral blood is used, the blood is collected at appropriate times after treatment and smear preparations are made and stained. Preparations are analyzed for the presence of micronucleated polychromatic erythrocytes in treated animals is an indication of induced chromosome damage.	The test is used for the detection of cytogenetic damage to the chromosomes or the mitotic apparatus of erythroblasts by analysis of erythrocytes for formation of micronuclei (small nuclei, separate from and additional to the main nuclei of cells, produced during the telophase of mitosis (meiosis) by lagging chromosome fragments or whole chromosomes). When a bone marrow erythroblast develops into a polychromatic erythrocyte (immature erythrocyte), the main nucleus is extruded; any micronucleus that has been formed may remain behind in the otherwise anucleated cytoplasm.
The mammalian *in vivo* bone marrow chromosome aberration test OECD TG 475 US-EPA OPPTS 870.5385 EU Annex V B.11	Animals, usually rodents, are exposed to the test substance by an appropriate route of exposure, usually by gavage or by intraperitoneal injection, and are sacrificed at appropriate times after treatment. Each treated and control group must include at least five animals per sex. Positive controls should produce structural chromosome aberrations *in vivo* at exposure levels expected to give a detectable increase over background. The test substance is preferably administered as a single treatment but may also be administered as a split dose, i.e., two treatments on the same day separated by no more than a few hours, to facilitate administering a large volume of the substance. Three dose levels are generally used; these should cover a range from the maximum to little or no toxicity. Prior to sacrifice, animals are treated with a metaphase-arresting agent (e.g., colchicine). Chromosome preparations are then made from the bone marrow cells and stained, and metaphase cells are analyzed for chromosome aberrations.	The bone marrow test is used for the detection of structural chromosome aberrations induced by a test substance in bone marrow cells of animals. A structural chromosome aberration is a change in chromosome structure detectable by microscopic examination of the metaphase stage of cell division, observed as deletions and fragments, intrachanges or interchanges.

(*continued*)

TABLE 4.14 (continued)
Overview of *In Vivo* Mutagenicity/Genotoxicity Test Guideline Studies

Test	Design	Endpoints
The *in vivo* sister chromatid exchange (SCE) assay US-EPA OPPTS 870.5915	Animals, usually rodents, are exposed to a test substance by an appropriate route (usually oral or intraperitoneal injection, other routes may be appropriate) followed by administration of bromodeoxyuridine (BrdU). At least five female and five male animals per experimental and control group should be used. For an initial assessment, one dose of the test substance may be used, the dose being the maximum tolerated dose or that producing some indication of toxicity as evidenced by animal morbidity (including death) or target cell toxicity. For determination of dose–response, at least three dose levels should be used. A compound known to produce SCE *in vivo* should be employed as the positive control. A spindle inhibitor (e.g., colchicine) is administered prior to sacrifice. After sacrifice, tissue is obtained and metaphase preparations made, stained, and scored for SCE.	The test detects the ability of a chemical to enhance the exchange of DNA between two sister chromatids of a duplicating chromosome. SCEs are reciprocal interchanges of the two chromatid arms within a single chromosome. These exchanges are visualized during the metaphase portion of the cell cycle and presumably require enzymatic incision, translocation, and ligation of at least two DNA helices. The most commonly used assays employ bone marrow or lymphocytes from mammalian species such as mice, rats, or hamsters. Human lymphocytes may also be used.
The Unscheduled DNA Synthesis (UDS) test with mammalian liver cells *in vivo* OECD TG 486 EU Annex V B.39	Adult animals, usually rats, are administered the test substance usually by gavage and generally as a single treatment. Normally, at least two dose levels are used. Each treated and control group must include at least three animals per sex. Positive controls should be substances known to produce UDS when administered at exposure levels expected to give a detectable increase over background. Liver cells are prepared from treated animals normally 12–16 h after dosing. Freshly isolated mammalian liver cells are incubated for an appropriate length of time, then rinsed, fixed, and dried. Slides are developed, stained, and exposed silver grains are counted. The endpoint of UDS is measured by determining the uptake of labeled nucleosides in cells that are not undergoing scheduled (S-phase) DNA synthesis. The most widely used technique is the determination of the uptake of ^{3}H-TdR by autoradiography.	The test is used to identify substances that induce DNA repair in liver cells of treated animals. UDS is DNA repair synthesis after excision and removal of a stretch of DNA containing a region of damage induced by a chemical substance. The test is usually based on the incorporation of tritium-labeled thymidine (^{3}H-TdR) into the DNA of liver cells, which have a low frequency of cells in the S-phase of the cell cycle.
The mouse spot test OECD TG 484 EU Annex V B.24	Pregnant female mice are treated by a single oral gavage dose or intraperitoneal injection normally given on day 8, 9, or 10 of pregnancy. At least two appropriately spaced dose levels are used including one producing signs of toxicity or reduced litter size. The mice are scored for spots of changed color in the coat between 3 and 4 weeks after birth. The frequency of such spots in treated groups is compared with their frequency in the control group.	The test detects presumed somatic mutations in fetal cells following transplacental absorption of the test substance, i.e., the developing embryo is exposed to the substance. The target cells in the developing embryos are the melanoblasts, and the target genes are those which control the pigmentation of the coat hairs. The genotype of the parental animals is essential and must be chosen to produce the developing embryos, which are heterozygous for a number of

these coat color genes. A mutation in, or loss of (by a variety of genetic events), the dominant allele of such a gene in a melanoblast results in the expression of the recessive phenotype in its descendant cells, constituting a spot of changed color in the coat of the resulting mouse.

The *in vivo* mammalian spermatogonial chromosome aberration test OECD TG 483 US-EPA OPPTS 870.5380 EU Annex V B.23	Animals, usually male Chinese hamsters and mice, are exposed to the test substance by an appropriate route of exposure, usually by gavage or by intraperitoneal injection, and are sacrificed at appropriate times after treatment. Each treated and control group must include at least five males. Positive controls should produce structural chromosome aberrations *in vivo* in spermatogonial cells when administered at exposure levels expected to give a detectable increase over background. The test substance is preferably administered as a single treatment but may also be administered as a split dose, i.e., two treatments on the same day separated by no more than a few hours, to facilitate administering a large volume of the substance. Three dose levels are generally used; these should cover a range from the maximum to little or no toxicity. Prior to sacrifice, animals are treated with a metaphase-arresting agent (e.g., colchicine). Chromosome preparations are then made from the germ cells and stained, and metaphase cells are analyzed for chromosome aberrations.	The spermatogonial test is used to identify substances that cause structural chromosome aberrations in mammalian spermatogonial cells, and is designed to investigate whether somatic cell mutagens are also active in germ cells. This test is not designed to measure numerical aberrations and is not routinely used for this purpose.
The rodent dominant lethal test OECD TG 478 US-EPA OPPTS 870.5450 EU Annex V B.22	Generally, male animals are exposed to a test substance (usually oral or by intraperitoneal injection, a single exposure or on five consecutive days) and mated to untreated virgin females. The various germ cell stages can be tested separately by the use of sequential mating intervals. Normally, three dose levels should be used. The highest dose should produce signs of toxicity (e.g., slightly reduced fertility). The females are sacrificed after an appropriate period of time, and the contents of the uteri are examined to determine the numbers of implants and live and dead embryos. The calculation of the total dominant lethal effect is based on comparison of the live implants per female in the treated group to the live implants per female in the control group. The increase of dead implants per female in the treated group over the dead implants per female in the control group reflects the postimplantation loss. The postimplantation loss is calculated by determining the ratio of dead to total implants from the treated group compared to the ratio of dead to total implants from the control group. Preimplantation loss can be estimated based on corpora lutea counts or by comparing the total implants per female in treated and control groups. The total dominant lethal effect is the sum of pre- and postimplantation loss.	The test is used to detect dominant lethal mutations. A dominant lethal mutation is one occurring in a germ cell, which does not cause dysfunction of the gamete but which is lethal to the fertilized egg or developing embryo.

(continued)

TABLE 4.14 (continued)

Overview of *In Vivo* Mutagenicity/Genotoxicity Test Guideline Studies

Test	Design	Endpoints
The mouse heritable translocation test OECD TG 485 US-EPA OPPTS 870.5460 EU Annex V B.25	Two treatment schedules are available: (1) single administration of a test substance is most widely used; (2) administration of the test substance on 7 days/week for 35 days may also be used. One dose level is tested, usually the highest dose associated with the production of minimal toxic effects, but without affecting reproductive behavior or survival. To establish a dose–response relationship, two additional lower doses are required. Routes of administration are usually oral intubation or intraperitoneal injection. Other routes of administration may be appropriate. The number of matings following treatment is governed by the treatment schedule and should ensure that all treated germ cell stages are sampled. All male progeny are weaned and all female progeny are discarded unless they are included in the experiment. One of two possible methods is used for testing of translocation heterozygosity: (1) fertility testing of F1 progeny and subsequent verification of possible translocation carriers by cytogenetic analysis; (2) cytogenetic analysis of all male F1 progeny without prior selection by fertility testing. Translocations are cytogenetically observed in meiotic cells at diakinesis-metaphase I of male individuals, either F1 males or male offspring of F1 females. The XO-females are cytogenetically identified by the presence of only 39 chromosomes in bone marrow mitoses.	The test detects structural and numerical chromosome changes in mammalian germ cells as recovered in first generation progeny. No special genotype is needed in the parental animals. The types of chromosome changes detected are reciprocal translocations and, if female progeny are included, X-chromosome loss. Carriers of translocations and XO-females show reduced fertility, which is used to select F1 progeny for cytogenetic analysis. Complete sterility is caused by certain types of translocations (X-autosome and c/t type).
The Mouse Visible Specific Locus Test (MSLT) US-EPA OPPTS 870.5200	Generally, male mice of a certain hybrid genotype are treated with the test substance (gavage, inhalation, mixture with food or water, intraperitoneal or intravenous injections) and then mated to females, which are genetically homozygous for certain specific visible marker loci. Usually, only one dose level need to be tested. This should be the highest dose tolerated without toxic effects, provided that any temporary sterility induced due to elimination of spermatogonia is of only moderate duration, as determined by a return of males to fertility within 80 days after treatment. For evaluation of dose–response, it is recommended that at least two dose levels be tested. Offspring are examined for evidence that a new mutation has arisen.	The test is used to detect and quantitate mutations in the germ line of a mammalian species. A visible specific locus mutation is a genetic change that alters factors responsible for coat color and other visible characteristics of certain mouse strains. The principle of the MSLT is to cross individuals who differ with respect to the genes present at certain specific loci, so that a genetic alteration involving the standard gene at any one of these loci will produce an offspring detectably different from the standard heterozygote. The genetic change may be detectable by various means, depending on the loci chosen to be marked.

The Mouse Biochemical Specific Locus Test (MBSL) US-EPA OPPTS 870.5195	The test is used to detect and quantitate mutations originating in the germ line of a mammalian species. A biochemical specific locus mutation is a genetic change resulting from a DNA lesion causing alterations in proteins. The principle of the MBSL is that heritable damage to the genome can be detected by electrophoretic analysis of proteins in the tissues of the progeny of mice treated with germ cell mutagens.	Generally, male mice are treated with the test substance (gavage, inhalation, mixture with food or water, intraperitoneal or intravenous injections) and are then mated to untreated females to produce F1 progeny. Usually, only one dose needs to be tested. This should be the maximum tolerated dose (MTD), the highest dose tolerated without toxic effects. Any temporary sterility induced due to elimination of spermatogonia at this dose must be of only moderate duration, as determined by a return of males to fertility within 80 days after treatment. For evaluation of dose–response, it is recommended that at least two dose levels be tested. The parental animals must be of two different inbred strains to ensure uniform genotype in the F1 animals except for the possible substance-induced mutagenic effect. Both blood and kidney samples are taken from progeny for electrophoretic analysis. Mutants are identified by variations from the normal electrophoretic pattern. Presumed mutants are bred to confirm the genetic nature of the change.
The sex-linked recessive lethal (SLRL) test in *Drosophila melanogaster* OECD TG 477 US-EPA OPPTS 870.5275 EU Annex V B.20	The test detects mutations, both point mutations and small deletions, in the germ line of the insect. The test is a forward mutation assay capable of screening for mutations at about 800 loci on the X-chromosome; this represents about 80% of all X-chromosomal loci. The X-chromosome represents approximately one-fifth of the entire haploid genome. Mutations in the X-chromosome are phenotypically expressed in males carrying the mutant gene. When the mutation is lethal in the hemizygous condition, its presence is inferred from the absence of one class of male offspring out of the two that are normally produced by a heterozygous female, i.e., absence of the phenotypically males indicates that a sex-linked recessive lethal mutation has occurred in a germ cell of the parent male.	Wild-type males are treated with a test substance (oral, by injection or by exposure to gases or vapors) and mated to appropriate females. A single exposure is generally used, that exposure being the maximum-tolerated concentration or that producing some indication of toxicity, where possible. The offspring of these females are scored for lethal effects in the various germ cell stages. Heterozygous F1 females from the above crosses are mated individually to their brothers, and in the next generation the progeny from each separate cross is scored for the absence of wild-type males.

TABLE 4.15

Test Guidelines Adopted for *In Vitro* Testing of Mutagenicity/Genotoxicity

Title

OECD Test Guidelines	Year
471 Bacterial reverse mutation test	1997
473 *In vitro* mammalian chromosomal aberration test	1997
476 *In vitro* mammalian cell gene mutation test	1997
479 Genetic toxicology: *In vitro* sister chromatid exchange assay in mammalian cells	1986
480 Genetic toxicology: *Saccharomyces cerevisiae*, gene mutation assay	1986
481 Genetic toxicology: *Saccharomyces cerevisiae*, mitotic recombination assay	1986
482 Genetic toxicology: DNA damage and repair, unscheduled DNA synthesis in mammalian cells *in vitro*	1986
487 *In vitro* micronucleus test (Draft)	2004

US-EPA OPPTS Harmonized Test Guidelines	
870.5100 Bacterial reverse mutation test	1998
870.5140 Gene mutation in *Aspergillus nidulans*	1998
870.5250 Gene mutation in *Neurospora crassa*	1998
870.5300 *In vitro* mammalian cell gene mutation test	1998
870.5375 *In vitro* mammalian chromosome aberration test	1998
870.5500 Bacterial DNA damage or repair tests	1998
870.5550 Unscheduled DNA synthesis in mammalian cells in culture	1998
870.5575 Mitotic gene conversion in *Saccharomyces cerevisiae*	1998
870.5900 *In vitro* sister chromatid exchange assay	1998

EU Annex V Test Methods	
B.10 Mutagenicity - *In vitro* mammalian chromosome aberration test	2000
B.13/14 Mutagenicity - Reverse mutation test using bacteria	2000
B.15 Gene mutation - *Saccharomyces cerevisiae*	1988
B.16 Mitotic recombination - *Saccharomyces cerevisiae*	1988
B.17 Mutagenicity - *In vitro* mammalian cell gene mutation test	2000
B.18 DNA damage and repair - Unscheduled DNA synthesis - Mammalian cells *in vitro*	1988
B.19 Sister chromatid exchange assay *in vitro*	1988
B.21 *In vitro* mammalian cell transformation test	1988

replicates of or very similar to the corresponding OECD test guidelines. The US-EPA OPPTS test guidelines are, in principle, comparable to the corresponding OECD test guidelines.

The principle of the various *in vitro* mutagenicity/genotoxicity tests is that cells in culture are exposed to the test substance with and without an exogenous metabolic activation system. The number of dose levels differs for the various tests. The relevant endpoints for an evaluation of the mutagenicity/genotoxicity in the specific test are investigated. Concurrent positive and negative (solvent or vehicle) controls both with and without metabolic activation should be included in each experiment. When metabolic activation is used, the positive control chemical should be the one that requires activation to give a mutagenic response. Table 4.16 summarizes the various *in vitro* mutagenicity/genotoxicity test guideline studies in more detail.

4.8.4 GUIDANCE DOCUMENTS

4.8.4.1 WHO

WHO/IPCS has published "Summary Reports on the Evaluation of Short-Term Tests for Carcinogens" (WHO/IPCS 1985a, WHO/IPCS 1990) and a "Guide to Short-Term Tests for Detecting Mutagenic and Carcinogenic Chemicals" (WHO/IPCS 1985b).

TABLE 4.16

Overview of *In Vitro* Mutagenicity/Genotoxicity Test Guideline Studies

Test	Design	Endpoints
The bacterial reverse mutation test OECD TG 471 US-EPA OPPTS 870.5100 EU Annex V B.13/14	Suspensions of bacterial cells are exposed to the test substance in the presence and in the absence of an exogenous metabolic activation system. In the plate incorporation method, these suspensions are mixed with an overlay agar and plated immediately onto minimal medium. In the pre-incubation method, the treatment mixture is incubated and then mixed with an overlay agar before plating onto minimal medium. For both techniques, after 2 or 3 days of incubation, revertant colonies are counted and compared to the number of spontaneous revertant colonies on solvent control plates. At least five strains of bacteria should be used. These should include four strains of *Salmonella typhimurium* (TA1535; TA1537 or TA97a or TA97; TA98; and TA100) that have been shown to be reliable and reproducibly responsive between laboratories. These *Salmonella typhimurium* strains have GC base pairs at the primary reversion site and it is known that they may not detect certain oxidizing mutagens, cross-linking agents, and hydrazines. Such substances may be detected by *Escherichia coli* WP2 strains or *Salmonella typhimurium* TA102, which have an AT base pair at the primary reversion site. At least five different concentrations of the test substance should be used with approximately half log intervals between test points. The recommended maximum test concentration for soluble non-cytotoxic substances is 5 mg/plate or 5 μl/plate. For non-cytotoxic substances that are not soluble at 5 mg/plate or 5 μl/plate, one or more concentrations tested should be insoluble in the final treatment mixture. Test substances that are cytotoxic below 5 mg/plate or 5 μl/plate should be tested up to a cytotoxic concentration.	The test is commonly employed as an initial screen for genotoxic activity and, in particular, for point mutation-inducing activity. It detects point mutations, which involve substitution, addition or deletion of one or a few DNA base pairs. The reverse mutation test in either *Salmonella typhimurium* or *Escherichia coli* detects mutation in an amino acid requiring strain (histidine or tryptophan, respectively) to produce a strain independent of an outside supply of amino acid. The principle of the test is that it detects mutations, which revert mutations present in the test strains and restore the functional capability of the bacteria to synthesize an essential amino acid. The revertant bacteria are detected by their ability to grow in the absence of the amino acid required by the parent test strain.
The bacterial DNA damage or repair tests US-EPA OPPTS 870.5500	Several methods for performing the test have been described, the guideline describes the diffusion and suspension test. Bacteria should be exposed to the test substance both in the presence and absence of an appropriate metabolic activation system. The response is expressed in the preferential inhibition of growth or the preferential killing of the DNA repair deficient strain. *Escherichia coli polA* (W3110/p3478) or *Bacillus subtilis rec* (H17/M45) pairs are recommended. The test should initially be performed over a broad range of concentrations.	The tests are used to detect DNA damage, which is expressed as differential cell killing or growth inhibition of repair deficient bacteria in a set of repair proficient and deficient strains. These tests do not measure mutagenic events per se. Tests for differential growth inhibition of repair proficient and repair deficient bacteria measure differences in chemically induced cell killing between wild-type strains with full repair capacity and mutant strains deficient in one or more of the enzymes, which govern repair of damaged DNA.

(*continued*)

TABLE 4.16 (continued)
Overview of *In Vitro* Mutagenicity/Genotoxicity Test Guideline Studies

Test	Design	Endpoints
The *Saccharomyces cerevisiae* gene mutation assay OECD TG 480 EU Annex V B.15	Treatment of *Saccharomyces cerevisiae* is usually performed in a liquid test procedure using either stationary or growing cells. Initial experiments should be done on growing cells, which are exposed to the test substance with and without an exogenous mammalian metabolic activation system. At the end of the treatment, cells are seeded upon an appropriate culture medium, incubated, and scored for survival and the induction of gene mutation/mitotic recombination. If the first experiment is negative, then a second experiment should be carried out using stationary phase cells. The haploid strain XV 185-14C and the diploid strain D7 are the most widely used in gene mutation studies; other strains may also be appropriate. At least five adequately spaced concentrations of the test substance should be used.	The test is used to detect gene mutation in yeast, a eukaryotic microorganism.
The *Saccharomyces cerevisiae* mitotic recombination assay OECD TG 481 US-EPA OPPTS 870.5575 EU Annex V B.16	Treatment of *Saccharomyces cerevisiae* is usually performed in a liquid test procedure using either stationary or growing cells. Initial experiments should be done on growing cells, which are exposed to the test substance with and without an exogenous mammalian metabolic activation system. At the end of the treatment, cells are seeded upon an appropriate culture medium, incubated, and scored for survival and the induction of gene mutation/mitotic recombination. If the first experiment is negative, then a second experiment should be carried out using stationary phase cells. The most commonly used strains for the detection of mitotic gene conversion are D4, D7, BZ34, and JD1. Mitotic crossing-over producing red and pink homozygous sectors can be assayed in D5 or D7. At least five adequately spaced concentrations of the test substance should be used.	The test is used to detect mitotic recombination (gene conversion or crossing-over) in yeast, a eukaryotic microorganism. Mitotic recombination can be detected between genes (or more generally between a gene and its centromere) and within genes. Mitotic gene conversion is the unilateral transfer of DNA sequence information within a gene, resulting most frequently in nonreciprocal products. Mitotic crossing-over is the exchange of segments of DNA between genes (or more generally between a gene and its centromere), resulting in reciprocal products. Crossing-over is generally assayed by the production of recessive homozygous colonies or sectors produced in a heterozygous strain, whereas gene conversion is assayed by the production of prototrophic revertants produced in an auxotrophic heteroallelic strain carrying two different defective alleles of the same gene.
The gene mutation test in *Aspergillus nidulans* US-EPA OPPTS 870.5140	Tests for mutation in *Aspergillus nidulans* are performed in liquid suspension. Conidia are exposed to the test substance both with and without metabolic activation and plated on selective medium to determine changes in colonial morphology or nutritional requirements. At the end of a suitable incubation period, mutant colonies are counted and compared to the number of spontaneous mutants in an untreated control culture. Simultaneous determination of survival	The test is used to detect forward and reverse gene mutation in *Aspergillus nidulans*, a eukaryotic fungus. These mutations are detected by changes in colonial morphology or nutritional requirements in treated populations. The methionine and 2-thioxanthine forward mutation systems can be used to detect mutations in *Aspergillus nidulans*.

permits calculation of mutation frequency. The test should initially be performed over a broad range of concentrations selected on the basis of a preliminary assay. Each test should include five treatment points, two at fixed concentrations for different time periods, and three at varying concentrations for fixed periods of time.

Test	Method	Purpose
The gene mutation test in Neurospora crassa US-EPA OPPTS 870.5250	Conidia are exposed to the test substance both with and without metabolic activation. Forward mutations at the ad-3 locus can be detected using non-colonial strains of Neurospora crassa grown on media containing sorbose as well as glucose. Under these conditions, colonies are formed and reproducible colonial morphology results. Adenine-requiring mutants, which accumulate a reddish-purple pigment can be readily identified from the white (wild-type) colonies and counted. The test should initially be performed over a broad range of concentrations selected on the basis of a preliminary assay. Each test should include five treatment points, two at fixed concentrations for different time periods, and three at varying concentrations for fixed periods of time.	The test is used to detect forward and reverse gene mutation in Neurospora crassa, a eukaryotic fungus. These mutations are detected by biochemical or morphological changes in the treated population. The most commonly used mutation assay in Neurospora crassa measures forward mutation in the ad-3 region of the genome.
The in vitro mammalian cell gene mutation test OECD TG 476 US-EPA OPPTS 870.5300 EU Annex V B.17	Cells in suspension or monolayer culture are exposed to the test substance, both with and without metabolic activation, for a suitable period of time and subcultured to determine cytotoxicity and to allow phenotypic expression prior to mutant selection. At least four concentrations should be used. Cytotoxicity is usually determined by measuring the relative cloning efficiency (survival) or relative total growth of the cultures after the treatment period. Mutant frequency is determined by seeding known numbers of cells in medium containing the selective agent to detect mutant cells, and in medium without selective agent to determine the cloning efficiency (viability). After a suitable incubation time, the colonies are counted. The mutant frequency is derived from the number of mutant colonies in selective medium and the number of colonies in nonselective medium.	The test is used to detect gene mutations in mammalian cells induced by chemical substances. Suitable cell lines include L5178Y mouse lymphoma cells, the CHO, AS52, and V79 lines of Chinese hamster cells, and TK6 human lymphoblastoid cells. In these cell lines, the most commonly used genetic endpoints measure mutation at thymidine kinase (TK) and hypoxanthine–guanine phosphoribosyl transferase (HPRT), and a transgene of xanthine–guanine phosphoribosyl transferase (XPRT). Thymidine kinase proficient cells are sensitive to trifluorothymidine (TFT), which causes the inhibition of cellular metabolism and halts further cell division. Thus mutant cells are able to proliferate in the presence of TFT, whereas normal cells, which contain thymidine kinase, are not. Similarly, cells deficient in HPRT or XPRT are selected by resistance to 6-thioguanine (TG) or 8-azaguanine (AG).
The in vitro chromosome aberration test OECD TG 473 US-EPA OPPTS 870.5375 EU Annex V B.10	A variety of cell lines, strains, or primary cell cultures, including human cells, may be used (e.g., Chinese hamster fibroblasts, human or other mammalian peripheral blood lymphocytes). Cell cultures are exposed to the test substance both with and without metabolic activation, and at predetermined intervals after exposure, they are treated with a metaphase-arresting substance (e.g., colchicine), harvested, stained, and metaphase cells are analyzed microscopically for the presence of structural chromosome aberrations. At least three concentrations should be used.	The test is used to identify substances that cause structural chromosome aberrations in cultured mammalian cells. A structural chromosome aberration is a change in chromosome structure detectable by microscopic examination of the metaphase stage of cell division, observed as deletions and fragments, intrachanges or interchanges. This test is not designed to measure numerical aberrations and is not routinely used for that purpose.

(continued)

TABLE 4.16 (continued)
Overview of *In Vitro* Mutagenicity/Genotoxicity Test Guideline Studies

Test	Design	Endpoints
The *in vitro* sister chromatid exchange (SCE) assay in mammalian cells OECD TG 479 US-EPA OPPTS 870.5900 EU Annex V B.19	Mammalian cells *in vitro* are exposed to the test chemical with and without an exogenous mammalian metabolic activation system and cultured for two rounds of replication in bromodeoxyuridine (BrdU) containing medium. After treatment with a spindle inhibitor (e.g., colchicine) to accumulate cells in a metaphase-like stage of mitosis (c-metaphase), cells are harvested, stained, and metaphase cells analyzed for SCEs. Primary cultures (e.g., human lymphocytes) or established cell lines (e.g., Chinese hamster ovary or lung cells) may be used in the assay. At least three adequately spaced concentrations of the test substance should be used.	The test is used to detect the ability of a substance to enhance the exchange of DNA between two sister chromatids of a duplicating chromosome. SCEs are reciprocal interchanges of the two chromatid arms within a single chromosome. These exchanges are visualized during the metaphase portion of the cell cycle and presumably require enzymatic incision, translocation, and ligation of at least two DNA helices.
The DNA damage and repair, unscheduled DNA synthesis in mammalian cells *in vitro* test OECD TG 482 US-EPA OPPTS 870.5550 EU Annex V B.18	Mammalian cells in culture are exposed to the test substance. Established cell lines are treated both with and without metabolic activation. Cells are incubated for an appropriate length of time, then rinsed, fixed, and dried. Slides are developed, stained, and exposed silver grains are counted. The endpoint of UDS is measured by determining the uptake of labeled nucleosides in cells that are not undergoing scheduled (S-phase) DNA synthesis. The most widely used technique is the determination of the uptake of ^{3}H-TdR by autoradiography. Primary cultures (e.g., rat hepatocytes), human lymphocytes, or established cell lines (e.g., human diploid fibroblasts) may be used in the assay. Multiple concentrations of the test substance over a range adequate to define the response, should be used.	The test is used to detect DNA repair synthesis after excision and removal of a stretch of DNA containing the region of damage induced by a chemical substance. The test is usually based on the incorporation of tritium-labeled thymidine (^{3}H-TdR) into the DNA of mammalian cells, which are not in the S-phase of the cell cycle.
The *in vitro* micronucleus test OECD TG 487 Draft	Cultured cells (from human peripheral blood lymphocytes or from Syrian hamster embryo (SHE)) or cell lines (CHO, V79, CHL/IU, L5178Y) are exposed to the test substances both with and without an exogenous source of metabolic activation unless primary cells with metabolizing capability are used. After exposure to the test substance, cell cultures are grown for a sufficient period to	The test is used to detect DNA repair synthesis and for the detection of micronuclei in the cytoplasm of interphase cells. These micronuclei may originate from acentric fragments (chromosome fragments lacking a centromere) or whole chromosomes that are unable to migrate with the rest of the chromosomes during the

anaphase of cell division. The assay detects the activity of both clastogenic and aneugenic substances in cells that have undergone cell division after exposure to the test substance. A clastogen causes chromosome breaks or sister chromatid exchange, an aneugen causes loss or gain of whole chromosomes. Development of the cytokinesis-block methodology, by addition of the actin polymerization inhibitor cytochalasin B during the targeted mitosis, allows the identification and selective analysis of micronucleus frequency in cells that have completed one cell division as such cells are binucleate.

allow chromosome or spindle damage to lead to the formation of micronuclei in interphase cells and to trigger the aneuploidy sensitive cell stage (G2/M). Harvested and stained interphase cells are then analyzed microscopically for the presence of micronuclei. In cultures that have been treated with a cytokinesis blocker, only binucleate cells are scored. At least three test concentrations should be used.

The *in vitro* mammalian cell transformation test EU Annex V B.21

The test is used to detect phenotypic changes *in vitro* induced by chemical substances associated with malignant transformation *in vivo*. Widely used cells include C3H10T1/2, 3T3, SHE, Fischer rat and the tests rely on changes in cell morphology, focus formation or changes in anchorage dependence in semisolid agar. Less widely used systems exist which detect other physiological or morphological changes in cells following exposure to carcinogenic chemicals.

Cells should be exposed for a suitable period of time. Cells without sufficient intrinsic metabolic activity should be exposed to the test substance in the presence and absence of an appropriate metabolic activation system. At the end of the exposure period, cells are cultured under conditions appropriate for the appearance of the transformed phenotype being monitored and the incidence of transformation determined. Cytotoxicity may be determined by measuring the effect of the test material on colony-forming abilities (cloning efficiency) or growth rates of the cultures. The measurement of cytotoxicity is to establish that exposure to the test chemical has been toxicologically relevant but cannot be used to calculate transformation: frequency in all assays since some may involve prolonged incubation and/or replating. Several concentrations of the test substance should be used. These concentrations should yield a concentration-related toxic effect, the highest concentration producing a low level of survival and the survival in the lowest concentration being approximately the same as that in the negative control.

The need for inter-laboratory collaborative studies on an international scale arose from the necessity to investigate the value of short-term tests for detecting mutagenic and carcinogenic chemicals. Short-term assays were proposed as alternatives or supplementary procedures to traditional long-term rodent bioassays. Concern about the choice of short-term tests and their reliability and sensitivity led to the proposal of the Collaborative Study on the Assessment and Validation of Short-Term Tests for Genotoxicity and Carcinogenicity (CSSTT) by the International Programme on Chemical Safety (IPCS) and the U.S. National Institute of Environmental Health Sciences (NIEHS). The first part of the project, dealing with *in vitro* studies, was summarized in the Environmental Health Criteria 47 (WHO/IPCS 1985a). The second part of the project, dealing with *in vivo* studies, was summarized in the Environmental Health Criteria 109 (WHO/IPCS 1990). The Environmental Health Criteria 51 (WHO/IPCS 1985b) contains guidance on short-term testing for mutagens and carcinogens with genetic activity and was published in an attempt to stimulate scientific discussion, as well as to provide guidance on the use of genotoxicity tests in chemical safety programs.

4.8.4.2   OECD

OECD has published a "Detailed Review Document on Classification Systems for Germ Cell Mutagenicity in OECD Member Countries" (OECD 1999b). The Review Document presents an overview of classification systems/guidelines used in OECD Member countries relating to the mutagenicity of chemicals. The existing classification systems make use of similar approaches to describe the various degrees of evidence for (germ cell) mutagenicity.

OECD has also published a document on a "Harmonised integrated classification system for human health and environmental hazards of chemical substances and mixtures" (OECD 2001b). Chapter 2.5 addresses a harmonized system for the classification of chemicals, which cause mutations in germ cells.

4.8.4.3   US-EPA

The series of Risk Assessment Guidelines includes a guideline for mutagenicity risk assessment (US-EPA 1986a). The Guidelines describe the procedures that the US-EPA will follow in evaluating the potential genetic risk associated with human exposure to chemicals. The central purpose of the health risk assessment is to provide a judgment concerning the weight of evidence that an agent is a potential human mutagen capable of inducing transmitted genetic changes, and, if so, to provide a judgment on how great an impact this agent is likely to have on public health.

4.8.4.4   EU

The TGD (EC 2003), Chapter 3.10, addresses mutagenicity and provides guidance on data requirements, evaluation of data, and dose–response assessment.

4.8.5   Use of Information on Mutagenicity in Hazard Assessment

For a comprehensive coverage of the potential mutagenicity of a substance, information on gene mutations, structural chromosome aberrations (clastogenicity), and numerical chromosome aberrations (aneugenicity) is required.

For some substances, a wealth of genotoxicity data may be available from studies conducted *in vitro* and/or *in vivo*. In addition to studies conducted according to test guideline methods, there may be nonstandard studies available in which "first site of contact" tissues, i.e., skin, epithelium of the respiratory or gastrointestinal tract, have been examined. In addition, data from plant, fungal, or insect (*Drosophila*) systems may be available. Occasionally, studies of genotoxic effects in humans may also be available. The validity and usefulness of each of the data sets to the overall assessment of genotoxicity should be individually assessed, taking protocol design and current expert views on the value of the test systems into account.

Useful data for the hazard assessment may also be obtained from studies on toxicokinetics (including metabolism), *in vitro* studies on macromolecule binding, from knowledge of the reactivity and electrophilicity of a substance, and from the presence or absence of "structural alerts" for genotoxicity.

Certain substances may need special consideration, such as highly electrophilic substances, which give positive results *in vitro*, particularly in the absence of metabolic activation. Although these substances may react with proteins and water *in vivo* and thus be rendered inactive toward many tissues, they may be able to express their mutagenic potential at the first site of contact with the body. Consequently, the use of test methods that can be applied to the respiratory tract, upper gastrointestinal tract, and skin may be appropriate.

Contradictory results between different test systems should be evaluated with respect to their individual significance. Examples of points to be considered are as follows:

- Conflicting results obtained in nonmammalian systems and in mammalian cell tests may be addressed by considering possible differences in metabolism or in the organization of genetic material. Additional data may be needed to resolve contradictions.
- If the results of indicator tests, e.g., DNA binding and SCE, are not supported by results obtained in tests for mutagenicity, the results of mutagenicity tests are generally of higher significance.
- If contradictory findings are obtained *in vitro* and *in vivo*, in general, the results of *in vivo* tests indicate a higher degree of reliability. However, for evaluation of "negative" results *in vivo*, it should be considered whether there is adequate evidence of target tissue exposure.

Conflicting results may also be available from the same test, performed by different laboratories or on different occasions. In this case, expert judgment should be used to reach an overall evaluation of the data. In particular, the quality of each of the studies and of the data provided should be evaluated, with consideration especially of the study design, reproducibility of data, dose–effect relationships, and biological plausibility of the findings. The purity of the test substance may also be a factor to take into account. Furthermore, studies compliant with GLP may be regarded as being of a higher quality.

Particular points to take into account when evaluating "negative" test results include:

- Doses or concentrations of test substance used, e.g., were they high enough?
- Volatility of the test substance, e.g., were the concentrations maintained in tests conducted *in vitro*?
- For studies *in vitro*, the possibility of metabolism not active in the system, including those in extrahepatic organs.
- Is the substance reaching the target organ?
- Reactivity of the substance, e.g., rate of hydrolysis, electrophilicity, presence or absence of structural alerts.
- If a mixture was tested, did the cytotoxicity of one component mask the mutagenicity of another?

Regarding "positive" findings, responses may be generated only at highly toxic/cytotoxic concentrations, and the presence or absence of a dose–response relationship should be considered.

Unless a threshold mechanism of action is clearly demonstrated, mutagenicity and genotoxicity are considered as being "non-threshold effects," i.e., a threshold of substance concentration cannot be identified below which the effects will not be manifested. For non-threshold mechanisms of genotoxicity, the dose–response relationship is generally linear. It should be kept in mind, however, that both direct and indirect mechanisms of genotoxicity can be nonlinear (threshold). Examples of mechanisms of genotoxicity that may lead to nonlinear dose–response relationships include extremes of pH, ionic strength, and osmolarity; inhibition of DNA synthesis; alterations in DNA repair; overloading of defense mechanisms (antioxidants or metal homeostasis); interaction with microtubule assembly leading to aneuploidy; topoisomerase inhibition; high cytotoxicity; metabolic overload; and physiological perturbations (e.g., induction of erythropoeisis) (EC 2003).

In general, several doses are tested in genotoxicity assays. Determination of experimental dose–response relationships may be used to assess the genotoxic potential of a substance, as indicated below (EC 2003):

- A dose-related increase in genotoxicity is one of the relevant criteria for identification of positive findings. In practice, this will be most helpful for *in vitro* tests, but care is needed to check for cytotoxicity or cell cycle delay, which may cause deviations from a dose–response related effect in some experimental systems.
- Routine genotoxicity tests are not designed in order to derive no-effect levels. However, the magnitude of the lowest dose with an observed effect (i.e., the LOEL) may, on certain occasions, be a helpful tool in the hazard assessment. Specifically, it can give an indication of the potency of the test substance. Modified studies, with additional dose levels and improved statistical power may be useful in this regard.
- Unusual shapes of dose–response curves may contribute to the identification of specific mechanisms of genotoxicity. For example, extremely steep increases suggest an indirect mode of action or metabolic switching.

4.8.5.1   Human Data

Human data and their relevance have to be assessed carefully on a case-by-case basis due to limitations of the techniques available. In particular, attention should be paid to the adequacy of the exposure information, confounding factors, and to sources of bias in the study design. The statistical power of the test may also be considered.

4.8.5.2   Animal Data

Animal tests will, in general, be needed for the clarification of positive findings in *in vitro* tests and in case of specific metabolic pathways that cannot be simulated adequately *in vitro*.

The information that can be obtained from the various types of *in vivo* mutagenicity/genotoxicity studies is briefly summarized below.

Positive results in the mammalian *in vivo* erythrocyte micronucleus test indicate that a substance produces micronuclei in the immature erythrocytes of the test species, which are the result of chromosomal damage or damage to the mitotic apparatus in the erythroblasts of the test species.

Positive results in the mammalian *in vivo* bone marrow chromosome aberration test indicate that a substance induces structural chromosome aberrations in the bone marrow of the species tested. An increase in polyploidy (a multiple of the haploid chromosome number (n) other than the diploid number, i.e., $3n$, $4n$ and so on) may indicate that a substance has the potential to induce numerical aberrations (change in the number of chromosomes from the normal number characteristic of the animals utilized).

Positive results in the *in vivo* mammalian spermatogonial chromosome aberration test indicate that a substance induces structural chromosome aberrations in the germ cells of the species tested. This test measures chromosome events in spermatogonial germ cells and is, therefore, expected to be predictive of induction of inheritable mutations in germ cells.

Structural chromosome aberrations may be of two types, chromosome or chromatid. A chromosome-type aberration is a structural chromosome damage expressed as breakage, or breakage and reunion, of both chromatids at an identical site. A chromatid-type aberration is a structural chromosome damage expressed as breakage of single chromatids or breakage and reunion between chromatids. With the majority of chemical mutagens, induced structural aberrations are of the chromatid-type, but chromosome-type aberrations also occur.

A positive response in the sex-linked recessive lethal (SLRL) test in *Drosophila melanogaster* indicates that the test substance causes mutations in germ cells of this insect as evaluated by an increase in the frequency of sex-linked recessive lethal mutations. A lethal mutation is a change in the genome, which, when expressed, causes death to the carrier; a recessive mutation is a change in the genome, which is expressed in the homozygous or hemizygous condition. Sex-linked genes are present on the X- or Y-chromosome; in this test, sex-linked genes refer only to those located on the X-chromosome.

A positive response in the rodent dominant lethal test indicates that the test substance may be genotoxic in the germ cells of the treated sex of the test species as evaluated by an increase in the number of dominant lethals evaluated as the sum of pre- and postimplantation loss. A dominant lethal mutation is one occurring in a germ cell, which does not cause dysfunction of the gamete but which is lethal to the fertilized egg or developing embryo.

A positive response in the mouse spot test indicates that exposure to the test substance results in an increased frequency of spots of changed color in the coat as a result of heritable gene mutations.

A positive response in the mouse heritable translocation test indicates that exposure to the test substance results in an increase in the number of translocations observed for at least one test point, or an increase in the number of translocations observed, as a result of heritable gene mutations.

A positive result in the unscheduled DNA synthesis (UDS) test with mammalian liver cells *in vivo* indicates that a substance induces DNA damage in mammalian liver cells *in vivo* that can be repaired by UDS *in vitro*. The liver is usually the major site of metabolism of absorbed compounds and is thus an appropriate site to measure DNA damage *in vivo*. The detection of a UDS response is dependent on the number of DNA bases excised and replaced at the site of the damage. Therefore, the UDS test is particularly valuable to detect substance-induced "long-patch repair" (20–30 bases). In contrast, "short-patch repair" (1–3 bases) is detected with much lower sensitivity. Furthermore, mutagenic events may result because of non-repair, mis-repair, or mis-replication of DNA lesions. The extent of the UDS response gives no indication of the fidelity of the repair process. In addition, it is possible that a mutagen reacts with DNA but the DNA damage is not repaired via an excision repair process. The lack of specific information on mutagenic activity provided by the UDS test is compensated for by the potential sensitivity of this endpoint because it is measured in the whole genome.

Positive results in the Mouse Biochemical Specific Locus Test (MBSL) and/or in the Mouse Visible Specific Locus Test (MSLT) indicate that the test substance induces heritable gene mutations in a mammalian species.

Positive results in the *in vivo* sister chromatid exchange (SCE) assay indicate that the test substance induces reciprocal chromatid interchanges in the bone marrow or lymphocytes of the test species. SCEs represent the interchange of DNA replication products at apparently homologous loci. The exchange process presumably involves DNA breakage and reunion, although little is known about its molecular basis.

4.8.5.3 *In Vitro* Data

In vitro tests are particularly useful for gaining an understanding of the potential mutagenicity of a substance and have a critical role in the hazard identification.

The information that can be obtained from the various types of *in vitro* mutagenicity/genotoxicity studies is briefly summarized below.

Positive results from the bacterial reverse mutation test (often referred to as the Ames test) indicate that a substance induces point mutations by base pair substitution or frameshift in the genome of either *Salmonella typhimurium* and/or *Escherichia coli*. Base pair substitution mutagens are substances that cause a base change in DNA; in a reversion test, this change may occur at the site of the original mutation, or at a second site in the bacterial genome. Frameshift mutagens are substances that cause the addition or deletion of one or more base pairs in the DNA, thus changing the reading frame in the RNA. Many of the test strains have several features that make them more sensitive for the detection of mutations, including responsive DNA sequences at the reversion sites, increased cell permeability to large molecules, and elimination of DNA repair systems or enhancement of error-prone DNA repair processes. The specificity of the test strains can provide some useful information on the types of mutations that are induced by genotoxic agents.

An extensive database has demonstrated that many chemicals that are positive in this test also exhibit mutagenic activity in other tests. There are, however, examples of mutagenic substances, which are not detected by this test; reasons for these shortcomings can be ascribed to the specific nature of the endpoint detected, differences in metabolic activation, or differences in bioavailability. On the other hand, factors which enhance the sensitivity of the bacterial reverse mutation test can lead to an overestimation of mutagenic activity. The bacterial reverse mutation test may not be appropriate for the evaluation of certain classes of chemicals; for example, highly bactericidal compounds (e.g., certain antibiotics) and those which are thought (or known) to interfere specifically with the mammalian cell replication system (e.g., some topoisomerase inhibitors and some nucleoside analogues). In such cases, mammalian mutation tests may be more appropriate.

Positive results from the *in vitro* chromosome aberration test indicate that the test substance induces structural chromosome aberrations in cultured mammalian somatic cells. More details about this endpoint are provided under the *in vivo* tests.

Positive results in the *in vitro* mammalian cell gene mutation test indicate that the test substance induces gene mutations in the cultured mammalian cells used. The thymidine kinase (TK), hypoxanthine–guanine phosphoribosyl transferase (HPRT), and xanthine–guanine phosphoribosyl transferase (XPRT) mutation tests detect different spectra of genetic events. The autosomal location of TK and XPRT may allow the detection of genetic events (e.g., large deletions) not detected at the HPRT locus on X-chromosomes. Cells deficient in TK due to the mutation $TK^{+/-} \rightarrow TK^{-/-}$ are resistant to the cytotoxic effects of the pyrimidine analogue trifluorothymidine (TFT) (see Table 4.16 for further details).

Positive results in the *in vitro* sister chromatid exchange (SCE) assay indicate that the test substance induces reciprocal chromatid interchanges in cultured mammalian somatic cells. For further description of SCE, see the *in vivo* test.

Positive results from the *Saccharomyces cerevisiae* gene mutation assay indicate that a substance induces point mutations by base pair substitution or frameshift in the genome. Base pair substitution and frameshift is explained under the *in vivo* test.

Positive results from the *Saccharomyces cerevisiae* mitotic recombination assay indicate that a substance induces DNA recombination in this test system, i.e., mitotic gene conversion (within a gene) and/or mitotic crossing-over (between a gene and its centromere).

A positive result in the DNA damage and repair, unscheduled DNA synthesis in mammalian cells *in vitro* test indicates that a substance induces DNA damage in cultured mammalian somatic cells that can be repaired by UDS. UDS is described in more detail under the *in vivo* test.

Positive results from the *in vitro* micronucleus test indicate that the test substance induces chromosome damage and/or damage to the cell division apparatus, in cultured mammalian somatic cells. Immunochemical labeling (FISH: fluorescence in situ hybridization) of kinetochores, or hybridization with general or chromosome specific centromeric/telomeric probes can provide useful information on the mechanism of micronucleus formation. Use of cytokinesis block facilitates the acquisition of the additional mechanistic information (e.g., chromosome nondisjunction) that can be obtained by FISH techniques. The micronucleus assay has a number of advantages over metaphase analysis performed to measure chromosome aberrations (see OECD TG 487 draft).

Positive results from the *in vitro* mammalian cell transformation test indicate that the test substance produces phenotypic changes in cultured mammalian cells associated with malignant transformation *in vivo*. None of the *in vitro* test endpoints has an established mechanistic link with cancer. Some of the test systems are capable of detecting tumor promoters.

Positive results from the gene mutation test in *Aspergillus nidulans* indicate that the test substance causes gene (point) mutations in the DNA of this organism caused by base pair changes and small deletions in the genome.

Positive results from the gene mutation test in *Neurospora crassa* indicate that the test substance causes mutations in the DNA of this organism.

Positive results from the bacterial DNA damage or repair tests indicate that the test substance causes DNA damage, which is expressed as differential cell killing or growth inhibition of repair deficient bacteria. The DNA damage tests do not measure DNA repair per se nor do they measure mutations. A positive result in a DNA damage test in the absence of a positive result in another system is difficult to evaluate in the absence of a better database.

4.8.5.4   Other Data

The chemical structure of a substance can provide information for an initial assessment of the mutagenic potential. By using expert judgment, it may be possible to identify whether a substance, or a potential metabolite of a substance, shares structural characteristics with known mutagens or nonmutagens. This can be used to justify a higher or lower level of priority for the characterization of the mutagenic potential of a substance. Where the level of evidence for mutagenicity is particularly strong, it may be possible to make a conclusive hazard identification without additional information on the basis of (Q)SAR alone.

The need for data in relation to photomutagenicity may be indicated by the structure of a molecule, its light absorbing potential, or its potential to be photoactivated.

Highly electrophilic substances may react with proteins and water *in vivo* and thus be rendered inactive toward many tissues; however, they may be able to express their mutagenic potential at the initial site of contact with the body.

The pH, solubility, volatility, and stability of a substance in test vehicles can affect the performance of mutagenicity tests and therefore influence the design of test protocols.

Use of (Q)SAR may be valuable in the evaluation of the mutagenicity/genotoxicity of a substance for which no data are available, see Sections 3.5.2 and 3.5.3.

4.9   CARCINOGENICITY

Adult tissues, even those composed of rapidly replicating cells, maintain a constant size and number of cells through regulation of the rate of cell replication, the differentiation of cells to assume specialized functions, and programmed cell death (apoptosis). Cancers are diseases in which somatic mutation of genes critical to the maintenance of control over cell division leads to loss of control over cell replication, differentiation, and death. Such uncontrolled cell replication can cause the growth of tumors (neoplasms), i.e., masses of abnormal disorganized cells that arise from preexisting tissue and are characterized by excessive and uncoordinated proliferation and abnormal differentiation.

Tumors are classified as either benign or malignant. Malignant tumors invade or infiltrate surrounding tissues, often damaging or destroying them. They may also spread by dissemination via the circulatory and vascular systems to distant sites, a process known as metastasis. Growth may be rapid. The morphology of malignant tumors is variable. Some well-differentiated tumors bear a resemblance to their parent tissues, but recognizable features are progressively lost in moderately and poorly differentiated malignancies. Undifferentiated, or anaplastic, tumors are composed of pleomorphic cells (of varying shape or size) that do not resemble normal tissue. Benign tumors, by comparison, show a close morphological resemblance to their tissue of origin, grow by slow expansion, and form circumscribed and (usually) encapsulated masses. They may stop growing or even regress, and do not metastatize or invade surrounding structures, although they may compress them. However, benign tumors may become malignant, and benign tumors that are considered to have the potential to progress to malignant tumors are generally considered along with malignant tumors. For example, the U.S. National Toxicology Program (US-NTP) considers a study to show clear evidence of carcinogenic activity if one of the following outcomes are found: (1) an increase of malignant neoplasms; (2) an increase of a combination of malignant and benign neoplasms; or (3) a marked increase of benign neoplasms if there is an indication from this or other studies of the ability of such tumors to progress to malignancy (US-NTP 2007).

The process of chemical carcinogenesis is a multistep process and involves the transition of normal cells into cancer cells via a sequence of stages that entail both genetic alterations (i.e., mutations) and nongenetic events (see Figure 4.9). Nongenetic events are those alterations/processes that are mediated by mechanisms that do not affect the primary sequence of DNA and yet increase the incidence of tumors or decrease the latency time for the appearance of tumors. For example, altered growth and death rates, (de)differentiation of the altered or target cells, and modulation of the expression of specific genes associated with the expression of neoplastic potential (e.g., tumor suppressor genes) are recognized to play an important role in the process of carcinogenesis and can be modulated by a chemical substance in the absence of genetic change to increase the incidence of cancer.

Carcinogenic substances have conventionally been divided into two categories according to the presumed mode of action: genotoxic and non-genotoxic (epigenetic).

Genotoxic modes of action involve genetic alterations caused by the substance interacting directly with DNA to result in a change in the primary sequence of DNA. A substance can also cause genetic alterations indirectly following interaction with other cellular processes (e.g., secondary to the induction of oxidative stress).

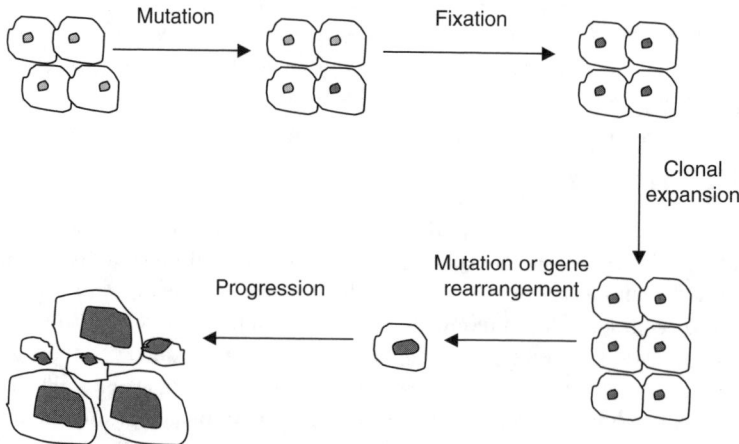

FIGURE 4.9 Stages in the development of cancer. A critical mutation is fixed in one of the cells, i.e., the cell is initiated. The initiation gives the cell a growth advantage, so that a whole clone of initiated cells grows from the first cell. The risk of yet another critical hit affecting an initiated cell is increased because the number of initiated cells has increased. The result of the second hit is conversion of the initiated cell into a cancer cell. The cancer cell quickly grows into a tumor, which develops at the expense of the normal cells (progression). (Adapted from DK NFA, *Quantitative Risk Analysis of Carcinogens*, National Food Agency of Denmark, Søborg, Denmark, 1990.)

Non-genotoxic modes of action include epigenetic changes, i.e., effects that do not involve alterations in DNA but that may influence gene expression, altered cell–cell communication, or other factors involved in the carcinogenic process. For example, chronic cytotoxicity with subsequent regenerative cell proliferation is considered a mode of action by which tumor development can be enhanced. The induction of urinary bladder tumors may, in certain cases, be due to persistent irritation/inflammation, tissue erosion, and regenerative hyperplasia of the urothelium following the formation of bladder stones. Other modes of non-genotoxic action can involve specific receptors (e.g., PPARα, which is associated with liver tumors in rodents; or tumors induced by various hormonal mechanisms). As with other non-genotoxic modes of action for toxicity, these can all be presumed to have a threshold.

Genotoxic carcinogens may possess both genotoxic and non-genotoxic properties that contribute to the carcinogenic effect. Such substances are sometimes called complete carcinogens since they may act as "initiators" (i.e., cause a change in DNA that allows cancer to develop) as well as "promoters" (i.e., cause cells with DNA changes to multiply and become tumors) in two-step/multistep experimental models of carcinogenicity as illustrated in Figure 4.9 (see also Section 6.2.1).

4.9.1 Definitions

Generally, a substance is defined as carcinogenic if it induces tumors (benign or malignant), increases tumor incidence and/or malignancy, or shortens the time to tumor occurrence (OECD 2002a, EC 2003).

According to OECD (2001b), "the term carcinogen denotes a chemical substance or a mixture of chemical substances, which induce cancer or increase its incidence. Substances which have induced benign and malignant tumors in well-performed experimental studies on animals are considered also to be presumed or suspected human carcinogens unless there is strong evidence that the mechanism of tumor formation is not relevant for humans."

In the US-EPA test guidelines for carcinogenicity and combined chronic toxicity/carcinogenicity (OPPTS 870.4200 and OPPTS 870.4300, respectively), the following definition is provided: "Carcinogenicity is the development of neoplastic lesions as a result of the repeated daily exposure of experimental animals to a chemical by the oral, dermal, or inhalation routes of exposure."

4.9.2 Objectives for Assessing the Carcinogenicity of a Substance

The objective of assessing the carcinogenicity of a substance is to identify potential human carcinogens, their mode(s) of action, and their potency.

Carcinogenicity testing is intended to differentiate carcinogens from noncarcinogens. Associated targeted and mechanistic studies derive information on their mode of action. Carcinogens acting without a threshold mode of action or lacking an acceptable evidence for a threshold mode of action have to be discriminated from those carcinogens for which the evidence for a threshold mode of action is plausibly and convincingly demonstrated to be of relevance for humans.

4.9.3 Test Guidelines

4.9.3.1 *In Vivo* Tests

The OECD, US-EPA, and EU have developed specific test guidelines for *in vivo* testing of carcinogenicity, see Table 4.17. The US-EPA OPPTS and EU Annex V test guidelines are, in principle, similar to the corresponding OECD test guidelines. The carcinogenicity part of the combined chronic toxicity/carcinogenicity test guidelines are, in principle, similar to the carcinogenicity test guidelines.

TABLE 4.17

Test Guidelines Adopted for Carcinogenicity Testing

Title

OECD Test Guidelines	Year
451 Carcinogenicity studies	1981
453 Combined chronic toxicity/carcinogenicity studies	1981
US-EPA OPPTS Harmonized Test Guidelines	
870.4200 Carcinogenicity	1998
870.4300 Combined chronic toxicity/carcinogenicity	1998
EU Annex V Test Methods	
B.32 Carcinogenicity test	1988
B.33 Combined chronic toxicity/carcinogenicity test	1988

The principle of the carcinogenicity studies is that the test substance is administered daily in graduated doses or concentrations to several groups of experimental animals, one dose level per group for a major part of their life span. Generally, the termination of the study should be at 18 months for mice and 24 months for rats. Survival of all groups at these time points should be no less than 50% (OECD TG 451, EU Annex B.32), or below 25% (US-EPA OPPTS 870.4200). During the period of administration, the animals are observed each day for signs of toxicity. Animals, which die or are killed during the test are necropsied and at the conclusion of the test surviving animals are killed and necropsied. The macroscopic and histopathological examination is the main endpoint in the carcinogenicity study.

Advances in genetic engineering have created opportunities for improved understanding of the molecular basis of carcinogenesis. Through selective introduction, activation, and inactivation of specific genes, investigators can produce mice of unique genotypes and phenotypes that afford insights into the events and mechanisms responsible for tumor formation. It has been suggested that such animals might be used for routine testing of chemicals to determine their carcinogenic potential because the animals may be mechanistically relevant for understanding and predicting the human response to exposure to the chemical being tested. Information related to the animal line to be used, study design, and data analysis and interpretation must be carefully considered before transgenic and knockout mice can be used as an adjunct or alternative to the conventional 2-year rodent bioassay. A number of transgenic and knockout models are available. Gulezian et al. (2000) have identified and reviewed information relative to four animal lines: the Tg.AC and rasH2 transgenic mice, and the p53+/− (heterozygous) and XPA−/− knockout mice, all of which have been proposed for use in chemical carcinogenicity testing. In addition, the implications of finding of tumors in transgenic and knockout animals exposed to chemicals are discussed in the context of human health risk assessment.

4.9.3.2 *In Vitro* Tests

The OECD, US-EPA, and EU have no adopted test guidelines for *in vitro* testing of carcinogenicity.

4.9.4 GUIDANCE DOCUMENTS

4.9.4.1 WHO

WHO/IPCS have published Summary Reports on the Evaluation of Short-Term Tests for Carcinogens (WHO/IPCS 1985a, 1990) and a Guide to Short-Term Tests for Detecting Mutagenic and Carcinogenic Chemicals (WHO/IPCS 1985b), see Section 4.8.4.1.

4.9.4.2   OECD

OECD has published "Guidance Notes for Analysis and Evaluation of Chronic Toxicity and Carcinogenicity Studies" (OECD 2002a). The aim of this OECD project was to develop harmonized guidance on conducting independent evaluations of, and writing reviews of, chronic oral toxicity and carcinogenicity tests. These Guidance Notes contain much material identical to or derived from the "Guidance Notes for Analysis and Evaluation of Repeat-dose Toxicity Studies" (Section 4.7.4.2), because many concepts apply to both types of study.

The objective of the Guidance Notes is to outline core concepts in order to avoid the need to make reference to large numbers of textbooks, but to refer the reader to other useful sources when more detailed and specific information is required. They are intended to complement OECD Test Guidelines and other publications by the OECD, including the Guidance for Industry Data Submissions (Dossier Guidance) and Guidance for Country Data Review Reports (Monograph Guidance) on Plant Protection Products and their Active Substances. However, whereas the latter publication provides guidance on the format and presentation of entire review reports (monographs), the Guidance Notes place emphasis on data interpretation, scientific judgment, and report writing in the context of regulatory toxicology evaluations.

OECD has also published a document on a "Harmonised integrated classification system for human health and environmental hazards of chemical substances and mixtures" (OECD 2001b). Chapter 2.6 addresses a harmonized system for the classification of chemicals, which cause cancer.

4.9.4.3   US-EPA

The series of Risk Assessment Guidelines includes a guideline for carcinogen risk assessment (US-EPA 2005). The guideline does not establish any substantive "rules" under any other law and has no binding effect on US-EPA or any regulated entity, but instead represents a nonbinding statement of policy. Deviations from the guideline may occur as a result of new information, new scientific understanding, or new science policy judgment.

4.9.4.4   EU

The TGD (EC 2003), Chapter 3.11, addresses carcinogenicity and provides guidance on data requirements, evaluation of data, and dose–response assessment.

4.9.4.5   IARC

The preamble to the IARC (International Agency for Research on Cancer) monographs (IARC 2006) describes the scientific principles and procedures used in the evaluation of the potential carcinogenicity of a substance, the types of evidence considered, and the scientific criteria that guide the evaluations (Section 3.6.1.2).

In October 1997, an expert meeting discussed the use in carcinogen evaluation of emerging short- and medium-term carcinogenicity tests, including transgenic and knockout mouse models, and results from the use of multistage cancer models. The significance of specific gene alterations found in human and rodent tumors was considered, as were the significance of genetic and related properties of chemical agents. The IARC Scientific Publication No. 146 (IARC 1998) includes a consensus document setting out the weight that data derived from such tests can be given when making evaluations of carcinogenic risks to humans.

4.9.5   Use of Information on Carcinogenicity in Hazard Assessment

The identification of the carcinogenic potential of substances often requires the consideration of a large set of data. An important part of the assessment of the available data concerns the evaluation of the mode of action underlying the carcinogenic activity, as this information also allows an evaluation of possible human relevance, existence of thresholds, and comparability with structurally related carcinogens. Genotoxicity data play an important role for predicting carcinogenicity as well as in the assessment of the mode of action of the carcinogen. Expert judgment and a weight of evidence approach are required for the evaluation.

Unless a threshold mechanism of action is clearly demonstrated, genotoxic carcinogenicity is considered as being "non-threshold effects," i.e., a threshold of substance concentration cannot be identified below which the effects will not be manifested. In the case of non-threshold carcinogens, it is thus assumed that there is no level of exposure without a potential effect and, in theory, there will always be a risk associated with exposure even to the lowest exposure levels. It should be kept in mind, however, that both direct and indirect mechanisms of genotoxicity can be nonlinear (threshold) and thus it is also recognized that for certain genotoxic carcinogens a threshold may exist for the underlying genotoxic effect; this issue is addressed in detail in Chapter 6.

For non-threshold mechanisms of genotoxic carcinogenicity, the dose–response relationship is considered to be linear. The observed dose–response curve in some cases represents a single rate-determining step; however, in many cases it may be more complex and represent a superposition of a number of dose–response curves for the various steps involved in the tumor formation (EC 2003). Because of the small number of doses tested experimentally, i.e., usually only two or three, almost all data sets fit equally well various mathematical functions, and it is generally not possible to determine valid dose–response curves on the basis of mathematical modeling. This issue is addressed in further detail in Chapter 6.

For threshold carcinogens, it is possible to identify a NOAEL for the underlying toxicity responsible for tumor formation. The following general guidance is provided for the dose–response assessment for non-genotoxic (threshold) carcinogens (EC 2003). The dose–response assessment for the relevant tumor types is performed in a two-step process.

The first step of the dose–response assessment is an evaluation within the range of tumor observations. In addition to the tumor-data analysis, attempts are also made to identify and determine the significance and dose–response for toxicological effects possibly underlying the tumor formation, i.e., their role in the induction and/or promotion of the carcinogenic process. If appropriate, the analyses of tumor incidence and of "precursor steps" may be combined, using precursor data to extend the dose–response curve beyond that of the tumor data. In this respect, it is important that the data on "precursor steps" come from *in vivo* repeated dose toxicity studies where the exposure is over an extended period of time. Moreover, it is desirable to have data on the "precursor steps" in the same target organ, sex, animal strain, and species as the tumor data. Since a substance may induce multiple tumor types, the dose–response assessment may include analysis of several types, followed by an overall synthesis, which includes an analysis and comparison of the risk estimates across tumor types, an evaluation of the strength of the information on the mode of action of the tumor type considered, and the relevance of the tumor types to humans. Normally, the most sensitive tumor endpoint is used in the hazard and risk assessment.

The second step of dose–response assessment is the extrapolation to lower dose levels, which are relevant in relation to human exposures. For some threshold carcinogens, the mechanisms underlying the tumor formation have been well characterized, and a NOAEL for the underlying toxicity can be derived according to the principles described for repeated dose toxicity (Section 4.7.5). In practice, the mode of action of carcinogenesis for a given substance is often not well understood. Thus, neither a genotoxic nor a non-genotoxic mode of action, i.e., threshold or non-threshold, can be derived with scientific certainty. In such cases, the substance under evaluation may be treated as a threshold or a non-threshold carcinogen, or both in the risk assessment.

4.9.5.1   Human Data

Human data may provide direct information on the potential carcinogenicity of a substance. Human data may also reveal the carcinogenic potential of a substance for which experiments in animals either do not exist or have failed to indicate the carcinogenic potential of the substance.

The degree of reliability for each study on the carcinogenic potential of a substance should be evaluated using generally accepted causality criteria, such as those of Bradford Hill (1965) (Section 3.2.5). Often a significant uncertainty exists about identifying a substance unequivocally as being carcinogenic, because of inadequate reporting of exposure data and because potential errors such as chance, bias, and confounding factors can frequently not be ruled out. A series of studies revealing similar excesses of the same tumor type, even if not statistically significant, may suggest a positive association, and an appropriate joint evaluation (meta-analysis) may be used in order to increase the sensitivity, provided that the studies are sufficiently similar for such an evaluation. When the results of different studies are inconsistent, possible explanations should be sought and the various studies judged on the basis of the methods employed.

When several epidemiological studies show little or no indication of an association between an exposure and cancer, a judgment may be made that they show evidence of lack of carcinogenicity. Neither individual studies nor the pooled results of all the studies should show any consistent tendency that the relative risk of cancer increases with increasing level of exposure. It is important to note that evidence of lack of carcinogenicity obtained from several epidemiological studies can apply only to the type(s) of cancer studied, to the dose levels reported, and to the intervals between first exposure and disease onset observed in these studies. Experience with human cancer indicates that the period from first exposure to the development of clinical cancer is sometimes longer than 20 years; latent periods substantially shorter than 30 years cannot provide evidence for lack of carcinogenicity.

When human data of sufficient quality are available, they are preferable to animal data as no interspecies extrapolation is necessary and exposure scenarios are likely to be more realistic. However, the relatively low sensitivity of epidemiological studies implies that it is very difficult to demonstrate the noncarcinogenicity of a substance, unless exposure conditions are exceptional and well documented. Negative human data cannot be used to override positive findings in animals, unless it has been demonstrated that the mode of action of a certain toxic response observed in animals is not relevant for humans, see Section 4.9.6. In such a case, a full justification is required.

4.9.5.2 Animal Data

4.9.5.2.1 Information To Be Obtained from Carcinogenicity Studies
In the absence of adequate human data, animal carcinogenicity tests may be used to differentiate carcinogens from noncarcinogens.

All known human carcinogens that have been studied adequately for carcinogenicity in experimental animals have produced positive results in one or more animal species (IARC 2006). IARC (2006) also states that although this association cannot establish that all substances that cause cancer in experimental animals also cause cancer in humans, it is biologically plausible that substances for which there is sufficient evidence of carcinogenicity in experimental animals also present a carcinogenic hazard to humans.

The primary focus on the carcinogenicity studies is to provide information on neoplastic lesions. An assessment of carcinogenicity involves several considerations of qualitative importance, including (1) the experimental conditions under which the test was performed, including route, schedule and duration of exposure, species, strain (including genetic background where applicable), sex, age and duration of follow-up; (2) the consistency of the results, e.g., across species and target organ(s); (3) the spectrum of neoplastic response, from pre-neoplastic lesions and benign tumors to malignant neoplasms; and (4) the possible role of modifying factors.

For a negative test result to be acceptable, survival of all groups is no less than 50% at 18 months for mice and at 24 months for rats (OECD TG 451, EU Annex B.32), or below 25% (US-EPA 870.4200).

When data are available from several different studies, all of which are assessed as being of an adequate quality, the results should be analyzed for their consistency. It is seldom problematic to reach a conclusion about the carcinogenic potential of a substance, where there are consistent results from a number of studies, particularly if the studies were conducted in more than one species, or where there is a treatment-related incidence of malignant tumors in a single study.

However, the results from the experimental studies may not unequivocally demonstrate the carcinogenic potential of a substance under consideration. For instance, there may be an increase in the incidence of benign tumors only, or of tumors which have a high spontaneous (background) incidence. Although such an outcome is less convincing than an increase in malignant and rare tumors, a detailed and substantiated rationale should be given before such positive findings can be dismissed as not relevant.

If a single adequate study demonstrates no carcinogenic effects, expert judgment is needed to decide on whether a second study is needed to further support the noncarcinogenicity of the substance, based on all available data in addition to the carcinogenicity study.

Positive carcinogenic findings in animals require careful evaluation to determine their relevance to humans. Of key importance is the mode of action of tumor induction. The WHO/IPCS has developed a conceptual framework to provide a structured and transparent approach for the assessment of the overall weight of evidence for a postulated mode of induction for each tumor type observed (Sonich-Mullin et al. 2001). The framework promotes confidence in the conclusions reached by the use of a defined procedure, which mandates clear and consistent documentation of the reasoning used and inconsistencies and uncertainties in the available data.

In general, tumors induced by a genotoxic mechanism are considered to be relevant to humans even when observed in tissues with no direct human equivalent. Tumors induced by a non-genotoxic mode of action are, in principle, also considered relevant to humans. However, there is a scientific consensus that some tumors seen in rodents arising by specific non-genotoxic mechanisms are not relevant for humans. This consensus exists for some mode of actions of tumor formation, for instance specific types of rodent kidney, thyroid, urinary bladder, forestomach and glandular stomach tumors. For some of these mechanisms, the IARC has provided detailed characterization and has identified the key biochemical and histopathological events, which should be observed in order to conclude that the tumors arose via one of these mechanisms and can therefore be dismissed as not relevant for humans (IARC 1999a,b). This issue is addressed further in Section 4.9.6.

4.9.5.2.2 Information To Be Obtained from Other Studies

Experimental data not directly detecting carcinogenicity as an endpoint may also be informative about the potential of a substance to induce cancer.

Genotoxicity studies may provide information on whether or not the substance is likely to be a genotoxic carcinogen. Also, positive results in cell transformation or intercellular gap junction communication tests should be taken as alerts for potential carcinogenicity.

Repeated dose toxicity studies may indicate that the substance is able to induce hyperplasia, either through such mechanisms as cytotoxicity and mitogenicity, or interference with cellular control mechanisms, and/or pre-neoplastic lesions giving cause for concern for potential carcinogenicity by non-genotoxic mechanisms, i.e., depending on the outcome of genotoxicity tests. Repeated dose toxicity studies may also indicate a strong immunosuppressive activity of a substance, a condition favoring tumor development under conditions of chronic exposure.

Toxicokinetic data may reveal the generation of metabolites with structural alerts for genotoxicity and/or carcinogenicity (based on clear evidence for carcinogenicity of structural analogues), and their possible species-specificity. It may also give important information as to the relevance of carcinogenicity and related data on one species to another, based upon differences in absorption, distribution, metabolism, and/or excretion of the substance either directly or by the application of toxicokinetic modeling.

4.9.5.3 *In Vitro* Data

Currently, there are no validated and regulatory accepted *in vitro* methods for assessing carcinogenicity.

4.9.5.4 Other Data

The chemical structure of a substance may contain structural alerts for genotoxicity and/or carcinogenicity, based on clear evidence for carcinogenicity of structural analogues.

Use of (Q)SAR may be valuable in the evaluation of the carcinogenicity of a substance for which no data are available, see Sections 3.5.2 and 3.5.3.

4.9.6 EFFECTS IN EXPERIMENTAL ANIMALS OF DISPUTED RELEVANCE FOR HUMANS

A basic assumption for toxicological risk assessment is that effects observed in laboratory animals are relevant for humans, i.e., would also be expressed in humans. This is also assumed for tumor development, although the type and site of tumor will not always be the same in animals and humans.

In general, tumors induced by a genotoxic mechanism (non-threshold) are considered to be relevant for humans even when observed in tissues with no direct human equivalent, e.g., the Zymbal's gland of rats.

Tumors induced by a non-genotoxic mode of action (epigenetic, threshold) are in principle also considered relevant to humans. Non-genotoxic mechanisms or mode(s) of action include chronic cell damage, immune suppression, increased secretion of trophic hormones, and receptor activation. In addition, other mechanisms, such as CYP450 induction, may cause tumor development. Such non-genotoxic mechanisms or mode(s) of action can be relevant for humans as well as for laboratory animals. However, there is a scientific consensus that some tumors seen in rodents arising by specific non-genotoxic mechanisms or mode(s) of action are not relevant for humans.

As a general rule, tumor morphology in rodents is similar for tumors at a given site, irrespective of the nature of the inducing agent. Thus, agents that may be acting by fundamentally different carcinogenic mechanisms may not be distinguishable by histopathology alone (IARC 1999a). For a given chemical-induced tumor in animals, the developmental history must be considered before it can be determined whether the tumor is relevant for human risk assessment or not.

The WHO/IPCS has developed a conceptual framework to provide a structured and transparent approach for the assessment of the overall weight of evidence for a postulated mode of induction for each tumor type observed (Sonich-Mullin et al. 2001). The framework promotes confidence in the conclusions reached by the use of a defined procedure, which mandates clear and consistent documentation of the reasoning used and inconsistencies and uncertainties in the available data.

In this section, a number of tumor types and mechanisms/mode(s) of action, which are believed to be of limited relevance for humans, are addressed. It must be stressed that it is important that each individual case must be thoroughly evaluated, and to take into consideration that the response of various laboratory species and strains may vary greatly and may change considerably over time.

4.9.6.1 Leukemia (Mononuclear Cell Type) in the Fischer Rat

Mononuclear Cell Leukemia (MNCL) is unique to the rat, and is only common in the F-344 (common name: Fischer rat) inbred rat strain, which is the strain used by the U.S. National Toxicology Program (NTP). Elevated incidences of MNCL have been observed in a number of chronic bioassays in the F-344 rat. The frequency differs between males and females, with an incidence in males around 50%, and an incidence in females around 30%, with a large variation from study to study. It has been shown for some genotoxic carcinogens that exposure does not lead to an increase in MNCL in the F-344 rat, while a number of substances, which are believed to be noncarcinogens, do cause an increase. Based on this, an increased incidence of MNCL in studies with F-344 rats is considered of limited relevance for humans (Caldwell 1999).

4.9.6.2 Kidney Tumors in Male Rats

Certain kidney tumors occurring in male rats are considered not relevant for humans, because there is strong evidence that the underlying mode of action for tumor induction is specific to the male rat. These male rat–specific tumors occur as a result of accumulation of a single, major male rat–specific protein, $\alpha_{2\mu}$-globulin in phagolysosomes of renal proximal tubule cells. This protein is synthesized in the liver in male rats, while female rats and other animal species, including humans produce no or only very low levels of $\alpha_{2\mu}$-globulin. Lysosomal accumulation of $\alpha_{2\mu}$-globulin leads to death of individual renal cells and compensatory cell proliferation, which may result in atypical tubule hyperplasia and ultimately renal tubule tumors. Mechanistic studies have demonstrated that the requisite step in the development of the syndrome is the ability of a chemical (or metabolite(s)) to bind reversibly, and specifically, to $\alpha_{2\mu}$-globulin. Binding of chemicals to $\alpha_{2\mu}$-globulin appears to alter the lysosomal degradation of the protein, leading to its accumulation in phagolysosomes. There is no evidence that a mechanism similar to $\alpha_{2\mu}$-globulin nephropathy in male rats occurs in humans. In case the available data are sufficient to show that $\alpha_{2\mu}$-globulin nephropathy is the underlying mechanism for formation of renal tumors in an animal study, the tumor development is regarded to be of limited relevance for humans. The IARC has formulated a list of criteria (all of which must be met) for concluding that an agent causes kidney tumors through an $\alpha_{2\mu}$-globulin-associated response and thus are of limited relevance for humans (IARC 1999a).

4.9.6.3 Liver Tumors in Mice and Rats

Liver tumors are a common finding in rodent carcinogenicity studies. In mice, the background occurrence of liver tumors is very high. The historical control data reported by U.S. NTP (US-NTP 2006) shows an incidence in hepatocellular adenomas, carcinomas, or hepatoblastoma in B6C3F1mice in oral feeding studies of 52.1% in males and 30.7% in females. The occurrence of liver tumors in mice is strongly influenced by a number of factors such as stress and nutrition. An increased incidence of liver tumors in a mouse study may therefore very well be caused by other factors than the test substance. In addition, mice are very prone to develop liver tumors following long-term stimulation of cell proliferation, or long-term enzyme induction caused by the high exposure levels of the test substance. This mode of action is not relevant for humans, where the exposure levels usually are much lower than the doses used in laboratory animal studies. A mouse study, in which the only increased tumor type is liver tumors, is often regarded of limited relevance to humans (Carmichael et al. 1997, Williams 1997).

A number of substances, e.g., hypolipidemic fibrates and certain phthalates, have been shown to induce peroxisome proliferation in the rodent liver. There is a strong concordance between this effect and development of liver cancer in rats and mice. There is evidence that the peroxisome proliferators induce cancer via a non-genotoxic, receptor-mediated mechanism, the peroxisome proliferator-activated receptor alpha (PPARα). Both oxidative stress as a consequence of peroxisome proliferation and preferential growth of pre-neoplastic lesions following hepatocyte proliferation have been proposed as underlying processes in the neoplastic development. Humans also possess the PPARα; however, this receptor is expressed at a much lower level than in rodents and peroxisome proliferation does not seem to occur in human liver to any significant extent. Therefore, tumors, which are associated with peroxisome proliferation apparently represents little, if any, human carcinogenic hazard. However, the mechanism for this type of tumor development is not fully known, and the relation between peroxisome proliferation and the development of liver tumors is not unequivocal. In order to reject the relevance of liver tumors associated with peroxisome proliferation, a large data material is necessary (IARC 1994, Williams 1997).

4.9.6.4 Leydig Cell Tumors in Rats

Leydig cells are located in the testis and their primary function is the production of the male sex hormone, testosterone. In cancer studies with rats, Leydig cell tumors (most often adenoma, rarely carcinoma) are a frequent finding in older animals. There is a large variation in the background incidence among rat strains, the Sprague-Dawley rat has a background incidence of 1%–5%, while the Fischer rat almost has a 100% incidence. The Fischer rat is therefore regarded as having an abnormally high tendency to develop Leydig cell tumors. In mice, the background incidence is generally low. The frequency of this tumor type is very low in man, where only 2% of all testis tumors are of Leydig cell origin.

There are plausible mechanisms (non-genotoxic) for the chemical induction of Leydig cell tumors; most of these ultimately involve increased concentration of serum LH (luteinising hormone) and associated stimulation of Leydig cells to growth and proliferation, or an increased sensitivity to LH in the Leydig cells. Regarding human relevance, the pathways for regulation of the hypothalamo–pituitary–testis (HPT) axis (see Figure 4.10) of rats and humans are similar and the mechanism is relevant for humans. Hence, chemicals that induce Leydig cell tumors in rats by disruption of the HPT axis pose a cancer risk to humans. Therefore, the central issue becomes what is the relative sensitivity between rat and human Leydig cells in their response to increased LH levels. There is evidence suggesting that human Leydig cells are quantitatively less sensitive than rats in their proliferative response to LH, and hence in their sensitivity to chemically induced Leydig cell tumors. Men suffering from endocrine diseases, which cause a constantly increased LH level, develop Leydig cell adenomas with a frequency of 2%–3%. In comparison, the incidence of Leydig cell tumors in rats receiving the androgen receptor antagonist flutamide, which causes increased LH secretion via the hypothalamo–gonadal axis (see Figure 4.10), is almost 100%. Furthermore, a genetic defect in men, which causes a constant LH receptor activation, does not lead to an increased incidence in Leydig cell tumors. In addition, several epidemiological studies are available on a number of substances, which induce Leydig cell tumors in rats (1,3-butadiene, cadmium, ethanol, lactose, lead, nicotine), that demonstrate no association between human exposure to these compounds and induction of Leydig cell hyperplasia or adenomas. The occurrence of Leydig cell tumors in a rat study may therefore have limited relevance for man (Cook et al. 1999).

4.9.6.5 Thyroid Tumors in the Rat

Spontaneous thyroid tumors derived from the thyroid follicular cell occur in 1%–3% of laboratory rats (adenomas and carcinomas combined in a variety of strains of rats aged 2 or more years).

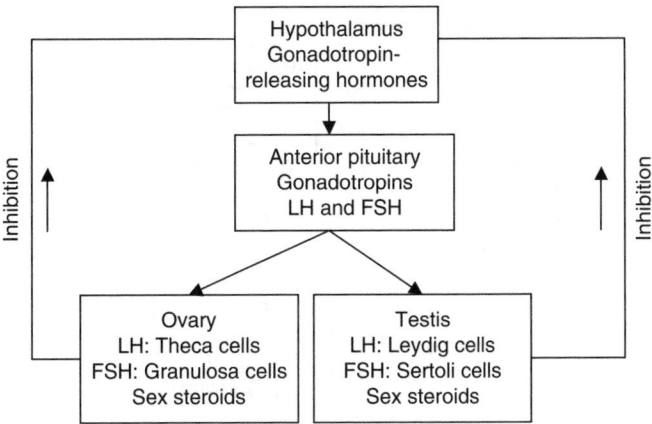

FIGURE 4.10 Negative feedback loop regulating the function of the gonads.

In general, male rats show a higher incidence of follicular-cell tumors than female rats. Both genotoxic and non-genotoxic agents have been shown to induce thyroid follicular-cell tumors. Non-genotoxic agents can be divided into those which have effects directly on the thyroid (blocking uptake of iodine into the follicular cell, e.g., perchlorate; inhibiting thyroid peroxidase, e.g., thioureas; or inhibiting hormone release, e.g., lithium) and those which have effects on thyroid hormone catabolism and excretion (e.g., agents such as lupiditine which increase uptake into the hepatic cell, or those such as phenobarbital which increase thyroid hormone loss from the liver through enzyme induction). The only known common pathway through which these agents act is the pituitary-thyroid feedback mechanism involving the Thyroid Stimulating Hormone (TSH) (IARC 1999a).

TSH, also known as thyrotropin, is secreted from cells in the anterior pituitary called thyrotrophs, finds its receptors on epithelial cells in the thyroid gland, and stimulates the gland to synthesize and release thyroid hormones. The most important controller of TSH secretion is the Thyroid-Releasing Hormone (TRH). The TRH is secreted by hypothalamic neurons into hypothalamic–hypophyseal portal blood, finds its receptors on thyrotrophs in the anterior pituitary, and stimulates the secretion of TSH. Secretion of TRH, and hence, TSH, is inhibited by high blood levels of thyroid hormones in a classical negative feedback loop. See also Figure 4.11.

There are several ways by which TSH secretion can be increased. An increased hepatic enzyme activity may cause an increased metabolism of thyroid hormones, leading to lower serum hormone levels, which in turn leads to increased secretion of TRH, and subsequently increased TSH secretion. Regarding human relevance, the pathways for regulation of the hypothalamo–pituitary–thyroid axis of rats and humans are similar and the mechanism is relevant for humans, but the human system is far more resistant to perturbation.

There are several species differences in thyroid physiology. Thyroxine-Binding Globulin (TBG) is the predominant plasma protein in humans that binds and transports thyroid hormone in the blood. The lack of TBG in the adult rat is one important difference. Major differences are also present in the half-life of the thyroid hormone, thyroxine (12 h in the rat versus 5–9 days in humans), and in the serum level of TSH, which is 25 or more times higher in the rodent than in humans. The rat also exhibits enhanced thyroid hormone elimination. Thus, both the physiological parameters indicate that the rodent thyroid gland is more active and operates at a higher level with respect to thyroid hormone turnover as compared to the human gland. The weight of the evidence suggests that rodents are more sensitive than humans to thyroid tumor induction due to hormonal imbalances that cause elevated TSH levels. Therefore, the relevance of rat thyroid tumors for humans is considered

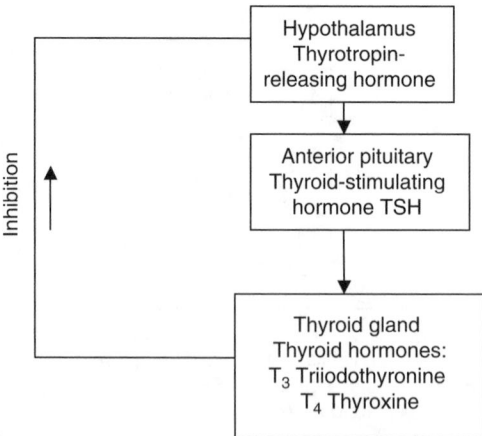

FIGURE 4.11 Negative feedback loop regulating the function of the thyroid gland.

to be limited if the chemical is non-genotoxic, and it has been demonstrated that the mechanism is a disturbance of the hormonal balance that causes increased TSH levels (IARC 1999a).

4.9.6.6 Urinary Bladder Tumors in the Rat and Mouse

Many non-genotoxic chemicals have been shown to induce formation of microcrystals, amorphous precipitates, and/or calculi in the urinary bladder of mice and rats. This is associated with irritation and may cause increased cell proliferation and subsequent tumor formation in the urothelium lining the urinary bladder. Urinary bladder calculi also cause irritation and cell proliferation in humans and there is some epidemiological evidence that urinary tract cancer in humans is associated with a history of calculi in the bladder. The proliferative and carcinogenic effects observed in the bladder of rodents require that the concentration of the chemical in the urine is sufficiently high to lead to precipitate formation and ultimately to calculi, a situation that will normally not occur in humans. Thus, although there are quantitative differences in the carcinogenic response to calculi between rodents and humans, the effect is not species-specific. Tumors in rats and mice related to the formation of urinary bladder calculi are thus relevant for humans, but will not develop if the exposure of humans is below the threshold concentration for precipitate formation (IARC 1999a).

In rats, ingestion of high doses of several sodium salts of moderate to strong organic acids produces a calcium phosphate-containing precipitate in the urine and increased urothelial proliferation. Increased bladder tumor incidences are seen when these sodium salts are administered after treatment with bladder carcinogens. Formation of this precipitate appears to require high urinary concentrations of protein and a high osmolality of the urine, as found in rats. Male rats are more susceptible than female rats, presumably because of the higher concentration of protein in their urine. Healthy humans have very low concentrations of urinary protein and much lower urinary osmolalities than rodents. In humans with renal diseases, the urinary protein concentrations can be nearly as high as those found in rats; however, the osmolality does not increase above normal levels in humans. Thus, the data suggest that the carcinogenic response in the urinary bladder of rats generated by these sodium salts is a species- and dose-specific phenomenon that does not occur in humans (IARC 1999a).

4.9.6.7 Forestomach Tumors in Mice and Rats

The stomach of rats and mice is anatomically different from the human stomach. A part of the stomach in rodents is devoid of glands; this part is called the forestomach. The squamous epithelium in the non-glandular rodent forestomach resembles the epithelium covering the esophagus. Therefore, the human esophagus must be regarded as similar with respect to the epithelium of the non-glandular rodent forestomach.

A considerable number of compounds induce forestomach tumors in different species. Naturally occurring forestomach tumors are rare (an incidence of less than 1%) in rats and mice, but they do occur in hamsters with incidences up to 12%. Most forestomach carcinogens are known to be genotoxic, although a few non-genotoxic substances are also able to induce this type of tumor. A possible mechanism of action of non-genotoxic forestomach carcinogens is related to irritation of the epithelium and hyperplasia caused by direct contact to the chemical substance. Since humans have no squamous epithelium in the stomach, the potential risk from non-genotoxic forestomach carcinogens involves exposure of the mouth, pharynx, and esophagus at dose levels that exert irritating action. Since the time of epithelial contact is much shorter in the human esophagus compared to the rodent forestomach, it seems very unlikely that exposure to concentrations far below those having an irritating potential is hazardous to man. In cases, where this specific mode of action can be shown to be the cause of forestomach tumors in rodent studies, the animal findings are evaluated as not being relevant for humans (Kroes and Wester 1986).

4.9.6.8   Other Types of Tumors

In addition to the above-mentioned examples of tumors with disputed relevance to humans, mammary tumors in mice have been regarded as being of uncertain relevance for humans. The incidence of mammary tumors in mice has been shown to be highly influenced by environmental factors such as stress (Riley 1975), which complicates the interpretation of increased mammary tumor incidence in a carcinogenicity study.

Phaeochromocytoma (a tumor in the adrenal medulla) is not uncommon in rats, but rare in humans. Pheochromocytomas are induced in rats by a variety of non-genotoxic substances that may act indirectly by stimulating chromaffin cell proliferation. They are not known to be similarly inducible in other species. In the rat, a mechanism for the development may be hypercalcaemia (Tischler et al. 1999; Capen et al., in Haschek et al. 2001).

Chemical substances, which are injected into laboratory animals, may cause tumors locally at the injection site, e.g., in the muscle, irrespective of the mechanism for the tumor formation. It is now well established that smooth-surfaced foreign bodies, regardless of their chemical composition, will produce sarcomas when transplanted subcutaneously into rodents (Moore 1991). It is difficult to evaluate the relevance for humans of such site-related tumor formation; this issue has been further addressed by IARC (1999b).

4.9.7   CATEGORIZATION FOR CARCINOGENICITY

Carcinogenic substances may be classified in various ways for regulatory purposes. The most common classification systems use the strength of the evidence for carcinogenicity arising from human and experimental animal data to classify substances, and standard terms for the various classes are used. These categories refer only to the strength of the evidence that a substance or exposure scenario is carcinogenic and not to the potency of the carcinogenic activity, i.e., the categories are evidence-based, not potency-based.

4.9.7.1   IARC

The preamble to the IARC monographs (IARC 2006) describes the evaluation and rationale for classification of a substance into one of four defined categories, called "groups":

Group 1: The agent is carcinogenic to humans.
This category is used when there is sufficient evidence of carcinogenicity in humans. Exceptionally, an agent may be placed in this category when evidence of carcinogenicity in humans is less than sufficient but there is sufficient evidence of carcinogenicity in experimental animals and strong evidence in exposed humans that the agent acts through a relevant mechanism of carcinogenicity.

Group 2:
This category includes agents for which, at one extreme, the degree of evidence of carcinogenicity in humans is almost sufficient, as well as those for which, at the other extreme, there are no human data but for which there is evidence of carcinogenicity in experimental animals. Agents are assigned to either Group 2A or Group 2B on the basis of epidemiological and experimental evidence of carcinogenicity and mechanistic and other relevant data. The terms "probably carcinogenic" and "possibly carcinogenic" have no quantitative significance and are used simply as descriptors of different levels of evidence of human carcinogenicity, with probably carcinogenic signifying a higher level of evidence than possibly carcinogenic.

Group 2A: The agent is probably carcinogenic to humans.
This category is used when there is limited evidence of carcinogenicity in humans and sufficient evidence of carcinogenicity in experimental animals. In some cases, an agent may be classified in this category when there is inadequate evidence of carcinogenicity in humans and sufficient evidence of carcinogenicity in experimental animals and strong evidence that the carcinogenesis

is mediated by a mechanism that also operates in humans. Exceptionally, an agent may be classified in this category solely on the basis of limited evidence of carcinogenicity in humans. An agent may be assigned to this category if it clearly belongs, based on mechanistic considerations, to a class of agents for which one or more members have been classified in Group 1 or Group 2A.

Group 2B: The agent is possibly carcinogenic to humans.

This category is used for agents for which there is limited evidence of carcinogenicity in humans and less than sufficient evidence of carcinogenicity in experimental animals. It may also be used when there is inadequate evidence of carcinogenicity in humans but there is sufficient evidence of carcinogenicity in experimental animals. In some instances, an agent for which there is inadequate evidence of carcinogenicity in humans and less than sufficient evidence of carcinogenicity in experimental animals together with supporting evidence from mechanistic and other relevant data may be placed in this group. An agent may be classified in this category solely on the basis of strong evidence from mechanistic and other relevant data.

Group 3: The agent is not classifiable as to its carcinogenicity to humans.

This category is used most commonly for agents for which the evidence of carcinogenicity is inadequate in humans and inadequate or limited in experimental animals.

Exceptionally, agents for which the evidence of carcinogenicity is inadequate in humans but sufficient in experimental animals may be placed in this category when there is strong evidence that the mechanism of carcinogenicity in experimental animals does not operate in humans.

Agents that do not fall into any other group are also placed in this category.

An evaluation in Group 3 is not a determination of noncarcinogenicity or overall safety. It often means that further research is needed, especially when exposures are widespread or the cancer data are consistent with differing interpretations.

Group 4: The agent is probably not carcinogenic to humans.

This category is used for agents for which there is evidence suggesting lack of carcinogenicity in humans and in experimental animals. In some instances, agents for which there is inadequate evidence of carcinogenicity in humans but evidence suggesting lack of carcinogenicity in experimental animals, consistently and strongly supported by a broad range of mechanistic and other relevant data, may be classified in this group.

4.9.7.2 US-EPA

A classification scheme to describe the nature of the cancer database and evidence supporting the carcinogenicity of an agent was provided in the US-EPA's 1986 Cancer Guidelines (US-EPA 1986b). This classification system, which is based on a similar scheme developed by the IARC, describes the evaluation and rationale for classification of a substance into one of five defined categories, called "groups" by using an alphanumerical classification system. The 1986 classification system has been replaced by a new approach since 2005, see below.

- Group A: Human carcinogen; sufficient evidence from epidemiological studies
- Group B: Probable human carcinogen; sufficient evidence in animals or limited evidence in humans
- Group C: Possible human carcinogen; limited evidence of carcinogenicity in animals in the absence of adequate human data
- Group D: Not classifiable; inadequate data or no data
- Group E: No evidence of carcinogenicity in adequate studies in at least two species or in both epidemiological and animal studies

Within Group B, there are two subgroups: B1 and B2. According to the 1986 Cancer Guidelines, Group B1 is usually reserved for agents for which there is limited evidence of carcinogenicity from epidemiological studies. It is reasonable, for practical purposes, to regard an agent for which there is sufficient evidence of carcinogenicity in animals as if it presented a carcinogenic risk to humans. Therefore, agents for which there is "sufficient" evidence from animal studies and for which there is "inadequate evidence" or "no data" from epidemiological studies would usually be categorized under Group B2 (US-EPA 1986b).

In the revised Cancer Guidelines (US-EPA 2005), the use of a "weight of evidence narrative" that includes both a conclusion about the weight of evidence of carcinogenic potential and a summary of the data on which the conclusion rests *in toto* replaces the alphanumerical classification system used in US-EPA's 1986 Cancer Guidelines (US-EPA 1986b).

In order to provide some measure of clarity and consistency in an otherwise free-form, narrative characterization, standard descriptors are used as part of the hazard narrative to express the conclusion regarding the weight of evidence for carcinogenic hazard potential. There are five recommended standard hazard descriptors:

- Carcinogenic to humans
- Likely to be carcinogenic to humans
- Suggestive evidence of carcinogenic potential
- Inadequate information to assess carcinogenic potential
- Not likely to be carcinogenic to humans

Each standard descriptor may be applicable to a wide variety of data sets and weights of evidence and is presented only in the context of a weight of evidence narrative. Furthermore, more than one conclusion may be reached for an agent (US-EPA 2005).

4.9.7.3   EU

According to the classification criteria provided in Annex VI of Directive 67/548/EEC (EU 2001), carcinogenic substances are divided into three categories:

Category 1: Substances known to be carcinogenic to man.

There is sufficient evidence to establish a causal association between human exposure to a substance and the development of cancer.

Category 2: Substances, which should be regarded as if they are carcinogenic to man.

There is sufficient evidence to provide a strong presumption that human exposure to a substance may result in the development of cancer, generally on the basis of appropriate long-term animal studies, other relevant information.

Category 3: Substances, which cause concern for man.

Owing to possible carcinogenic effects but in respect of which the available information is not adequate for making a satisfactory assessment. There is some evidence from appropriate animal studies, but this is insufficient to place the substance in Category 2.

Category 3 actually comprises two subcategories: (a) Substances which are well investigated but for which the evidence of a tumor-inducing effect is insufficient for classification in Category 2. Additional experiments would not be expected to yield further relevant information with respect to classification. (b) Substances which are insufficiently investigated. The available data are inadequate, but they raise concern for man. This classification is provisional; further experiments are necessary before a final decision can be made.

4.9.8   The Carcinogenic Potency Database

The Carcinogenic Potency Database (CPDB) is a unique and widely used international database containing results from 6153 chronic, long-term animal cancer studies on 1485 chemicals. CPDB provides a standardized and easily accessible database with qualitative and quantitative analyses of both positive and negative studies that have been published in the general literature through 1997, and by the U.S. National Cancer Institute/U.S. National Toxicology Program (Section 2.3.3) through 1998 (CPDB 2007).

For each study, information is included on species, strain, and sex of test animal; features of the experimental protocol such as route of administration, duration of dosing, dose level(s) in mg/kg body weight/day, and duration of experiment; target organ, tumor type, and tumor incidence; carcinogenic potency (TD_{50}, see Section 6.2.2) and its statistical significance; shape of the dose–response, author's opinion as to carcinogenicity, and literature citation.

4.10 REPRODUCTIVE TOXICITY

Normal human reproduction is regulated by a finely tuned system of coordinated signals that direct the activity of multiple interdependent target cells, leading to the formation of gametes, their transport, release, fertilization, implantation and gestation, and, ultimately, the development of offspring that is eventually capable of successfully repeating the entire process under similar or different environmental conditions. Sexual function and fertility reflect a wide variety of functions that are necessary for reproduction and may be affected by exposure to environmental factors. Any disturbance in the integrity of the reproductive system may affect these functions. Patterns of reported infertility vary around the world, but approximately 10% of all couples experience infertility at some time during their reproductive years (WHO/IPCS 2001b).

Throughout the entire life cycle, all aspects of reproductive function are dependent on various endocrine communicating systems that employ a wide variety of protein/peptide and steroid hormones, growth factors and other signaling molecules that affect target cell gene expression and/or protein synthesis. In particular, development and gametogenesis are regulated by a myriad of signals delivered in appropriate strength at precisely defined times. Although recent animal studies demonstrate that the developing fetus may be more sensitive to the effects of exposure to environmental chemicals than the adult system, effects may not be manifest until adulthood. Further characterization of the molecular mechanisms regulating the various aspects of normal reproduction and development is critical to our understanding of the variety of mechanisms through which exogenous chemicals may disrupt normal reproduction and development (WHO/IPCS 2001b).

Developmental toxicity, defined in its widest sense to include any adverse effect on normal development either before or after birth, has become of increasing concern in recent years. Developmental toxicity can result from exposure of either parent prior to conception, from exposure of the embryo or fetus in utero or from exposure of the progeny after birth. Adverse developmental effects may be detected at any point in the life span of the organism. In addition to structural abnormalities, examples of manifestations of developmental toxicity include fetal loss, altered growth, functional defects, latent onset of adult disease, early reproductive senescence, and shortened life span (WHO/IPCS 2001b).

The term "reproductive toxicity" is used to describe the adverse effects induced by a substance on any aspect of mammalian reproduction and covers all phases of the reproductive cycle, including impairment of male or female reproductive function or capacity and the induction of nonheritable adverse effects in the progeny such as death, growth retardation, structural and functional effects (EC 2003).

4.10.1 DEFINITIONS

The term reproductive toxicity is sometimes used exclusively to describe toxic effects on male and female sexual function and fertility. More commonly, and in this book, reproductive effects are considered to include adverse effects on sexual function and fertility in males and females as well as developmental toxicity.

Human sexual function and fertility disorders include, e.g., spontaneous abortions, impaired spermatogenesis, menstrual disorders, impotence, and early menopause.

Developmental toxicity is defined by the WHO as "any effect interfering with normal development both before and after birth resulting from exposure of either parent prior to conception, exposure during prenatal development or exposure postnatally to the time of sexual maturation" (WHO/IPCS 2001b). Effects can include birth defects, low birth weight, biological dysfunctions, or psychological or behavioral deficits that become manifest as the child grows. Maternal exposure to toxic chemicals during pregnancy can disrupt the development or even cause the death of the fetus. Death of a fetus may result in abortion. Developmental toxicity can also result from paternal exposures. Early postnatal contact with toxicants can also affect normal development.

US-EPA, in its Guidelines for Reproductive Toxicity Risk Assessment (US-EPA 1996), defines reproductive toxicity as "The occurrence of biologically adverse effects on the reproductive systems of females or males that may result from exposure to environmental agents. The toxicity may be expressed as alterations to the female or male reproductive organs, the related endocrine system, or pregnancy outcomes. The manifestation of such toxicity may include, but not be limited to, adverse effects on onset of puberty, gamete production and transport, reproductive cycle normality, sexual behavior, fertility, gestation, parturition, lactation, developmental toxicity, premature reproductive senescence, or modifications in other functions that are dependent on the integrity of the reproductive Systems."

EU, in Annex VI to Directive 67/548 (EU 2001) defines reproductive toxicity as including "impairment of male and female reproductive functions or capacity and the induction of non-inheritable harmful effects on the progeny." The definition of fertility effects is similar to that of the US-EPA, while more detail is provided on developmental toxicity, which "is taken in its widest sense to include any effect interfering with normal development, both before and after birth. It includes effects induced or manifested prenatally as well as those manifested postnatally. This includes embryotoxic/foetotoxic effects such as reduced body weight, growth and developmental retardation, organ toxicity, death, abortion, structural defects (teratogenic effects), functional defects, peri-postnatal defects, and impaired postnatal mental or physical development up to and including normal pubertal development."

4.10.2 Objectives for Assessing the Reproductive Toxicity of a Substance

The general objectives of the testing are to establish:

- Whether exposure of humans to the substance has been associated with adverse effects on reproductive function or capacity
- Whether, in studies in animals, administration of the substance to males and/or females prior to conception and during pregnancy and lactation, causes adverse effects on reproductive function or capacity
- Whether, in studies in animals, administration of the substance during the period of pre- or postnatal development induces nonheritable adverse effects in the progeny
- Whether the pregnant female is potentially more susceptible to general toxicity
- The dose–response relationship for any adverse effects on reproduction

Substance-related adverse effects on reproduction are always of potential concern, but it is important, where possible, to distinguish between a specific effect on reproduction as a consequence of an intrinsic property of the substance and an adverse reproductive effect, which is a nonspecific consequence of general toxicity (e.g., reduced food or water intake, maternal stress). Hence, reproductive toxicity should be assessed alongside parental toxicity in the same study (EC 2003).

However, the presence of concurrent maternal toxicity does not automatically mean that effects in the offspring have no significance. Mother and fetus could very well be equally sensitive to the toxicity of the substance, and effects in the offspring should not be considered to be secondary to maternal effects unless there is justification for this assumption.

4.10.3 TEST GUIDELINES

4.10.3.1 *In Vivo* Tests

Specific reproductive toxicity test guidelines have been published by OECD, US-EPA, and EU, see Table 4.18. In the following text, the principles of the various OECD tests are briefly described. Table 4.19 summarizes the various OECD test guideline studies for reproductive toxicity in more detail, including the parameters examined in each test, in order to provide a brief overview of the similarities and differences between the various reproductive toxicity studies.

In the prenatal developmental toxicity study (OECD TG 414, US-EPA OPPTS 870.3700, EU Annex V B.31), the test substance is administered to pregnant animals at least from the day of implantation to 1 day prior to the day of scheduled sacrifice, which should be as close as possible to the normal day of delivery without risking loss of data resulting from early delivery. The test is not intended to examine solely the period of organogenesis (e.g., days 5–15 in the rodent and days 6–18 in the rabbit), but also effects from the period of preimplantation, when appropriate, through the entire period of gestation to the day before caesarean section. Shortly before caesarean section, the females are sacrificed (rodents will, as a natural behavior, consume abnormal pups at birth, and therefore, allowing dams to give natural birth means a great risk of missing abnormally developed pups), the uterine contents are examined, and the fetuses are evaluated for soft tissue and skeletal changes.

In the one-generation study (OECD TG 415, EU Annex V B.34, the test substance is administered in graduated doses to several groups of males and females (P generation). Males of the P generation should be dosed during growth and for at least one complete spermatogenic cycle. Females of the P generation should be dosed for at least two complete estrous cycles in order to elicit any adverse effects on estrus by the test substance. The animals are then mated. The test substance is administered to both sexes during the mating period and thereafter only to females during pregnancy and for the duration of the nursing period. Pregnancy and litter data are collected. Histopathology is performed on the sexual organs of the P animals if considered necessary.

TABLE 4.18
Test Guidelines Adopted for Reproductive Toxicity Testing

Title

OECD Test Guidelines	Year
414 Prenatal developmental toxicity study	2001
415 One-generation reproduction toxicity study	1983
416 Two-generation reproduction toxicity study	2001
421 Reproduction/developmental toxicity screening test	1995
422 Combined repeated dose toxicity study with the reproduction/developmental toxicity screening test	1996
426 Developmental neurotoxicity study, draft	2003

US-EPA OPPTS Harmonized Test Guidelines	
870.3550 Reproduction/developmental toxicity screening test	1999
870.3650 Combined repeated dose toxicity study with the reproduction/developmental toxicity screening test	1999
870.3700 Prenatal developmental toxicity study	1998
870.3800 Reproduction and fertility effects	1998
870.6300 Developmental neurotoxicity study	1998

EU Annex V Test Methods	
B.31 Teratogenicity test - Rodent and non-rodent	2004
B.34 One-generation reproduction toxicity test	1988
B.35 Two-generation reproduction toxicity test	2004

TABLE 4.19

Overview of *In Vivo* Reproductive Toxicity Test Guideline Studies

Test	Design	Endpoints (General Toxicity)
OECD TG 414 Prenatal developmental toxicity study (Teratology study)	Exposure at least from implantation to 1 or 2 days before expected birth At least three dose levels plus control At least 20 pregnant females per group	Clinical observations Body weight and food/water consumption Macroscopical examination of all dams for any structural abnormalities or pathological changes, which may have influenced the pregnancy Implantation, resorptions Foetal growth Morphological variations and malformations
OECD TG 415 One-generation reproduction toxicity study	Exposure before mating for at least one spermatogenic cycle until weaning of first generation At least three dose levels plus control At least 20 parental males and females per group	Clinical observations Body weight and food/water consumption Gross necropsy (all parental animals) Fertility Growth, development, and viability of offspring Organ weights (reproductive organs, brain, liver, kidneys, spleen, pituitary, thyroid, adrenal glands, and known target organs) Histopathology (reproductive organs, previously identified target organ(s) - at least control and high-dose groups)
OECD TG 416 Two-generation reproduction toxicity study	Exposure before mating for at least one spermatogenic cycle until weaning of second generation At least three dose levels plus control At least 20 parental males and females per group	Clinical observations Body weight and food/water consumption Fertility Estrus cyclicity and sperm quality Pregnancy outcome, e.g., dystocia Growth, development, and viability of offspring Anogenital distance if triggered Sexual maturation Gross necropsy (all parental animals) Organ weights (reproductive organs, brain, liver, kidneys, spleen, pituitary, thyroid, adrenal glands, and known target organs) Histopathology (reproductive organs, previously identified target organ(s) - at least control and high-dose groups)
OECD TG 421 Reproduction/ developmental toxicity screening test	Exposure from 2 weeks prior to mating until at least postnatal day 4 At least three dose levels plus control At least 8–10 parental males and females per group	Clinical observations Body weight and food/water consumption Fertility Pregnancy length and birth Fetal and pup growth and survival until day 4 Gross necropsy (adult animals, special attention to reproductive organs) Organ weights (all adult males: testes, epididymides) Histopathology (reproductive organs - at least control and high-dose groups)
OECD TG 422 Combined repeated dose toxicity study with the reproduction/ developmental toxicity screening test	Exposure for a minimum of 4 weeks (males) or from 2 weeks prior to mating until at least postnatal day 4 (females - at least 6 weeks of exposure At least three dose levels plus control At least 10 males and females per group	Clinical observations as in TG 407 Functional observations as in TG 407 Body weight and food/water consumption Haematology as in TG 407 Clinical biochemistry Urinalysis

TABLE 4.19 (continued)

Overview of *In Vivo* Reproductive Toxicity Test Guideline Studies

Test	Design	Endpoints (General Toxicity)
		Fertility
		Pregnancy length and birth
		Fetal and pup growth and survival until day 4
		Gross necropsy (full, detailed, all adult animals)
		Organ weights (testes and epididymides - all males; liver, kidneys, adrenals, thymus, spleen, brain, heart - in five animals of each sex per group, i.e., as in TG 407)
		Histopathology (ovaries, testes, epididymides, accessory sex organs, all gross lesions - all animals in at least control and high-dose groups; brain, spinal cord, stomach, small and large intestines, liver, kidneys, adrenals, spleen, heart, thymus, thyroid, trachea and lungs, urinary bladder, lymph nodes, peripheral nerve, a section of bone marrow - in five animals of each sex in at least control and high-dose groups, i.e., as in TG 407)
OECD TG 426 Developmental neurotoxicity study (Draft)	Exposure at least from implantation throughout lactation (PND 20) At least three dose levels plus control At least 20 pregnant females per group	Clinical observations Body weight and food/water consumption Birth and pregnancy length Growth, development, and viability of offspring Physical and functional maturation Behavioral changes due to CNS and PNS effects Brain weights and neuropathology

In the two-generation study (OECD TG 416, US-EPA OPPTS 870.3800, EU Annex V B.35), male and female rats (the P generation) give rise to the next generation (F1 generation), which is allowed to live unto adulthood and produce the next generation (F2 generation). The test substance is administered in graduated doses to several groups of males and females.

Males of the P generation should be dosed during growth and for at least one complete spermatogenic cycle. The P male sperm is examined. Females of the P generation should be dosed during growth and for several complete estrous cycles in order to detect any adverse effects on estrous cycle normality by the test substance. The test substance is administered to P animals during their mating, during the resulting pregnancies, and through the weaning of their F1 offspring. At weaning, the administration of the substance is continued to F1 offspring during their growth into adulthood, mating, and production of an F2 generation, until the F2 generation is weaned. Clinical observations and pathological examinations are performed on all animals for signs of toxicity with special emphasis on effects on the integrity and performance of the male and female reproductive systems, and on the growth and development of the offspring.

In the reproduction/developmental toxicity screening test (OECD TG 421, US-EPA OPPTS 870.3550), the test substance is administered in graduated doses to several groups of males and females. Males should be dosed for a minimum of 4 weeks and up to and including the day before scheduled sacrifice (this includes a minimum of 2 weeks prior to mating, during the mating period and, approximately, 2 weeks postmating). In view of the limited premating dosing period in males, fertility may not be a particularly sensitive indicator of testicular toxicity and therefore a detailed histological examination of the testes is essential. Females should be dosed throughout the study, i.e., 2 weeks prior to mating with the objective of covering at least two complete estrous cycles, the variable time to conception, the duration of pregnancy and at least 4 days after delivery, up to and including the day before scheduled sacrifice. During the period of administration, the animals are observed closely each day for signs of toxicity. Animals, which die or are killed during the test period are necropsied and, at the conclusion of the test, surviving animals are killed and necropsied.

The combined repeated dose toxicity study with the reproduction/developmental toxicity screening test (OECD TG 422, US-EPA OPPTS 870.3650) comprises a basic repeated dose toxicity study and a fertility/developmental toxicity screening test and, therefore, can be used to provide initial information on possible effects on a limited number of reproductive performance parameters. The test does not provide complete information on all aspects of reproduction, has a relatively short period of exposure, and does not provide evidence for definite claims of no reproductive effects, while positive results are useful for initial hazard assessment. Furthermore, results regarding repeated dose toxicity are influenced by the pregnant state of the female animals (see also Sections 4.7.3.1 and 4.7.5.2.2).

The developmental neurotoxicity test (OECD TG 426 Draft, US-EPA OPPTS 870.6300) can be conducted as a separate study, incorporated into a reproductive toxicity study and/or adult neuro-toxicity study, or added onto a prenatal developmental toxicity study.

The test substance is administered to animals during gestation and lactation. Dams are tested to assess effects in pregnant and lactating females and to provide comparative information (dams versus offspring). Offspring are randomly selected from within litters for neurotoxicity evaluation. The evaluation consists of observations to detect gross neurological and behavioral abnormalities, including the assessment of physical development, behavioral ontogeny, motor activity, motor and sensory function, and learning and memory, and the evaluation of brain weights and neuropathology during postnatal development and adulthood. When the test method is conducted as a separate study, additional available animals in each group could be used for specific neurobehavioral, neuropathological, neurochemical, or electrophysiological procedures as a supplement to the core battery. The supplemental procedures can be particularly useful when empirical observation, anticipated effects, or mechanism/mode of action indicate a specific type of neurotoxicity. These supplemental procedures may be used in the dams as well as in the pups.

4.10.3.2 *In Vitro* Tests

The OECD, US-EPA, and EU have no adopted test guidelines for *in vitro* testing of reproductive toxicity.

4.10.4 GUIDANCE DOCUMENTS

4.10.4.1 WHO

WHO has published a monograph titled "Principles for Evaluating Health Risks to Reproduction Associated with Exposure to Chemicals" (WHO/IPCS 2001b). The monograph summarizes current scientific knowledge on hazard identification and risk assessment for reproductive toxicity and is intended as a tool for use by public health officials, research and regulatory scientists, and risk managers. It seeks to provide a scientific framework for the use and interpretation of reproductive toxicity data from human and animal studies. It also discusses emerging methodology and testing strategy in reproductive toxicity. The monograph offers a discussion of basic reproductive physiol-ogy and the relative vulnerability of specific reproductive structures and processes and provides the scientific background for understanding specific methods and procedures used in reproductive toxicology. It describes on methods for assessing and evaluating altered sexual function and fertility and methodologies for assessing developmental toxicity. Finally, the monograph deals with the general principles of risk assessment for reproductive toxicity and identifies areas where research is needed.

4.10.4.2   OECD

OECD is in the process of developing a guidance document on reproductive toxicity testing and assessment (OECD 2004d). The guidance document is intended to provide guidance on methodological aspects, interpretation of data, and strategy for testing of chemicals for potential reproductive toxicity, with the primary objective to ensure that necessary and sufficient data are obtained to enable adequate evaluation of the risk of reproductive toxicity arising from exposure to a chemical. The guidance paper recommends a stepwise assessment/testing strategy. To minimize animal usage and optimize allocation of resources, data should be assessed following each step of testing to decide if they are adequate for the evaluation of the risk arising from the intended use of the chemical, or if further testing is needed. The document provides a definition of reproductive toxicity, and discusses different types of effects including adverse effects on sexual function and fertility in adult males and females as well as developmental toxicity. The guidance document also gives advice on test methods.

4.10.4.3   US-EPA

The series of Risk Assessment Guidelines includes a guideline for reproductive toxicity risk assessment and a guideline for developmental toxicity risk assessment.

The Guidelines for Reproductive Toxicity Risk Assessment (US-EPA 1996) discuss the scientific basis for concern about exposure to agents that cause reproductive toxicity and describe the principles and procedures to be followed in conducting risk assessments for reproductive toxicity. They include the female (nonpregnant and pregnant) and male reproductive systems.

The Guidelines for Developmental Toxicity Risk Assessment (US-EPA 1991) outline principles and methods for evaluating data from animal and human studies, exposure data, and other information to characterize risk to human development, growth, survival, and function because of exposure prior to conception, prenatally, or to infants and children.

4.10.4.4   EU

The TGD (EC 2003), Chapter 3.12, addresses reproductive toxicity and provides guidance on data requirements, evaluation of data, and dose–response assessment.

4.10.5   Use of Information on Reproductive Toxicity in Hazard Identification

Many countries have developed risk assessment processes for reproductive and developmental toxicity in order to set standards and regulate exposures. Experimental testing protocols are largely based on identifying structural anomalies and/or functional deficits following chemical exposure during critical windows of the reproductive cycle. All available sources of animal and human data should be considered to assess specific reproductive and developmental toxic effects. Approaches for evaluating and summarizing reproductive toxicity data have improved. Nevertheless, assumptions must often be made in the risk assessment process because of gaps in knowledge about underlying biological processes and species differences. Risk assessment test methods and strategies need to be continually refined as new data and technologies become available (WHO/IPCS 2001b).

The adverse reproductive effects are considered as being "threshold effects," i.e., effects for which there are expected to be a threshold of substance concentration below which the effects will not be manifested. For the hazard and risk assessment, it is important to identify those dose levels at which adverse reproductive effects are observed, and the dose level at which adverse reproductive effects are not observed, i.e., to derive a NOAEL for reproductive toxicity. Crucial in the derivation of the NOAEL and/or LOAEL, is the definition of "adverse effects" (Section 4.2.2). In the derivation of the NOAEL and/or LOAEL, a number of factors need to be considered; these issues are addressed in detail in Sections 4.2.3 and 4.2.4. An alternative approach to the derivation of the

NOAEL, the BMD concept, and the limitations of its applicability are addressed in Section 4.2.5. Attention is brought to the fact that the BMD approach has been studied most widely in the contexts of developmental toxicity endpoints (Falk Filipsson et al. 2003).

4.10.5.1 Human Data

Human studies on altered sexual function/fertility provide the most direct means of assessing risk, but data are often unavailable (WHO/IPCS 2001b).

The most feasible endpoints for evaluating developmental toxicity in humans are vital status at birth (including embryo/fetal loss), readily identifiable congenital anomalies, gestational length, birth weight, and sex ratio. Measurable postnatal developmental effects include changes in growth, behavior, and organ or system function, as well as cancer. Both prenatal and postnatal effects may not be apparent until well after birth, and some may not appear until adulthood. For example, some congenital anomalies are not immediately apparent, and the long-term sequelae of intrauterine growth retardation are just now being appreciated. Chemical exposure during development may also affect the later reproductive function of the offspring. For example, chemicals could damage female germ cells in utero and affect the mature female's fertility; similarly, male stem cells or Sertoli cells could be depleted, potentially affecting sperm production (WHO/IPCS 2001b).

Well-designed and well-reported human studies, such as epidemiological studies and workplace monitoring data, in which both reproductive and relevant nonreproductive effects are described will contribute to the weight of evidence for whether or not the substance is evaluated as being toxic to reproduction. It is often very difficult unequivocally to associate human exposure to a specific substance with adverse effects on reproduction unless the adverse effect is a rare birth defect and the exposure is very well characterized. This is partly because it is difficult to identify a cause–effect relationship for reproductive effects, which have a high "natural" incidence (e.g., spontaneous abortion) and which may predispose to unreliable results due to recall bias. Also, some effects can be very subjective, e.g., effects on libido. As with much human data, there may be mixed exposures and/or lifestyle related confounding factors.

If it is possible to identify a NOAEL from well-reported and reliable human studies, this value may be used preferentially in the risk assessment. However, it is expected that this will rarely be the case.

4.10.5.2 Animal Data

For many environmental chemicals, it is still necessary to rely on information derived from experimental animal models and laboratory studies (WHO/IPCS 2001b).

In order to fully assess the hazardous properties of a substance with respect to reproductive toxicity, the key data requirements are a two-generation study and a prenatal developmental toxicity (teratogenicity) study in two species (EC 2003).

In vivo animal studies for reproductive toxicity risk assessment typically utilize standard laboratory rodents. Fertility assessments in male animals have limited sensitivity as measures of reproductive injury, because, unlike humans, males of most test species produce sperm in numbers that greatly exceed the minimum requirements for fertility. Histopathological data on reproductive tissues play an important role in male reproductive toxicity risk assessment. Chemicals with estrogenic or anti-androgenic activity have been identified that are capable of causing reproductive effects in males. While sensitivity may differ, it is likely that mechanisms of action for these endocrine disrupting agents will be consistent or similar across mammalian species. For females, all functions of the reproductive system are under endocrine control and can be susceptible to disruption by effects on the reproductive endocrine system. However, single measurements of hormonal changes may be insensitive indicators of any damage because of large normal variability in females (WHO/IPCS 2001b).

In vivo studies on pregnant experimental animals and their progeny have been widely used in developmental toxicity assessment. The aim of the maternal observations is to assess the relative contribution of maternal toxicity to any observed embryo/fetal toxicity. Observations on progeny include early and late embryonic deaths (resorptions), fetal weight, external malformations, visceral and skeletal anomalies, and sex determination. Background information and historical records on abnormal development of the experimental animals are important for adequate interpretation of such toxicity studies. Functions that can be evaluated postnatally include neurological development, simple and complex behaviors, reproduction, endocrine function, immune competence, xenobiotic metabolism, and physiological function of different organ systems. Latent manifestations of toxicity may include transplacental carcinogenicity (neoplasia in the progeny resulting from maternal exposure to chemical agents during pregnancy) and shortened life span (WHO/IPCS 2001b).

The reproductive/developmental toxicity screening test can provide initial information on possible effects on reproduction and/or development and may make it possible to identify a substance as being toxic to reproduction, i.e., the test gives a clear "positive" result. However, this test offers only limited means of detecting postnatal manifestations of prenatal exposure or effects that may be induced during postnatal exposure. In addition, because of the study design (e.g., relatively small numbers of animals per dose level, relatively short study duration), the test will not provide evidence for definite claims of no effects.

Similar criteria apply when evaluating data obtained from the combined repeated dose toxicity study with the reproductive/developmental toxicity screening test.

Clearly, well-reported two-generation or developmental toxicity studies of international guideline standard, particularly if conducted in accordance with the principles of GLP, can be used to identify substances as being specifically toxic to reproduction. These tests can also be used to identify substances as being of no concern in relation to the endpoints that they address. Non-GLP studies and studies not having been performed according to test guideline protocols may also be used in the same way to decide whether a substance is toxic to reproduction when sufficient animals of an appropriate species have been used and have survived, when the dose levels used are sufficient in number and sufficiently high, and the relevant observations have been made.

If peri-postnatal tests, developmental neurotoxicity studies, or specific male or female fertility studies are available they can be used to identify a substance as being toxic to reproduction. Data from such studies alone cannot be used to identify a substance as being of no concern in relation to reproduction.

Data from repeated dose toxicity studies in which there are marked adverse effects on the reproductive organs (usually the testes) can also be used to identify a substance as being toxic to reproduction. Data from such studies cannot be used to identify a substance as being of no concern in relation to reproduction.

4.10.5.3 *In Vitro* Data

Currently, there are no validated and regulatory accepted *in vitro* methods for assessing reproductive toxicity.

A variety of *in vitro* test systems, including isolated perfused testis/ovary, primary cultures of gonadal cells, investigation of subcellular fractions of different organs and cell types and *in vitro* fertilization techniques, are available that can be used in supplementary investigational studies of different aspects of the reproductive system. *In vitro* testing systems are especially useful for screening for toxicity potential and for identifying potential mechanisms of action of potential toxicants. However, these tests are limited in their ability to assess complex, integrative reproductive functions (WHO/IPCS 2001b).

A wide range of *in vitro* systems, ranging from whole embryo culture through organ and tissue culture to a variety of nonmammalian systems, has also been developed for the study of developmental toxicity. *In vitro* tests are useful in investigation of mechanisms of normal and abnormal development to obtain information on dose–response relationships and specific organ toxicity, and perhaps as screening systems for selection or prioritization of chemicals for further *in vivo* studies (WHO/IPCS 2001b).

In vitro studies will not, in the absence of more definitive data, provide a basis for a firm decision about the reproductive toxicity of a substance. "Positive" results from such studies indicate that there may be some concern in relation to the potential for reproductive toxicity, but they can be overridden by clearly negative data from well-conducted test guideline studies for reproductive toxicity. "Negative" data from *in vitro* studies, if well conducted, may contribute to the weight of evidence.

4.10.5.4 Other Data

The chemical structure of a substance may contain structural alerts for reproductive toxicity, based on clear evidence for reproductive toxicity of structural analogues. For example, glycol ethers, which are metabolized to a reproductive toxic metabolite (an oxyacetic acid) should be considered as reproductive toxicants unless data are available showing no reproductive toxicity of the substance under evaluation.

Use of (Q)SAR may be valuable in the evaluation of the reproductive toxicity of a substance for which no data are available, see Sections 3.5.2 and 3.5.3.

4.11 ENDOCRINE DISRUPTERS

Global concerns have been raised in recent years over the potential adverse effects that may result from exposure to chemicals that have the potential to interfere with the endocrine system. Wildlife and human health effects of endocrine disrupting chemicals (EDCs) were first proclaimed by Rachel Carson in 1962, and based on a growing body of knowledge those concerns have increased. This concern regarding EDCs is directed at both humans and wildlife (WHO/IPCS 2002).

A variety of chemicals have been found to disrupt the endocrine systems of animals in laboratory studies, and there is strong evidence that chemical exposure has been associated with adverse developmental and reproductive effects in fish and wildlife in particular locations. There are also associations between exposure to the endocrine disrupting chemicals so far evaluated and human health disturbances such as testicular, breast and prostate cancers, thyroid dysfunction, as well as intelligence and neurological problems, although a causative role has not been verified. It should be noted that endocrine disruption is not considered a toxicological endpoint in its own right, but a functional change or mode of action that may lead to adverse effects.

Endocrine systems, also referred to as the hormone systems, are found in all mammals, birds, fish, and many other types of living organisms. The endocrine system consists of a set of glands such as the thyroid, gonads, and the adrenal glands; the hormones they produce such as thyroxine, estrogen, testosterone, and adrenaline; and receptors in various organs and tissues that recognize and respond to the hormones. The endocrine system regulates all biological processes in the body from conception through adulthood and into old age, including the development of the brain and nervous system, the growth and function of the reproductive system, as well as the metabolism and blood sugar levels. Table 4.20 provides an overview of the mammalian endocrine system.

Disruption of the endocrine system can occur in various ways (EC 2003, US-EPA 2007c):

- By mimicking the action of a naturally produced hormone such as estrogen or testosterone and thereby setting off similar chemical reactions in the body
- By blocking the receptors in cells receiving the hormones (hormone receptors), thereby preventing the action of normal hormones
- By affecting the synthesis, transport, metabolism, and excretion of hormones, thus altering the concentration of normal hormones

TABLE 4.20

Overview of the Mammalian Endocrine System

Endocrine Gland	Function/Hormones (Major)
Hypothalamus	Drives the endocrine system
	Links the nervous system and the endocrine system
	Produces releasing hormones, e.g., for corticotropin, growth hormone, gonadotropin, thyrotropin
	Produces inhibiting hormones, e.g., for the growth hormone
Pituitary	Posterior part secretes hormones produced by the hypothalamus, e.g., antidiuretic hormone, oxytocin
	Anterior part produces its own hormones in response to hypothalamic releasing hormones,
	e.g., adrenocorticotropic hormone ACTH, luteinizing hormone LH, follicle-stimulating
	hormone FSH, prolactin, growth hormone, thyroid-stimulating hormone TSH
Thyroid	Regulation of metabolism, development, and maturation
	Produces tyroxine and triiodothyronine hormones
Adrenals	Stress response, blood pressure, glucose metabolism, salt and water balance
	Cortical hormones include glucocorticoids (cortisol/corticosterone) and mineralocorticoids
	(aldosterone); medullar hormones are adrenaline and noradrenaline
Pancreas	Regulates glucose in blood via production of the hormones glucagon and insulin
Ovaries	Sexual development and function. Produce the hormones androgens, estrogens, and progestins
Testes	Sexual development and function. Produce the hormones androgens, estrogens, and progestins

4.11.1 DEFINITIONS

In the WHO/IPCS monograph "Global Assessment of the State-of-the-Science of Endocrine Disruptors" (WHO/IPCS 2002), endocrine disruptors are defined in a generic sense as follows: "An endocrine disruptor is an exogenous substance or mixture that alters function(s) of the endocrine system and consequently causes adverse health effects in an intact organism, or its progeny, or (sub)populations. A potential endocrine disruptor is an exogenous substance or mixture that possesses properties that might be expected to lead to endocrine disruption in an intact organism, or its progeny, or (sub)populations."

In the OECD "Draft Guidance Document on Reproductive Toxicity Testing and Assessment" (OECD 2004d), the term "endocrine disruption" is defined as above.

4.11.2 STRATEGIES FOR ASSESSMENT OF ENDOCRINE DISRUPTERS

Endocrine disruption is a relatively new concept in risk assessment, and there are currently no test methods available, which specifically detect all effects that have been linked to the endocrine disruption mechanism(s) and mode(s) of action.

With respect to endocrine disruption, the two-generation study (OECD TG 416, US-EPA OPPTS 870.3800, EU Annex V B.35) is currently the most complete study available. Both in this study and in the developmental toxicity study (OECD TG 414, US-EPA OPPTS 870.3700, EU Annex V B.31), additional endocrine-sensitive parameters may be studied on a case-by-case basis when endocrine disruption is an issue of concern.

Test methods specifically directed toward endocrine disrupting endpoints are still in the phase of being standardized and validated. Many *in vivo* and *in vitro* tests have been proposed for screening, and strategies for assessment of endocrine disrupters have been proposed by the WHO, OECD, US-EPA, and EU, see below.

Many of the proposed toxicity tests have been criticized for nonspecificity and lack of reproducibility. Concern has also been raised about their relevance for generating useful data for hazard and risk assessment purposes. The diversity of the possible modes of action (e.g., receptor binding, steroidogenesis, and modulation of the homeostatic processes which regulate endogenous responses to hormones) makes it difficult to develop tests, especially screening tests, which typically focus on a single or few endpoints (Combes 2000).

4.11.2.1   WHO

In response to the increasing concerns regarding endocrine disrupters, the Intergovernmental Forum on Chemical Safety in 1997 made a number of recommendations, to the Member Organizations of the IOMC (Section 2.1.5), notably, IPCS and OECD, concerning approaches and means for coordinating and/or supporting efforts to address the issues internationally, including the development of an international inventory of research and coordinated testing and assessment strategies. The 50th World Health Assembly adopted a resolution in 1997 (WHO 50.13), which called upon the Director-General of WHO to "take the necessary steps to reinforce WHO leadership in undertaking risk assessment as a basis for tackling high priority problems as they emerge, and in promoting and coordinating related research, e.g., on potential endocrine-related health effects of exposure to chemicals." In response to these recommendations, the IPCS assumed responsibility for developing this global assessment of the current state of scientific knowledge relative to environmental endocrine disruption. Concurrently, the IPCS assisted in the development of a Global Endocrine Disruptor Research Inventory (see http://endocrine.ei.jrc.it), which serves as a tool to foster complementary research efforts and identify strengths and weaknesses of current global research efforts. The IPCS (in collaboration with the OECD) convened an informal consultation in 1997 and a Scoping Meeting in 1998 to outline the objectives, scope, and development process for the assessment document. A preliminary draft of the document was circulated to numerous scientific experts and IPCS contact points for their review. The final document entitled "Global Assessment of the State-of-the-Science of Endocrine Disruptors" was published in 2002 (WHO/IPCS 2002).

The document builds on existing reviews and documents, and focuses on the global peer-reviewed scientific literature where the associations between environmental exposures and adverse outcomes have been demonstrated or hypothesized to occur via mechanisms of endocrine disruption. The document provides a framework for the assessment of causality between exposures to endocrine disruptors and selected outcomes. It summarizes critical generic issues (e.g., exposure–outcome associations, dose–response relationships, role of natural hormones and phytoestrogens, etc.), several of which are particularly relevant to endocrine disruptors. It provides background information on the endocrine system, the role of hormones, and potential mechanisms of endocrine disruption along with specific chemical examples of multiple modes of action. The emphasis is on the vertebrate endocrine system and on the hypothalamic–pituitary–gonad, hypothalamic–pituitary–adrenal, and hypothalamic–pituitary–thyroid axes (WHO/IPCS 2002).

4.11.2.2   OECD

In 1996, the OECD established a Special Activity on Endocrine Disrupter Testing and Assessment (EDTA) with the objectives of:

- Providing information and coordinating activities
- Developing new and revised existing Test Guidelines to detect endocrine disrupters
- Harmonizing hazard and risk characterization approaches

This activity was launched at the request of the Member countries and the Business and Industry Advisory Commitee (BIAC) to the OECD to ensure that testing and assessment approaches for endocrine disrupters would not substantially differ among countries. An overview of the extent and nature of current OECD activities on endocrine disrupters can be found at the OECD Web site for endocrine disrupters (OECD 2007b).

The issue of developing new and revised test guidelines to detect endocrine disrupters has been given very high priority because of the concern of the OECD Member countries that currently no existing test guidelines are fully sufficient to identify the potential effects of endocrine disrupting substances. The current test guidelines may enable the identification of endocrine disruptor-related effects if conducted with specific attention to this issue.

The work includes the development of test guidelines for both human health and the environment, and is guided by a conceptual framework developed to provide a framework for the testing and assessment of potential endocrine disrupters. Initially it was developed to guide the EDTA Task Force's deliberations in deciding which tests were suitable for OECD test development and validation work. The initial framework has been revised and the conceptual framework agreed by the EDTA in 2002 is not a testing scheme but rather a toolbox in which the various tests that can contribute information for the detection of the hazards of endocrine disruption are placed. The toolbox is organized into five compartments or levels, each corresponding to a different level of biological complexity (for both toxicological and ecotoxicological areas). It is stressed that the tests listed in the conceptual framework constitute many tools, which can be used independently of each other and do not each represent data requirements and even though the conceptual framework may be full of testing tools this does not imply that they will all be needed for assessment purposes. The conceptual framework is subject to further elaboration and discussion as the work on endocrine disrupters proceeds and tools (assays and tests) will be added as they are validated in future (OECD 2007b).

The OECD has published a Detailed Review Paper (DRP) "Appraisal of Test Methods for Sex Hormone Disrupting Chemicals" (OECD 2002b), which was intended to make an important contribution to the work on testing by serving as the basis for the first step in the consideration and development of OECD Test Guidelines for the testing of chemicals for endocrine-disrupting effects. The focus of the DRP is on test methods for sex hormone-disrupting chemicals capable of affecting the reproductive process. Other hormone systems, which are also important in the control of reproduction, such as the thyroid and adrenal systems, have not been considered in the DRP. In addition, test methods for the effects of sex hormones on nonreproductive processes such as brain development and behavior were considered to be beyond the scope of the DRP.

The OECD "Draft Guidance Document on Reproductive Toxicity Testing and Assessment" (OECD 2004d) is intended to provide guidance on methodological aspects, interpretation of data, and strategy for testing of chemicals for potential reproductive toxicity (see Section 4.9.4.2). The Guidance document includes considerations on endocrine disrupter-related effects but endocrine disruption is not devoted a specific paragraph. The guidance document has specifically pointed out that current test guidelines for toxicological testing could be improved with the aim of increasing sensitivity and specificity of detection of endocrine disrupter-related effects.

Newly proposed test guidelines to specifically address effects on hormone homeostasis and on male and female reproductive organs include the rodent uterotrophic assay to detect estrogenic effects (OECD 2007a) and the rodent Hershberger assay to detect androgenic effects (OECD 2007b).

The rodent uterotrophic assay is based on the principle that the uterus is under the control of estrogens to stimulate and maintain growth. If endogenous sources of this hormone are not available, the animal will require an exogenous source to initiate and/or restore uterine growth. The objective of the OECD work on the uterotrophic bioassay is to develop and validate a new Test Guideline for the detection of chemicals having the potential to act like, and consequently interfere with, endogenous female sex hormones. More specifically the rodent uterotrophic bioassay is intended to identify chemicals that act like estrogen agonists or antagonists. The assay, once validated, is intended to be used as a short-term assay within an overall testing strategy for the detection and assessment of potential endocrine disrupters. The OECD has published several reports regarding the development and validation of the rodent uterotrophic assay (OECD 2003b, 2006a–c) and a draft test guideline is now available (OECD 2007a). The test substance is administered daily by oral gavage or subcutaneous injection for a minimum administration period of three consecutive days. A statistically significant increase in the mean uterine weight of a test group relative to the vehicle group indicates a positive response for estrogen agonists.

The rodent Hershberger assay is intended to detect androgen agonists, antagonists, and 5α-reductase inhibitors. The principle of the assay is that there are organs and tissues in the animal that are under the control of androgens, which stimulate and maintain growth. If the endogenous source of this hormone is not available, either because of immaturity of the animals or because the animals have been castrated, the animal requires an exogenous source to initiate or restore growth of these tissues, and for normal sexual development. Chemicals that act as agonists may be identified as potential endocrine disrupters if they cause an increase in the weights of these androgen-dependent tissues, or as antagonists if they cause a relative decrease when coadministered with a potent androgen. The rodent Hershberger assay may also serve as a tool for the prioritization of chemicals for further testing.

The validation of the Hershberger assay is now being reviewed by an international panel of reviewers. The peer review package, submitted to the panel and available at the OECD Web site for endocrine disrupters (OECD 2007b), includes three validation reports, the draft test guideline (OECD 2007c), and a Secretariat document to support the peer review panel. A detailed draft background review document was published in 2006 (OECD 2006d).

In addition to these newly proposed test guidelines, suggestions are being considered for new parameters to be included in the present repeated dose oral toxicity test (OECD TG 407) with more emphasis to be placed on detection of endocrine effects. The validation of the enhanced OECD TG 407 is now being reviewed by an international panel of reviewers. The peer review package, submitted to the panel and available at the OECD Web site for endocrine disrupters (OECD 2007b), includes the validation report, the draft and the current test guidelines, and a Secretariat document to support the peer review panel.

Also an *in vitro* test is currently being developed. The Stably Transfected Transcriptional Activation (TA) Assay is intended to detect estrogenic activity in order to provide mechanistic information. The assay is an *in vitro* screening assay that has long been used to evaluate the specific gene expression regulated by specific nuclear receptors. It is based on the production of reporter gene product induced by a chemical following the ligand-receptor binding followed by a transcriptional activation. Therefore, it can evaluate the ability of a chemical to activate estrogenic responses. The test uses the human cervical tumor HeLa cell line with an inserted construct: human ERα expression vector with a firefly luciferase reporter construct bearing five tandem repeats of a vitellogenin estrogen-responsive element (ERE) driven by a mouse metallothionein promoter TATA element. Test and control chemicals are incubated with the HeLa cells. Transcriptional activity is evidenced as luminescence. Validation studies of this test have been performed by Japan. The validation is now reviewed by an international panel of reviewers. The peer review package, submitted to the panel and available at the OECD home page for endocrine disrupters (OECD 2007b), includes the draft validation report and a draft test guideline (OECD 2006e).

The OECD has also published a Detailed Review Paper on Thyroid Hormone Disruption Assays (OECD 2006f). This document reviews the state of assays for thyroid toxicants across four vertebrate classes (fish, amphibians, birds, and mammals) within the context of a thorough review of thyroid endocrinology across these classes. Considering the similarity of the endocrinology of the thyroid gland across these classes, a major feature of the document is the integration of assay comparisons across vertebrates. By assessing the state of the assays amongst mammalian (rodent), fish, amphibian, and avian species, the state of thyroid assays, redundancies, and information gaps presented themselves.

4.11.2.3   US-EPA

The US-EPA's Endocrine Disruption Screening Program (EDSP) is mandated to use validated methods for the screening and testing of chemicals to identify potential endocrine disrupters, determine adverse effects, dose–response, assess risk, and ultimately manage risk under current laws (US-EPA 2007c). The "EDSP Team" is currently conducting the studies needed to validate the endocrine disruptor screening and testing methods, using the general principles and processes developed by the Interagency Coordinating Committee on the Validation of Alternative Methods (ICCVAM). The intention is to identify and characterize the endocrine activity (specifically estrogen, androgen, and thyroid) of pesticides, commercial chemicals, and environmental contaminants by these methods; using a two-tiered screening and testing process. In Tier 1, chemicals that have the potential to interact with the endocrine system should be identified. In Tier 2, the specific effect caused by each endocrine disrupter should be determined and the dose at which the effect occurs should be established. The Screening batteries are based on the Endocrine Disruptor Screening and Testing Advisory Committee's (EDSTAC) recommendations. The EDSTAC was a federal advisory committee formed in 1996 to make recommendations to the US-EPA on how to develop the screening and testing program called for by Congress. Its final report was presented to the US-EPA in September 1998 (EDSTAC 1998). For the tests under validation, review papers are available at the US-EPA Web site for the EDSP (US-EPA 2007d).

The Tier 1 screening battery is intended to identify chemicals affecting the estrogen, androgen, or thyroid hormone systems through any of several recognized modes of action. The EDSTAC selected Tier 1 assays on the basis of (1) maximum sensitivity which serves to minimize false negatives; (2) inclusion of a range of organisms representing differences in metabolism; (3) detection of all known modes of action for the endocrine endpoints of concern; (4) inclusion of a sufficient range of taxonomic groups among the test organisms; and (5) incorporation of sufficient diversity among the endpoints, permitting weight-of-evidence conclusions. The Tier 1 screening assays under consideration by the US-EPA are summarized in Table 4.21.

The Tier 2 testing assays are intended to confirm, characterize, and quantify effects for estrogen, androgen, and thyroid active substances. The EDSTAC selected Tier 2 to include the most sensitive developmental lifestage, to identify the specific hazard caused by the chemical and establish a dose–response relationship, and to include a range of taxa. The Tier 2 screening assays under consideration by the US-EPA are summarized in Table 4.22.

Finally, the In Utero through Lactation Assay involves the use of pregnant rats to assess postnatal development of the neonate after in utero and lactational exposure. This assay has not yet been assigned to either Tier 1 or 2.

4.11.2.4   EU

In December 1999, the EU Commission published the "Community Strategy for Endocrine Disrupters" which set out a general framework for studying endocrine disrupters (ED) (EC 1999). This strategy focused on short, medium- and long-term actions that would contribute to ensure a better environment and health of people within the EU.

The short-term actions aim at information gathering, to provide background information for medium- and long-term actions, and to identify knowledge gaps that may need to be addressed in the future.

The short-term actions include: (1) establishment of a priority list of substances for further evaluation of their role in endocrine disruption; monitoring levels of suspect chemicals in food and the environment; (2) identification of vulnerable groups of people (such as children) who need to be given special consideration; (3) establishment of an international network to enable information exchange and coordination of research and testing; and (4) communication with the public and continuing consultation with stakeholders.

TABLE 4.21

US-EPA Tier 1 Screening Assays for Endocrine Disruption

Assay	Test System/Process	Endpoint(s)
Amphibian (Frog) Metamorphosis assay	Tadpoles, metamorphosis	Developmental effects resulting from affection of the thyroid
Androgen receptor Binding assays	Rat prostate cytosol or rat recombinant androgen receptor	Ability of a test chemical to bind with androgen receptors
Estrogen receptor Binding assays	Rat uterine cytosol or alpha isoform of the human recombinant estrogen receptor	Ability of a test chemical to bind with estrogen receptors
Aromatase *in vitro* assay	Cell culture or cellular fractions (e.g., human placental microsomes)	Inhibition of aromatase activity
Fish screen assay	Fish (e.g., fathead minnow, Japanese medaka, zebrafish)	Abnormalities associated with survival, reproductive behavior, secondary sex characteristics, histopathology, and fecundity (i.e., number of spawns, number of eggs/spawn, fertility, and development of offspring)
Hershberger assay	Please see Section 4.11.2.1	
Pubertal female assay	Female rats during sexual maturation	Abnormalities associated with sex organs and puberty markers, as well as thyroid tissue
Pubertal male assay	Male rats during sexual maturation	Abnormalities associated with sex organs and puberty markers, as well as thyroid tissue
Steroidogenesis *in vitro* assay	Human cells	Interference with the cellular production of male and female steroid sex hormones
Uterotrophic assay	Please see Section 4.11.2.1	
Adult male assay	Male rats	Abnormalities associated with primary and secondary sex organs, systemic hormone concentrations, and thyroid; resulting from anti-androgenic and thyroid activity

The medium-term actions center on the practical and experimental activities needed to ensure that suspected chemicals are tested in a speedy and accurate way. The test development process is directed by the OECD, and the EU Commission contributes by coordinating the input of Member States. These actions include: (1) the development and validation of internationally agreed test

TABLE 4.22

US-EPA Tier 2 Screening Assays for Endocrine Disruption

Assay	Test System/Process	Endpoint(s)
Amphibian development, reproduction assay	Frogs	Dose–response characteristics and adverse reproductive and developmental effect
Avian two-generation assay	Japanese quail	Dose–response characteristics and adverse reproductive and developmental effect
Fish lifecycle assay		Dose–response characteristics and adverse reproductive and developmental effect
Invertebrate lifecycle assay	Mysid shrimp	Dose–response characteristics and adverse reproductive and developmental effect
Mammalian two-generation assay	Rats	Dose–response characteristics and adverse reproductive and developmental effect

methods to assess endocrine disruption in people and wildlife; (2) the development of a European test strategy for identifying and assessing endocrine disrupters that is consistent with similar strategies in other countries such as the United States and Japan; and (3) the coordination and funding of international research into the underlying mechanisms of endocrine disruption and understanding how these mechanisms can impact on human health.

The long-term actions relate to updating, amending, or adapting the legislative instruments that protect the health of humans and wildlife in the EU, and include addition of new or adaptation of existing toxicity tests for hazard assessment and the adaptation of risk assessment methodology; updating the way in which chemicals are classified, packaged, labeled, used, or marketed in order to ensure safe usage and disposal within the EU; and review of legislation relating to the testing, assessment, use, and disposal of specific substance groups such as pesticides, biocides, and consumer products.

Endocrine disrupters not addressed by specific legislation (e.g., natural substances and by-products such as hormones and dioxins) will be dealt with under environmental legislative instruments such as the EU Water Framework Directive or through the adaptation of existing international legislation.

In June 2001, the EU Commission published the first progress report following the adoption of the Community Strategy for Endocrine Disrupters, covering the time period from 1999 to 2001 (EC 2001b); and in October 2004, the second progress report was published covering the time period from 2001 to 2003 (EC 2004).

More information on the EU activities within this area can be found at the DG Environment's Endocrine Disrupters Web site (EU 2007).

4.12 HORMESIS

Hormesis is the term used for the phenomenon of stimulatory effects at low-level exposure, and inhibition at high-level exposure. The term derives from the Greek word *Hormo* which means "excite" or "set in motion," and which is also the root of the word "hormone." The concept of hormesis dates back to the 1920s. A substance showing hormesis has the opposite effect in small doses compared to effects at large doses. The definition of hormesis does not imply that low-dose effects are necessarily beneficial, only that they are opposite to high-dose effects.

Hormetic effects of low-dose stimulatory and high-dose inhibitory response include parameters of good health such as growth rate, fecundity, and longevity. The dose–response curve for this type of hormetic effect is "the inverted U," see Figure 4.3.

For adverse effects such as carcinogenicity, mutagenicity, and disease incidence, a hormetic effect means low-dose reduction and high-dose enhancement of response. The dose–response curve is the "J-shape," see Figure 4.4.

According to systematic investigations of peer-reviewed published scientific literature, thousands of examples of likely hormetic action exist (Calabrese 2005).

The maximum hormetic stimulatory response is typically a 30%–60% increase of the control value, and this appears to be the case across systems, whether the system is a plant, fish, cell line, mammal, or bacteria. The hormetic response is typically observed at dose levels at $1/10$–$1/5$ of the NOAEL and up to just below the NOAEL. The frequency with which hormesis occurs in toxicity studies may be quite high. An analysis of several hundred articles selected from a large database of published toxicity studies showed a frequency of 40% (Calabrese 2005).

A controversy exists because hormesis implies that small doses of toxic substances may actually have a beneficial effect on living organisms, an idea which is difficult to perceive intellectually. Hormetic effects have been reported in cases that are too diverse for any single mechanism to be able to account for the phenomenon. In pharmacology, biphasic responses have been shown to occur via high and low affinity subsets of receptors with opposite action. In toxicology, no attempts have yet been made to assess mechanisms by authors reporting hormetic dose responses (Calabrese 2005).

Hormesis has implications for risk assessment. The traditional risk characterization involves establishing "safe" exposure levels based on extrapolation from high-level exposure. In the threshold model, anything above a certain dose is considered dangerous, and anything below it is considered safe. For non-threshold effects, there is no safe dose. However, if the effects at low exposure levels are opposite to effects at high exposure levels, the traditional approaches may not make sense. Furthermore, in hormetic models, a treatment-related response occurs below the NOAEL. The application of a hormesis model in regulatory risk assessment could change exposure standards for contaminants in air, water, food, and soil.

Regulatory toxicology attempts to protect the entire, or the majority of, the human population, or particular exposed subgroups of the population. Concern has been raised that hormesis may be a population phenomenon rather than an individual phenomenon. That is, the beneficial effects of low-dose exposure may be restricted to part of the population, while other parts of the population may experience no, or even harmful effects. If risk assessment takes hormesis into consideration, individuals unable to exhibit a hormetic response may be insufficiently protected. According to a review by Calabrese and Baldwin (2002), there are sufficient examples which show that some strains/individuals lack the capacity to produce the low-dose stimulatory response. Examples of species- and strain-related differences include pesticide effects on growth in various algal species, alcohol effects on longevity in various *Drosophila* strains, vanadate effects on cell proliferation in rat versus mouse cell lines, and ethanol effects on locomotor patterns in rat sub-strains. Differences in hormetic response within a species (related to age) were indicated in rats in response to anticonvulsants on drug-induced convulsions, in response to irradiation of the brain cortex, or in response to epinephrine on memory; and in dogs in response of isolated coronary arteries to adrenergic agonists. Different human breast cancer cell lines exhibited different quantitative and qualitative hormetic response to a chemotherapeutic agent. Hormetic response to a toxic substance, like a traditional toxic response, may also be limited to the particular species in which the phenomenon was observed.

The design and conduct of toxicity studies and the selection of statistical models that estimate risk would need to be changed to accommodate the concept of hormesis. Toxicity studies are normally conducted at high exposure levels, and the aim is to identify a LOAEL and a NOAEL for harmful effects. An OECD test guideline method for repeated dose toxicity typically includes three dose groups, of which the lowest dose group should turn out to be the NOAEL. The inclusion of doses below the presumed NOAEL is not encouraged. Possible hormetic effects are difficult to distinguish from normal background variation, taking into consideration the typical hormetic stimulatory response of 30%–60% increase of the control value. In order to establish hormetic action for a toxic substance, study designs would need to be changed to include dose levels below the NOAEL (1/10–1/5 NOAEL), and to use group sizes sufficient to pick up a 30%–60% increase of the control value.

A group of scientists representing several federal agencies, the International Society of Regulatory Toxicology and Pharmacology, the private sector, and academia have formed the BELLE (Biological Effects of Low Level Exposures) Advisory Committee. The initial goal of BELLE is the scientific evaluation of the existing literature and of ways to improve research and assessment methods for low-level exposure. The BELLE Advisory Committee is committed to the enhanced understanding of low-dose responses of all types, whether of an expected nature (e.g., linear, sublinear) or of a paradoxical nature, including U-shaped dose–response curves, hormesis, and biphasic dose–response curves. BELLE considers toxic agents, pharmaceuticals, and natural products over wide dosage ranges in *in vitro* and *in vivo* systems, including human populations (BELLE 2007).

4.13 THRESHOLD OF TOXICOLOGICAL CONCERN

For most types of toxic effects (e.g., organ-specific, neurological, immunological, non-genotoxic carcinogenicity, reproductive, developmental), it is generally considered that there is a dose or concentration below which adverse effects will not occur, i.e., a threshold exists (Section 4.2). For other types of effects (e.g., sensitization, mutagenicity, genotoxicity, genotoxic carcinogenicity) it is assumed that there is some probability of harm at any level of exposure, i.e., no threshold exists. From this, the concept of the Threshold of Toxicological Concern (TTC) has arisen, which refers to the establishment of a human exposure threshold value for all chemicals, below which there would be no appreciable risk to human health.

The establishment of TTC is based on the analysis of the toxicological and structural data of a broad range of different substances. It might be used as a substitute for substance-specific information in situations where there are limited or no information on the toxicity of a given substance to which the human exposure is so low that undertaking toxicity studies is considered not warranted, because of the costs incurred in the use of animals, manpower, and laboratory resources as well as for animal welfare reasons.

So far, two principal approaches exist in the thresholds developed to date: the general TTC concept and the TTC concept in relation to structural information and/or metabolic and toxicological data of substances.

4.13.1 GENERAL TTC CONCEPT

The general TTC concept, covering also carcinogenic effects, was introduced by Rulis (1986, 1989) as a "Threshold of Regulation." Rulis used data on a subset of 343 oral carcinogens from animal studies compiled in the Carcinogenic Potency Database (CPDB) (Gold et al. 1984).

The CPDB is an international resource of results from several thousand chronic, long-term animal cancer tests on chemicals. CPDB provides a standardized and easily accessible database with qualitative and quantitative analyses of both positive and negative experiments that have been published in the general literature through 1997 and by the National Cancer Institute/National Toxicology Program through 1998. For each experiment, information is included on species, strain, and sex of test animal; features of experimental protocol such as route of administration, duration of dosing, dose level(s) in mg/kg body weight/day, and duration of experiment; target organ, tumor type, and tumor incidence; carcinogenic potency (TD_{50}) and its statistical significance; shape of the dose–response, author's opinion as to carcinogenicity, and literature citation (CPDB 2007).

From the CPDB, Rulis (1986, 1989) initially proposed, for illustration, a threshold value of 0.15 µg/person/day. Subsequently, Munro (1990) confirmed the observation of Rulis and included more rodent carcinogens in the original database, bringing the total to 492 rodent carcinogens (Gold et al. 1989). Munro (1990) concluded that a threshold value of 1.5 µg/person/day would provide a high degree of health protection. The robustness of the database was confirmed by Cheeseman et al. (1999) who expanded the data set to 709 carcinogens based on the continuously updated CPDB.

The Threshold of Regulation approach has been applied by US-FDA to food contact materials since 1995 (Federal Register 1995) for substances that are not known to be carcinogens and that do not contain structural alerts indicative of carcinogenicity. Substances meeting the threshold of regulation criteria, i.e., the use in food-contact articles result in a dietary concentration of the substance of 0.5 µg/kg food (corresponding to an intake of 1.5 µg/person assuming the consumption of 3000 g food and liquid) or less would not require toxicological testing, although a toxicity profile based on available data is expected. Above this threshold, the degree of required testing increases as exposure increases (Federal Register, 1993, 1995).

4.13.2 STRUCTURE-BASED, TIERED TTC CONCEPT

Munro et al. (1996) explored the relationship between chemical structure and toxicities through the compilation of a large reference database consisting of 613 chemical substances tested for a variety of noncarcinogenic toxicological endpoints in rodents and rabbits in oral toxicity tests, including subchronic, chronic, reproductive, and developmental toxicity. For many of the substances, more than one NOEL was identified and in all, the reference database contained 2941 NOELs. The substances were grouped into one of three general classes (Classes I, II, and III) based on the chemical structure using the decision tree of Cramer et al. (1978). The structural classification was based on the assumption that inherent toxicity is dependent on chemical structure.

The decision tree method of Cramer et al. (1978) was based on the toxicological data then available and used a series of 33 questions, each leading either to another question or to classification into one of three classes of presumptive toxicity. The questions were primarily based on chemical structure, but natural occurrence in body tissues or fluids as well as natural occurrence in traditional foods was also considered. The three classes of substances were defined as follows:

- Class I substances are those with structures and related data suggesting a low order of oral toxicity. They have simple chemical structures and are efficiently metabolized by high-capacity pathways.
- Class II substances are simply "intermediate" substances with less clearly innocuous structures than those of Class I substances, but without structural features suggestive of toxicity.
- Class III substances are those that have chemical structures that permit no strong initial presumptions of safety, or that may even suggest significant toxicity. They thus deserve the highest priority for investigation.

Cumulative distributions of the logarithms of NOELs were plotted separately for each of the structural classes. The 5th percentile NOEL was estimated for each structural class and this was in turn converted to a human exposure threshold by applying the conventional default safety factor of 100 (Section 5.2.1). The structure-based, tiered TTC values established were 1800 μg/person/day (Class I), 540 μg/person/day (Class II), and 90 μg/person/day (Class III). Endpoints covered include systemic toxicity except mutagenicity and carcinogenicity. Later work increased the number of chemicals in the database from 613 to 900 without altering the cumulative distributions of NOELs (Barlow 2005).

The structure-based, tiered TTC approach as outlined by Munro et al. (1996, 1999) is used by the Joint FAO/WHO Expert Committee on Food Additives (JECFA) in a procedure for the evaluation of flavoring substances in food, including an acceptance of the general TTC of 1.5 μg/person/day, i.e., the US-FDA Threshold of Regulation (Section 4.13.1). The European Food Safety Authority (EFSA) also uses this approach for evaluation of flavoring substances, except that the general TTC of 1.5 μg/person/day is not accepted (Larsen 2006).

4.13.3 ENHANCED STRUCTURE-BASED, TIERED TTC CONCEPT

Cheeseman et al. (1999) has further extended the TTC concept with a TTC of 1.5 μg/person/day (0.025 μg/kg body weight/day), used by US-FDA in the Threshold of Regulation policy, by incorporation of acute and short-term toxicity data, the results of genotoxicity testing, and structural alerts to identify potent and nonpotent carcinogens. This work confirmed the validity of 1.5 μg/person/day as an appropriate threshold for most carcinogens. Cheeseman et al. (1999) also concluded that there may be some chemicals with a very high carcinogenic potency that may not be covered by this TTC approach.

Kroes et al. (2000, 2004) and Barlow (2005) combined the general TTC concept and the structure-based, tiered TTC concept to develop an enhanced structure-based, tiered TTC approach.

Neurotoxicants, immunotoxicants, and teratogens were also further explored, and it was concluded that, except for the organophosphorous neurotoxicants, such compounds would be covered by the structure-based, tiered TTC approach. For organophosphates, a human exposure threshold of

18 µg/person/day was derived. This threshold for organophosphates was not intended to replace the normal regulatory assessments and controls for organophosphates used as pesticides, but could be used to evaluate the risk should a nonapproved or unregulated organophosphate be detected as a contaminant in food.

Five groups of compounds were identified having a significant fraction of their members that may still be of concern at an intake of 0.15 µg/person/day (0.0025 µg/kg body weight/day), which is 10-fold below the Threshold of Regulation figure. Three of these groups are genotoxic compounds: the aflatoxin-like compounds, azoxy-compounds, and *N*-nitroso-compounds, while two groups were non-genotoxic compounds: the 2,3,7,8-tetrachlorodibenzo-*p*-dioxin (TCDD, the "Seveso dioxin") and its analogues, and the steroids. Compounds with these structural alerts for high carcinogenic potency are excluded from the TTC approach. A TTC of 0.15 µg/person/day could be used for all other substances with structural alerts for genotoxicity (Kroes et al. 2004).

Specific considerations of metabolism and accumulation are not necessary in the application of a TTC provided that the substances are not likely to show very large species differences in accumulations such as, e.g., polyhalogenated-dibenzo-*p*-dioxins, dibenzofurans, and biphenyls and related compounds, as well as nonessential heavy metals in elemental, ionic or organic forms. Such substances are known to accumulate in the body, and the traditionally employed safety factors (Section 5.2.1) may not be high enough to account for species differences in rates of elimination of such chemicals. Therefore, the TTC approach should not be used for such substances.

In addition, because it is still not known whether endocrine disrupters are active at very low exposures, it would be premature to include low-dose, endocrine-mediated effects in the TTC approach.

While thresholds probably exist for sensitization and elicitation of allergic responses, they have not been established yet even for common allergens, and are known to vary between individuals and within an individual over time. Thus, although the TTC approach does take account for substances causing immunotoxicity other than allergenicity, it cannot be used to assess the concern for allergenicity. In addition, proteins should be excluded from the TTC approach because of their potential for allergenicity and because some peptides have potent biological activities, and because they were not included in the original database.

It should be noted that the TTC concept at present is limited to oral exposure as the databases behind the concept do not include studies using dermal application or exposure by inhalation.

The human exposure TTC values suggested by Kroes et al. (2004) to be used for individual types of chemicals are summarized in Table 4.23.

TABLE 4.23

Human Exposure Threshold of Toxicological Concern (TTC) Values

Type of Chemical	µg/person/day	µg/kg body weight/day
Genotoxic compounds	0.15	0.0025
Non-genotoxic compounds	1.5	0.025
Organophosphates	18	0.3
Cramer class III	90	1.5
Cramer class II	540	9
Cramer class I	1800	30

Source: Adapted from Kroes, R., Renwick, A., Cheeseman, M., et al., *Food Chem. Toxicol.*, 42, 65, 2004.

4.13.4   TTC Concept, Industrial Chemicals within REACH

In order to reduce animal testing under REACH, and also because of the ban on the marketing of cosmetic products/ingredients tested on animals, a broader strategy for risk assessment has been suggested. One element suggested in this strategy is the TTC concept.

4.13.4.1   Tiered TTC Concept: ECETOC

ECETOC (2004) has proposed a concept of generic threshold values based on hazard categories primarily intended to be used in the risk assessment procedure of industrial chemicals within REACH. The hazard categories are based on EU classification limits and for each substance to be risk assessed, inclusion in hazard categories depends on the substance's specific classification (or no classification) according to the Commission Directive 67/548/EC (EC 1967). Three hazard categories have been suggested:

Low hazard category:
Substances classified as harmful (Xn;R20/21/22) for acute toxicity
Substances classified as irritating to eyes, respiratory system and skin (Xi;R36/37/38)

Medium hazard category:
Substances classified as toxic (T;R23/24/25) for acute toxicity
Substances classified as harmful (Xn;R48/20/21/22) for repeated dose toxicity
Substances classified as carcinogens, mutagens, and reproductive toxins in Category 3
Substances classified as skin sensitizers (Xi;R43)
Substances classified as severe eye irritants (Xi;R41)

High hazard category:
Substances classified as very toxic (Tx;R26/27/28) for acute toxicity
Substances classified as toxic (T;R48/23/24/25) for repeated dose toxicity
Substances classified as carcinogens, mutagens, and reproductive toxins in Categories 1 and 2
Substances classified as respiratory sensitizers (Xi;R42)

Generic Exposure Values (GEVs) are generic threshold values for occupational exposure (and derived dermal values) derived from OELs (Occupational Exposure Limits). The effects used to estimate GEVs are acute and repeated dose toxicity for a total of 63 organic and nonorganic substances, both volatile and nonvolatile.

For inhalation of solids, the GEVs are: 0.005, 0.1, and $1 \, mg/m^3$, while for inhalation of volatiles, the GEVs are 0.05, 1, or 10 ppm for high, medium, and low hazard category, respectively.

A margin of exposure of 2 has been selected as a basis for distinguishing scenarios that are of concern from those which are unlikely to be of concern. Consequently, for inhalation of solids, the GEVs are corresponding to an intake of 25, 500, and 5000 µg/person/day for high, medium, and low hazard category, respectively, based on a respiratory volume of $10 \, m^3/day$.

It should be noted that, in addition to the scientific toxicological information, the OELs include socioeconomic and technical arguments.

Generic Lowest Exposure Values (GLEVs) are suggested to be used in tiered processes of consumer risk assessment as an estimate of the actual LOAEL for the substance's repeated dose toxicity. The GLEVs are based on the classification limit (50 mg/kg body weight/day; R48 "Danger of serious damage to health by prolonged exposure;" based on a 90-day study) for repeated dose toxicity according to the Commission Directive 67/548/EC (EC 1967). None of the values includes carcinogens, mutagens, and reproductive toxins in Categories 1 and 2.

An assessment factor of 240 is applied to take into consideration: extrapolation from LOAEL to NOAEL (a factor of 6), extrapolation from subchronic to chronic study (a factor of 2), and inter- and intraspecies variation (a factor of 4 and 5, respectively).

For oral intake, the GLEVs are 0.5, 5, and 50 mg/kg body weight/day, corresponding to an intake of 150, 1,500, and 15,000 µg/person/day for high, medium, and low hazard category, respectively, for a 70 kg person.

For inhalation of solids, the GLEVs are 2.5, 25, and 250 mg/m^3, corresponding to an intake of 210, 2,100, and 21,000 µg/person/day for high, medium, and low hazard category, respectively, based on a respiratory volume of 20 m^3/day.

4.13.4.2 Nordic Project: Applicability of TTC within REACH

A Nordic project has evaluated how different TTC-like concepts have been used, and assessed their potential usability in risk assessment of industrial chemicals within REACH (NCM 2005). The Nordic group considered that if the TTC concept is appropriately derived and used, it might imply a better focus of chemicals at risk. However, it was also stated that, independent of the approach used in risk assessment of industrial chemicals, it is important to maintain a sufficient level of protection and that application of the TTC concept in REACH would imply that limited data may be generated and thus that the level of protection might be influenced.

The Nordic group noted a number of limitations or drawbacks that should be taken into consideration in deciding whether the TTC concept in general might be applicable for use within REACH (NCM 2005):

- In general, the consequences of the assumptions of toxicological, statistical, and/or uncertainty factors made in the derivations of the TTC concepts are difficult to overview since there are uncertainties and drawbacks in more or less all of the available TTC approaches.
- A number of studies have been undertaken to investigate whether different endpoints of concern, which might give rise to effects at low doses like immunotoxic, endocrinologic, neurotoxic, and developmentally toxic effects could be included in the TTC values. The Nordic group considered that not all these endpoints have been adequately covered by the analyses performed today.
- From the investigations performed it has been concluded that all types of substances cannot be included in a concept using TTC; in this respect, the Nordic group pointed out that industrial chemicals are diverse and often of complex nature.
- Use pattern of industrial chemicals can often be characterized as wide and dispersive. The Nordic group noted that this is different from other groups of chemicals for which the use patterns are considered to be more specific. As an example, the Nordic group pointed at food contact materials and flavoring substances.
- Up to now, the TTC concept has only been developed and used for systemic effects following oral exposure (dietary uptake). For industrial chemicals, the predominant exposure is to workers and consumers via inhalation and/or by skin contact. Toxic endpoints of concern for industrial chemicals such as irritation and sensitization relevant for skin and lung are therefore not covered by the TTC concepts developed up to now.
- Exposure of industrial chemicals also includes, in addition to workers and consumers, man exposed via the environment. TTC values are intended to be used for the general population; however, the Nordic group considered that up to now, no considerations to vulnerable subgroups such as children, the elderly, and pregnant women, etc. have been made. The Nordic group also stated that the problem with exposure to the same substance from multiple sources is not solved by the use of TTC.
- Use of the TTC approach is dependent on rather precise quantitative exposure estimates. Experience from the EU Risk Assessment Program for Existing Substances is that it is very difficult to get sufficient information on the different uses and related exposure to make precise exposure estimates. The Nordic group considered that for substances where only

very limited toxicological data is available, it seems very unlikely that high quality exposure data exist. Furthermore, in relation to industrial chemicals, many different and changing uses of a substance make it very difficult to obtain a robust overall exposure estimate for the substance.

- The Nordic group considered that in order to be sure of protective TTC values, the values would be rather small. Using rather crude or conservative exposure estimates (e.g., worst-case scenarios and modeling), as is the case for risk characterization of industrial chemicals, would usually be at a quantitative higher level and thus this combination would probably lead to limit the use of the TTC approach to a great extent within REACH.

The Nordic group has questioned the ECETOC (2004) approach described in Section 4.13.4.1:

- Reasoning for using the classification limit for R48 as the numerical starting point for calculating TTC levels is rather unclear and may not be especially relevant as a starting point.
- Up to now, there is no experience of hazard categorization.
- Classification limits are effect values - not "no-effect" values.
- Use of assessment factors seems rather controversial in this approach.
- There is an obvious risk of misuse if the concept of generic threshold values derived for a specific use (food contact materials and flavorings) is expanded to be used for all kinds of substances, including industrial chemicals, and all possible exposure situations (workers, consumers, and man via the environment). For example, the intended use of GLEV/GEV means use outside the original applicability domain of the concept.

The Nordic group concluded that the TTC concept is not applicable within REACH at tonnage levels below 100 tons/year because of the classification requirements and because waiving from testing is possible only at tonnage levels at or above 100 tons/year.

In the decision whether toxicity studies may be omitted at tonnage levels at or above 100 tons/year is appropriate or not, a TTC value might be used in the comparison with the available exposure information. However, due to limitations and uncertainties in the derivation of TTC values, as well as the fact that the TTC concept has not yet been evaluated for the diverse group of industrial chemicals and for different routes of exposure other than dietary, the Nordic group concluded that it is too premature to use the TTC concept within REACH.

Finally, the Nordic group considered it of utmost importance to obtain a sufficient level of protection in risk assessment of industrial chemicals and found it doubtful whether this is possible to achieve by the use of the TTC concept.

4.13.4.3   TTC Concept within EU REACH: Dutch Document

Veenstra and Kroese (2005) have discussed the concept of TTC for use within REACH.

It is mentioned that the TTC concept has been incorporated in the risk assessment processes in a number of regulatory schemes as a scientifically sound tool to justify waiving or generation of animal data. It is also stressed that, in contrast to approaches such as read-across or chemical categorization, the use of the TTC is not focused or limited to the identification of potential hazards but also provides a quantitative estimate of potency.

It is furthermore noted that the TTC concepts, including the structure-based approaches, are derived from databases covering substances used as direct and indirect food additives, pesticides, and industrial chemicals, and cover toxic effects related to systemic exposure to these chemicals. In addition, it is underlined that TTC has not been developed for endpoints associated with direct contact such as irritation or sensitization.

It is concluded that the TTC concept for systemic toxicity is suitable as a starting point for a tiered testing and risk assessment strategy within REACH. However, it is also underlined that the

TTC concepts require a minimum set of information in order to be applied successfully. A draft generic TTC concept under REACH is proposed and it is stressed that the information requirements for the draft generic TTC concept under REACH are consistent with the tiered approach proposed by Kroes et al. (2004) (Section 4.13.3).

According to Veenstra and Kroes (2005), the following structural characteristics or properties needs special attention:

- Polyhalogenated dibenzo-*p*-dioxins, -dibenzofurans, or -biphenyls and similar substances; nonessential, heavy metals
- Genotoxic carcinogens
- Organophosphates
- Proteins

Taken together, the information necessary for an initial assessment of a substance using the TTC concept is:

- Potential to persist and bioaccumulate
- Potential for genotoxic carcinogenic action
- Potential for neurotoxicity and cholinesterase inhibition
- Potential for inducing allergies, hypersensitivity, intolerances, or local effects

4.14 PROBABILISTIC METHODS FOR HAZARD ASSESSMENT

Individual tolerances to chemicals may vary extensively in humans depending on genetics, coincident exposures, nutritional status, and various other susceptibility factors (Section 5.4). In a population, each individual may have an exposure threshold below which there is no excess risk of that individual experiencing a particular adverse effect following exposure to a chemical. However, the existence of individual dose–response thresholds does not necessarily mean that it is possible, or practical to define a population dose–response threshold below which no individual in the population is at excess risk. For risk assessment, it could be more useful to include an estimation of the variability in dose–response thresholds in the description of the population. Techniques such as probabilistic analysis attempt to predict a range and likelihood of plausible risk estimates rather than a single estimate of the magnitude of risk. Although formal quantitative uncertainty analysis techniques are commonly applied in exposure assessments (Section 7.5) and in pharmacokinetic modeling parts of chemical risk assessments, they are not yet widely used for dose–response modeling.

Probabilistic methods can be applied in dose–response assessment when there is an understanding of the important parameters and their relationships, such as identification of the key determinants of human variation (e.g., metabolic polymorphisms, hormone levels, and cell replication rates), observation of the distributions of these variables, and valid models for combining these variables. With appropriate data and expert judgment, formal approaches to probabilistic risk assessment can be applied to provide insight into the overall extent and dominant sources of human variation and uncertainty.

When for a particular endpoint, data are available that allow for fitting a regression function, the BMD or CED may be estimated (Section 4.2.5). Depending on the quality of the data, this estimate has a certain degree of imprecision. To take uncertainty of the data into consideration, it has been proposed to calculate the lower 95% confidence limit of the estimated BMD, the BMDL. In the probabilistic approach, the complete uncertainty distribution of this estimate is needed. One way of obtaining this distribution is by the bootstrap method. Once a regression model has been fitted to the data, Monte Carlo sampling (a repeated random sampling from the distribution of values for each of the parameters in a calculation to derive a distribution of estimates in the population) is used to generate a large number of new data sets from this regression model, each time with the same number of data points per dose group as observed in the real study. For each generated data set, the BMD is reestimated. Taking all these BMDs together results in the required distribution (Slob and Pieters 1998).

REFERENCES

Alarie, Y. 1973. Sensory irritation in the upper airways by airborne chemicals. *Toxicol. Appl. Pharmacol.* 24:279–297.

Alarie, Y. 1981. Bioassay for evaluating the potency of airborne sensory irritants and predicting acceptable levels of exposure in man. *Food Cosmet. Toxicol.* 19:623–626.

Alarie, Y. 2007. Yves Alarie (Alarie test) website. www.yvesalarie.com

Andersen, M.E. 2003. Toxicokinetic modeling and its applications in chemical risk assessment. *Toxicol. Lett.* 18:9–27.

Arts, J.H.E. and C.F. Kuper. 2007. Animal models to test respiratory allergy of low molecular weight chemicals: A guidance. *Methods* 41:61–71.

Barlow, S. 2005. Threshold of Toxicological Concern (TTC). A tool for assessing substances of unknown toxicity present at low levels in the diet. ILSI Europe Concise Monograph Series. Europe, Brussels, Belgium: ILSI. http://europe.ilsi.org/publications/Monographs/ThresholdToxicologicalConcern.htm

Barnes, D., G. Daston, J. Evans, et al. 1995. Benchmark dose workshop: Criteria for use of a benchmark dose to estimate a reference dose. *Regul. Toxicol. Pharmacol.* 21:296–306.

BELLE. 2007. Biological effects of low level exposures website. http://www.belleonline.com/

Bos, P.M.J., M. Busschers, and J.H.E. Arts. 2002. Evaluation of the sensory irritation test (Alarie Test) for the assessment of respiratory tract irritation. *J. Occup. Environ. Med.* 44:968–976.

Bradford Hill, A. 1965. The environment and diseases: Association or causation? *Proc. R. Soc. Med.* 58:295–300.

Calabrese, E.J. 2005. Paradigm lost, paradigm found: The re-emergence of hormesis as a fundamental dose response model in the toxicological sciences. *Environ. Pollut.* 138:379–412.

Calabrese, E.J. and L.A. Baldwin. 2002. Hormesis and high-risk groups. *Regul. Toxicol. Pharmacol.* 35:414–428.

Caldwell, D.J. 1999. Review of mononuclear cell leukemia in F-344 rat bioassays and its significance to human cancer risk: A case study using alkyl phthalates. *Regul. Toxicol. Pharmacol.* 30:45–53.

Carmichael, N.G., H. Enzmann, I. Pate, and F. Waechter, 1997. The significance of mouse liver tumor formation for carcinogenic risk assessment: results and conclusions from a survey of ten years of testing by the agrochemical industry. *Environ. Health Perspect.* 105:1196–1203.

Cheeseman, M.A., E.J. Machuga, and A.B. Bailey. 1999. A tiered approach to threshold of regulation. *Food Chem. Toxicol.* 37:387–412.

Combes, R.D. 2000. Endocrine disruptors: A critical review of *in vitro* and *in vivo* testing strategies for assessing their toxic hazard to humans. *ATLA* 28:81–118.

Conolly, R.B., J.S. Kimbell, D. Janszen, et al. 2004. Human respiratory tract cancer risks of inhaled formaldehyde: Dose–response predictions derived from biologically-motivated computational modeling of a combined rodent and human dataset. *Toxicol. Sci.* 82:279–296.

Cook, J.C., G.R. Klinefelter, J.F. Hardisty, R.M. Sharpe, and P.M. Foster. 1999. Rodent Leydig cell tumorigenesis: A review of the physiology, pathology, mechanisms, and relevance to humans. *Crit. Rev. Toxicol.* 29:169–261.

CPDB. 2007: The carcinogenic potency database website. http://potency.berkeley.edu/cpdb.html

Cramer, G.M., R.A. Ford, and R.A. Hall. 1978. Estimation of toxic hazard - A decision tree approach. *Food Cosmet. Toxicol.* 16:255–276.

Crump, K.S. 1984. A new method for determining allowable daily intakes. *Fundam. Appl. Toxicol.* 4:854–871.

Crump, K. 2002. Critical issues in benchmark calculations from continuous data. *Crit. Rev. Toxicol.* 32:133–153.

Dearman, R.J. and I. Kimber. 2007. A mouse model for food allergy using intraperitoneal sensitization. *Methods* 41:91–98.

DK NFA. 1990. *Quantitative Risk Analysis of Carcinogens*. Søborg, Denmark: National Food Agency of Denmark, Institute of Toxicology, Ministry of Health.

Draize, J.H., G. Woodard, and H.O. Calvery. 1944. Methods for the study of irritation and toxicity of substances applied topically to the skin and mucous membranes. *J. Pharmacol. Exp. Ther.* 82:377–390.

EC. 1967. Council Directive 67/548/EEC of 27 June 1967 on the approximation of laws, regulations and administrative provisions relating to the classification, packaging and labelling of dangerous substances. *Off. J. Eur. Communities* L 196, 16.8.1967, pp. 1–98.

EC. 1999. Communication from the Commission to the Council and the European Parliament on Community strategy for endocrine disruptors, a range of substances suspected of interfering with the hormone systems of humans and wildlife. COM (1999) 706 Final. http://ec.europa.eu/environment/docum/99706sm.htm

EC. 2001a. Annex VI to Directive 67/548. http://ecb.jrc.it/classification-labelling/

EC. 2001b. Communication from the Commission to the Council and the European Parliament on implementation of the Community strategy for endocrine disrupters - A range of substances suspected of interfering with the hormone systems of humans and wildlife (COM (1999) 706). http://eur-lex.europa.eu/LexUriServ/site/en/com/2001/com2001_0262en01.pdf

EC. 2003. Technical guidance document on risk assessment in support of Commission Directive 93/67/EEC on risk assessment for new notified substances, Commission Regulation (EC) No 1488/94 on Risk Assessment for existing substances and Directive 98/8/EC of the European Parliament and of the Council concerning the placing of biocidal products on the market. http://ecb.jrc.it/tgd

EC. 2004. Commission staff working document on implementation of the Community strategy for endocrine disrupters - A range of substances suspected of interfering with the hormone systems of humans and wildlife (COM (1999) 706). http://ec.europa.eu/environment/endocrine/documents/sec_2004_1372_en.pdf

ECETOC. 1997. ECETOC Document No.37, EC Classification of Eye Irritancy. Brussels: ECETOC.

ECETOC. 2000. *Skin sensitisation testing for the purpose of hazard identification and risk assessment.* ECETOC Monograph No. 29. Brussels: ECETOC.

ECETOC. 2002a. *Recognition of, and differentiation between, adverse and non-adverse effects in toxicology studies.* ECETOC Technical Report No. 85. Brussels: ECETOC.

ECETOC. 2002b. *Use of human data in hazard classification for irritation and sensitisation.* ECETOC Monograph No. 32. Brussels: ECETOC.

ECETOC. 2004. *Targeted risk assessment.* ECETOC Technical Report No. 93. Brussels: ECETOC.

ECETOC. 2006. *Toxicological modes of action: Relevance for human risk assessment.* ECETOC Technical Report No. 99. Brussels: ECETOC.

EDSTAC. 1998. Endocrine Disruptor Screening and Testing Advisory Committee (EDSTAC) Final Report. http://www.epa.gov/scipoly/oscpendo/pubs/edstac/exesum14.pdf

EU. 2001. Commission Directive 2001/59/EC of 6 August 2001 adapting to technical progress for the 28th time Council Directive 67/548/EEC on the approximation of the laws, regulations and administrative provisions relating to the classification, packaging and labelling of dangerous substances (text with EEA relevance). *Off. J. Eur. Communities* L 225, 21.8.2001, pp. 1–333.

EU. 2006. The DG Environment REACH website. http://ec.europa.eu/environment/chemicals/reach

EU. 2007. The DG Environment Endocrine Disrupters website. http://ec.europa.eu/environment/endocrine/index_en.htm

Falk Filipsson, A., S. Sand, J. Nilsson, and K. Victorin. 2003. The benchmark dose method - Review of available models, and recommendations for application in health risk assessment. *Crit. Rev. Toxicol.* 33:505–542.

FAO. 1998. Pesticide residues in food. *Report of the Joint Meeting of the FAO Panel of Experts on Pesticide Residues in Food and the Environment and the WHO Core Assessment Group.* Rome: FAO.

FAO. 1999. Plant production and protection paper, No. 148, 17–19. Rome: FAO.

Federal Register. 1993. Food additives: Threshold of regulation for substances used in food-contact articles: Proposed rule. 58:52719–52729.

Federal Register. 1995. Food additives: Threshold of regulation for substances used in food-contact articles: Final rule. 60:36582–36596. http://www.cfsan.fda.gov/~lrd/t36582.html

Gephart, L.A., W.F. Salminen, M.J. Nicolich, and M. Pelekis. 2001. Evaluation of subchronic toxicity data using the benchmark dose approach. *Regul. Toxicol. Pharmacol.* 33:37–59.

Gold, L.S., C.B. Sawyer, R. Magaw, et al. 1984. A carcinogenic potency database of the standardized results of animal bioassay. *Environ. Health Perspect.* 58:9–319.

Gold, L.S., T.H. Slone, and L. Bernstein. 1989. Summary of carcinogenic potency and positivity for 492 rodent carcinogens in the carcinogenic potency database. *Environ. Health Perspect.* 79:259–272.

Gulezian, D., D. Jacobson-Kram, C.B. McCullough, et al. 2000. Use of transgenic animals for carcinogenicity testing: Considerations and implications for risk assessment. *Toxicol. Pathol.* 28:482–499.

Haschek, W.M., C.G. Rousseaux, and M.A. Wallig. 2001. *Handbook of Toxicologic Pathology. Second edition.* San Diego, San Fransisco, New York, Boston, London, Sydney, Tokyo: Academic Press.

Hester, S.D., G.B. Benavides, M. Sartor, L. Yoon, K.T. Morgan, and D.C. Wolf. 2001. Studies of normal gene expression in the rat nasal epithelium using cDNA array technology. *Nature Genetics* 27:59–60.

IARC. 1994. *Peroxisome proliferation and its role in carcinogenesis.* IARC Technical Publication No. 24. Lyon, France: IARC.

IARC. 1998. *The use of short- and medium-term tests for carcinogens and data on genetic effects in carcinogenic hazard evaluation.* IARC Scientific Publications No 146. Lyon, France: IARC.

IARC. 1999a. *Species differences in thyroid, kidney and urinary bladder carcinogenesis.* International Agency for Research on Cancer (IARC), IARC Scientific Publications No 147. Lyon, France: IARC.

IARC. 1999b. *Surgical implants and other foreign bodies*, IARC Summary & Evaluation, Volume 74. Lyon, France: IARC.

IARC. 2006. Preamble to the IARC Monographs (amended January 2006). http://monographs.iarc.fr/ENG/Preamble/index.php

Kelly, J.T. and B. Asgharian. 2003. Nasal molds as predictors of fine and coarse particle deposition in rat nasal airways. *Inhal. Toxicol.* 15:859–875.

Kimber, I. and R.J. Dearman. 2002. Immune responses: Adverse versus non-adverse effects. *Toxicol. Pathol.* 30:54–58.

Kroes, R. and W. Wester. 1986. Forestomach carcinogens: Possible mechanisms of action. *Food Chem. Toxicol.* 24:1083–1089.

Kroes, R., C. Galli, I. Munro, et al. 2000. Threshold pf toxicological concern for chemical substances in the diet: A practical tool for assessing the need for toxicity testing. *Food Chem. Toxicol.* 38:255–312.

Kroes, R., A. Renwick, M. Cheeseman, et al. 2004. Structure-based thresholds of toxicological concern (TTC): Guidance for application to substances present at low levels in the diet. *Food Chem. Toxicol.* 42:65–83.

Larsen, J.C. 2006. Risk assessment of chemicals in European traditional foods. *Trends Food Sci. Tech.* 17:471–481.

Mainwaring, G., J.R. Foster, V.J. Lund, and T. Green. 2001. Methyl methacrylate toxicity in rat nasal epithelium: Studies of the mechanism of action and comparisons between species. *Toxicology* 158:109–118.

Moore, G.E. 1991. Foreign body carcinogenesis. *Cancer* 67:2731–2732.

Munro, I.C. 1990. Safety assessment procedures for indirect food additives: An overview. Report of a workshop. *Regul. Toxicol. Pharmacol.*, 12:2–12.

Munro, I.C., R.A. Ford, E. Kennepohl, and J.G. Sprenger. 1996. Correlation of structural class with no-observed-effect levels: A proposal for establishing a threshold of concern. *Food Chem. Toxicol.* 34:829–867.

Munro, I.C., E. Kennepohl, and R. Kroes. 1999. A procedure for the safety evaluation of flavouring substances. *Food Chem. Toxicol.* 37:207–232.

NCM. 2005. *Threshold of toxicological concern (TTC). Literature review and applicability.* TemaNord 2005, 559, Copenhagen: Nordic Council of Ministers http://www.norden.org/pub/miljo/miljo/sk/TN2005559.pdf

OECD. 1999a. *Detailed review document on classification systems for sensitising substances in OECD member countries.* OECD Series on Testing and Assessment No. 13. Environment Directorate, Joint Meeting of the Chemicals Committee and the Working Party on Chemicals. ENV/JM/MONO(99)3. Paris: OECD.

OECD. 1999b. *Detailed review document on classification systems for germ cell mutagenicity in OECD member countries.* OECD Series on Testing and Assessment No. 12. Environment Directorate, Joint Meeting of the Chemicals Committee and the Working Party on Chemicals. ENV/JM/MONO(99)2. Paris: OECD.

OECD. 2000. *Guidance notes for analysis and evaluation of repeat-dose toxicity studies.* OECD Series on Testing and Assessment No. 32. Environment Directorate, Joint Meeting of the Chemicals Committee and the Working Party on Chemicals, Pesticides and Biotechnology. ENV/JM/MONO(2000)18. Paris: OECD.

OECD. 2001a. *Guidance document on acute oral toxicity testing.* OECD Series on Testing and Assessment No. 24. Environment Directorate, Joint Meeting of the Chemicals Committee and the Working Party on Chemicals, Pesticides and Biotechnology. ENV/JM/MONO(2001)4. Paris: OECD.

OECD. 2001b. *Harmonised integrated classification system for human health and environmental hazards of chemical substances and mixtures*. OECD Series on Testing and Assessment No. 33. Environment Directorate, Joint Meeting of the Chemicals Committee and the Working Party on Chemicals, Pesticides and Biotechnology. ENV/JM/MONO(2001)6. Paris: OECD.

OECD. 2002a. *Guidance notes for analysis and evaluation of chronic toxicity and carcinogenicity studies*. OECD Series on Testing and Assessment No. 35. Environment Directorate, Joint Meeting of the Chemicals Committee and the Working Party on Chemicals, Pesticides and Biotechnology. ENV/JM/MONO(2002)19. Paris: OECD.

OECD. 2002b. *Detailed review paper. Appraisal of test methods for sex hormone disrupting chemicals*. OECD Series on Testing and Assessment No. 21. Environment Directorate, Joint Meeting of the Chemicals Committee and the Working Party on Chemicals, Pesticides and Biotechnology. ENV/JM/MONO (2002)8. Paris: OECD.

OECD. 2003a. *Descriptions of selected key generic terms used in chemical hazard/risk assessment*. Joint project with IPCS on the harmonisation of hazard/risk assessment terminology. OECD Series on Testing and Assessment No. 44. Environment Directorate, Joint Meeting of the Chemicals Committee and the Working Party on Chemicals, Pesticides and Biotechnology. ENV/JM/MONO(2003)15. Paris: OECD.

OECD. 2003b. *Detailed background review of the uterotrophic assay. Summary of the available literature in support of the project of the OECD task force on endocrine disrupters testing and assessment (EDTA) to standardise and validate the uterotrophic assay*. OECD Series on Testing and Assessment No. 38. Environment Directorate, Joint Meeting of the Chemicals Committee and the Working Party on Chemicals, Pesticides and Biotechnology. ENV/JM/MONO(2003)1. Paris: OECD.

OECD. 2004a. *Manual for Investigation of HPV Chemicals*. Paris: OECD. http://www.oecd.org/document/7/0,2340,en_2649_34379_1947463_1_1_1_1,00.html

OECD. 2004b. *Draft guidance document on acute inhalation toxicity testing*. OECD Series on Testing and Assessment No. 39B. Environment Directorate, Joint Meeting of the Chemicals Committee and the Working Party on Chemicals, Pesticides and Biotechnology. December 8, 2004 (1st version). Paris: OECD.

OECD. 2004c. *Guidance document for neurotoxicity testing*. OECD Series on Testing and Assessment No. 20. Environment Directorate, Joint Meeting of the Chemicals Committee and the Working Party on Chemicals, Pesticides and Biotechnology. ENV/JM/MONO(2004)25. Paris: OECD.

OECD. 2004d. *Draft guidance document on reproductive toxicity testing and assessment*. OECD Series on Testing and Assessment No. 43. Environment Directorate, Joint Meeting of the Chemicals Committee and the Working Party on Chemicals, Pesticides and Biotechnology. November 10, 2004 (1st version). Paris: OECD.

OECD. 2005. Environment, Health & Safety News, No. 18, November 2005.

OECD. 2006a. *OECD report of the initial work towards the validation of the rodent uterotrophic assay - Phase 1*. OECD Series on Testing and Assessment No. 65. Environment Directorate, Joint Meeting of the Chemicals Committee and the Working Party on Chemicals, Pesticides and Biotechnology. ENV/JM/MONO(2006)33. Paris: OECD.

OECD. 2006b. OECD report of the validation of the rodent uterotrophic bioassay: Phase 2. Testing of potent and weak oestrogen agonists by multiple laboratories. OECD Series on Testing and Assessment No. 66. Environment Directorate, Joint Meeting of the Chemicals Committee and the Working Party on Chemicals, Pesticides and Biotechnology. ENV/JM/MONO(2006)34. Paris: OECD.

OECD. 2006c. *Summary report of the uterotrophic bioassay peer review panel, including agreement of the working group of national coordinators of the test guidelines programme on the follow-up of this report*. OECD Series on Testing and Assessment No. 68. Environment Directorate, Joint Meeting of the Chemicals Committee and the Working Party on Chemicals, Pesticides and Biotechnology. ENV/JM/MONO(2006)37. Paris: OECD.

OECD. 2006d. *Draft Hershberger background review document*. Paris: OECD, http://www.oecd.org/dataoecd/18/57/37880949.pdf

OECD. 2006e. *Stably transfected transcriptional activation (TA) assay for detecting estrogenic activity of chemicals - The human estrogen receptor alpha mediated reporter gene assay using HeLa-hER-9903 cell line*. Version October 2006. Paris: OECD, http://www.oecd.org/dataoecd/49/55/37531918.pdf

OECD. 2006f. *Detailed review paper on thyroid hormone disruption assays*. OECD Series on Testing and Assessment No. 57. Environment Directorate, Joint Meeting of the Chemicals Committee and the Working Party on Chemicals, Pesticides and Biotechnology. ENV/JM/MONO(2006)24. Paris: OECD.

OECD. 2007a. *The uterotrophic bioassay in rodents: A short-term screening test for oestrogenic properties*. Draft OECD guideline for the testing of chemicals. Revised 23 January 2007. Paris: OECD, http://www.oecd.org/dataoecd/48/55/37994719.doc

OECD. 2007b. Endocrine disrupter testing and assessment website. http://www.oecd.org/document/62/0,2340,en_2649_34377_2348606_1_1_1_1,00.htm

OECD. 2007c. *The Hershberger bioassay in rats*. Draft OECD guideline for the testing of chemicals. Paris: OECD, http://www.oecd.org/dataoecd/49/60/37478355.pdf

Prieto, P., C. Clemedson, A. Meneguz, W. Pfaller, U.G. Sauer, and C. Westmoreland. 2005. 3.6 Subacute and subchronic toxicity. *ATLA* 33:109–116.

Prieto, P., A.W. Baird, B.J. Blaauboer, J.V. Castell Ripoll, et al. 2006. The assessment of repeated dose toxicity *in vitro*: A proposed approach. The report and recommendations of ECVAM Workshop 56. *ATLA* 34:315–341.

Riley, V. 1975. Mouse mammary tumors: Alteration of incidence as apparent function of stress. *Science* 189:465–467.

Rulis, A.M. 1986. *De minimis* and the threshold of regulation. In: Felix, C.W. (Ed.) *Food Protection Technology* 29–37. Chelsea, Michigan: Lewis Publishers Inc.

Rulis, A.M. 1989. Establishing a threshold of concern. In: Bonin, J.J. and Donald, E.S., eds. *Risk Assessment in Setting National Priorities, Vol. 7.*, 271–278. New York: Plenum Press.

Slob, W. 1999. Thresholds in toxicology and risk assessment. *Int. J. Toxicol.* 18:259–268.

Slob, W. and M.N. Pieters. 1998. A probabilistic approach for deriving acceptable human intake limits and human risks from toxicological studies: General framework. *Risk Anal.* 18:787–798.

Sonich-Mullin, C., R. Fielder, J. Wiltse, et al. 2001. IPCS conceptual framework for evaluating a mode of action for chemical carcinogenesis. *Regul. Toxicol. Pharmacol.* 34:146–152.

Sunderman, F.W. Jr. 2001. Nasal toxicity, carcinogenicity, and olfactory uptake of metals. *Ann. Clin. Lab. Sci.* 31:3–24.

Tischler, A.S., J.F. Powers, M. Pignatello, P. Tsokas, J.C. Downing, and R.M. McClain. 1999. Vitamin D3-induced proliferative lesions in the rat adrenal medulla. *Toxicol. Sci.* 51:9–18.

US-EPA. 1986a. Guidelines for mutagenicity risk assessment. Federal Register 51(185):34006–34012. http://cfpub.epa.gov/ncea/raf/rafguid.cfm

US-EPA. 1986b. Guidelines for carcinogen risk assessment. Federal Register 51(185):33992–34003. http://cfpub.epa.gov/ncea/raf/rafguid.cfm

US-EPA. 1991. Guidelines for developmental toxicity risk assessment. Federal Register 56(234): 63798–63826. http://cfpub.epa.gov/ncea/raf/rafguid.cfm

US-EPA. 1995. The use of the benchmark dose approach in health risk assessment. US Environmental Protection Agency (EPA), Office of Research and Development, Doc. EPA/630/R-94/007, Washington, DC.

US-EPA. 1996. Guidelines for reproductive toxicity risk assessment. Federal Register 61(212): 56274–56322. http://cfpub.epa.gov/ncea/raf/rafguid.cfm

US-EPA. 1998. Guidelines for neurotoxicity risk assessment. U.S. Environmental Protection Agency, Risk Assessment Forum, Washington, DC, 630/R-95/001F, 1998.

US-EPA. 2005. Guidelines for carcinogen risk assessment. U.S. Environmental Protection Agency, Risk Assessment Forum, Washington, DC, 630/P-03/001F, 2005.

US-EPA. 2007a. Integrated risk information system (IRIS). http://www.epa.gov/iris/index.html

US-EPA. 2007b. Glossary of IRIS terms. http://www.epa.gov/iris/gloss8.htm

US-EPA. 2007c. US-EPA Endocrine Disruptor Screening Program - Overview website. http://www.epa.gov/scipoly/oscpendo/pubs/edspoverview/index.htm

US-EPA. 2007d. US-EPA Endocrine Disruptor Screening Program - Assay status website. http://www.epa.gov/scipoly/oscpendo/pubs/assayvalidation/status.htm

US-NTP. 2006. NTP Historical controls. Report. All routes and vehicles. Mice. June 2006. http://ntp.niehs.nih.gov/files/HIST_CONTROLS_2006_TUMOR_MICE1.pdf

US-NTP. 2007. Study results and research projects website, definition of carcinogenicity results. http://ntp.niehs.nih.gov/index.cfm?objectid = 07027D0E-E5CB-050E-027371D9CC0AAACF

Veenstra, G. and D. Kroese. 2005. Threshold of toxicological concern. A concept in toxicological risk assessment. Discussion Document, March 2005. RIP 3.3 SEG1–19.

WHO/IPCS. 1985a. Summary report on the evaluation of short-term tests for carcinogens (collaborative study on *in vitro* tests). Environmental Health Criteria 47. Geneva: WHO. http://www.inchem.org/documents/ehc/ehc/ehc47.htm

WHO/IPCS. 1985b. Guide to short-term tests for detecting mutagenic and carcinogenic chemicals. Environmental Health Criteria 51. Geneva: WHO. http://www.inchem.org/documents/ehc/ehc/ehc051.htm

WHO/IPCS. 1986a. Principles of toxicokinetic studies. Environmental Health Criteria 57. Geneva: WHO. http://www.inchem.org/documents/ehc/ehc/ehc57.htm

WHO/IPCS. 1986b. Principles and methods for the assessment of neurotoxicity associated with exposure to chemicals. Environmental Health Criteria 60. Geneva: WHO. http://www.inchem.org/documents/ehc/ehc/ehc060.htm

WHO/IPCS. 1990. Summary report on the evaluation of short-term tests for carcinogens (collaborative study on *in vivo* tests). Environmental Health Criteria 109. Geneva: WHO. http://www.inchem.org/documents/ehc/ehc/ehc109.htm

WHO/IPCS. 1991. Principles and methods for the assessment of nephrotoxicity associated with exposure to chemicals. Environmental Health Criteria 119. Geneva: WHO. http://www.inchem.org/documents/ehc/ehc/ehc119.htm

WHO/IPCS. 1994. Assessing human health risks of chemicals: Derivation of guidance values for health-based exposure limits. Environmental Health Criteria 170. Geneva: WHO. http://www.inchem.org/documents/ehc/ehc/ehc170.htm

WHO/IPCS. 1996. Principles and methods for assessing direct immunotoxicity associated with exposure to chemicals. Environmental Health Criteria 180. Geneva: WHO. http://www.inchem.org/documents/ehc/ehc/ehc180.htm

WHO/IPCS. 1999a. Principles for the assessment of risks to human health from exposure to chemicals. Environmental Health Criteria 210. Geneva: WHO. http://www.inchem.org/documents/ehc/ehc/ehc210.htm

WHO/IPCS. 1999b. Principles and methods for assessing direct allergic hypersensitization associated with exposure to chemicals. Environmental Health Criteria 212. Geneva: WHO. http://www.inchem.org/documents/ehc/ehc/ehc212.htm

WHO/IPCS. 2001a. Neurotoxicity risk assessment for human health: Principles and approaches. Environmental Health Criteria 223. Geneva: WHO. http://www.inchem.org/documents/ehc/ehc/ehc223.htm

WHO/IPCS. 2001b. Principles for evaluating health risks to reproduction associated with exposure to chemicals. Environmental Health Criteria 225. Geneva: WHO. http://www.inchem.org/documents/ehc/ehc/ehc225.htm

WHO/IPCS. 2002. Global assessment of the state-of-the-science of endocrine disruptors. International Programme on Chemical Safety. Geneva: WHO. http://www.who.int/ipcs/publications/en/toc.pdf

WHO/IPCS. 2005. Chemical-specific adjustment factors for interspecies differences and human variability: Guidance document for use of data in dose/concentration–response assessment. Harmonization Project Document No. 2. http://whqlibdoc.who.int/publications/2005/9241546786_eng.pdf

WHO/IPCS. 2007. IPCS Project on the Harmonization of Approaches to the Assessment of Risk from Exposure to Chemicals. General Conclusions and Recommendations of an IPCS International Workshop on Skin Sensitization in Chemical Risk Assessment. http://www.who.int/ipcs/methods/harmonization/areas/sensitization_summary.pdf

Williams, G.M. 1997. Chemicals with carcinogenic activity in the rodent liver; mechanistic evaluation of human risk. *Cancer Lett.* 117:175–188.

Worth, A.P. and M. Balls, eds. 2002. Alternative (non-animal) methods for chemicals testing: Current status and future prospects. A report prepared by ECVAM and the ECVAM working group on chemicals. *ATLA* 30:71.

5 Standard Setting: Threshold Effects

Adverse health effects can be considered to be of two types (see Section 4.2): those considered to have a threshold, known as "threshold effects" (effects such as, e.g., organ-specific, neurological, immunological, non-genotoxic carcinogenicity, reproductive, developmental), and those for which there is considered to be some risk at any exposure level, known as "non-threshold effects" (effects such as, e.g., mutagenicity, genotoxicity, genotoxic carcinogenicity). Though it is not possible to demonstrate experimentally the presence or absence of a threshold, differences in the approach to the hazard assessment of threshold versus non-threshold effects have been adopted widely. The distinction in approaches is based primarily on the premise that simple events such as *in vitro* activation and covalent binding may be linear over many orders of magnitude, i.e., that these events occur even at very low exposure levels. However, a simple pragmatic distinction on this basis is increasingly problematic as it is likely that there is a threshold for a number of genotoxic effects; this is addressed in detail in Chapter 6.

In the hazard assessment process, described in detail in Chapter 4, all effects observed are evaluated in terms of the type and severity (adverse or non-adverse), their dose–response relationship, and the relevance for humans of the effects observed in experimental animals. For threshold effects, a No- or a Lowest-Observed-Adverse-Effect Level (N/LOAEL), or alternatively a Benchmark Dose (BMD), is derived for every single effect in all the available studies provided that data are sufficient for such an evaluation. In the last step of the hazard assessment for threshold effects, all this information is assessed in total in order to identify the critical effect(s) and to derive a NOAEL, or LOAEL, for the critical effect(s).

The approach of deriving a tolerable intake by dividing the N/LOAEL, or alternatively a BMD for the critical effect(s) by an assessment factor has been described and discussed extensively in the scientific literature. It is beyond the scope of this book to review all these references. This chapter presents an overview of published extrapolation methods for the derivation of a tolerable intake based on the assessment factor approach, i.e., limited to address effects with threshold characteristics, and is not meant to be exhaustive. The main focus is on the rationale for and the use of the assessment factors. Pertinent guidance documents and reviews for the issues addressed in this chapter include WHO/IPCS (1994, 1996, 1999), US-EPA (2002, 2004), IGHRC (2003), ECETOC (2003), KEMI (2003), Kalberlah and Schneider (1998), Vermeire et al. (1999), and Nielsen et al. (2005).

The approach of standard setting for non-threshold effects is addressed in Chapter 6.

The development of regulatory standards derived from a standard such as, e.g., the Tolerable Daily Intake or a Reference Dose, is addressed in Chapter 9.

5.1 INTRODUCTION

According to the OECD/IPCS definitions listed in Annexure 1 of Chapter 1 (OECD 2003):

Threshold is "Dose or exposure concentration of a substance below that a stated effect is not observed or expected to occur."

Tolerable Intake is "Estimated maximum amount of an agent, expressed on a body mass basis, to which each individual in a (sub) population may be exposed over a specified period without appreciable risk."

A tolerable intake may have different units depending on the route of administration upon which it is based, and is generally expressed on a daily or weekly basis. For the oral and dermal routes, a tolerable intake is generally expressed on a body weight basis, e.g., mg/kg body weight per day. Though not strictly an "intake," tolerable intakes for inhalation are generally expressed as an airborne concentration, e.g., mg/m^3.

According to the OECD/IPCS definitions listed in Annexure 1 of Chapter 1 (OECD 2003):

Acceptable/Tolerable Daily Intake is "Estimated maximum amount of an agent, expressed on a body mass basis, to which an individual in a (sub) population may be exposed daily over its lifetime without appreciable health risk."

Reference Dose is "An estimate of the daily exposure dose that is likely to be without deleterious effect even if continued exposure occurs over a lifetime."

Related terms: Acceptable/Tolerable Daily Intake.

The term "acceptable" is used widely to describe "safe" levels of intake and is applied for chemicals to be used in food production such as, e.g., food additives, pesticides, and veterinary drugs. The term "tolerable" is applied for chemicals unavoidably present in a media such as contaminants in, e.g., drinking water and food. The term "PTWI" (Provisional Tolerable Weekly Intake) is generally used for contaminants that may accumulate in the body, and the weekly designation is used to stress the importance of limiting intake over a period of time for such substances. The tolerable intake is similar in definition and intent to terms such as "Reference Dose" and "Reference Concentration" (RfD/RfC), which are widely used by, e.g., the US-EPA. For some substances, notably pesticides, the "ARfD" (Acute Reference Dose), is also established, often from shorter-term studies than those that would support the ADI. The ARfD is defined as the amount of a substance in food that can be consumed in the course of a day or at a single meal with no adverse effects.

In inhalation studies, laboratory animals are generally exposed to an airborne chemical for a limited period of time, e.g., 6 h a day, 5 days per week. Adjustment of such an intermittent exposure to a continuous exposure scenario is regularly applied as a default procedure to inhalation studies with repeated exposures but not to single-exposure inhalation toxicity studies. Operationally, this is accomplished by a correction for both the number of hours in a daily exposure period and the number of days per week that the exposures were performed. In an inhalation study in which animals were exposed to an airborne concentration of a substance at 5 mg/m^3 for 6 h a day, for 5 days per week, the adjustment of this intermittent exposure concentration to a continuous exposure concentration would consider both hours per day and days per week: 5 mg/m$^3 \times 6/24$ h $\times 5/7$ days/week $= 0.9$ mg/m^3, with 0.9 mg/m^3 being the concentration adjusted to continuous exposure.

For systemic effects observed in inhalation studies, the determining factor for effects to occur at the systemic target is generally the total dose rather than the concentration of the chemical in the air. In such cases, a tolerable intake (expressed as mg/kg body weight per day, or mg/m^3 depending on the standard to be derived, i.e., a tolerable intake in its strict meaning, or a tolerable concentration) is established from the NOAEC, or LOAEC, derived in the inhalation study and adjusted for continuous exposure.

For local effects, in contrast, the determining factor for effects to occur at the site of first contact (mucous membrane of the respiratory tract, the eyes, or the skin) is generally the concentration of the chemical in the air rather than the total dose at the site of first contact. In such cases, a tolerable concentration (expressed as mg/m^3) is established from the NOAEC, or LOAEC, derived in the inhalation study without an adjustment to a continuous exposure.

The overall principles for the derivation of a tolerable intake are equal irrespective of chemical class (e.g., food additives, pesticides, veterinary drugs, contaminants) although it should be recognized that the available database for chemicals deliberately added to, e.g., food is generally more comprehensive than for contaminants. This is because there are extensive regulatory demands for toxicity data in relation to marketing of chemicals, which are intentionally applied to food, etc. For threshold effects, a tolerable intake is generally derived from the NOAEL, or LOAEL, for the critical effect(s) by dividing the NOAEL/LOAEL, by an overall assessment factor.

According to the OECD/IPCS definitions listed in Annexure 1 of Chapter 1 (OECD 2003):

Assessment Factor is "Numerical adjustment used to extrapolate from experimentally determined (dose–response) relationships to estimate the agent exposure below which an adverse effect is not likely to occur."

Related terms: Safety Factor, Uncertainty Factor, Extrapolation Factor, Adjustment Factor, Conversion Factor.

There is an enormous variability in the extent and nature of different databases for chemical substances. For example, in some cases, the evaluation of a chemical must be based on limited data in experimental animals, whereas in other cases detailed information on the various endpoints, toxicokinetics, and mode(s) of action may be available. In some cases, the evaluation can be based on data on effects in exposed human populations. Clearly as the amount of information available increases, the degree of understanding of the hazards expressed also increases, and the uncertainties due to lack of information decrease. However, even with complex databases, uncertainties still remain.

The assessment factors generally applied in the establishment of a tolerable intake from the NOAEL, or LOAEL, for the critical effect(s) are applied in order to compensate for uncertainties inherent to extrapolation of experimental animals data to a given human situation, and for uncertainties in the toxicological database, i.e., in cases where the substance-specific knowledge required for risk assessment is not available. As a consequence of the variability in the extent and nature of different databases for chemical substances, the range of assessment factors applied in the establishment of a tolerable intake has been wide (1–10,000), although a value of 100 has been used most often. An overview of different approaches in using assessment factors, historically and currently, is provided in Section 5.2.

The key areas of uncertainty when using data from experimental animals include uncertainty related to:

- Extrapolation from animal species to humans (Section 5.3)
- Variability in the human population (Section 5.4)
- Route-to-route extrapolation (Section 5.5)
- Duration of exposure in experimental studies (Section 5.6)
- Dose–response curve/NOAEL not established (Section 5.7)
- Nature and severity of the effects (Section 5.8)
- Gaps or other deficiencies in the database (Section 5.9)

5.2 ASSESSMENT FACTORS: GENERAL ASPECTS

In the context of assessment factors, it is important to distinguish between the two terms "variability" and "uncertainty." Variability refers to observed differences attributable to true heterogeneity or diversity, i.e., inherent biological differences between species, strains, and individuals. Variability is the result of natural random processes and is usually not reducible by further measurement or study although it can be better characterized. Uncertainty relates to lack of knowledge about, e.g., models, parameters, constants, data, etc., and can sometimes be minimized, reduced, or eliminated if additional information is obtained (US-EPA 2003).

It should be recognized that a lack of knowledge of variability is a source of uncertainty.

The terminology within this area is not standardized. Other terms include "safety factor," "uncertainty factor," "extrapolation factor," "adjustment factor," and "conversion factor." None of these terms are ideal. For example, the term safety factor has implications of absolute safety, whereas the term uncertainty factor, although being broader, may be interpreted differently in relation to variability and uncertainty. For the sake of clarity in this book, the term assessment factor is used and is meant as a general term to cover all factors designated in the literature as safety factor, uncertainty factor, extrapolation factor, adjustment factor, conversion factor, etc. The other mentioned terms are not used unless reference is made to a specific term or method. The assessment factor can cover both variability and uncertainty.

The following section gives an overview of different approaches in using assessment factors, historically and currently, beginning with the introduction of the so-called "safety factor approach" in the mid-1950s and reflecting the development up to the regulatory approaches currently used by international and federal bodies. The overview does not attempt to cover all publications in this field, but includes the approaches suggested by different scientific groups and international and federal bodies, which are considered as being the most central ones in the development of the approaches currently used regulatory. Default assessment factors used or suggested in the various approaches are summarized in Table 5.1.

5.2.1 Assessment Factors: Various Approaches

Historically, the so-called safety factor approach was introduced in the United States in the mid-1950s in response to the legislative needs in the area of the safety of chemical food additives (Lehman and Fitzhugh 1954). This approach proposed that a "safe level" of chemical food additives could be derived from a chronic NOAEL from animal studies divided by a 100-fold safety factor. The 100-fold safety factor as proposed by Lehman and Fitzhugh was based on a limited analysis of subchronic/chronic data on fluorine and arsenic in rats, dogs, and humans, and also on the assumption that the human population as a whole is heterogeneous. Initially, Lehman and Fitzhugh reasoned that the safety factor of 100 accounted for several areas of uncertainty:

- Intraspecies (human-to-human) variability
- Interspecies (animal-to-human) variability
- Allowance for sensitive human populations due to illness when compared with healthy experimental animals
- Possible synergistic action of the many intentional and unintentional food additives or contaminants

In 1961, the Joint FAO/WHO Expert Committee on Food Additives (JECFA) and the Joint Meeting of Experts on Pesticides Residues (JMPR) adopted this approach in a slightly modified form: The safe level was called the Acceptable Daily Intake (ADI) and expressed in mg/kg body weight per day (Vermeire et al. 1999, ECETOC 2003). Usually, a safety factor of 100 is used by JECFA and JMPR for establishing ADIs by this ADI approach; however, the procedures adopted by JECFA and JMPR do not generate a clear justification for deviation from the factor of 100, but in some individual cases, an expert explanation is given for the use of factors other than 100 (Vermeire et al. 1999).

It is apparent that the factor of 100 has no quantitative bases, and the choice of the value 100 is more or less arbitrary (Vermeire et al. 1999). Retrospectively, some attempts have been made to support a 100-fold factor (Bigwood 1973, Lu 1979, Vettorazzi 1977 as reviewed in Vermeire et al. 1999 and KEMI 2003), and the 100-fold factor was found to be justified.

The 100-fold safety factor has traditionally been interpreted as the product of two factors with default values of 10. For example, according to WHO/IPCS (1987), the safety factor is intended to provide an adequate Margin of Safety (MOS) by assuming that the human being is 10 times more sensitive than the test animal and that the difference of sensitivity within the human population is in a 10-fold range.

TABLE 5.1
Default Assessment Factors Used or Suggested for the Establishment of a Regulatory Standard or Health-Based Guidance Value for Threshold Effects

Factor	JECFA JMPR	US-EPA (2002)	Renwick (1993)	WHO	LLN (1990)	ECETOC (2003)	TNO (1996)	Kalberlah and Schneider (1998)	KEMI (2003)[a]	D-EPA (2006)
Interspecies	10	10			10					10
Toxicokinetic		A^b	4	4		A^b			4 (rat)	
Toxicodynamic		$10^{0.5}$	2.5	2.5		1			2.5	
Oral							$A^b \times 3$	$A^b \times 2\text{--}3$		
Inhalation							3			
Interindividual	10	10			10	5	10	25		10
Toxicokinetic			4	3.16				8	3–5	
Toxicodynamic			2.5	3.16				3	3.16	
Occupational						3	3			
Route-to-route		10								
Duration exposure							10			
Subacute-to-subchronic						6			3	
Subacute-to-chronic					10	2	10		24	10
Subchronic-to-chronic					10	3	1		8	10
LOAEL-to-NOAEL		10			10		1		3–10	up to 10
Nature and severity							1		1–10	1–10
Confidence database		10			10				1–10	
Nonscientific					1					

a See also Table 5.2.

b A is a calculated adjustment factor allowing for the differences in caloric requirement.

5.2.1.1 US-EPA Approach

In 1988, the US-EPA adopted the ADI approach in its regulatory measures against environmental pollution; with a number of modifications (US-EPA 1988, 1993). Instead of the terms ADI and safety factor, the terms Reference Dose (RfD) and uncertainty factor (UF), respectively, were selected. The RfD is derived from the NOAEL by dividing by the overall UF. The overall UF originally suggested and reconfirmed in 2002 (US-EPA 2002) generally consists of a 10-fold factor for each of the following:

- Human variation in sensitivity (UF_H)
- Interspecies extrapolation (UF_A)
- Use of the NOAEL obtained from a less than lifetime study (UF_S)
- Use of a LOAEL in the absence of a NOAEL (UF_L)
- Adequacy of the total database (UF_D)

According to US-EPA (1993), the first four of the above-mentioned factors are adapted from Dourson and Stara (1983).

The exact value of the UFs chosen should depend on the quality of the studies available, the extent of the database, and scientific judgment (US-EPA 2002). The default factors typically used cover a single order of magnitude (i.e., 10^1). By convention, in the US-EPA, a value of 3 is used in place of one-half power (i.e., $10^{0.5}$) when appropriate. These half-power values should be factored as whole numbers when they occur singly but as powers or logs when they occur in tandem. A composite UF of 3 and 10 would thus be expressed as 30 (3×10^1), whereas a composite UF of 3 and 3 would be expressed as 10 ($10^{0.5} \times 10^{0.5} = 10^1$). It should be noted, in addition, that rigid application of log (i.e., 10^1) or ½ log (i.e., $10^{0.5}$) units for UFs could lead to an illogical set of reference values; therefore, it has been emphasized that application of scientific judgment is critical to the overall process.

It is also noted that there is overlap in the individual UFs and that the application of five UFs of ten for the chronic reference value (yielding a total UF of 100,000) is inappropriate. In fact, in cases where maximum uncertainty exists in all five areas, it is unlikely that the database is sufficient to derive a reference value. Uncertainty in four areas may also indicate that the database is insufficient to derive a reference value. In the case of the RfC, the maximum UF would be 3,000, whereas the maximum would be 10,000 for the RfD. This is because the derivation of RfCs and RfDs has evolved somewhat differently. The RfC methodology (US-EPA 1994) recommends dividing the interspecies UF in half, one-half ($10^{0.5}$) each for toxicokinetic and toxicodynamic considerations, and it includes a Dosimetric Adjustment Factor (DAF, represents a multiplicative factor used to adjust an observed exposure concentration in a particular laboratory species to an exposure concentration for humans that would be associated with the same delivered dose) to account for toxicokinetic differences in calculating the Human Equivalent Concentration (HEC), thus reducing the interspecies UF to 3 for toxicodynamic issues. RfDs, however, do not incorporate a DAF for deriving a Human Equivalent Dose (HED), and the interspecies UF of 10 is typically applied, see also Section 5.3.4. It is recommended to limit the total UF applied for any particular chemical to no more than 3000, for both RfDs and RfCs, and avoiding the derivation of a reference value that involves application of the full 10-fold UF in four or more areas of extrapolation.

In addition, a modifying factor (MF) could be applied (US-EPA 1993). The MF is in reality an additional UF that is greater than 0 and less than or equal to 10; the default value is 1. The MF should account for uncertainties of the study and database not explicitly handled by the use of the general UFs; e.g., the completeness of the overall database and the number of species tested. In the 2002 review of the RfD and RfC processes (US-EPA 2002), it was recommended that use of the MF be discontinued as it was considered that the uncertainties accounted for by the MF is sufficiently accounted for by the general UF.

The US-EPA staff paper from 2004 titled "An Examination of EPA Risk Assessment Principles and Practices" (US-EPA 2004) provides comprehensive and detailed information on the practices employed in risk assessment, including use of UFs and use of default and extrapolation assumptions.

5.2.1.2 Calabrese and Gilbert Approach

Calabrese and Gilbert (1993) have demonstrated the lack of independence of the interspecies and intraspecies UFs, as well as of the intraspecies and the less-than-lifetime UFs. Based on their analyses, the authors concluded that most of the recommended US-EPA standards based on animal models needed to have some of their UFs modified. They recommended the following modifications of the intraspecies UF, see also Section 5.4.2:

- Significantly less-than-lifetime animal study: 5
- When the animal study is for a normal experimental lifetime (2 years in rodents): 4
- Occupational epidemiological study: 10
- Environmental epidemiological study, if study was for a normal human life span: 5

5.2.1.3 Renwick Approach

The approach proposed by Renwick (1991, 1993) is also based on the 100-fold factor. It attempts to give a scientific basis to the default values of 10 for the interspecies and 10 for the intraspecies (interindividual human) differences. Renwick also proposed a division of each of these UFs into sub-factors to allow for separate evaluations of differences in toxicokinetics and toxicodynamics. The advantage of such a subdivision is that components of these UFs can be addressed where data are available; for example, if available data show similar toxicokinetics of a given chemical in experimental animals and humans, then only an interspecies extrapolation factor would be needed to account for differences in toxicodynamics. Renwick examined the relative magnitude of toxicokinetic and toxicodynamic variations between and within species in detail. He found that toxicokinetic differences were generally greater than toxicodynamic differences resulting in the proposal that the 10-fold factors (for inter- and intraspecies variation) should, by default, be subdivided into factors of 4 for toxicokinetics and 2.5 for toxicodynamics. It should be noted that the proposed default values were derived from limited data.

The WHO/IPCS (1994) has adopted the approach set forth by Renwick (1993) with one deviation, see Figure 5.1. While the UF for interspecies (animal-to-human) extrapolation should be split into default values of 4 for toxicokinetics and 2.5 for toxicodynamics, the UF for intraspecies (human-to-human) extrapolation should be split evenly between both aspects, i.e., a sub-factor of 3.16 for both toxicokinetics and toxicodynamics. The reason for this deviation from Renwick's initial suggestion was that the WHO/IPCS considered that the slightly greater variability in the kinetics in humans compared with dynamics was not sufficient to warrant an unequal subdivision of the 10-fold factor into a toxicokinetic factor of 4 and a toxicodynamic factor of 2.5. Actual data should be used to replace the default values if available. It was furthermore noted that precise default values for kinetics and dynamics cannot be expected on the basis of a subdivision of the imprecise 10-fold composite factor for interspecies as well as for the interindividual variation. According to WHO/IPCS, the default values suggested above were considered as being reasonable since they provide a positive value greater than 2 for both aspects and are compatible with the species differences in physiological parameters such as renal and hepatic blood flow. It was also noted that since the database examined was limited, the default values suggested for subdivision of interspecies and interindividual variation should be adopted on an interim basis.

5.2.1.4 Lewis–Lynch–Nikiforov Approach

In 1990, Lewis et al. published a new approach introducing flexibility such that both new information and expert judgment could be readily incorporated. The Lewis–Lynch–Nikiforov (LLN) method, and its refinements, are extensions of established principles and procedures, and

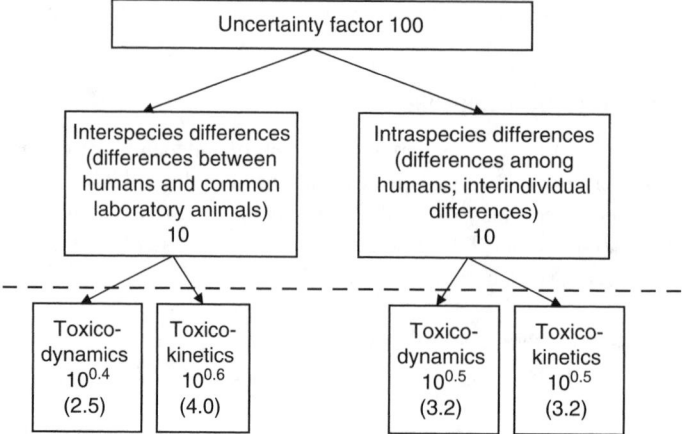

FIGURE 5.1 Subdivision of the 100-fold UF showing the relationship between the use of UFs (above the dashed line), and the proposed subdivisions (below the dashed line) based on toxicokinetics and toxicodynamics. (From Renwick, A.G., *Food Addit. Contam.*, 10, 275, 1993; WHO/IPCS Assessing human health risks of chemicals: Derivation of guidance values for health-based exposure limits. Environmental Health Criteria 170. Geneva, 1994. Available at http://www.inchem.org/documents/ehc/ehc/ehc170.htm)

guides the data evaluator to adjust experimentally determined "no-effect" (or "minimum effect") levels from experimental animal studies taking the following aspects into account:

- Known differences between laboratory animals and humans and between experimental conditions and the real world
- Sensitivity of the exposed human populations
- Strength of evidence that the chemical presents a real hazard to human health
- General quality of the experimental database
- Uncertainties in extrapolating from laboratory animals to humans
- Potency of the toxic agent
- Type and severity of the putative adverse effect

According to Lewis et al. (1990), a step-by-step sequence is used. Initially, a qualitative determination is made as to the strength of evidence that the putative toxic agent presents an actual health hazard to humans, i.e., how likely is this agent to produce the suspected adverse effect in humans? In contrast to the ADI and RfD method where no specific consideration is given to judging the likelihood that a chemical presents a real health hazard, the "strength of the qualitative evidence" is scored explicitly and separately in the LLN approach. The $NAEL_{human}$ is estimated from laboratory research results, using the following algorithm:

$$NAEL_{human} = \frac{NOAEL_{animal}[S]}{[I][R][Q_1][Q_2][Q_3][U][C]}$$

[S] is the aggregate "scaling factor" to account for known quantitative differences between species and between laboratory experimental conditions and the real world. The default value is 1, indicating that animals and humans are equivalent in these dimensions.

[I] is the adjustment factor to account for anticipated greater susceptibility among members of the test animal population than was observed in the experiment, i.e., to account for intraspecies variability. The default value is 10, indicating that extremely high variability was observed (or would be expected) among animals.

[R] is the adjustment factor to account for anticipated differences in susceptibility between humans and the laboratory animals, i.e., to account for interspecies variability. The default value is 10, indicating that humans are much more susceptible.

$[Q_{1-3}]$ and [U] are adjustment factors to account for variations in the reliability of the database (data quality) and other sources of uncertainty in the data evaluation process.

$[Q_1]$ reflects the data evaluator's certainty that the agent actually causes the specific "critical effect" in humans. The default value is 1, indicating that the agent causes similar toxic effects in animals and humans.

$[Q_2]$ is employed when extrapolating data from subchronic studies to estimate risk from lifelong exposures. The default value 10, indicating great uncertainty in estimating the $NOAEL_{chronic}$ from the $NOAEL_{subchronic}$.

$[Q_3]$ is employed when extrapolating LOAELs to NOAELs. The default value 10, indicating extremely great uncertainty associated with using a $LOAEL_{animal}$ to estimate a $NAEL_{human}$.

[U] is used to account for residual uncertainty in estimates of [S], [I], and [R]. The default value is 10 indicating very great overall uncertainty, which has not already been accounted for in $[Q_{1-3}]$.

[C] is a nonscientific, judgmental safety factor, i.e., a social or political value judgment. The default value is 1, indicating that no additional MOS is needed over that provided by the inherently conservative procedure above.

An aggregate adjustment factor of about 250 is typical; the theoretical maximum value is 100,000.

By application of factors $[Q_{1-3}]$ and [U], this approach attempted to separate scientific judgments from policy/value judgments. According to the authors, there are three distinguishing features of the LLN approach. The first is the emphasis on careful discrimination among the adjustments. The second is on discrimination between "best estimates" of the correct adjustments for [S], [I], and [R] and the completely separate adjustment for overall uncertainty. The third is on securing scientific consensus on the adjustment values. It should be recognized, however, that in practice, it will not be possible to distinguish all these factors, and that some factors may not be independent of each other. It could also be questioned whether a nonscientific factor [C] should be discussed in a scientific risk assessment.

5.2.1.5 EU TGD Approach

The process of human health risk assessment has been extensively addressed within the EU framework of Risk Assessment of New and Existing Chemical Substances. According to the EU Technical Guidance Document (TGD) on Risk Assessment of New and Existing Chemical Substances (EC 1996), the risk characterization is carried out by quantitatively comparing the outcome of the effects assessment to the outcome of the exposure assessment, i.e., a comparison of the NOAEL, or LOAEL, and the exposure estimate, see Section 8.3.3. The ratio resulting from this comparison is called the Margin of Safety (MOS). The TGD recommends the following parameters to be considered in assessing the MOS:

- Uncertainty arising, among other factors, from the variability in the experimental data and intra- and interspecies variation
- Nature and severity of the effect
- Human population to which the quantitative and/or qualitative information on exposure applies
- Differences in exposure (route, duration, frequency, and pattern)
- Dose–response relationship observed
- Overall confidence in the database

These parameters are parallel to those being considered in the evaluation of the assessment factors to be applied in the establishment of a tolerable intake.

The TGD has been revised and the second edition was published in 2003 (EC 2003). However, the human health risk characterization part was not included in this second edition. A final draft version of the human health risk characterization part was released in 2005 with a detailed guidance on, among others, the main issues to be included in derivation of the "reference MOS" (MOSref), which is analogous to an overall assessment factor. The individual factors contributing to the MOSref are described separately and guidance is given on how to combine these into the MOSref. The guidance provided in this draft version has been extensively used in relation to the risk assessment of prioritized substances carried out since the draft version was released; however, this version is not publicly available.

In the new EU chemicals regulation REACH, which entered into force on 1 June 2007, detailed guidance documents on different REACH elements, including risk characterization and the use of assessment factors, are currently in preparation (spring 2007). These documents will probably be available on the EU DG Environment REACH Web site (EU 2006) when published.

5.2.1.6 ECETOC Approach

The approach recommended by the ECETOC (1995) is to derive the best scientific estimate of a Human No-Adverse-Effect Level, referred to in the report as the Predicted No-Adverse-Effect Level (PNAEL). The approach distinguishes three stages:

- Application of a scientifically derived adjustment factor to the NOAEL, or LOAEL, of the critical effect established in the pivotal study. It is stated that if the database is inadequate, then human PNAELs cannot be derived scientifically.
- Application of a UF to the PNAEL to take into account the degree of scientific uncertainty involved. The following degrees of confidence in the human PNAEL are suggested: high = 1, medium = 1–2, low = larger UF.
- Application of a nonscientifically based safety factor to take into account political aspects, socioeconomic aspects (cost–benefit considerations), or risk perception factors (the nature of the effect may justify the use of an additional factor).

The scientifically derived adjustment factors include the following elements:

- Experimental exposure in relation to the expected human exposure: a default value of 3 for extrapolation from short-term to subchronic exposure; a default value of 2–3 for extrapolation from subchronic to chronic exposure
- Extrapolation from LOAEL to NOAEL: a default value of 3
- Route-to-route extrapolation: no default value
- Interspecies extrapolation (animal-to-human): a default value of 4 for oral exposure (for the rat with a body weight of 250 g and based on caloric demands); a default value of 1 for inhalation
- Intraspecies extrapolation (human-to-human): a default value of 3 for the general population; a default value of 2 for workers

This approach discriminates factors to a large extent in order to distinguish between the single adjustments and to separate best estimates from uncertainty. It should be noted that the ECETOC approach does not mention the establishment of an overall factor and although they mention that all discriminated aspects introduce uncertainties, they do not give guidance on how to account for this. It could also be questioned here whether a nonscientific factor should be discussed in a scientific risk assessment.

In a more recent report, ECETOC (2003) has further developed many of the principles established in the previous report (ECETOC 1995) and replaced the guidance provided therein on the use of assessment factors in human health risk assessment. The report provides a step-by-step guidance for deriving an approximation of a safe exposure level for humans from the appropriate NOAEL, or LOAEL, observed in animal studies, including guidance on assessment factors. The most scientifically supportable values for default assessment factors recommended to be used in the absence of substance-specific information include:

- LOAEL-to-NOAEL extrapolation: a default value of 3
- Duration extrapolation: subacute to chronic, a default value of 6; subchronic to chronic, a default value of 2; local effects by inhalation, a default value of 1
- Route-to-route extrapolation: oral to inhalation, no default value; oral to dermal, no default value
- Interspecies extrapolation (animal-to-human): systemic effects (scaling): mouse, a default value of 7; rat, a default value of 4; monkey, a default value of 2; dog, a default value of 2. Local effects by inhalation: a default value of 1
- Intraspecies extrapolation (human-to-human): systemic and local effects: a default value of 5 for the general population; a default value of 3 for workers

Similarly to the previous ECETOC approach, this revised approach does not mention the establishment of an overall factor.

5.2.1.7 Dutch Approaches

TNO (the Netherlands Organisation for Applied Scientific Research) has set up a method for setting Health-based Occupational Reference Values (HBORVs) (Hakkert et al. 1996). The HBORV is derived from the selected NOAEL by application of assessment factors compensating for uncertainties inherent to extrapolation of experimental data to a given human situation and for uncertainties in the toxicological database. The assessment factors should be derived considering the toxicity profile of the substance; if the available data are insufficient, an overall assessment factor is used comprising various sub-factors related to:

- Interspecies differences (animal-to-human): mouse, a default value of 7×3; rat, a default value of 4×3; rabbit, a default value of 2.4×3; dog, a default value of 1.4×3. The first factor for each species is a calculated adjustment factor, allowing for differences in basal metabolic rate (proportional to the 0.75th power of body weight). The second factor of 3 is the assessment factor applied for remaining uncertainties (Section 5.3.3), for which the default value is 3. For local skin and respiratory tract effects, the assessment factor is 3, as adjustment for differences in body size is inappropriate.
- Intraspecies differences (human-to-human): a default value of 3 (workers), a default value of 10 (general population)
- Differences between experimental conditions and exposure patterns for workers: chronic-to-chronic exposure, a default value of 1; subacute-to-subchronic exposure, a default value of 10; subchronic-to-chronic exposure, a default value of 10; other aspects, a default value of 1
- Type of critical effect: a default value of 1
- Dose–response curve: a default value of 1
- Confidence of the database: a default value of 1
- Route-to-route: no default value

Principally, the overall assessment factor is established by multiplication of the separate factors. The authors note that in practice it is not possible to distinguish all above-mentioned factors, and some factors are not independent of each other. Therefore, straightforward multiplication may lead to an unreasonably high overall factor. Discussion and weighing of the individual factors are therefore essential to establish a reliable and justifiable overall assessment factor.

Vermeire et al. (1999) have published a discussion paper with focus on assessment factors for human health risk assessment. The status quo with regard to assessment factors is reviewed and the paper discusses the development of a formal, harmonized set of assessment factors. Options are presented for a set of default values and probabilistic distributions for assessment factors based on the state of the art. Methods of combining default values or probabilistic distributions of assessment factors (Section 5.11) are also described. In relation to assessment factors, the authors recommended:

- For interspecies (animal-to-human) extrapolation, allometric scaling on the basis of caloric demands (the 0.75th power of body weight) is considered preferable above scaling on body weight.
- Traditional extrapolation approach, based on more or less arbitrarily chosen factors of 10, is considered simple to apply but obscures the relative contributions of scientific arguments and policy judgments. The other default approaches, including the application of a toxicity profile-derived factor as suggested by TNO, make better use of the data available.
- Worst-case character of the traditional default assessment factors is considered doubtful as the 95th percentile for the proposed distributions for the interspecies (animal-to-human) factor and the subchronic-to-chronic duration factor are considerably higher than 10. In addition, the limited data on intraspecies (human-to-human) variation is also considered to indicate that a default factor of 10 may not be sufficient.
- Derivation of approximations of the distribution of assessment factors from historical data (based on NOAEL ratios) has limitations as the use of the NOAEL instead of the True No-Adverse-Effect Level brings along the variation (error) in the NOAELs.
- Application of assessment factors derived from currently estimated distributions of assessment factors may lead to very wide distributions of the overall assessment factor.
- Probabilistic multiplication of distributions of assessment factors is preferred above the simple multiplication of percentiles to avoid extreme conservatism.

A more recent Dutch report (Vermeire et al. 2001) provides a practical guide for the application of probabilistic distributions of default assessment factors in human health risk assessments, and it is stated that the proposed distributions will be applied in risk assessments of new and existing substances and biocides prepared at RIVM (the National Institute of Public Health and the Environment) and TNO. The report concentrated on the quantification of default distributions of the assessment factors related to interspecies extrapolation (animal-to-human), intraspecies extrapolation (human-to-human), and exposure duration extrapolation.

5.2.1.8   Kalberlah and Schneider Approach

In a report on a research project "quantification of extrapolation factors" (Kalberlah and Schneider 1998), it is noted that extrapolation factors are intended to replace lack of knowledge by a plausible assumption, and that institutions with responsibility for establishing the rules must decide which level of statistical certainty, e.g., applicable for 50% or for 90% of a representative selection of substances, is desired for the selection of a standard value. It is furthermore noted that extrapolation factors are required for: (1) time extrapolation, e.g., from a subchronic to a chronic duration of exposure; (2) extrapolation from the LOAEL to the NAEL; (3) interspecies extrapolation, i.e., from experimental animals to humans; and (4) intraspecies extrapolation, i.e., from groups of persons with average sensitivity to groups of persons characterized by special sensitivity. In addition to these extrapolations, route-to-route extrapolation, e.g., oral-to-inhalation or dermal-to-oral must also be discussed.

If no substance-specific knowledge at all is available for one of the extrapolation steps, the extrapolation factor in each case is used in unaltered form; this factor is described as the standard value.

On average, a factor of 2–3 was considered sufficient for time extrapolation from a subchronic to a chronic duration of exposure. A higher factor is required in order to cover the 90th percentile.

Extrapolation from the LOAEL to the NAEL using standard factors should not be undertaken; the benchmark method should be used instead.

If physiologically based pharmacokinetic (PBPK) models cannot be used, interspecies extrapolation is best undertaken by means of scaling according to basal metabolic rate, see Section 5.3.2.3. A second aspect, interspecies variability, should be considered in cases where a higher than average level of safety (achieved by consideration of a higher percentile of the substances) is desired.

In general, an intraspecies factor of 10 should be sufficient to reflect the toxicokinetic variability between healthy adults; however, it is not sufficient with regard to toxicodynamic variability and possibly only considers risk groups to a limited extent.

If substance-specific knowledge regarding individual extrapolation steps is available but not sufficient to be able to dispense with the extrapolation entirely, the substance-specific information is used on a priority basis for the purpose of modifying the standard value, reduction or possibly also an increase.

A further problem lies in the combination of the individual extrapolation factors to form a total extrapolation factor. The type of combination results from the dependence or independence of the individual sub-factors. According to current knowledge, multiplicative combination of the individual factors is assumed. Substance-specific knowledge about the interdependencies among the sub-factors may lead to modification, i.e., a reduction of the total extrapolation factor.

5.2.1.9   UK Approach

The Interdepartmental Group on Health Risks from Chemicals (IGHRC) in the United Kingdom has published a document entitled "Uncertainty Factors: Their Use in Human Health Risk Assessment by UK Government" (IGHRC 2003). The document intended to lay out the principles used in the United Kingdom.

The uncertainties fall into two broad categories. Firstly, there are the uncertainties related to the extrapolation of the key data from experimental animal species to the "average" human (animal-to-human), and then from the "average" human to other members of the population with different characteristics (human-to-human), i.e., those with greater sensitivity. Secondly, there are then a number of uncertainties related to the available database including those arising from route-to route extrapolation, duration of exposure, NOAEL not established or not firmly established, and gaps or other deficiencies in the database.

Chapter 5 of the document reviews the UFs used by UK Government departments, agencies, and their advisory committees in human health risk assessment. Default values for UFs are provided in Table 3 in the UK document with the factors separated into four classes: (1) animal-to-human factor, (2) human variability factor, (3) quality or quantity of data factor, and (4) severity of effect factor. The following chemical sectors are addressed: food additives and contaminants, pesticides and biocides, air pollutants, drinking water contaminants, soil contaminants, consumer products and cosmetics, veterinary products, human medicines, medical devices, and industrial chemicals.

5.2.1.10   Swedish National Chemicals Inspectorate's Approach

The Swedish National Chemicals Inspectorate (KEMI) has published an extensive review on human health risk assessment with focus on the application of assessment factors in risk assessments for plant protection products, industrial chemicals, and biocidal products within the European Union (KEMI 2003).

One of the main conclusions drawn from the evaluation of the available data on default assessment factors was that the conventionally used factor of 100 (10 for animal-to-human and 10 for human-to-human variations) is probably an underestimate. It is stated that it is likely that the animal-to-human extrapolation is greatly underestimated, and in the case of human-to-human variability, an assessment factor of 10–16 is considered as a minimum.

Attention is also drawn to the fact that there are some other elements not included in the traditional assessment factor of 10 including adequacy of the database, nature of the effect, duration of exposure, route-to-route extrapolation, and considerations of extra-sensitive subpopulations such as children, the elderly, and patients under medical treatment.

The use of default assessment factors is recommended in risk assessments, when justifiable, although the scientific background for such factors in general was considered unsatisfactory. The default assessment factors suggested are summarized in Table 5.2. It is recommended to use assessment factors derived from probabilistic distributions in favor of deterministic assessment factors, see Table 5.2.

TABLE 5.2
Deterministic and Probabilistic Assessment Factors Suggested for Use in Human Health Risk Assessment

Area to be Extrapolated	Assessment Factor	
	Deterministic Approach	Probabilistic Approach
Adequacy of the toxicological database		
relevance, validity, reliability	1–5	—
children	1–10	—
Nature of the effect		
adversity, severity, potency	1–10	—
Duration of exposure		
subacute (1 month) to subchronic (3 months)	3	—
subchronic (3 months) to chronic (24 months)	8	10 (90th); 16 (95th); 37 (99th)
subacute (1 month) to chronic (24 months)	24	25 (90th); 39 (95th); 92 (99th)
Route-to-route extrapolation		
dermal NOAEL from oral NOAEL	100% or case-by-case	—
inhalation NOAEL from oral NOAEL	100% or case-by-case	—
Dose–response curve		
shape of the curve	Case-by-case	—
LOAEL to NOAEL	$BMDL_5$ or 3–10	—
Interspecies extrapolation		
rat to human	4 (TK) $\times$ 2.5 (TD) = 10	28 (90th); 48 (95th); 132 (99th)
mouse to human	—	49 (90th); 84 (95th); 231 (99th)
Intraspecies extrapolation	3–5 (TK) $\times$ 3.16 (TD) = 10–16	—

Source: Modified from KEMI, *Human health risk assessment. Proposals for the use of assessment (uncertainty) factors. Application to risk assessment for plant protection products, industrial chemicals and biocidal products within the European Union.* Report No. 1/03, Solna, Sweden, 2003.

TK: toxicokinetics.
TD: toxicodynamics.
$BMDL_5$: 5% lower confidence limit of the benchmark dose.

5.2.1.11 Danish EPA's Approach

In Denmark, health-based quality criteria are set for chemical substances in soil, drinking water, and ambient air according to principles laid down in a guidance document from the Danish Environmental Protection Agency (D-EPA 2006). The principles laid down in the guidance document are based on an extensive review addressing the hazard assessment of chemicals, including application of assessment factors (Nielsen et al. 2005).

For threshold effects, a Tolerable Daily Intake (TDI) is calculated by dividing the NOAEL (or LOAEL) for the critical effect(s) with an overall UF. The current practice according to the D-EPA in relation to the setting of quality criteria for chemical substances in soil, drinking water, and ambient air is to divide the overall UF into three categories (D-EPA 2006):

- UF_I accounts for the interspecies variation in susceptibility. The default value is 10 when correction for differences in body size between humans and experimental animals is based on the body weight.
- UF_{II} accounts for the differences in interindividual susceptibility. The default value is 10.
- UF_{III} accounts for the quality and relevance of the database, i.e., accounts for the uncertainties in the establishment of a NOAEL for the critical effect. The UF_{III} includes elements such as (1) the quality of the database, e.g., data on specific toxic endpoints are lacking or inadequate, default value of 1–10; (2) route-to-route extrapolation, e.g., no studies using the appropriate exposure route are available, no default value; (3) LOAEL-to-NOAEL extrapolation, e.g., a NOAEL cannot be established for the critical effect, default value of 10; (4) subchronic-to-chronic extrapolation, e.g., no chronic studies on which to establish the NOAEL are available, default value of 10; and (5) nature and severity of toxicity, e.g., the critical effect is toxicity to reproduction, carcinogenicity or sensitization, default value of up to 10. A default value for UF_{III} has not been recommended; however, a value from 1 to 100 is generally used. The value is evaluated case-by-case based on expert judgment.

The overall UF is derived by multiplication of the single UFs. Recognizing that the overall UF might be unrealistically high, a final review of the overall UF is performed in relation to the available data. If the magnitude of the overall UF is very high (e.g., above 10,000), the database is considered as being too limited in order to set a health-based quality criteria in soil, drinking water, and ambient air for the specific chemical substance.

5.2.1.12 Chemical-Specific Assessment Factors

A WHO/IPCS (2005) Harmonization Project Document has proposed using chemical-specific toxicological data instead of default assessment factors, when possible. The concept of Chemical-Specific Adjustment Factors (CSAFs) has been introduced to provide a method for the incorporation of quantitative data on interspecies differences or human variability in either toxicokinetics or toxicodynamics into the risk assessment procedure, by modifying the relevant default UF of 10. Incorporation of toxicokinetic or toxicodynamic data becomes possible if each factor of 10 is divided into appropriately weighted sub-factors as suggested by Renwick (1991, 1993) and adopted by WHO/IPCS (1994), see Section 5.2.1.3.

When appropriate chemical-specific data are available, a CSAF can be used to replace the relevant default sub-factor; for example, suitable data defining the difference in target organ exposure in animals and humans could be used to derive a CSAF to replace the uncertainty sub-factor for animal to human differences in toxicokinetics (a factor of 4). The overall UF would then be the value obtained on multiplying the CSAF(s), used to replace default sub-factor(s), by the remaining default sub-factor(s) for which suitable data were not available. In this way, chemical-specific data in one area could be introduced quantitatively into the derivation of a tolerable intake, and data would replace uncertainty.

The WHO/IPCS (2005) guidance document describes the types and quality of data that could be used to derive a CSAF. The guidance is separated into four main sections covering each of the four different areas where CSAFs can be introduced to replace a default sub-factor:

- Data related to interspecies differences in toxicokinetics
- Data related to interspecies differences in toxicodynamics
- Data related to human variability in toxicokinetics
- Data related to human variability in toxicodynamics

The combination of adjustment factors and default UFs to derive an overall UF is also addressed.

In the 2002 review of the RfD and RfC processes (US-EPA 2002), the growing support for the use of CSAFs in place of DAFs was noted, and this will provide an incentive to fill existing data gaps. The US-EPA has not yet established a guidance for the use of chemical-specific data for deriving UFs, but the division of UFs into toxicodynamic and toxicokinetic components is in the RfC methodology (US-EPA 1994). It was pointed out that, for many substances, there are relatively few data available to serve as an adequate basis to replace defaults for interspecies differences and human variability with more informative CSAFs. Currently, relevant data for consideration are often restricted to the component of uncertainty related to interspecies differences in toxicokinetics.

5.2.1.13 Children-Specific Assessment Factor

Concern has been raised that infants and children are at higher risk than adults from exposure to environmental chemicals. The question of an extra assessment factor in the hazard and risk assessment for chemicals of concern for children has therefore been raised and the rationale for such a children-specific assessment factor has been discussed.

Renwick et al. (2000) have performed an analysis of the need for an additional UF for infants and children. They considered that the proposal to introduce an additional 10-fold factor when exposure of infants and children is anticipated implies either age-related differences between species or differences within humans, which exceed those present in adults. Alternatively, the extra factor could be related to deficiencies of current testing methods or concerns over irreversibility in developing organ systems. They concluded that the available data did not provide a scientific rationale for an extra factor due to inadequacy of inter- and intraspecies UFs. Justification for the factor therefore must relate to the adequacy and sensitivity of current methods or concern about irreversible effects in the developing organism. They also pointed out that when adequate reproduction, multigeneration, or developmental studies are conducted, there will be no need for an additional 10-fold factor.

In setting pesticide tolerances, the U.S. Food Quality Protection Act (FQPA) adopted in 1996 directed the US-EPA to apply an extra safety factor of 10 in assessing the risks to infants and children (US-EPA 1996). This additional 10-fold MOS should take into account the potential for pre- and postnatal toxicity, and the completeness of the toxicology and exposure databases recognizing that maturing organ systems of infants and children may be susceptible to injury by chemicals. There may be developmental periods, i.e., windows of vulnerability, when endocrine, reproductive, immune, and nervous systems are particularly sensitive to certain chemicals, see Section 5.4.1.1. When data are missing or inadequate for an evaluation of the age group or of a window of vulnerability during development, the application of the extra factor, in addition to the general default factor of 10 for intraspecies variation (human-to-human), was considered appropriate.

The FQPA authorizes the US-EPA to replace this additional 10-fold factor with a factor of a different value (higher or lower, including 1) only if, on the basis of reliable data, the resulting level of exposure would be safe for infants and children. In practice, factors of 3 and 10 have been used, and the factor has also been used in cases when data have been sufficient but there were reasons for concern (US-EPA 2002).

In addition to considering the FQPA-relevant areas of uncertainty, assessments of pesticide risk to children also consider applying part or all of the FQPA factors in certain situations to account for areas of residual uncertainty that the traditional UFs do not address or for which they are believed to be insufficient. These areas of residual uncertainty include exposure uncertainties and high concern for an observed susceptibility (US-EPA 2002).

The US-EPA has concluded that in many cases, concerns regarding pre- and postnatal toxicity can be addressed by calculating an RfD by using pre- or postnatal developmental endpoints and applying the UFs (interspecies (Section 5.3), intraspecies (Section 5.4), LOAEL-to-NOAEL (Section 5.7), subchronic-to-chronic (Section 5.6), and database-deficiency (Section 5.9)) to account for deficiencies in the toxicity data when there are gaps considered essential for setting a reference value, including lack of data on children (US-EPA 2002).

The overlap of areas covered by the FQPA factor and those addressed by the traditional UFs was recognized, and it was concluded that the current UFs, if appropriately applied using the approaches recommended in the review (i.e., US-EPA 2002), will be adequate in most cases to cover concerns and uncertainties regarding the potential for pre- and postnatal toxicity and the completeness of the toxicology database. In other words, an additional UF is not needed in the RfC/RfD methodology because the currently available factors are considered sufficient to account for uncertainties in the database from which the reference values are derived (and it does not exclude the possibility that these UFs may be decreased or increased from the default value of 10).

In a report prepared for the Danish Environmental Protection Agency (Nielsen et al. 2001) with the purpose of reviewing the knowledge on the exposure and vulnerability of humans to chemical substances during the embryonic, fetal, and postnatal periods, it was strongly recommended to perform child-specific risk assessments for chemical substances in products and foods intended for children (e.g., in cosmetics, toys, child care products, food additives in preferred foods, and pesticide residues in processed baby foods and infant formulae). In addition, in the risk assessment of chemical substances in other use categories than the above-mentioned, it was recommended specifically to focus on children, including the unborn child, if a potential exposure to a given substance may occur to these age groups. Furthermore, it was recommended that the risk assessment should be performed by experts on a case-by-case basis for each substance and for each exposure scenario. In cases where the available data are insufficient to evaluate the susceptibility of children, including the unborn child, it was strongly recommended that additional safety measures (choice of safety factors) should be considered when tolerable intakes are established for chemical substances in products and foods intended for children.

In conclusion, the traditional assessment factors (interspecies, intraspecies, subchronic-to-chronic, LOAEL-to-NOAEL, and database-deficiency) are considered to cover the concerns and uncertainties for children adequately, i.e., no children-specific assessment factor is needed when setting tolerable intakes. However, it is recommended to perform children-specific risk assessments for chemical substances in products and foods intended for children, based on specific exposure assessments for children.

5.3 INTERSPECIES EXTRAPOLATION (ANIMAL-TO-HUMAN)

Data from studies in experimental animals are the typical starting points for hazard and risk assessments of chemical substances and thus differences in sensitivity between experimental animals and humans need to be addressed, with the default assumption that humans are more sensitive than experimental animals. The rationale for extrapolation of toxicity data across species is founded in the commonality of anatomic characteristics and the universality of physiological functions and biochemical reactions, despite the great diversity of sizes, shapes, and forms of mammalian species.

This section gives a short introduction regarding the biological variation between mammalian species (Section 5.3.1) as a basis for the subsequent section on allometric scaling (Section 5.3.2). Then a number of analyses performed regarding the validity of the default assessment factor of 10 are reviewed (Sections 5.3.3 and 5.3.4). Finally, the key issues are summarized and our recommendations are presented (Section 5.3.5).

5.3.1 Biological Variation

It is often postulated that extrapolation of biological data from animals to man must be viewed with extreme caution because of man's biological uniqueness.

Olson et al. (2000) examined the strengths and weaknesses of animal studies to predict human toxicity. The examination was based on the results of a multinational pharmaceutical company survey, which covered compounds where human toxicity was identified during clinical development of new pharmaceuticals, determining whether animal toxicity studies identified concordant target organ toxicities in humans. Data were compiled for 150 compounds with 221 human toxicity events reported; multiple human toxicity was reported in 47 cases. The results showed a positive human toxicity concordance rate of 71% for rodent and non-rodent species, with non-rodents alone being predictive for 63% of human toxicity and rodents alone for 43%. The highest incidence of overall concordance was seen in hematological, gastrointestinal, and cardiovascular effects, and the least was seen in cutaneous effects.

Although testing of chemicals in experimental animals to a great extent is predictive for human toxicity, humans might be more or less susceptible to the effect(s) exerted by a toxic chemical compared with other mammals, as also is the case between animal species. Absorption, distribution and storage, excretion, metabolism, site or target organ, and mechanism(s) of action are all involved in the toxicological response to a chemical. Thus, interspecies differences result from variation in the sensitivity of species due to differences in toxicokinetics as well as in toxicodynamics. Some of the toxicokinetic differences can be explained by differences in body size and related differences in basal metabolic rate (caloric requirement).

In general, absorption of chemicals is comparable among vertebrate species for the oral and inhalation route, whereas differences in dermal absorption are much more pronounced because of differences in skin morphology between vertebrates. The distribution and storage of chemicals, once they have been absorbed, also tend to be comparable across vertebrates, although there are differences related to, e.g., protein binding. In terms of renal and pulmonary excretion, the differences between the common laboratory animals and humans are minimal. The metabolism of chemicals is, however, generally far from comparable from species to species. Not only are different metabolites sometimes formed, but the rate of formation of identical metabolites may also be species specific. The mechanism(s) and sites of action may also differ across species, both qualitatively and quantitatively. A classical example is the thalidomide-induced teratogenicity in humans where the variability between species in susceptibility is thought to be largely explained by different metabolic pathways in humans and laboratory animals, since the ultimate human teratogen is a metabolite of thalidomide.

This highlights the importance of uncertainties in interspecies extrapolation in cases where the expression of chemical toxicity is related to its metabolism, and it is widely believed that interspecies differences in metabolism of xenobiotics is usually the most significant explanatory factor for observed interspecies differences (Davidson et al. 1986, Voisin et al. 1990, Calabrese et al. 1992).

RIVM, the Dutch National Institute for Public Health and the Environment, has launched a Web site in January 2006 with information of physiological and anatomical parameter values in various species frequently used in toxicity testing. The parameters are focused on organs and tissues relevant for pharmacokinetics following oral exposure. The aim of the Web site is to gain insight into the impact of anatomical and physiological differences between species on the pharmacokinetics. This insight may lead to improved species selection and subsequently to improved animal-to-human extrapolation (RIVM 2007).

In addition to the toxicokinetic and toxicodynamic differences mentioned above, other aspects of differences between experimental animals and humans include different types of organs and tissues, differences in digestion, and differences in the structure of the upper respiratory tract. Furthermore, animal studies are performed in homogenous groups of animals, but the results have to be applied for the protection of all individuals in a heterogeneous population of humans. In consequence of this, interspecies variation must also be expected.

Extrapolation of data from studies in experimental animals to the human situation involves two steps: a first step is to adjust the dose levels applied in the experimental animal studies to human equivalent dose levels, i.e., a correction for differences in body size between laboratory animals and humans. A second step involves the application of an assessment factor to compensate for uncertainties inherent in toxicity data as well as the interspecies variation in biological susceptibility. These two steps are addressed in the following sections.

5.3.2 Adjustment for Differences in Body Size: Allometry/Scaling

One aspect in the extrapolation of data from studies in experimental animals to the human situation is, as mentioned above, a correction of the dose levels in experimental animal studies to equivalent human dose levels, e.g., a NOAEL derived from an animal study to the equivalent human NOAEL.

Adolph (1949, as cited in Davidson et al. 1986, Voisin et al. 1990, ECETOC 2003) compiled a list of 34 morphological, physiological, and biochemical parameters, which correlated with interspecies body weight in accordance with the following general allometric equation:

$$Y = aW^n$$

where

    Y is a biological function
    W is body weight
    a, n are species-independent constants for the biological function Y

Values obtained for the exponent n ranged from 0.08 to 1.31. The geometric mean of all n values was 0.82 and a frequency distribution indicated that values from about 0.67 to 0.75 were most prominent (Adolph 1949, as cited in ECETOC 2003).

Today, well over 100 biological parameters of mammals are known to be linearly related to body weight and highly predictable on an interspecies basis (Davidson et al. 1986, Voisin et al. 1990, Calabrese et al. 1992). The allometric equation has traditionally been used for extrapolation of experimental data concerning physiological and biochemical functions from one mammalian species to another. In addition, the allometric equation has also been used extensively as the basis for extrapolation, or scaling, of e.g., a NOAEL derived for a chemical from studies in experimental animals to an equivalent human NOAEL, i.e., a correction for differences in body size between humans and experimental animals.

Where an interspecies correlation is assumed to exist between Y (biological effect) and body weight, such that if $Y = aW^n$ and the dose (mg) associated with Y in an experimental animal equals X, then

$$Y = f(X) = f(aW^n)$$

For the observed dose in an experimental animal study, X_{animal}, and the equivalent dose in man, X_{human}, the scaling factor for man from the experimental animal is

$$\text{Scaling} = X_{human}/X_{animal} = a(W_{human})^n/a(W_{animal})^n$$

and

$$X_{human} = X_{animal} \times [W_{human}/W_{animal}]^n$$

To correct for differences in body size between humans and experimental animals, three measures of body size are used in practice as the basis for the extrapolation: body weight, body surface area, and caloric requirement (Feron et al. 1990, Vermeire et al. 1999, KEMI 2003).

The reasons for using these three measures and the advantages and disadvantages of their use have been described by Davidson et al. (1986) and Vocci and Farber (1988). In these papers, it is also explained why the body weight can be used in all three cases. However, the body weight should be taken to the power of 1, 0.67, and 0.75 for the body weight approach, the body surface area approach, and the caloric requirement approach, respectively. These figures indicate that the approach used to correct for differences in body size will clearly affect the value of the NOAEL adjusted to the body size of humans.

5.3.2.1 Adjustment for Differences in Body Size: Body Weight Approach

Body weight is considered as being the most easily and accurately measurable of the three measures of body size used in practice as the basis for the extrapolation, and most often provides the quantitative basis for the correction of doses for differences in body size between experimental animals and humans.

When correction for differences in body size is based on body weight, the exponent n in the allometric equation is 1 and the human dose X_{human} (expressed in mg) can be calculated as follows:

$$X_{human} = X_{animal} \times [W_{human}/W_{animal}]^1$$

The scaling factor $[W_{human}/W_{animal}]^1$ between a man weighing 70 kg and a rat weighing 250 g is 280 when correction for differences in body size is based on body weight. Similarly, the scaling factor between a man weighing 70 kg and a mouse weighing 35 g is 2000. It should be recognized that the scaling factor and thus the uncertainty in extrapolating doses from experimental animals to equivalent human doses is heavily dependent on the choice of body weight for man as well as for experimental animals.

When the observed dose in an experimental animal study is expressed in mg/kg body weight, then the equivalent human dose (in mg/kg body weight) is equal to the dose in the experimental animal study as the scaling factor is 1. This is illustrated by the following example: a NOAEL of 1 mg has been derived from an experimental study with rats. By assuming a body weight of 250 g for the rat, the NOAEL is 4 mg/kg body weight. The equivalent human NOAEL can be calculated to 280 mg based on a human body weight of 70 kg (1 mg $\times$ [70 kg/0.25 kg]), or 4 mg/kg body weight (280 mg/70 kg), see also Table 5.3.

TABLE 5.3

Adjustment of Dose Levels for Differences in Body Size between Humans and Experimental Animals. Examples of Deriving a Human NOAEL by the Various Approaches for a Chemical with a NOAEL of 1 mg in a Rat Study

	Rat NOAEL (mg)	Rat NOAEL (mg/kg bw)	Human NOAEL (mg)	Human NOAEL (mg/kg bw)
Body weight approach[a]	1 mg	4 mg/kg bw (1 mg/0.25 kg)	280 mg (1 mg $\times$ [70 kg/0.25 kg]^1)	4 mg/kg bw (280 mg/70 kg)
Body surface area approach[b]	1 mg	4 mg/kg bw (1 mg/0.25 kg)	43.6 mg (1 mg $\times$ [70 kg/0.25 kg]$^{0.67}$)	0.62 mg/kg bw (43.6 mg/70 kg)
Caloric requirement approach[c]	1 mg	4 mg/kg bw (1 mg/0.25 kg)	68.4 mg (1 mg $\times$ [70 kg/0.25 kg]$^{0.75}$)	0.98 mg/kg bw (68.4 mg/70 kg)

Note: Assumed body weight (bw) of rat 0.25 kg, of human 70 kg.

[a] $X_{human} = X_{animal} \times [W_{human}/W_{animal}]^1$.

[b] $X_{human} = X_{animal} \times [W_{human}/W_{animal}]^{0.67}$.

[c] $X_{human} = X_{animal} \times [W_{human}/W_{animal}]^{0.75}$.

According to Voisin et al. (1990), physiological and metabolic processes such as renal function, metabolic rate, and cardiac function are not directly proportional to body weight and thus the toxic effects influenced by these physiological processes are not proportional to body weight, especially when extrapolating from smaller to larger animals. Consequently, interspecies comparisons based directly on body weight are likely to be very inaccurate in predicting chemical-induced toxicity across species.

5.3.2.2 Adjustment for Differences in Body Size: Body Surface Area Approach

The surface area approach has been proposed as an alternative to correction for differences in body size based on body weight. This approach is founded on the notion that the basal metabolic rate of vertebrates is a fundamental biological parameter, i.e., a final common expression of physiological and biochemical functions, which is remarkably well related to the body surface area across species and within species (Davidson et al. 1986).

The most comprehensive attempt to assess interspecies differences in susceptibility to toxic responses, based on two different dose correction approaches (body weight versus body surface area), was published in the classic paper by Freireich et al. (1966, as reviewed in Davidson et al. 1986, Calabrese et al. 1992, Grönlund 1992). The authors attempted to standardize various toxicological studies for 18 anticancer drugs performed in adult mice, rats, hamsters, dogs, monkeys, and humans. The findings of this study led to the conclusion that the toxic effects of an agent were similar across species when the dose was measured on the basis of the body surface area.

Dourson and Stara (1983) have noted that dose conversions based on body surface in general more accurately reflect differences among species in several biological parameters when compared to conversions based on the body weight.

Calabrese et al. (1992) have also noted that the use of the body surface area approach for correction for differences in body size between experimental animals and humans is a more conservative approach than use of the body weight approach. The authors also noted that as the animal model approaches human dimensions of weight and surface area, the differential in dose correction between body weight and body surface area is minimized. Furthermore, the body surface area approach was considered likely to account for certain toxicokinetic differences, especially those associated with interspecies differences in blood flow to organs or enzymatic parameters of importance for the metabolism of substances, which typically scale according to surface area. However, the body surface area approach did not appear to address issues of interspecies variation due to, e.g., differences in absorption efficiencies, thickness of epidermal tissue, number of hairs per square centimeter of skin, the presence and quantity of the gut microflora, the relative dominance of oxidative and conjugative metabolic pathways, and the rate of biliary excretion. It was also stated that the relative importance of these factors will differ from compound to compound and from species to species, ranging from unimportant to critical and thus standard dose correction practice does not eliminate the reality of variability with respect to how the different species handle and respond to agents over a wide range of doses.

Renwick (1999) has noted that most physiological and many biochemical processes correlate better with body surface area than with body weight. For compounds, which are metabolized by processes of intermediary metabolism, or for which the clearance is determined largely by blood flow to the organ(s) of elimination, there is a significant discrepancy between doses expressed on the basis of body surface area and those based on body weight, when comparing rodents with humans. Interspecies factors of about 3–4 and 8–10 would be necessary for rats and mice, respectively, to convert the external dose expressed in mg/kg body weight into a dose based on body surface area, and therefore, more closely related to the species differences in basal metabolic rate and organ blood flows between humans and these species.

The above-mentioned references thus indicate that the body surface area approach apparently is a more feasible approach than the body weight approach in terms of dose correction for differences in body size between experimental animals and humans.

The allometric equation relating the body surface area (BSA) to the body weight (W) is as follows:

$$BSA = aW^{2/3}$$

where the surface area is a function of body weight to the power $2/3$, and a species-specific constant, a.

When correction for differences in body size is based on body surface area, the power n in the allometric equation is $2/3$, or 0.67, and the human dose X_{human} (expressed in mg) can be calculated as follows:

$$X_{human} = X_{animal} \times [W_{human}/W_{animal}]^{0.67} \tag{5.1}$$

The scaling factor $[W_{human}/W_{animal}]^{0.67}$ between a man weighing 70 kg and a rat weighing 250 g is 43.6 when correction for differences in body size is based on body surface area. Similarly, the scaling factor between a man weighing 70 kg and a mouse weighing 35 g is 163. The corresponding scaling factors obtained based on the body weight approach are 280 and 2000, respectively (Section 5.3.2.1). Thus, the difference between the two extrapolation bases, W^1 and $W^{0.67}$, for "to man from rat" and for "to man from mouse" is 6.4-fold and 12.3-fold, respectively, greater for W^1 compared to $W^{0.67}$. Therefore, scaling by the body weight approach provides equivalent human doses of roughly an order of magnitude greater than scaling by the body surface area approach. Hence, scaling by the body weight approach without a biologically justifiable reason may overestimate the equivalent human dose.

When the observed dose in an experimental animal study is expressed in mg/kg body weight, then the equivalent human dose (in mg/kg body weight) is equal to the dose in the experimental animal study divided by a scaling factor according to the following equation:

$$X_{human} = X_{animal}/[W_{human}/W_{animal}]^{0.33} \tag{5.2}$$

According to this equation, the scaling factor $[W_{human}/W_{animal}]^{0.33}$ is 6.4 "to man (70 kg) from rat (250 g)" and 12.3 "to man (70 kg) from mouse (35 g)." This is illustrated by the following example: a NOAEL of 1 mg has been derived from an experimental study with rats. By assuming a body weight of 250 g for the rat, the NOAEL is 4 mg/kg body weight. According to Equation 5.1, the equivalent human NOAEL can be calculated to 43.6 mg based on a human body weight of 70 kg (1 mg $\times$ [70 kg/0.25 kg]$^{0.67}$), or 0.62 mg/kg body weight (43.6 mg/70 kg). According to Equation 5.2, the equivalent human NOAEL can be calculated to 0.62 mg/kg body weight (4 mg/kg $\times$ [70 kg/0.25 kg]$^{0.33}$), i.e., the animal dose (in mg/kg body weight) divided by the scaling factor, in this case 6.4. See also Table 5.3.

The surface area for different species can be calculated by empirically derived equations using a species-specific "shape factor," which depends on the ration of weight to height (Voisin et al. 1990). According to Voisin et al. (1990), there are several limitations in the accuracy of conversions based on body surface area: (1) the surface area appears to be difficult to estimate; (2) some analyses have indicated that the exponent of $2/3$ may be inaccurate; (3) some physiological parameters are not so well related to body surface area; (4) body surface area conversions are inaccurate when the mode of administration is different across species; and (5) not all types of toxicity correlate with body surface area, e.g., skin toxicity.

5.3.2.3   Adjustment for Differences in Body Size: Caloric Requirement Approach

The caloric requirement, or metabolic rate approach, has also been proposed as an alternative to correction for differences in body size based on body weight.

A number of parameters such as, e.g., renal clearance, basal oxygen consumption (metabolic rate), area under the curve (AUC), maximum metabolic velocity, or cardiac output correlate to the body weight to the power of 0.75 ($W^{0.75}$). Further support for the power of 0.75 comes from a more theoretical approach based on fractal geometric and energy conservation rules for mammalian species (West et al. 1997, 1999, as cited in ECETOC 2003).

It is important to note that extrapolation using allometric scaling based on metabolic rate assumes that the parent compound is the toxic agent and that the detoxification is related to the metabolic rate and thus controls the tissue level. This is relevant for oral exposure only (ECETOC 2003).

Feron et al. (1990) have concluded that, in general, adjustment for differences in body size between experimental animals and humans should be based on caloric requirement (energy metabolism) as this was considered to be both scientifically sound and of practical significance.

In 1992, the US-EPA has adopted the caloric requirement approach for oral exposures (US-EPA 1992, as cited in US-EPA 2005), and it is stated that doses should be scaled from animals to humans on the basis of equivalence of milligrams of the agent normalized by the 3/4 power of body weight ($W^{0.75}$) per day (US-EPA 2005). The 3/4 power is considered as being consistent with current science, including empirical data that allow comparison of potencies in humans and animals, and it is also supported by analysis of the allometric variation of key physiological parameters across mammalian species. It is generally more appropriate at low doses, where sources of nonlinearity such as saturation of enzyme activity are less likely to occur. This scaling is intended as an unbiased estimate rather than a conservative one. It is furthermore noted that equating exposure concentrations in food or water is an alternative version of the same approach, because daily intakes of food or water are approximately proportional to $W^{0.75}$.

Vermeire et al. (1999) have noted that scaling on the basis of surface area or caloric demand can be considered more appropriate compared to extrapolation based on body weight; however, they also noted that experimental work did not answer the question regarding which of these two methods is the most correct. Based on theoretical grounds, and supported by their own analyses, Vermeire et al. (1999) concluded that scaling on the basis of caloric demand to adjust oral NOAELs for metabolic size can be considered more appropriate compared with extrapolation based on body weight. It was also noted that an allometric exponent of 0.67, i.e., the body surface area approach, seems to better describe intraspecies relations.

Based on theoretical grounds, the TNO (Hakkert et al. 1996) and Kalberlah and Schneider (1998) consider the interspecies extrapolation based on caloric demands ($W^{0.75}$) as preferable above scaling on body weight.

The allometric equation relating the caloric requirement (CR) to the body weight (W) is as follows:

$$CR = aW^{3/4}$$

where the caloric requirement is a function of body weight to the 3/4 power, and a species-specific constant, a.

When correction for differences in body size is based on caloric requirement, the exponent n in the allometric equation is thus 3/4, or 0.75, and the human dose X_{human} (expressed in mg) can be calculated as follows:

$$X_{human} = X_{animal} \times [W_{human}/W_{animal}]^{0.75} \tag{5.3}$$

The scaling factor $[W_{human}/W_{animal}]^{0.75}$ between a man weighing 70 kg and a rat weighing 250 g is 68.4 when correction for differences in body size is based on caloric requirement. Similarly, the scaling factor between a man weighing 70 kg and a mouse weighing 35 g is 299. The corresponding scaling factors obtained based on the body weight approach are 280 and 2000, respectively.

TABLE 5.4

**Scaling Factors for Adjusting the Dose Expressed Per Unit Body
Weight to the Dose Expressed Per Unit Caloric Requirement Taking
70 kg as the Body Weight for an Adult Human**

Animal Species	Body Weight (kg)	Scaling Factor
Rat	0.200	4.3
Rat	0.250	4.1
Rat	0.300	3.9
Mouse	0.025	7.3
Mouse	0.035	6.7
Mouse	0.050	6.1
Guinea pig	0.500	3.4
Dog	10	1.6
Dog	15	1.5

Note: When the dose for a given species is expressed in mg/kg body weight, the equivalent
human dose (in mg/kg body weight) is obtained by dividing the animal dose by the
scaling factor.

Thus, the difference between the two extrapolation bases, W^1 and $W^{0.75}$, for "to man from rat" and
for "to man from mouse" is 4.1-fold and 6.7-fold, respectively, greater for W^1 compared to $W^{0.75}$.
Therefore, scaling by the body weight approach provides equivalent human doses greater than
scaling by the caloric requirement approach. Hence, scaling by the body weight approach without a
biologically justifiable reason may overestimate the equivalent human dose. As can be seen from the
two scaling factors derived for rat and mouse, respectively, the greater the difference in body weight
between animal and man, the greater the scaling factor; this is illustrated in Table 5.4.

When the observed dose in an experimental animal study is expressed in mg/kg body weight,
then the equivalent human dose (in mg/kg body weight) is equal to the dose in the experimental
animal study divided by the scaling factor according to the following equation:

$$X_{\text{human}} = X_{\text{animal}} / [W_{\text{human}} / W_{\text{animal}}]^{0.25} \qquad (5.4)$$

According to this equation, the scaling factor $[W_{\text{human}}/W_{\text{animal}}]^{0.25}$ is 4.1 "to man (70 kg) from rat
(250 g)" and 6.7 "to man (70 kg) from mouse (35 g)." This is illustrated by the following example:
a NOAEL of 1 mg has been derived from an experimental study with rats. By assuming a body
weight of 250 g for the rat, the NOAEL is 4 mg/kg body weight. According to Equation 5.3, the
equivalent human NOAEL can be calculated to 68.4 mg based on a human body weight of 70 kg
(1 mg × [70 kg/0.25 kg]$^{0.75}$), or 0.98 mg/kg body weight (68.4 mg/70 kg). According to
Equation 5.4, the equivalent human NOAEL can be calculated to 0.98 mg/kg body weight
(4 mg/kg × [70 kg/0.25 kg]$^{0.25}$), i.e., the animal dose (in mg/kg body weight) divided by the
scaling factor, in this case 4.1. See also Table 5.3.

5.3.2.4   Adjustment for Differences in Body Size: Exposure Route

According to Feron et al. (1990), simplicity is probably the main reason for applying the caloric
requirement approach in extrapolating inhalation toxicity data from animals to humans. This method
is based on the assumption that (small) animals and humans breathe at a rate related to their need for
oxygen, thus automatically at a rate depending on their caloric requirement (energy metabolism),
and thus, most importantly, they are automatically being exposed to chemicals occurring in the
breathing atmosphere at a rate similar to that of the caloric requirement.

In practice, this means that no adjustment for difference in body size is needed for a NOAEC obtained for systemic effects in an inhalation toxicity study (van Genderen 1988, Feron et al. 1990, Vermeire et al. 1999, KEMI 2003). For example, a NOAEC of 50 mg/m^3 observed for laboratory animals is also the equivalent human NOAEC (note that so far species-specific sensitivity has not been taken into account).

As mentioned in Section 5.3.2.3, extrapolation using allometric scaling based on metabolic rate assumes that the parent compound is the toxic agent and that the detoxification is related to the metabolic rate and thus controls the tissue level. This is relevant for oral exposure only. With regard to inhalation of substances, which act systemically, the lower detoxification (metabolic) rate in larger animals is balanced by a lower uptake (lower respiratory rate) and thus no scaling factor is needed (ECETOC 2003).

For substances with local effects on the respiratory tract, no general approach for interspecies scaling can be given. Anatomical and physiological differences in the airways between experimental animals and humans contribute to interspecies differences in local effects observed between animals and humans, see Section 4.7.8. It should be noted, however, that for local effects the determining factor for effects to occur in the respiratory tract is generally the concentration of the chemical in the air rather than the total dose and thus allometric scaling is not relevant.

5.3.2.5 Adjustment for Differences in Body Size: PBPK Models

Extrapolation between species should ideally take into account metabolic routes, i.e., the absence or presence of metabolites, as well as the relative rate of formation of the individual metabolites. In PBPK models (Section 4.3.6), both aspects (nonlinearity, formation of active metabolites) are incorporated. This modeling technique uses compartments that correspond to actual tissues or tissue groups of the body. Size, blood flow, air flow, etc. are taken into account, in addition to specific compound-related parameters such as partition coefficients and metabolic rate data. Based on such studies, target-organ concentrations of active metabolites can be predicted in experimental animals and humans, thus providing the best possible basis for extrapolation (Feron et al. 1990).

According to Clewell et al. (2002a), PBPK modeling provides important capabilities for improving the reliability of the extrapolations across dose, species, and exposure route that are generally required in chemical risk assessment regardless of the toxic endpoint being considered. The authors have described an approach, which provides a common template for incorporating pharmacokinetic modeling to estimate tissue dosimetry (e.g., tissue concentrations, body burdens, area under the curve, see Section 4.3.5) into chemical risk assessment. They noted that chemical risk assessments typically depend upon comparisons across species that often simplify to ratios reflecting the differences, and have described the uses of this ratio concept and discussed the advantages of a pharmacokinetic-based approach as compared to the use of default dosimetry. Based on their analyses, they concluded that the correct relationship for cross-species dosimetry depends on whether the toxicity is due to the parent chemical or a metabolite, and in the case of toxicity from a metabolite, whether the metabolite is highly reactive or sufficiently stable to enter the circulation. Moreover, the nature of the cross-species relationship for each of these possibilities is different for oral exposure than for inhalation. Therefore, PBPK modeling is required to improve the reliability of cross-species extrapolation that considers the nature of the toxic entity. Thus, the availability of information on the parent compound and its metabolism may allow modification of the default assessment factor.

5.3.3 REMAINING SPECIES-SPECIFIC DIFFERENCES

As mentioned earlier, the interspecies differences can be divided into differences in metabolic size (Section 5.3.2) and remaining species-specific differences. The average sensitivity of humans to the adverse effects of chemicals (after scaling for caloric requirement) is comparable to that of other species (KEMI 2003). However, an extra assessment factor is needed to account for the remaining interspecies differences, which include differences in toxicodynamics as well as in species-specific toxicokinetics such as different expressions of metabolizing enzymes.

To account for the remaining interspecies uncertainties, a default factor is usually used. In theory, the remaining uncertainty could be assessed by comparing NOAELs in test animals with estimates of human NOAELs. However, in practice, such an assessment must rely on data from studies derived experimentally for the same substance in different animal species because human data are lacking. The degree of remaining interspecies uncertainty may be obtained by examining the differences (ratios) of the NOAELs established for the same substance in different species. The uncertainty in extrapolating from animals to humans is likely to be at least as large as the uncertainty in extrapolating among mice, rats, and dogs (Vermeire et al. 1999).

For the purpose of assessing the remaining interspecies uncertainty, Vermeire et al. (1999) collected and analyzed data for 184 chemicals tested in different species and via different exposure routes. NOAELs were selected from studies with mice, rats, and dogs exposed to the same chemical via the same exposure route and with the same duration of exposure. Two categories of exposure duration were defined, subacute and (sub)chronic, in order to increase the comparability of the different studies. The definition of these exposure categories is species specific, partly depending on their maximum lifetime. Subacute exposure was defined as 21–50 days for the mouse and rat, and as 28–90 days for the dog; (sub)chronic exposure was defined as 90–730 days for the mouse and rat, and as 365–730 days for the dog. The oral NOAELs were adjusted to account for differences in metabolic size, i.e., by the caloric requirement approach (Section 5.3.2.3).

In order to increase the comparability of the derived factors to the actual uncertainty (animal-to-human), the ratios were calculated by dividing the NOAELs derived in the smaller animal by the NOAEL derived in the larger animal. The following ratios for oral exposure were calculated: $NOAEL_{mouse}/NOAEL_{rat}$, $NOAEL_{mouse}/NOAEL_{dog}$, and $NOAEL_{rat}/NOAEL_{dog}$. For respiratory toxicity data, only the ratios $NOAEC_{mouse}/NOAEC_{rat}$ were analyzed, as insufficient data for statistical analyses were available with respect to the other ratios. For dermal toxicity, insufficient data were available for further analyses. The ratios, both adjusted and unadjusted for metabolic size, were evaluated by examining their distributions, see Table 5.5.

The results suggest that the distribution of the ratios can be described sufficiently by a lognormal distribution. If the interspecies differences would depend only on the differences in metabolic size,

TABLE 5.5
Distribution of Parameters Derived from the NOAEL Ratios

Ratio	N	GM	GSD	P_{90}	P_{95}
$NOAEL_{rat}/NOAEL_{dog}$ (oral, unadjusted)	63	1.3	5.1	10.4	18.8
$NOAEL_{rat}/NOAEL_{dog}$ (oral, adjusted)	63	0.5	5.1	3.6	6.6
$NOAEL_{mouse}/NOAEL_{rat}$ (oral, unadjusted)	67	4.2	5.7	39.3	73.9
$NOAEL_{mouse}/NOAEL_{rat}$ (oral, adjusted)	67	2.4	5.7	22.5	42.2
$NOAEL_{mouse}/NOAEL_{dog}$ (oral, unadjusted)	40	6.4	6.1	64.7	124.6
$NOAEL_{mouse}/NOAEL_{dog}$ (oral, adjusted)	40	1.3	6.1	12.9	24.9
$NOAEL_{mouse}/NOAEL_{rat}$ (respiratory)	21	3.1	7.8	43.6	91.8

Source: Modified from Vermeire, T., Stevenson, H., Pieters, M.N., Rennen, M., Slob, W., and Hakkert, B.C., *Crit. Rev. Toxicol.*, 29, 439, 1999.

N: number of ratios.
GM: geometric mean.
GSD: geometric standard deviation.
P_{90}: 90th percentile.
P_{95}: 95th percentile.

and if NOAELs were perfect estimates of the true no-effect levels (which they clearly are not), the geometric mean (GM) and the geometric standard deviation (GSD) of the ratio distributions would be unity. The GMs of the ratios of adjusted NOAELs for mouse/rat and mouse/dog, but not for rat/dog, were closer to 1 than the means of the unadjusted NOAELs giving some support to the idea of accounting for the differences in metabolic size (scaling based on caloric requirement). In the absence of equivalent human NOAELs, it was suggested that this lognormal distribution (GM 1; GSD 6) would also characterize the remaining interspecies differences between animals and humans. If the 90th, 95th, and 99th percentiles are calculated from this distribution of remaining interspecies differences, default values for the assessment factor adjusted for metabolic size would be 10, 19, and 65, respectively; the often applied default factor of 12 (adjustment for metabolic size 4, remaining uncertainty 3) coincided with the 73rd percentile.

A reanalysis and extension of this database (now 198 substances) led to a lower standard deviation (GSD 4.5) (Rennen et al. 2001, Vermeire et al. 2001). If the 90th, 95th, and 99th percentiles are calculated from this distribution of remaining interspecies differences, default values for the assessment factor adjusted for metabolic size would be 7, 12, and 33, respectively. The percentiles for a default factor of 10 (not adjusted for allometric scaling) were 77 for extrapolating oral data for mouse-to-rat, 81 for rat-to-dog, and 66 for mouse-to-dog; for a default factor of 3 (adjusted for allometric scaling), the percentiles were 50 for mouse-to-rat, 52 for rat-to-dog, and 59 for mouse-to-dog. For extrapolation from the rat to human, the traditional factor of 10 coincided with the 73rd percentile when an allometric scaling factor for differences in metabolic size is included, in this case 4 for extrapolation from a rat study (KEMI 2003).

These analyses thus indicate that the default factor of 3 for remaining uncertainty, as well as the traditional factor of 10 for interspecies differences in general, in many cases does not sufficiently account for the remaining interspecies differences.

5.3.4 Assessment Factor for Interspecies Variation (Animal-to-Human): Default Value

The interspecies assessment factor is generally recognized as providing an extrapolation from the average animal studied to an average human being, assuming that humans are 10 times more sensitive to a chemical's toxic effects than experimental animals. When dose correction for differences in body size between experimental animals and humans is performed by the body weight approach (Section 5.3.2.1), the traditionally used default interspecies assessment factor is 10; however, the rationale for this value is not known.

In the following text, various studies will be described, which attempt to establish a scientific rationale for the selection of the interspecies assessment factor. Based on these studies, it can be concluded that a species-specific default factor based on differences in caloric requirement (see Table 5.4) should be used for interspecies extrapolation regarding metabolic size. The remaining interspecies differences should preferentially be described probabilistically, or a deterministic default factor of 2.5 could be used for extrapolation of data from rat studies to the human situation.

Dourson and Stara (1983) plotted experimental animal weights versus an interspecies adjustment factor, calculated as the cube root of the assumed average human body weight (70 kg) divided by the experimental animal body weight. These interspecies adjustment factors were stated to account for differences in doses expressed as mg/kg body weight due to different body surface areas between experimental animals and humans, based on the assumption that different species are equally sensitive to the effects of a toxic substance on a dose per unit body surface area. When this surface area dose is converted to corresponding units of mg/kg body weight, species with greater body weight (e.g., humans) appeared to be more sensitive to the toxicity of a compound than species of smaller body weight (e.g., rodents); the factors varied from 1 (humans) up to 15 (mice). The factors were thought of as reductions in experimental animal dose (in mg/kg body weight) needed to estimate a comparable human dose (in mg/kg body weight). The factors were also viewed as support of a 10-fold UF to account for interspecies variability to the toxicity of a chemical when estimating an Acceptable Daily Intake (ADI) from animal doses expressed as mg/kg body weight. The authors found that the 10-fold factor in that way appeared to incorporate a MOS if the underlying assumption of dose equivalence among species per unit of surface area is correct, i.e., if dose correction for differences in body size between experimental animals and humans is performed by the body surface area approach (Section 5.3.2.2).

Feron et al. (1990) concluded that the sensitivity of humans to chemicals is probably not very different from that of other mammals, and that a systematic error is made by carrying out extrapolation by using the body weight approach. For metabolizable compounds, the authors strongly recommended a procedure that takes the metabolic rate into account ($W^{0.75}$) for scaling across species, i.e., dose correction for differences in body size between experimental animals and humans by the caloric requirement approach (Section 5.3.2.3). This approach was also considered to provide a contribution to reducing the size of the traditional safety factor in a justifiable way.

Calabrese et al. (1992) noted that the animal-to-human UF is supposed to account for all possible factors that could result in interspecies differences in susceptibility, regardless of non-carcinogenic endpoint. The value of the interspecies UF is traditionally 10 and was stated as being founded on a reasonable public-health-based protective philosophy that assumes that an average group of humans may be as much as 10-fold more susceptible than the average group of animals under study. It has been argued that the apparent conflict between the use of body weight and that of surface area for dose correction is not real since the animal-to-human UF actually incorporates the interspecies variation factor with respect to scaling as well as other interspecies differences. Consequently, it may be inferred that the use of surface area scaling amounts to "double counting" when the animal-to-human UF of 10 is also employed. This argument was investigated by the authors. Their assessment led to the conclusion that, under ideal circumstances, dose adjustment using body surface area may lead to a close similarity across species in the blood level of the agent reaching potential target organs. However, this adjustment will not address some of the principal causes of interspecies variation in response to xenobiotics, i.e., certain metabolic and toxicodynamic factors. Thus, the dose correction using the body surface approach was not considered to eliminate all major causes on interspecies differences and had no obvious consistent quantitative interspecies relationship to numerous other factors that affect susceptibility to toxic substances. Therefore, on theoretical grounds, the authors recommended that dose correction using the body surface approach should be considered independent of the animal-to-human UF for the purpose of risk management.

Grönlund (1992) has investigated methods used for quantitative risk assessment of non-genotoxic substances, with special regard to the selection of assessment factors. Grönlund found that humans, in most cases, seem to be more sensitive to the toxic effects of chemicals than experimental animals, and that the traditional 10-fold factor for interspecies differences apparently is too small in order to cover the real variation. It was also noted that a general interspecies factor to cover all types of chemicals and all types of experimental animals cannot be expected. It was concluded that a 10-fold factor for interspecies variability probably protects a majority, but not all of the population, provided that the dose correction for differences in body size between experimental animals and humans is performed by the body surface area approach (Section 5.3.2.2). If the dose correction is based on the body weight approach (Section 5.3.2.1), the 10-fold factor was considered to be too small in most cases.

Renwick (1993) examined the relative magnitude of toxicokinetic and toxicodynamic variations between species in detail and found that toxicokinetic differences were generally greater than toxicodynamic differences. In order to allow for separate evaluations of differences in toxicokinetics and toxicodynamics, he proposed that the default interspecies UF of 10 should, by default, be subdivided into a sub-factor of 4 for toxicokinetics and a sub-factor of 2.5 for toxicodynamics. The suggested factor of 4 for differences in toxicokinetics was largely based on the extent of absorption and the rate of elimination or clearance in different experimental animals. The suggested factor of 2.5 for toxicodynamic differences was not scientifically based, but mainly the remaining value to fit the traditional default interspecies assessment factor of 10.

Kalberlah and Schneider (1998) have analyzed the information on the quantification of extrapolation factors. They noted that in interspecies extrapolation, two variables must be differentiated: The *systematic* differences between different species, and the *variability* in the sensitivity of the species. Systematic differences can, e.g., be recorded by means of allometric approaches, "scaling" (Section 5.3.2). The reasons for the variability in sensitivity may be due to both toxicokinetic and toxicodynamic characteristics of a species.

In their analyses, statistics on the relevant extrapolation factor from animals to humans, as reported in the literature, were considered synoptically, and distinctions were made between: (1) publications which focused on allometrically justifiable differences; (2) publications which examined the toxicodynamic or toxicokinetic variability; and (3) publications which considered the total (gross) interspecies factor. In addition, consideration of PBPK models was discussed as a possible alternative.

If sufficient data are available, substance-specific PBPK models should always be given preference over the use of general scaling factors. However, PBPK models were considered not to replace all of the sub-factors in the interspecies comparison and should, by definition, only include toxicokinetic differences. A further extrapolation factor for toxicodynamic differences between the species needs to be discussed.

If PBPK models cannot be used, scaling is the recommended approach. Theoretical considerations and the evaluation of numerous publications supported scaling according to basal metabolic rate ($W^{0.75}$). When doses from experimental animal studies are expressed as mg/kg body weight, this scaling approach means an extrapolation factor of 7 for mouse-to-man, 3.9 for rat(Fischer) -to-man, 3.6 for rat(Sprague–Dawley)-to-man, 1.6 for dog(Beagle)-to-man, 3.9 for monkey (marmoset)-to-man, and 1.6 for monkey(rhesus)-to-man. It was noted that these extrapolation factors only account for toxicokinetic differences in the basal metabolic rate. If an interspecies extrapolation with an average degree of statistical certainty is undertaken (for approximately 50% of substances, i.e., the 50th percentile), this extrapolation step is sufficient.

A second extrapolation step for toxicokinetic and toxicodynamic variability was recommended if interspecies comparisons involve, with a certain statistical probability, the occurrence of an above-average sensitivity in humans as compared with the test animal species.

The authors noted that, in comparison with the scaling factor, the traditional 10-fold factor contains an additional extrapolation factor for possible additional toxicokinetic or toxicodynamic variability apart from the basal metabolic rate scaling. This additional factor, which can be interpreted as the traditional 10-fold factor divided by the scaling factor, ranges from approximately 1.5 for the mouse ($10/7 = 1.4$) to approximately 6 for the rhesus monkey ($10/1.6 = 6.3$). The authors considered that the additional factor thus comprises levels of safety, which are currently nonuniform, and this inhomogeneity is not supported toxicologically.

The authors also noted that the database for the derivation of a standard value is very limited. The few available data support a factor of 2–3 additional (multiplicatively) to the scaling factor in order to cover, for approximately 95% of the substances (i.e., the 95th percentile), a possibly greater sensitivity of humans compared with experimental animals. The overall interspecies assessment factor would then be 8–12 for a rat study and 14–21 for a mouse study.

When dose levels are expressed as the concentration in the medium (mg/m^3 air or mg/kg feed), a scaling factor is not relevant. In such cases, only the toxicokinetic and toxicodynamic variability, i.e., an interspecies factor of 2–3 should be applied.

Finally, it was emphasized that substance-specific modifications are always possible when data on toxicokinetics and toxicodynamics permit more precise statements on interspecies differences.

Renwick (1999) noted that, from the perspective of the late 1990s, it is naïve to expect a single 10-fold assessment factor to allow for differences between different test animals and humans. It was also noted that the 10-fold interspecies factor is applied to an intake expressed in mg/kg body weight, but most physiological and many biochemical processes correlate better with body surface area. It was therefore concluded that interspecies factors of about 3–4 and 8–10 would be necessary for rats and mice, respectively, to convert the external dose expressed in mg/kg body weight into a dose based on body surface area, see also Section 5.3.2.2.

Vermeire et al. (1999) have assessed the remaining interspecies uncertainty, i.e., the uncertainty not related to differences in metabolic size that can be accounted for by allometric scaling (Section 5.3.3). A reanalysis and extension of the database was published a few years later (Rennen et al. 2001, Vermeire et al. 2001). The reanalyses indicated that the default factor of 3 for remaining uncertainty, as well as the traditional factor of 10 for interspecies differences in general, in many cases does not sufficiently account for the remaining interspecies differences.

Analyses of species differences in the toxicokinetics of compounds eliminated by a single major metabolic pathway in humans have been performed by Renwick and coworkers using published data for compounds in four test species (dog, rabbit, rat, and mouse).

Walton et al. (2001a) examined data for compounds eliminated by the cytochrome P450 isoenzymes CYP1A2 in humans. Absorption, bioavailability, and route of excretion were generally similar between humans and the test species for each of the substances (caffeine, paraxanthine, theobromine, and theophylline). However, interspecies differences in the route of metabolism, and the enzymes involved in this process, were identified. The magnitude of difference in the internal dose, between species, showed that values for the mouse (10.6) and rat (5.4) exceeded the fourfold default factor for toxicokinetics, whereas the rabbit (2.6) and the dog (1.6) were below this value.

In a second study (Walton et al. 2001b), the magnitude of the interspecies differences in the internal dose of compounds for which glucuronidation is the major pathway of metabolism in either humans or in the test species was determined. There were major interspecies differences in the nature of the biological processes that influence the internal dose including route of metabolism, the extent of pre-systemic metabolism, and enterohepatic recirculation. There was also a wide variability in the magnitude of differences in the internal dose for all of the test species. The mean values for the clearance ratios compared to humans were 4.5 for the mouse, 9.1 for the rat, 8.7 for the rabbit, and 9.7 for the dog. Thus, the fourfold default factor was exceeded for all the species.

Walton et al. (2004) determined the extent of interspecies differences in the internal dose of compounds, which are eliminated primarily by renal excretion in humans. Renal excretion was also the main route of elimination in the test species for most of the compounds. Interspecies differences were apparent for both the mechanism of renal excretion (glomerular filtration, tubular secretion, and/or reabsorption), and the extent of plasma protein binding. Both of these may affect renal clearance and therefore the magnitude of species differences in the internal dose. For compounds which were eliminated unchanged by both humans and the test species, the average difference in the internal dose between humans and animals were 1.6 for dogs, 3.3 for rabbits, 5.2 for rats, and 13 for mice. This suggests that for renal excretion the differences between humans and the rat, and especially the mouse, may exceed the fourfold default factor for toxicokinetics.

The analyses thus indicated that pathway-related factors for different species could be derived for some pathways; however, according to Walton et al. (2001c), the pathway of elimination in humans for most compounds did not reliably predict the pathway in animals. Thus, there would be considerable uncertainty in using a species-specific pathway-related factor, unless there are detailed data for both humans and the animal species.

ECETOC (2003) recommended that in the absence of any substance- or species-specific mechanism or PBPK modeling (Section 4.3.6), allometric scaling based on metabolic rate ($W^{0.75}$) (caloric requirement approach, Section 5.3.2.3) is considered to provide an appropriate default for an assessment factor for interspecies differences with respect to systemic effects. Allometric scaling was stated as being a tool for estimating interspecies differences of internal exposure or body burden and to provide indirectly information on differences in sensitivity between species. Typical scaling factors for interspecies adjustment were noted as 7 for mouse, 4 for rat, and 2 for dog; however, adjustments of these scaling factors may be necessary especially for directly acting and metabolically activated/inactivated compounds.

ECETOC also considered that the scaling approach might not account completely for interspecies variation in biological sensitivity and might not address special cases of higher sensitivity in humans due to toxicokinetic or toxicodynamic differences between animals and humans. The database for determination of this (additional) assessment factor for interspecies sensitivity was considered as small and most likely confounded by intraspecies variability. It was concluded that, although residual interspecies variability may remain following allometric scaling, this is largely accounted for in the default assessment factor for intraspecies variability reflecting the inherent interdependency of the inter- and intraspecies factors.

For local effects, a default assessment factor of 1 for interspecies extrapolation for water-soluble gases and vapors was considered to be sufficiently conservative, as well as for aerosols since the respiratory rate of rodents leads to a greater respiratory tract burden as compared to humans.

The WHO/IPCS (1994, 1996, 1999) has adopted the approach set forth by Renwick (1993), i.e., the UF for interspecies (animal-to-human) extrapolation should be split into default values of 4 for toxicokinetics and 2.5 for toxicodynamics, see Section 5.2.1.3. In situations where appropriate toxicokinetic and/or toxicodynamic data are available for a particular compound, the relevant UF should be replaced by the data-derived factor. If a data-derived factor is introduced, then the commonly used 10-fold factor would be replaced by the product of that data-derived factor and the remaining default factor. It is also noted that for some classes of compounds, a data-derived factor for one member of the class may be applicable to all members thereby producing a group-based data-derived factor.

In 1988, the US-EPA adopted the ADI approach with respect to the derivation of RfDs and RfCs with a 10-fold UF to account for interspecies extrapolation (US-EPA 1988, 1993), see Section 5.2.1.1. It was noted, in the 2002 review of the RfD and RfC processes (US-EPA 2002), that the interspecies UF is generally presumed to include both toxicokinetic and toxicodynamic aspects.

Much of the RfC methodology (US-EPA 1994) focused on improving the science underlying the animal-to-human UF, segregating it into toxicokinetic and toxicodynamic components and providing generalized procedures to derive DAFs (domimetric adjustment factors, represents a multiplicative factor used to adjust an observed exposure concentration in a particular laboratory species to an exposure concentration for humans that would be associated with the same delivered dose). Application of a DAF in the calculation of a human equivalent concentration (HEC) was considered to address the toxicokinetic aspects of the animal-to-human UF, i.e., to estimate from animal exposure information the human exposure scenario that would result in the same dose to a given target tissue. The RfC methodology recommended dividing the default interspecies UF of 10 in half, one-half ($10^{0.5}$) each for toxicokinetic and toxicodynamic considerations. The methodology included a DAF to account for toxicokinetic differences in calculating the HEC, thus reducing the uncertainty about the remaining toxicodynamic component through application of the partial animal-to-human UF ($10^{0.5}$, which is typically rounded to 3). It was also noted that seldom are there data available to inform toxicodynamic differences. One-half the default 10-fold interspecies UF (i.e., $10^{0.5}$) was assumed to account for such differences, but more specific data should be used when available and the flexibility for applying a factor greater than 10 should be recognized. Unless data support the conclusion that the test species is more or equally susceptible to the pollutant as are humans, and in the absence of any other specific toxicokinetic or toxicodynamic data, a default factor of 3 (in conjunction with HEC derivation) or 10 is applied.

It was also noted, in the 2002 review of the RfD and RfC processes (US-EPA 2002), that currently, no procedures parallel to the inhalation RfC methodology exist for deriving either oral or dermal human equivalents from animal data. Default factors (usually of 10) are routinely applied to address the issue of animal-to-human extrapolation. Thus, no parallel to the HEC, i.e., a human equivalent dose (HED), is derived nor are other adjustments applied to the animal oral or dermal dose. Instead, assumptions are made regarding the comparability of ingested or applied dose, based on a mg/kg body weight basis. The caloric requirement approach for dose correction was analyzed and this process was recommended as a possible candidate for estimating cross-species toxicokinetic relationships in the absence of adequate toxicokinetic information. That is, $W^{0.75}$ factors could be applied as DAFs for deriving a human equivalent dose (HED). This procedure would parallel the one used for deriving the HEC. It was noted that, as with the HEC, this process applies only to toxicokinetic aspects of cross-species extrapolation and does not address the toxicodynamic differences that may exist between species. As with the HEC, consideration of toxicodynamic differences was proposed to be through application of a portion of the animal-to-human extrapolation ($10^{0.5}$, which is typically rounded to 3).

TNO has suggested a default interspecies assessment factor composed of a scaling factor allowing for differences in basal metabolic rate ($W^{0.75}$) for oral studies depending on the species (mouse 7, rat 4, rabbit 2.4, dog 1.4), and a factor of 3 for remaining variability (Hakkert et al. 1996). For local skin and respiratory tract effects, an assessment factor of 3 was suggested, as an adjustment for differences in body size is inappropriate for local effects, see also Section 5.2.1.7.

KEMI (2003) has suggested that a species-specific default factor should be used for interspecies extrapolation regarding metabolic size. This factor should be based on differences in caloric demand ($W^{0.75}$) and the factor for extrapolation from rats is 4, from mice 7, from guinea pigs 3, from rabbits 2.4, and from dogs 1.4. The remaining variability can be described by a distribution, i.e., a probabilistic approach, see Section 5.3.3. Which percentile of the distribution, i.e., percent of substances to be covered by the factor should be chosen is a matter of judgment. The composite interspecies assessment factor would be 48 (12×4) if the 95th percentile (i.e., 95% of substances to be covered) is chosen in the case of extrapolation from a study in rats. The traditional default factor of 10 only covers the 73rd percentile of the distribution, i.e., 73% of the substances are covered.

5.3.5 INTERSPECIES EXTRAPOLATION (ANIMAL-TO-HUMAN): SUMMARY AND RECOMMENDATIONS

The rationale for extrapolation of toxicity data across species is founded in the commonality of anatomic characteristics and the universality of physiological functions and biochemical reactions, despite the great diversity of sizes, shapes, and forms of mammalian species.

For extrapolation of data from animal studies to humans, account should be taken of species-specific differences between animals and humans.

Ideally, the interspecies extrapolation should be based on substance-specific information; however, for most substances, only limited or no data are available. Therefore, an assessment factor is usually applied in the interspecies extrapolation. The traditionally used default interspecies assessment factor is 10, possibly divided into a sub-factor of 4 for differences in toxicokinetics and a sub-factor of 2.5 for differences in toxicodynamics as proposed by Renwick (1993) and adopted by the WHO/IPCS (1994). The validity of the interspecies default factor of 4 for toxicokinetics has been assessed by Walton et al. (2001a,b) for each of the test species (dog, rabbit, rat, and mouse); the authors concluded that their assessment supports the need to replace the generic default factor by a compound-related value derived from specific, relevant, quantitative data.

Two variables must be differentiated in interspecies extrapolation:

- Consideration of the systematic differences between different species
- Consideration of the variability in the sensitivity of the species

Systematic differences can be accounted for by means of allometric "scaling." The reasons for the variability in sensitivity may be due to both toxicokinetic and toxicodynamic characteristics of a species and may imply both higher and lower sensitivity for man when compared to experimental animals. These interspecies differences can thus be seen as differences in metabolic size and remaining species-specific differences.

Differences in metabolic size, the major part of toxicokinetic differences, can be accounted for by allometric scaling (Section 5.3.2). To account for differences in metabolic size (differences in body size between humans and experimental animals), three measures of body size are used in practice as the basis for the extrapolation: body weight, body surface area, and caloric requirement. These methods can be described by an allometric equation in which body weight has to be raised to the power of 1, 0.67, and 0.75, respectively. Scaling on the basis of body weight (Section 5.3.2.1) most often provides the quantitative basis for the correction of doses for differences in body size between experimental animals and humans. However, scaling on the basis of body surface area (Section 5.3.2.2) or caloric requirement (Section 5.3.2.3) is considered as more appropriate compared to extrapolation based on body weight. There is a general, international, and scientific recommendation to the caloric requirement approach, i.e., scaling according to the basal metabolic rate ($W^{0.75}$); scaling factors for the most commonly used experimental animals are provided in Table 5.4. If sufficient data are available, substance-specific PBPK models should always be given preference over the use of general scaling factors. However, PBPK models only account for toxicokinetic differences and thus cannot replace all of the sub-factors in the interspecies extrapolation. It should be remembered that allometric scaling is relevant for oral exposure (and dermal exposure if systemic effects occur) only as (small) animals and humans breathe at a rate related to their need for oxygen, i.e., at a rate depending on their caloric requirement; therefore, they are automatically being exposed to chemicals occurring in the breathing atmosphere at a rate similar to that of the caloric requirement (Section 5.3.2.4).

The average sensitivity of humans (after scaling for caloric requirement) is considered to be comparable to other species. However, an extra assessment factor is needed to account for the remaining interspecies differences, which include differences in toxicodynamics as well as in species-specific toxicokinetics. Usually, a default factor is used; however, there is no general, international, and scientific recommendation to a standard value for the remaining interspecies differences (Section 5.3.4). The TNO in the Netherlands has proposed a standard value of 3 (Hakkert et al. 1996), and Kalberlah and Schneider (1998) a factor of 2–3 additional (multiplicatively) to the scaling factor. Consideration of toxicodynamic differences was proposed by the WHO/IPCS (1994) through application of a default factor of 2.5 as proposed by Renwick (1993), and by the US-EPA (2002) through application of a portion of the animal-to-human extrapolation default factor of 10, i.e., $10^{0.5}$, which is typically rounded to 3. ECETOC (2003) concluded that, although residual interspecies variability may remain following allometric scaling, this is largely accounted for in the default assessment factor for intraspecies variability, reflecting the inherent interdependency of the inter- and intraspecies factors. However, the analyses performed by Vermeire et al. (1999, 2001) and Rennen et al. (2001) indicate that a default factor of 3 for remaining uncertainty, in many cases, does not sufficiently account for the remaining interspecies differences (Section 5.3.3). For these reasons, Vermeire et al. (1999, 2001) and KEMI (2003) have recommended that the remaining variability should be described by a distribution, i.e., a probabilistic approach; which percentile of the distribution, i.e., percent of substances to be covered by the factor, should be chosen is a matter of judgment.

In conclusion, if no substance-specific data are available, it is recommended as a default to correct for differences in metabolic size (differences in body size between humans and experimental animals) by using allometric scaling based on the caloric requirement approach (see Table 5.4). The assessment factor accounting for remaining interspecies differences should preferentially be described probabilistically as suggested by Vermeire et al. (1999, 2001) and KEMI (2003), or a deterministic default factor of 2.5 could be used for extrapolation of data from rat studies to the human situation.

5.4 INTRASPECIES EXTRAPOLATION (INTERINDIVIDUAL, HUMAN-TO-HUMAN)

Risk assessments are usually based on data from studies in animals of similar age. In addition, the animals are initially healthy and are fed with the same feed, etc. The NOAEL from animal studies is extrapolated to a tolerable intake that is considered to be without appreciable health risk for the general population. This raises the questions whether it is possible to generalize to the average human population or whether there is any particular vulnerable subpopulation that should be taken into consideration in the risk assessment.

This section gives an overview regarding the biological variation between human individuals (Section 5.4.1). Then a number of analyses performed regarding the validity of the default assessment factor of 10 are reviewed (Section 5.4.2). Finally, the key issues are summarized and our recommendations are presented (Section 5.4.3).

5.4.1 BIOLOGICAL VARIATION

In comparison with the genetically relatively homogenous inbred and outbred strains of experimental animals used for toxicological testing, a considerably greater variability in the responses to chemicals can be expected in the heterogeneous human population. This is due in part to genetic factors, but also to acquired susceptibility factors as well as to previous or simultaneous exposure to multiple compounds (industrial chemicals, food additives, pesticides, drugs), all of which may have an impact on the NOAEL for the different individuals of a population. The following factors may play a role for the marked differences with respect to the interindividual sensitivity in the responses to chemicals:

- Genetic factors (enzyme polymorphisms, hereditary metabolic disorders)
- Age and development (physiology, organ sensitivity)
- Gender
- Health and disease status (diet, stress, lifestyle)
- Specific constitution and situation (weight, proportion of fat, pregnancy)

The interindividual variability reflects differences in the extent of exposure, in toxicokinetics as well as in toxicodynamics. The variability due to factors which influence the extent of exposure (physiological differences in the intake, e.g., inhalation rates) can be considered by means of suitable parameters for the internal exposure (absorbed dose, area under the curve AUC, plasma concentration) if sufficient information is available. With respect to toxicokinetic factors, interindividual differences in the metabolism of chemicals are generally considered as the most significant explanatory factor. Hardly any knowledge is available with respect to the factors that influence toxicodynamics. In the following, a brief overview of the factors playing a role for the toxicokinetic and toxicodynamic differences is presented.

5.4.1.1 Age and Development

A human being changes, anatomically, biochemically, and physiologically, during its lifetime from conception to death, and there may be windows of increased vulnerability or periods of human development when chemical exposures may substantially alter organ structure or function. Some organ systems show specific vulnerability to chemical toxicity during development, as organ maturation is an ongoing process throughout the embryo-fetal period and during infancy, childhood, pubertal period, and adolescence, and will not be completed until adulthood. For example, the reproductive and endocrine systems mature slowly, reaching a peak immediately prior to adulthood, the immune system develops both pre- and postpartum, and the nervous system develops both late during pregnancy and in the infant and child (Nielsen et al. 2001).

The concept that infants and children may be a sensitive subgroup relates to their relative immaturity compared to adults. Children, as well as the unborn child, have in some cases appeared to be uniquely vulnerable to toxic effects of chemicals because periods of rapid growth and development render them more susceptible to some specific toxic effects when compared to adults. In addition to such toxicodynamic factors, differences in toxicokinetics may contribute to an increased susceptibility during these periods. It should be noted, however, that during the developmental and maturational periods the susceptibility to exposure to xenobiotics in children may be higher, equal, or even lower than in adults. Except for a few specific substances, not very much is known about whether and why the response to a substance may differ between age groups. It should also be borne in mind that, in terms of risk assessment, children are not simply small adults, but rather a unique population (Nielsen et al. 2001).

In general, the fetus is not protected against xenobiotics that circulate in the maternal blood as chemicals, which pass maternal membranes, are also likely to pass the placental barrier. The human fetus and the placenta possess metabolic capacity, but the contribution of these metabolizing entities to the total kinetics is probably minimal. The period of organogenesis (the embryonic stage from approximately 2 to 8 weeks of gestation) is considered to be the developmental phase most sensitive to exogenously induced classical malformations in single organs. During the fetal stage (from the 9th week of pregnancy until birth), which is characterized by differentiation, growth, and physiological maturation, the fetus becomes increasingly resistant to the actions of teratogens. Exposure to chemicals during this period is most likely to result in effects on growth and functional maturation. Receptors and other molecular targets for chemicals affecting future functions are developing continuously, so that the fetus may be even more sensitive than the embryo to some toxic effects (Nielsen et al. 2001).

Infancy is the period from birth up to 12 months of age. In early infancy (0 to 4 months), the organs are still rather immature and various maturation processes take place. The complexity of all these factors makes it difficult to predict the net effect of xenobiotics on toxicokinetics; however, the maturation of the gastrointestinal system, liver, and kidneys has generally taken place within 6–12 months after birth. By late infancy (4 to 12 months), most processes related to metabolic activity and excretion are probably comparable to those of adults for most substances. Because of the immature function of the organs, neonates and young infants may have lower biotransformation and elimination capacities. This may render these individuals less able to detoxify and excrete xenobiotics and thereby more vulnerable to toxicants. On the other hand, if toxicity is caused by a toxic intermediate produced via biotransformation, young infants may be less sensitive. In childhood (1 to 12 years), metabolism and excretion of xenobiotics may be equal to or even higher than in adults due to the higher basal metabolic rate and larger relative liver size. It is difficult to generalize about age-dependent deficiencies in the metabolism of xenobiotics because the various enzyme systems mature at different time points. The age at which metabolism is similar to the adult value may be different for each substance. In general, the most prominent differences in toxicokinetics are seen in children less than 1 year of age, especially in the first few days and weeks of life. However, children do not seem to represent a special group from a toxicokinetic viewpoint regarding variability among children, as the toxicokinetic variability among children generally appears to be of a similar magnitude as the variability among adults (Nielsen et al. 2001).

Generally, it appears that effects of xenobiotics on organs or endpoints may be similar in children and adults, e.g., liver necrosis observed in adults will also be observed in children. As regards toxicodynamics, age-dependent differences are primarily related to the specific and unique effects that substances may have on the development of the embryo, fetus, and child in that the physiological development of the nervous, immune, and endocrine/reproductive systems continues until adolescence (12 to 18 years). Furthermore, receptors and other molecular targets for various xenobiotics are continuously developing during the embryonic, fetal, and infant periods. This may cause age-dependent differences in the outcome of receptor–xenobiotic interactions and even result in opposite effects of xenobiotics in infants and adults. The available data are insufficient to evaluate the toxicodynamic variability among children as well as the differences between children and adults (Nielsen et al. 2001).

Children's exposure pattern differs from that of adults and children may be more heavily exposed than adults to certain chemicals in the environment as they, on a body weight basis, breathe more air, drink more water, and eat more food than adults; additionally, their behavior patterns, such as play close to the ground and hand-to-mouth activities, can increase their exposure. The differences in exposure patterns between children and adults are often used as an argument for increased susceptibility of children to chemicals; however, it should be recognized that such differences are not related to increased vulnerability to chemicals but are purely related to an increased internal exposure (Nielsen et al. 2001).

Irrespective of other differences, there might be different conditions during adulthood, such as pregnancy, that might change the susceptibility to chemicals. During pregnancy, many physiological changes occur in the maternal organism as a consequence of, and in order to support, the rapid growth of the fetus and reproductive tissues. These changes may in different ways influence the toxicokinetic handling of a chemical. Absorption of chemicals from the gastrointestinal tract increases and hepatic metabolism decreases during pregnancy; this may favor retention of chemicals leading to enhanced toxicity. Also, the higher fat content during pregnancy increases the potential for greater body burden of lipophilic chemicals. There are also signs of metabolic changes and increasing activity in certain endocrine organs (Nielsen et al. 2001).

The numerous physiological and biochemical changes occurring during aging can modify the toxicokinetics and toxicodynamics of chemicals in the elderly, resulting in either higher or lower toxicity. Enhanced effects caused by xenobiotics in the elderly (from 60–65 years), may be due to decreases in renal excretion, hepatic extraction, plasma protein binding, or volume of distribution for water-soluble chemicals. Many elderly people have impaired renal function and chemicals that are excreted primarily by the kidneys will persist longer in the blood of older individuals than in younger people. Elderly people have decreased hepatic mass and blood flow. Moreover, phase I enzymes often decline in rate of function. The decrease in cardiac output that accompanies aging gradually changes the regional distribution of blood flow in the body. As a result, transport of chemicals to the liver and kidneys is slower, delaying inactivation of many chemicals. Elderly people also have a higher percentage of body fat, on average, than younger adults, so lipophilic chemicals are more extensively stored in older people and accumulate to a greater extent. Furthermore, various organ systems alter functionally with increasing age. In the nervous system, the density of neurons and dendrites is reduced, the receptor densities on the cell surfaces are reduced, the immune system reacts more slowly to xenobiotics, the cardiovascular system is limited functionally as a result of arteriosclerotic symptoms, and the respiratory tract has a smaller area for gas exchange. The responses of target tissues in the elderly may also be enhanced because of a less unused functional capacity in reserve. As a result, chemical exposures impact more keenly on tissue systems already stretched to the limit. In addition, aging can potentially alter not only the structure of genes, but also the way in which they function. Changes in the DNA are often thought to be integral to aging. It is clear that not only mutations but also chromosomal rearrangements accumulate with age (WHO/IPCS 1993, Kalberlah and Schneider 1998, Dybing and Søderlund 1999, Beltoft et al. 2001).

Clewell et al. (2002b) have reviewed and evaluated the potential impact of age-specific pharmacokinetic differences on tissue dosimetry. A large number of age-specific quantitative differences in pharmacokinetic parameters were identified. The majority of these differences were identified between neonates/children and adults, with fewer differences being identified between young adults and the elderly.

5.4.1.2 Gender

Women and men differ from each other in some constitutive and physiological parameters. For example, weight, tidal volume, and the water and fat content of the body differ between genders.

However, the differences become smaller when normalized according to body weight or body surface area. Many physiological parameters are altered in women during pregnancy due to specific processes of adaptation to the needs of the circulatory system of pregnant women and fetuses, e.g., the blood output of the heart increases by 50% for only a slight increase in the body surface area. Data indicate that differences in physiological parameters that may influence the toxicokinetics of chemicals are generally below a factor of 2 (Kalberlah and Schneider 1998).

Clewell et al. (2002b) have reviewed and evaluated the potential impact of gender-specific pharmacokinetic differences on tissue dosimetry. A large number of gender-specific quantitative differences in pharmacokinetic parameters were identified. The majority of these differences were identified between neonates/children and adults, with fewer differences being identified between young adults and the elderly.

Differences are also obvious if chemicals have specific mechanisms of action that affect gender-specific differences, i.e., as a result of interaction with hormonal regulation, specific damage to the sex organs, or adverse effects on organs in the development of the infantile organism (Kalberlah and Schneider 1998).

5.4.1.3 Genetic Polymorphism

Enzyme levels and activities within the human population can vary considerably and many of the enzymes involved in the metabolism of xenobiotics are polymorphically distributed in the human population. Genetic polymorphism (from Greek: *poly* "many", *morph* "form") is defined as the occurrence of at least two different alleles, with allele frequencies exceeding 1% at a particular locus. The allelic variants include point mutations as well as deletions and insertions and genetic polymorphism may cause an increase, a decrease, or no change in enzymatic activity.

Genetic polymorphism may result in "poor metabolizers" (i.e., individuals who have only a limited or no capacity to metabolize a given chemical via a specific enzymatic pathway), and "extensive metabolizers" (i.e., individuals who have a sufficient capacity to metabolize a given chemical via a specific enzymatic pathway) and individuals of a particular group may therefore respond differently to exposure to chemicals.

The polymorphisms in genes encoding xenobiotic metabolizing enzymes are far more prominent than those seen in genes encoding enzymes of endogenous importance with defined physiological functions. Because of genetic drift, where a small population has migrated and then expanded, interethnic differences in the distribution of the genes encoding xenobiotic metabolizing enzymes are sometimes large (KEMI 2003).

It should be noted that a large variation in metabolic capacity not necessarily corresponds to an equal variation in toxicity.

There are many examples in the literature showing a more than 10-fold variation in metabolic capacity depending on genetic polymorphism in the involved enzymes (Kalberlah and Schneider 1998).

Data reviewed in Beltoft et al. (2001) show interindividual levels and differences in the activities of cytochrome P450 (CYP) enzymes within the human population. Polymorphisms in various CYP enzymes have been described and may result in poor metabolizers. For example, approximately 2%–7% of the Caucasian population and 18%–23% of Japanese are reported to be poor metabolizers due to genetic alterations. The CYP enzymes are haemeprotein oxidoreductases that utilize electrons (from NADPH) to reduce molecular oxygen and oxidize molecules containing, e.g., carbon, nitrogen, oxygen, sulphur, phosphorus, and/or halogens. The CYP enzymes belong to three main P450 gene families: CYP1, CYP2, and CYP3. The highest concentration of CYP enzymes involved in biotransformation of xenobiotics is found in the liver, but CYP enzymes are present in virtually all tissues.

The glutathione S-transferases (GST) enzymes are dimeric enzymes that catalyze the conjugation of glutathione (GSH) to electrophilic xenobiotics in order to inactivate them and facilitate their excretion from the body. The isoenzymes are divided into at least seven classes, and marked interindividual differences exist in the expression of some of these classes. Genetic polymorphisms, deletion of the whole gene (null allele), have been detected in two of the classes (GSTM1 and GSTT1) of this enzyme family and ethnic differences have been reported with the frequency of the GSTM1 and GSTT1 genotypes ranging from 5.8% to 58% and from 12% to 62%, respectively, among different ethnic groups (Beltoft et al. 2001).

The paraoxonase enzyme (PON1), involved in the metabolism of organophosphate pesticides and aryl ester compounds, is widely distributed in mammals, including humans. Differences in observed rates of hydrolysis of the toxic metabolite (paraoxon) of the pesticide parathion between individuals have been reported to vary by at least 20-fold. There is also a large interindividual variability of the level of circulating PON1. Human PON1 exists in two polymorphic forms: $PON1_{R192}$, which hydrolyzes paraoxon at a high rate; and $PON1_{Q192}$, which hydrolyzes paraoxon at a low rate. Significant differences in gene frequencies between different ethnic groups have been reported with a frequency for the $PON1_{R192}$ allele of 0.31 in Caucasian populations, 0.41 in Hispanic populations, and 0.66 in Japanese populations (Beltoft et al. 2001).

Other metabolic enzymes that show polymorphic differences in that they can occur as genetic high-activity and low-activity variants include acetylcholinesterase, butyrylcholinesterases, flavin-dependent monooxygenase, alcohol dehydrogenase, epoxide hydrolase, and arylesterase (Beltoft et al. 2001).

Altered enzyme levels and activities may thus render some individuals more susceptible to exposure to chemicals than the general population. It could therefore be hypothesized that even a very low exposure to a chemical may be associated with various biological responses in such susceptible individuals as altered enzyme levels and activities may influence the individual's ability to detoxify a chemical or increase the conversion of a chemical to a toxic metabolite. Whether and to what extent an altered enzyme level or activity will increase an individual's risk of experience adverse effects from exposure to chemicals is generally not known.

In addition to polymorphisms in biotransformation enzymes, some genetically determined variations in toxicodynamic processes have also been described. Genetic polymorphisms are present in many receptor genes, for example the D2 dopamine receptor or the β_2-adrenoceptor, but the functional significance of these variations and the importance for cell signaling is uncertain (KEMI 2003). The extreme polymorphism of the immune system can explain why chemicals alone or conjugated with tissue macromolecules are recognized very differently by different individuals (Weigle 1997, as cited in Dybing and Søderlund 1999).

Humans also display marked interindividual variability in the capacity to repair damaged DNA. This variation is partly due to genetic factors, and genetic polymorphisms have been found in several proteins involved in DNA repair (Kalberlah and Schneider 1998, KEMI 2003).

5.4.1.4 Health and Disease

A poor nutritional state can have a considerable influence on the metabolism of xenobiotics (Kalberlah and Schneider 1998).

Nutritional factors such as a poor or an unbalanced diet may influence the intake and probably also the biotransformation of food contaminants. Ethnic differences due to variations in, e.g., dietary habits have also been reported (KEMI 2003).

Starvation may be associated with an increase in the permeability of the blood–brain barrier (Hawkins 1986, as cited in Dybing and Søderlund 1999). Tissue antioxidant status may be compromised under nutritional deficiencies and starvation (Godin and Wohaieb 1988, as cited in Dybing and Søderlund 1999).

In general, a diseased state can be expected to influence the sensitivity to chemicals. This is of special concern when the target organ of the toxic effects is affected by the disease or when the metabolism and elimination are disturbed.

Liver disease may decrease hepatic metabolism resulting in enhanced responses to parent chemicals; however, for many compounds, metabolism is only slightly impaired in moderate to severe liver disease. Disease-induced alterations in clearance and volume of distribution often act in opposite directions with respect to their effect on half-life. Bioavailability may be markedly increased in liver disease with portal/systemic anastomosis (the connection of normally separate parts so they intercommunicate) so that orally administered chemicals bypass hepatic first-pass metabolism. Altered receptor sensitivity has been observed for some chemical substances in liver cirrhosis. When liver tissue repair is inhibited by chemical co-exposure, even an inconsequential level of liver injury may lead to fulminating liver failure from a nonlethal exposure of hepatotoxicants. (Several articles, as reviewed by Dybing and Søderlund 1999.)

Renal impairment results in reduced clearance of many chemicals or their metabolites that are eliminated largely in the urine; the decline in excretion is directly related to the glomerular filtration rate (Dybing and Søderlund 1999). A number of diseases, including preexisting renal disease, systemic hypertension, and diabetes, are significant risk factors for chemical-induced renal disease (Ritter et al. 1995, Bennett 1997, as cited in Dybing and Søderlund 1999).

Individuals with lung diseases such as asthma and chronic obstructive lung disease are generally regarded as sensitive groups with regard to air pollutants (KEMI 2003).

In addition to diseases of important organs such as the lungs, the liver, and the kidneys, hereditary or acquired characteristics such as immunodeficiency and hypersensitivity may also influence sensitivity to xenobiotics. For example, atopics (individuals with immunologically mediated allergy) may develop life-threatening reactions to a chemical at an exposure level that is insignificant for the population in general (KEMI 2003).

5.4.1.5 Lifestyle

The toxic effects of alcohol, tobacco smoke, and drugs may modify the toxic responses to chemicals.

Ethanol can increase the levels of many enzymes involved in metabolism of xenobiotics. Prolonged ethanol intake causes irreversible damage in the central nervous system and in the liver, resulting in marked decreased capacity for detoxification of xenobiotics and thereby increased sensitivity to a number of chemicals (KEMI 2003).

Reduced antioxidant capacity has been found in several tissues of alcoholics (Bjorneboe and Bjorneboe 1993, as cited in Dybing and Søderlund 1999).

Tobacco smoke is considered to be one of the more severe confounders in epidemiological studies, due, e.g., to its ability to affect enzyme activities and to cause various health effects (KEMI 2003).

Tobacco smoke contains more than 3800 different compounds. About 10% of these constitute the particulate phase, which contains nicotine and tar. The remaining 90% contains volatile substances such as carbon monoxide, carbon dioxide, cyanides, various hydrocarbons, aldehydes, and organic acids. Although all of these substances affect the smoker to some degree, nicotine is generally considered to be the primary substance responsible for the pharmacological responses to smoking (Nielsen et al. 2001).

5.4.2 ASSESSMENT FACTOR FOR INTRASPECIES VARIATION (HUMAN-TO-HUMAN): DEFAULT VALUE

The intraspecies (interindividual) assessment factor is generally recognized as providing an extrapolation from the average human being to the sensitive human being, assuming that most (not necessarily all) human responses to a chemical fall within a 10-fold range. The rationale for the traditionally used default interindividual assessment factor of 10 is not known.

If the N/LOAEL has been derived from an animal study, animal intraspecies differences have already to some extent been accounted for in that N/LOAEL. Ideally therefore, the assessment factor for interindividual variation should reflect the additional interspecies variability, i.e., the difference between variability in the human population and variability in the animal population. The variability within the experimental animals is, however, assumed to be small and in addition, difficult to quantify. Therefore the interindividual assessment factor is generally not corrected for animal variation.

In the following text, various studies will be described which attempt to establish a scientific rationale for the selection of the interindividual assessment factor. Based on these studies, it can be concluded that the factor for interindividual variability should preferentially be described probabilistically. However, at present there is no database-derived distribution of the interindividual factor and thus a deterministic default factor of 10, split evenly into a sub-factor of 3.16 for both toxicokinetics and for toxicodynamics, respectively, could be used to account for the interindividual variability in the human population. Alternatively, a pathway-related UF could be applied in case the pathway(s) of the metabolism of the chemical in humans and the particular enzyme(s) are known.

Calabrese (1985) examined the range of human responses with respect to (1) the degree of variation in the metabolism of xenobiotics, (2) the binding of toxic substances to molecules such as hemoglobin and DNA, (3) the activity level of selected cellular enzymes, and (4) the differential risks to diseases in the human population. Considerable differences were found among human subjects in their capacity to metabolize xenobiotics. The range of responses varied widely, depending on the substance, enzyme, and organ considered. In addition, it was apparent that human variation may range up to two or three orders of magnitude indicating that human variation in the metabolism of various xenobiotics can exceed a factor of 10. The author concluded that the commonly used safety factor of 10 appeared to provide protection for the majority of the population (80%–95%). The remaining 5%–20% exhibited responses outside the 10-fold range of variation.

Hattis et al. (1987) examined the variability in key pharmacokinetic parameters (elimination half-lives ($T_{1/2}$), area under the curve (AUC), and peak concentration (C_{max}) in blood) in healthy adults based on 101 data sets for 49 specific chemicals (mostly drugs). For the median chemical, a 10-fold difference in these parameters would correspond to 7–9 standard deviations in populations of normal healthy adults. For one relatively lipophilic chemical, a 10-fold difference would correspond to only about 2.5 standard deviations in the population. The authors remarked that the parameters studied are only components of the overall susceptibility to toxic substances and did not include contributions from variability in exposure- and response-determining parameters. The study also implicitly excluded most human interindividual variability from age and diseases. When these other sources of variability are included, it is likely that a 10-fold difference will correspond to fewer standard deviations in the overall population and thus a greater number of people at risk of toxicity.

According to Kalberlah and Schneider (1998), a factor of 10 ensures protection of much greater than 99% of the population if the average behavior of the chemicals is taken as the basis. In the case of the chemical for which a high degree of variability was observed, a factor of 10 signified protection of about 99% of the population.

Reanalysis of the data of Hattis et al. (1987) showed that the variation between individuals for the elimination half-life was quite small (Schaddelee 1997, as cited in Vermeire et al. 1999, 2001). Defining the interindividual factor as the ratio of the P_{50} (50th percentile) and P_{05} (5th percentile) resulted in a factor of 1.4. It was emphasized that although it appeared from this analysis that a 10-fold factor would be sufficient for pharmacokinetic variation, the real median to sensitive human variability is underestimated because variation also exists in pharmacodynamics and only data of healthy volunteers were available.

Grönlund (1992) has investigated methods used for quantitative risk assessment of non-genotoxic substances, with special regard to the selection of assessment factors. Grönlund found that the 10-fold factor suggested for interindividual variability probably protects a majority but not all of the population.

Renwick (1991, 1993) analyzed interindividual differences of healthy volunteers by comparing the maximum and mean values of pharmacokinetic parameters (7 substances) and pharmacodynamic parameters (6 substances). The data indicated that toxicokinetic differences were slightly greater than toxicodynamic differences. With one exception, the ratios between the maximum and mean value for a substance's kinetic parameter ranged from 1.8 to 4.2 with most values between 3 and 4, and it was concluded that a factor of 3–4 would be sufficient to consider toxicokinetic differences for 99% of the healthy, adult population and for 80% of the substances. The ratios between the maximum and mean value for a substance's dynamic parameter ranged from 1.5 to 6.9 with most values between 1.7 and 2.7. Based on the analyses, Renwick proposed to subdivide the interindividual factor of 10 into a factor of 4 for pharmacokinetic differences and a factor of 2.5 for pharmacodynamic differences. The aim of the subdivision of the 10-fold factor was to allow the incorporation of suitable compound-specific data for one particular aspect of uncertainty.

Reanalysis of the Renwick data by using distributions instead of ratios for max/min gave comparable results (Schaddelee 1997, as cited in Vermeire et al. 1999, 2001).

Given the assumption that most human responses fall within approximately a 10-fold range, Calabrese and Gilbert (1993) stated that the application of a 10-fold interindividual UF should begin with the average human and extend to cover the higher risk segments of the population. Consequently, a UF of 5 would be expected to protect most humans, see Figure 5.2. The application of a 10-fold UF for humans would be more justified if it were based on an occupational epidemiological study, because this type of study does not consider the most sensitive humans and is likely to involve principally healthy workers and a self-selection component that consists of the less sensitive members of the population, see Figure 5.3.

Dourson et al. (1996) considered that, in general, the default value of 10 for interindividual variability appears to be protective when starting from a median response, or by inference, from a NOAEL assumed to be from an average group of humans. When NOAELs are available in a known sensitive human subpopulation, or if human toxicokinetics or toxicodynamics are known with some certainty, this default value of 10 should be adjusted or replaced accordingly.

Kalberlah and Schneider (1998) considered that the usually applied extrapolation factor of 10 is probably sufficient to protect a large part of healthy adults in the human population with regard to toxicokinetic differences, whereas groups like children, elderly people, and individuals with diseases are not fully considered in the 10-fold factor. In addition, the 10-fold factor was not considered to take account of the toxicodynamic differences. The authors proposed an interindividual factor of 25 for the general population, consisting of a factor of 8 for toxicokinetic variation and

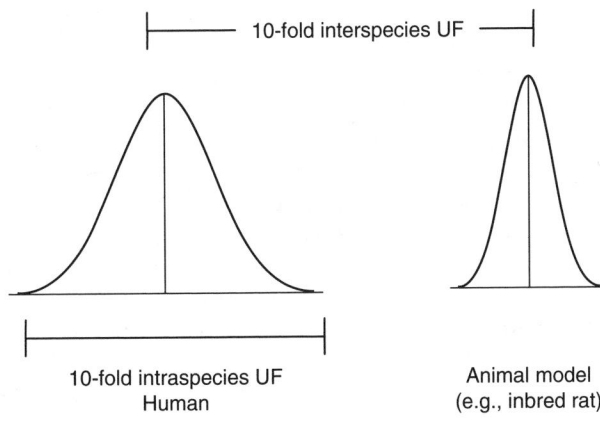

FIGURE 5.2 Interdependence of UFs. (Modified from Calabrese, E.J. and Gilbert, C.E., *Regul. Toxicol. Pharmacol.*, 17, 44, 1993.)

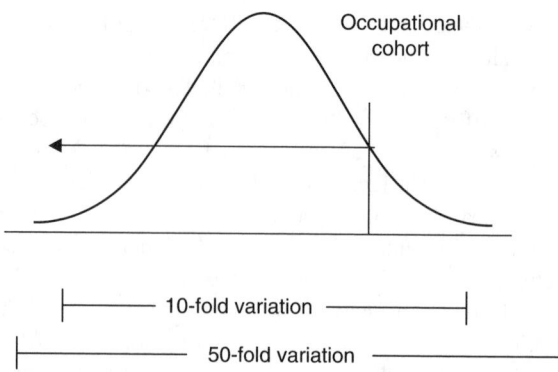

FIGURE 5.3 Interindividual UF based on occupational cohort. (Modified from Calabrese, E.J. and Gilbert, C.E., *Regul. Toxicol. Pharmacol.*, 17, 44, 1993.)

enzyme polymorphisms and a factor of 3 for toxicodynamic variation. The factor of 8 was derived from a factor of 5 based on the analyses of Hattis et al. (1987) and Renwick (1993) with healthy adult volunteers, and a factor of 3 for enzyme polymorphisms based on individual data in the range of 3–10. As the toxicokinetic variation and enzyme polymorphisms were considered to be interdependent, the factors are additive (5 + 3 = 8), not multiplicative. The toxicodynamic database was considered to be limited, consisting mainly of examples, but the majority of the examined substances could be considered by a factor of 3. Since toxicokinetic and toxicodynamic influences were regarded as largely independent parameters, these two factors are multiplicative (8 × 3 = 24, rounded to 25). The quantification of sub-factors for higher sensitivity in certain age groups (children and elderly people) as well as of a factor covering disease, pregnancy, gender, constitution, and physical activity was considered to require further validation, as the currently available examples were not considered representative in character.

Renwick and Lazarus (1998) analyzed the default UF for human variability based on the evaluation of an extensive database in relation to a subdivision of the 10-fold factor due to variability in toxicokinetics and toxicodynamics, as well as the adequacy of the 10-fold factor. Papers giving kinetic data were selected on the basis of the quality and/or size of the study, the interest of the results, and the physiological/metabolic process determining the kinetic parameter. Papers giving dynamic data were selected on the basis of the adequate separation of variability due to kinetics and dynamics. The data on kinetics and dynamics were tabulated, the coefficients of variation were averaged for different studies which measured a common endpoint, or for multiple doses which measured the same endpoint.

The data were analyzed for subdivision of the 10-fold factor into kinetic and dynamic aspects. Data for the kinetics of 60 compounds (drugs) were identified, which represented a range of pathways of metabolism or clearance; the mean coefficient of variation was 38% (range 9%–114%). The authors remarked that the selected studies represented only a small proportion of the total published literature on the kinetics of xenobiotics in humans. Concentration–effect data, mostly *in vivo* plasma concentration–response data (clinical treatment of patients), were identified for 49 compound-related effects. A variety of effects were included, but the majority of effects were short-term changes in the cardiovascular or central nervous system measurements, i.e., therapeutic rather than toxic effects. The mean coefficient of variation in dynamics was 51% (range 8%–137%). According to the authors, the slightly greater variability in the toxicodynamics compared with toxicokinetics is not sufficient to warrant an unequal subdivision weighted in favor of toxicodynamics and thus their analysis supported the subdivision of the 10-fold factor with an equal weighting for toxicokinetics and toxicodynamics (i.e., a factor of 3.16). The authors remarked that much of the toxicodynamic data, unlike the toxicokinetic data, were for the clinical treatment of patients and therefore aging and disease processes may have contributed to the greater variability in toxicodynamics compared to toxicokinetics.

The data were also analyzed using the reported standard deviation and the calculated geometric standard deviation (GSD) for each parameter to define the so-called Z-score in order to determine the proportion of a normally, or lognormally, distributed population, which would be covered by each of the sub-factors of 3.16. The number of subjects per million of the population not covered by the 3.16-fold factor was directly proportional to the standard deviation for the estimate. The choice of distribution model (normal or lognormal) had a greater impact when the population estimate was in the tail of the distribution, i.e., at 3.16 times the mean parameter estimate. For toxicokinetics, the average number of subjects not covered by a factor of 3.16 away from the mean parameter estimate was 685 per million of the population assuming a normal distribution, and 8,564 assuming a lognormal distribution. For toxicodynamics, the average number of subjects not covered by a factor of 3.16-fold away from the mean was 2,930 per million of the population assuming a normal distribution, and 18,896 assuming a lognormal distribution. The probability of the same individual falling outside the range for both toxicokinetics and toxicodynamics was also analyzed assuming that the toxicokinetic and toxicodynamic "risk factors" are independent variables. On average, only two persons in a million would not be covered by the combined factors (3.16 times 3.16) assuming a normal distribution, and 162 persons per million assuming a lognormal distribution. According to the authors, this demonstrates that the 10-fold factor is an adequate default assumption for the types of chemicals and biological effects considered in the performed analyses, and that the composite 10-fold factor would cover the vast majority (>99.9%) of the population assuming either normal or lognormal distribution.

The authors also analyzed the proportion of certain subgroups of the population that would not be covered if a 3.16-fold factor was applied to the mean for the major group in the population. As the majority of relevant data relate to the toxicokinetics of drugs in Caucasian adults, this group was taken as the reference subgroup for comparisons. According to the authors, young children frequently eliminate xenobiotics more rapidly by metabolism and excretion compared with adults, and children would therefore be covered adequately by a 3.16-fold factor for toxicokinetics applied to the mean data for adults. Different ethnic groups may show differences in both toxicokinetics and toxicodynamics, but according to the authors, the default value of 3.16 would cover adequately most ethnic groups in a population. However, the data also showed that ethnicity should be considered for some P450-mediated oxidation reactions, on a case-by-case basis. Of greater potential concern are undiagnosed and unrecognized sources of variability, such as genetically determined differences in enzymes affecting toxicokinetics, which must be covered by the default UF. According to the authors, it is clear that genetic polymorphisms can have a profound influence on the validity of the 3.16-fold default UF. The number of subjects in the whole population not covered by a factor of 3.16 applied to the mean for the extensive metabolizers would have to take into account the incidence of poor metabolizer status. For example, for fluoxetine (an antidepressive drug), the incidence outside a factor of 3.16 would be 85% of the incidence of poor metabolizers plus a small number of extensive metabolizers, i.e., up to about 8% of the whole population. Genetically determined differences were considered to be of greatest relevance to risk assessment when the polymorphic pathway represents the major route of elimination. Therefore, knowledge that a substance is substrate for a metabolic pathway which shows polymorphic expression raises questions about the validity of the 3.16-fold default factor for toxicokinetics and therefore also about the combined 10-fold factor for human variability, but does not automatically invalidate the default values. These observations led to a proposal for the generation of pathway-related UFs that would constitute an intermediate option between default UFs and CSAFs. The pathway-related UFs could be applied to chemicals for which the metabolic fate is known in humans, but for which chemical-specific toxicokinetic data were not available (Figure 5.4). The pathway-related UFs can be generated to cover a particular proportion of the population, e.g., the 95th percentile.

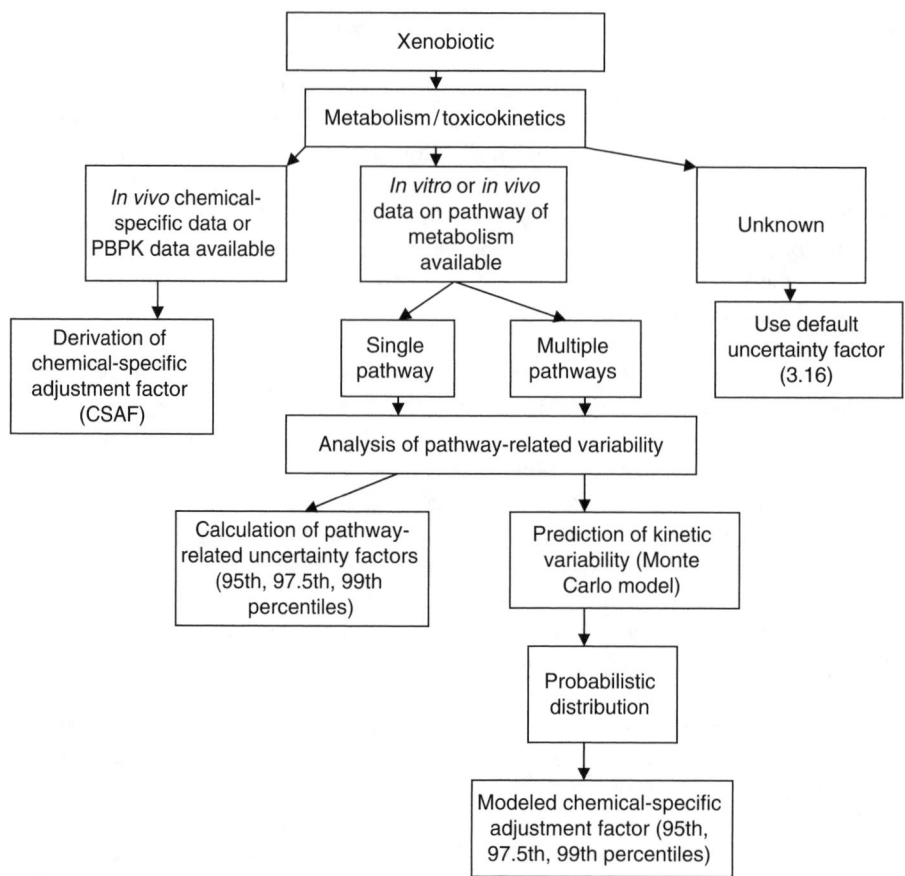

FIGURE 5.4 UFs, CSAFs, pathway-related UFs, and general default UFs. (Modified from Renwick, A.G. and Lazarus, N.R., *Regul. Toxicol. Pharmacol.*, 27, 3, 1998.)

In conclusion, the data and analyses performed by Renwick and Lazarus indicate that the 10-fold factor for human variability is an appropriate default value, but it has also identified a number of circumstances where this default value may be inadequate. For example, the 3.16 toxicokinetic factor could not cover human variability in the case of genetic polymorphisms.

Burin and Saunders (1999) have addressed the robustness of the interindividual UF. They considered that the best source of knowledge about human variability in the response to chemicals comes from clinical trials of pharmaceuticals. Based on these data, both qualitative and quantitative characterization of variability in pharmacokinetic and pharmacodynamic parameters can be performed for the general population and in subgroups such as children. The authors concluded that the preponderance of evidence in the areas of pharmacokinetics and pharmacodynamics supports the routine use of an interindividual UF in the range of 1–10 as being protective of greater than 99% of the human population. They also claimed that the interindividual UF is highly protective of various subpopulations, including infants and children.

According to Vermeire et al. (1999, 2001), several theoretical probabilistic distributions have been proposed. Distributions proposed by Price et al. (1997), Swartout et al. (1998), and Slob and Pieters (1998) as cited in Vermeire et al. (1999, 2001), were considered to be consistent with the current use of the default factor of 10 and these authors found the traditional factor 10 to be conservative, see also Section 5.11.

Vermeire et al. (1999, 2001) concluded that, currently, no proposal for a database-derived distribution of the interindividual factor can be made.

Vermeire et al. (1999) recommended to remain consistent with the traditional default value of 10 and to assume that this value protects the majority of the general human population. For workers, they recommended to remain consistent with the traditional default value of 3.

Vermeire et al. (2001) concluded that currently no adequate proposal for a database-derived distribution of the intraspecies factor can be made. Therefore, a distribution consistent with the default value of 10 as proposed by Slob and Pieters (1998) based on a theoretical distribution will be used for derivation of Human Limit Values. For workers, a distribution consistent with the default value for workers of 3, considered to be conservative, was proposed in parallel with the approach of Slob and Pieters (1998).

Ginsberg et al. (2002) have evaluated child/adult pharmacokinetic differences. Using published literature, a children's pharmacokinetic database has been compiled, which compares pharmacokinetic parameters between children and adults. The database contained 45 drugs covering a wide range of chemical structures, mechanisms of action, and metabolism and clearance pathways. The database has enabled comparison of child and adult pharmacokinetic function across a number of cytochrome P450 (CYP) pathways, as well as certain phase II conjugation reactions and renal elimination. The main common parameter was the half-life. The analysis indicated that premature and full-time neonates tend to have three to nine times longer half-life than adults for the drugs included in the database. The difference disappeared by 2–6 months of age. Beyond this age, the half-life can be shorter than in adults for specific drugs and pathways. The range of neonate/adult half-life ratios exceeded the 3.16-fold factor commonly ascribed to interindividual pharmacokinetic variability and thus this UF may not be adequate for certain chemicals in the early postnatal period.

The proposal of Renwick and Lazarus (1998) has been further developed by Renwick and coworkers using analyses of metabolism and pharmacokinetic data for probe substrates of phase I metabolism (CYP1A2, CYP2A6, CYP2C9, CYP2C19, CYP2D6, CYP2E1, CYP3A4, hydrolysis, alcohol dehydrogenase), phase II metabolism (N-acetyltransferases, glucuronidation, glycine conjugation, sulphation), and renal excretion for interspecies differences (Section 5.3.4) and human variability. Two papers by Dorne et al. (2005) and Dorne and Renwick (2005) describe the recent developments arising from a database that defined human variability in different routes of metabolism and was used to derive pathway-related UFs for chemical risk assessment that allow for human variability in toxicokinetics. Probe substrates for each pathway of elimination were selected on the basis that oral absorption was >95% and that the metabolic route was the primary route of elimination of the compound (60%–100% of a dose). Human variability in kinetics was quantified for each compound from published pharmacokinetic studies in healthy adults and other subgroups of the population using parameters relating to chronic exposure (metabolic and total clearances, area under the plasma concentration–time curve (AUC)). The analyses of human variability assessed critically the adequacy of the general 3.16-fold toxicokinetic default factor, and calculated sets of pathway-related UFs for each metabolic route to cover the 95th, 97.5th, and 99th percentile of the general healthy adult population and available subgroups of the population (healthy adults from different ethnic origins - African, Asian and South Asian, elderly, neonates, and children), respectively. The 3.16 factor would be too conservative to cover healthy adults to the 99th percentile for all monomorphic pathways, and this default could be replaced by the relevant pathway-related values. However, for polymorphic pathways, the 3.16 factor is insufficient to cover the majority of the general healthy adult population. The 3.16 factor would not be conservative enough to cover all subgroups of the population (due to variability and differences in internal dose) for compounds metabolized by highly variable monomorphic or polymorphic pathways. Neonates would be the most susceptible subgroup for all of these pathways because elimination processes are immature at and soon after birth. The analyses supported the need for a higher toxicokinetic factor for neonates, but not for infants and children.

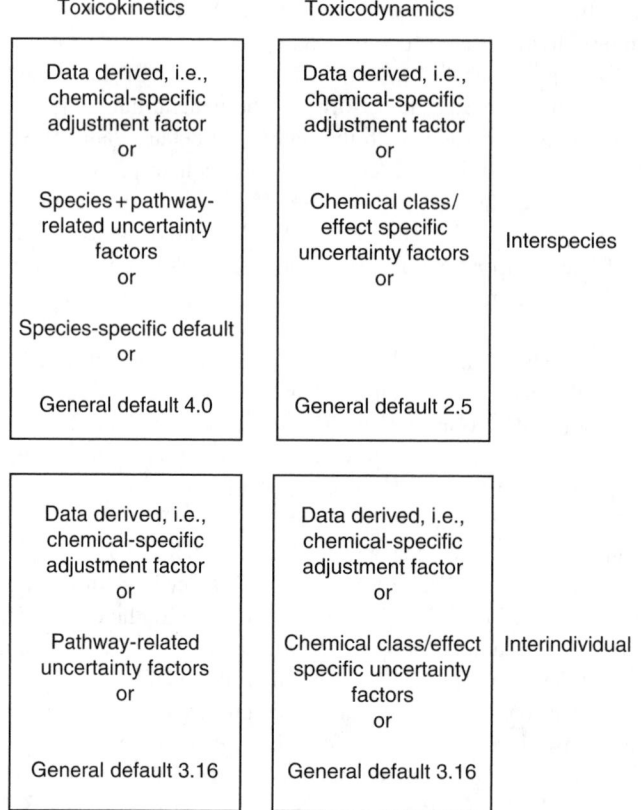

FIGURE 5.5 Applications of pathway-related UFs for chemical risk assessment and future refinements of toxicokinetic UFs. (Modified from Dorne, J.L., Walton, K., and Renwick, A.G., *Food Chem. Toxicol.*, 43, 203, 2005.)

Pathway-related UFs allow metabolism data to be incorporated into the derivation of, e.g., tolerable intakes. Both the enzyme involved in the metabolism as well as the toxicological consequences of metabolism need to be taken into account. Applications of pathway-related UFs for chemical risk assessment and future refinements of the approach are illustrated in Figure 5.5. In the ideal situation, databases describing the metabolism and/or toxicokinetics of a particular chemical in humans would be used to define the internal body burden, or to set up a PBPK model. In this situation, the data would be used to derive a CSAF. The worst-case scenario would be that no data on the metabolism or toxicokinetics of the compound are available and in this situation the general toxicokinetic default of 3.16 would be used. The pathway-related UFs approach lies in between these two extremes and constitutes an intermediate approach where the toxicokinetic default factor could be replaced by the appropriate UF when the pathway of the metabolism of the compound and the particular enzyme(s) are known. Recognizing the potential advantage of moving away from default factors and point estimates to probabilistic models, the authors suggested that probabilistic multiplication of the distributions of the pathway-related defaults available in the database combined with the distribution of the dynamic default derived from Renwick and Lazarus (1998) could be used to derive combined probability distributions to replace the 10-fold UF for human variability. The authors remarked that the pathway-related UFs have been derived for compounds handled by a single major route of elimination (>60%); however, this scenario is unusual as many chemicals are eliminated by several pathways. This situation could be addressed by the use of probabilistic approaches, such as Monte Carlo modeling, to combine the variability for different pathways, see Section 5.11.

ECETOC (2003) recommended a default assessment factor of 5 for the general population. This recommendation was based on the data from Renwick and Lazarus (1998) by considering that these data included both sexes, a variety of disease states, and ages. Thus, the use of the 95th percentile was considered sufficiently conservative to account for interindividual variability in the general population. A default value of 3 was recommended for the more homogeneous worker population, i.e., closer to the 90th percentile. Furthermore, ECETOC concluded that there is little scientific basis to support the need for an additional assessment factor for children in risk characterization, other than for substances that directly affect the developing fetus and which need to be considered on a case-by-case basis. In addition, attention should be given to substances affecting developing organ systems, such as reproductive development in prepuberty.

The WHO/IPCS (1994, 1996, 1999) noted that a factor of 10 is normally used to allow for differences in sensitivity *in vivo* between the population mean and highly sensitive subjects. WHO/IPCS has adopted the approach set forth by Renwick (1993) with one deviation: the UF for interindividual extrapolation should be split evenly between both aspects, i.e., a sub-factor of 3.16 for both toxicokinetics and toxicodynamics. The reason for this deviation from Renwick's initial suggestion was that the WHO/IPCS considered that the slightly greater variability in the kinetics compared with dynamics was not sufficient to warrant an unequal subdivision of the 10-fold factor into a toxicokinetic factor of 4 and a toxicodynamic factor of 2.5. In cases where there are appropriate data on the interindividual variability in toxicokinetic or toxicodynamics for a particular compound in humans, the relevant UF should be replaced by the data-derived factor. If a data-derived factor is introduced, then the commonly used 10-fold factor would be replaced by the product of the data-derived factor and the remaining default factor. It is also noted that for some compounds, it may be known that a subset of the population would be particularly sensitive, for example due to deficiencies in detoxication processes, and a specific reference was given to polymorphically distributed enzymes involved in xenobiotic biotransformation. In cases where the default factor would not adequately cover this additional variability, the default should be modified appropriately. In addition, it is noted that in cases where the risk assessment is based on *in vivo* data in the sensitive subgroup, the composite factor of 10 should be reduced to a much lower value. A value of 1 could be used if there is an extensive database in humans and the database adequately addresses any identified sensitive subgroups.

In 1988, the US-EPA adopted the ADI approach with respect to the derivation of RfD and RfC with a 10-fold UF to account for interindividual extrapolation (US-EPA 1988, 1993), see Section 5.2.1.1. It was noted in the 2002 review of the RfD and RfC processes (US-EPA 2002) that the interindividual UF is applied to account for variations in susceptibility within the human population and the possibility (given a lack of relevant data) that the database available is not representative of the dose/exposure–response relationship in the subgroups of the human population that are most sensitive to the health hazards of the chemical being assessed. As the RfC/RfD is defined to be applicable to "susceptible subgroups", the interindividual UF was established to account for uncertainty in that regard. In general, the importance of this UF was reconfirmed, and it was recommended that reduction of the UF from a default of 10 be considered only if data are sufficiently representative of the exposure/dose–response data for the most susceptible subpopulation(s). Given this, whether and how much the UF may be reduced must be linked to how completely the susceptible subpopulation has been identified and their sensitivity described (versus assumed). It was noted that Dourson et al. (1996) have documented the cases in the US-EPA IRIS database where the interindividual UF has been reduced from the default of 10-fold. At the other extreme, a 10-fold factor may sometimes be too small because of factors that can influence large differences in susceptibility, such as genetic polymorphisms.

The importance of development of data to support the selection of the appropriate size of this factor was emphasized, but it was also recognized that often there are insufficient data to support a factor other than the default.

TNO has suggested a default intraspecies assessment factor of 3 for workers instead of 10 as used for the general population, as it is assumed that the interindividual differences among workers are smaller than for the public at large because the worker population does not include the very young, the elderly, or the infirm. In case of embryotoxic and/or teratogenic effects, a factor of 10 should be used for workers, because no distinction should be made between the progeny of the occupational population and the general population (Hakkert et al. 1996), see also Section 5.2.1.7.

KEMI (2003) has suggested that for healthy adults, a factor of 3–5 may be considered for the toxicokinetic parameter, while for the general population, they did not find this factor protective enough. However, they did not consider that it is possible, based on present-day knowledge of interindividual variability, to suggest an alternative default value for toxicokinetics for the general population. A factor of 3 for genetic polymorphism has been suggested as additive to the toxico-kinetic factor of 5; however, KEMI did not consider this factor of 3 to be sufficiently protective since the genetic variation may be several orders of magnitude. However, they did not consider it possible to suggest a default assessment factor for genetic composition, but they stressed that it must be kept in mind that genetic composition really has to be taken into consideration. Toxicodynamic data were considered still to be limited and KEMI did not consider that it is possible, based on present-day knowledge of interindividual variability, to suggest an alternative to the default value of 3.2 for toxicodynamics for the general population. KEMI also noted that some probabilistic distributions have been proposed, but at present there is no database-derived distribution of the interindividual factor. They also questioned the use of a theoretical distribution based on the default value of 10. Use of an extra assessment factor for children has been considered in relation to the confidence in database (Section 5.9).

5.4.3 INTRASPECIES EXTRAPOLATION (HUMAN-TO-HUMAN): SUMMARY AND RECOMMENDATIONS

Humans differ in sensitivity due to biological factors such as metabolic polymorphism, age, gender, health status, and nutritional status. These differences can be the result of genetic and/or environ-mental/behavioral influences. This intraspecies (interindividual) variation is greater in humans than in the more inbred experimental animal population.

The interindividual variability reflects differences in toxicokinetics as well as in toxicody-namics. With respect to toxicokinetic factors, interindividual differences in the metabolism of chemicals are generally considered as the most significant explanatory factor. Hardly any know-ledge is available with respect to the factors that influence toxicodynamics. Thus, it is necessary to take such variation into account when extrapolating animal toxicity data to the human situation.

Ideally, the interindividual extrapolation should be based on substance-specific information; however, for most substances, only limited or no data are available. Therefore, an assessment factor is usually applied to account for interindividual human variation. A factor of 10 for the extrapolation from the average to the sensitive human being is traditionally used as a default value, eventually divided into two equal sub-factors of 3.16 for differences in toxicokinetics and toxicodynamics, respectively, as adopted by the WHO/IPCS (1994).

Children, as well as the unborn child, have in some cases appeared to be uniquely vulnerable to chemical's toxic effects because periods of rapid growth and development render them more susceptible to some specific toxic endpoints when compared to adults. Furthermore, there may be windows of vulnerability or periods of development when chemical exposures may substantially alter organ structure and function. In addition to such toxicodynamic factors, differences in toxicokinetics may contribute to an increased susceptibility during these periods. The greatest differences in comparison with adults occur in neonates and infants (<1 year). However, some organ systems, e.g., the nervous, immune, and endocrine/ reproductive systems continue to develop until adolescence (12 to 18 years) and may thus show specific vulnerability to chemical exposures until adulthood. The differences in sensitivity between infants/children and adults may be considerable; however, the available data are insufficient for a quantitative evaluation.

The numerous physiological and biochemical changes occurring during aging can modify the toxicokinetics and toxicodynamics of chemicals in the elderly, resulting in either higher or lower toxicity. Enhanced effects in the elderly (>60 years) are predominantly due to impaired renal function. The responses of target tissues in the elderly may also be enhanced because of a less unused functional capacity in reserve. The differences in sensitivity between elderly and adults in general are presumably less pronounced than those between infants and adults; however, the available data are insufficient for a quantitative evaluation.

Women and men differ from each other in some constitutive and physiological parameters. For example, weight, tidal volume, and the water and fat content of the body differ between genders; however, the differences become smaller when normalized according to body weight. Many physiological parameters are altered in women during pregnancy, as a consequence of, and in order to support, the rapid growth of the fetus and reproductive tissues. Absorption of chemicals from the gastrointestinal tract increases and hepatic metabolism decreases during pregnancy and this may favor retention of chemicals leading to enhanced toxicity. Data indicate that differences in physiological parameters that may influence the toxicokinetics of chemicals are generally below a factor of 2. However, greater differences could be expected if chemicals have specific mechanisms of action that affect other gender-specific differences, e.g., as a result of interaction with hormonal regulation or specific damage to the sex organs.

Genetically caused polymorphisms are known for a number of enzymes, which metabolize chemicals, and are important for the interindividual variability to chemical exposures, especially if the polymorphic pathway represents the major route of elimination. Altered enzyme levels and activities may thus render some individuals more susceptible to exposure to chemicals than the general population. It could therefore be hypothesized that even a very low exposure to a chemical may be associated with various biological responses in such susceptible individuals as altered enzyme levels and activities may influence the individual's ability to detoxify a chemical or increase the conversion of a chemical to a toxic metabolite.

There are many examples in the literature showing a more than 10-fold variation in metabolic capacity depending on genetic polymorphism in the involved enzymes. It should be noted, however, that a large variation in metabolic capacity not necessarily corresponds to an equal variation in toxicity.

In general, a poor nutritional state and a diseased state can be expected to influence the sensitivity to chemicals. This is of special concern when the target organ of the toxic effect is affected by the disease, or when the metabolism and elimination are disturbed. Also lifestyle factors such as alcohol, tobacco smoke, and drugs may modify the toxic responses of chemicals.

The interindividual assessment factor is generally recognized as providing an extrapolation from the average human being to the sensitive human being, assuming that most (not necessarily all) human responses to a chemical fall within a 10-fold range; however, there is no general, international, and scientific recommendation to a standard value. The available data generally support a default factor of 10 and a factor of 10 is considered to protect the majority (>99%) of the human population, including, e.g., children and the elderly. However, it is also obvious that in order to always cover the most sensitive individual exposed to any chemical would require a very high default assessment factor. It is recognized that there are differences between children and adults in toxicokinetics (especially babies in their first months) and toxicodynamics (especially at different stages of development); these differences may render children more or less susceptible to the toxic effects of a substance. A higher interindividual extrapolation factor for children from 10 to 100 could be considered when the following three criteria are fulfilled:

- Specific exposure of very young children
- Indications or suspicions of effects on organ systems and functions that are especially vulnerable during development and maturation in early life (in particular the nervous, reproductive, endocrine, and immune systems and also the metabolic pathways)
- Experimental data on such effects in young animals are not available

The US-EPA (1988, 1993) has adopted the 10-fold factor for interindividual extrapolation; this default value has later on been reconfirmed (US-EPA 2002) and is stated to account for "susceptible subgroups." The WHO/IPCS (1994, 1996, 1999) has recommended that the traditional UF for interindividual extrapolation should be split evenly into a sub-factor of 3.16 for both toxicokinetics and toxicodynamics, respectively. Calabrese and Gilbert (1993) have stated that the application of an interindividual assessment factor should begin with the average human and extend to cover the higher risk segments of the population (see Figure 5.2) and thus a factor of 5 would be expected to protect most humans. ECETOC (2003) recommended a default factor of 5 for the general population and a factor of 3 for the more homogeneous worker population. Kalberlah and Schneider (1998) have proposed an interindividual extrapolation factor of 25 for the general population, consisting of a factor of 8 for toxicokinetic variation and enzyme polymorphisms and a factor of 3 for toxicodynamic variation ($8 \times 3 = 24$, rounded to 25). The factor of 8 consists of a factor of 5 for toxicokinetic variation and a factor of 3 for enzyme polymorphism, additive to the factor of 5 because of interdependency ($5 + 3 = 8$). Some proposed theoretical probabilistic distributions have been considered to be consistent with the default factor of 10; however, Vermeire et al. (1999, 2001) concluded that, currently, no proposal for a database-derived distribution of the interindividual factor can be made. Vermeire et al. (1999) recommended to remain consistent with the traditional default value of 10 and to assume that this value protects the majority of the general human population. KEMI (2003) has suggested a default factor of 3–5 for healthy adults for the toxicokinetic parameters, while for the general population they did not find this factor protective enough; however, based on present-day knowledge, they did not consider it possible to suggest an alternative default value for toxicokinetics for the general population, as well as to suggest an alternative to the default value of 3.2 for toxicodynamics for the general population. In addition, KEMI did not consider it possible to suggest a default assessment factor for genetic composition, but they stressed that it must be kept in mind that genetic composition really has to be taken into consideration. Renwick and Lazarus (1998) have proposed the generation of pathway-related UFs that would constitute an intermediate option between default UFs and CSAFs. The pathway-related UFs could be applied to chemicals for which the metabolic fate is known in humans, but for which chemical-specific toxicokinetic data were not available, and can be generated to cover a particular proportion of the population, e.g., the 95th percentile. The proposal of Renwick and Lazarus (1998) has been further developed as reviewed by Dorne et al. (2005). Analyses of human variability in toxicokinetics for the main metabolic pathways enabled the derivation of pathway-related UFs for healthy adults and subgroups of the population, which can be used to replace the general toxicokinetic default factor of 3.16. Recognizing the potential advantage of moving away from default factors and point estimates to probabilistic models, the authors suggested that probabilistic multiplication of the distributions of the pathway-related defaults available in the database combined with the distribution of the dynamic default derived from Renwick and Lazarus (1998) could be used to derive combined probability distributions to replace the 10-fold UF for human variability.

In conclusion, the assessment factor for interindividual variability should preferentially be described probabilistically. However, at present there is no database-derived distribution of the interindividual factor and thus a deterministic default factor of 10, split evenly into a sub-factor of 3.16 for both toxicokinetics and toxicodynamics, respectively, is recommended in order to account for the interindividual variability in the human population. Alternatively, the pathway-related UF approach suggested by Renwick and Lazarus (1998) and further developed as reviewed by Dorne et al. (2005) could be applied in case the pathway(s) of the metabolism of the chemical in humans and the particular enzyme(s) are known. A higher interindividual extrapolation factor for infants (from 10 to 100) could be considered in some special situations, see above.

5.5 ROUTE-TO-ROUTE EXTRAPOLATION

Most toxicity studies are performed using oral exposure to the test chemical. However, in some situations, the predominant routes of exposure in humans are via inhalation or dermal contact, e.g., in the working environment. In the case where relevant data are lacking on the exposure route of interest for the derivation of a tolerable intake, a route-to-route extrapolation is used. A route-to-route extrapolation includes an examination of whether knowledge resulting from studies involving one exposure route can be applied to another exposure route. For example, it can be examined whether the internal dose after oral intake is identical to the internal dose following inhalation.

This section reviews the validity of route-to-route extrapolation. Then, the key issues are summarized and our recommendations are presented.

Sharratt (1988) has examined the possibilities and limitations of using toxicity data derived from one exposure route in assessing risks from other exposure routes. It was concluded that there is no simple and generally applicable way in which toxicity data derived from one route of exposure can be used to evaluate the effects of exposure by another route in the same species of animal, or for risk assessment of another exposure route in humans. It was further noted that reliable predictions may be made in some cases where substances act systemically and have relatively long half-lives, and where there are adequate data on toxicity by one route and on pharmacokinetics and metabolism by both routes. Extrapolation may also be possible where it can be shown that the nature and degree of toxicity are directly related to blood or tissue concentrations. The occurrence of toxic effects locally increases the uncertainties associated with extrapolation, usually making it infeasible. The use of safety factors to allow for uncertainties inherent in extrapolating between routes of exposure where data are inadequate was not advocated. Whether an extrapolation of data is justified in a given situation should be decided on a case-by-case basis, and generalizations cannot be made.

Kalberlah and Schneider (1998) evaluated the literature regarding route-to-route extrapolation and gave the general recommendation that, in the case of a total absence of information on the other route, a route-to-route extrapolation should generally not be undertaken. A route-to-route extrapolation should also not be undertaken in the case of:

- Local effects
- Wide variance in the exposure (intermittent, long interruptions) in the underlying study in combination with a short half-life of the substance under consideration
- Metal compounds, if the relevant absorption factors are not known, since the literature indicates a nonsystematic and widely varying absorption after oral or inhalation exposure
- Particulate bound substances
- A clear influence of dermal absorption on the total absorption in addition to the oral or inhalation absorption

According to Kalberlah and Schneider (1998), route-to-route extrapolations appear justifiable in principle whenever:

- Systemic effect represents the critical effect and local effects are of secondary importance.
- Half-life of the substance under consideration is not too short; it should be possible to assume a state of equilibrium in the body.
- Area under the curve (AUC) is of decisive importance for the effect under consideration.
- Relative absorption is known or a similar absorption can be assumed.
- There is no first-pass effect (this can lead to intoxication or detoxification); apart from the liver, first-pass effects may also be of relevance with regard to the lung and the skin.

Kalberlah and Schneider also noted that, in the case of studies with gavage, a route-to-route extrapolation is limited in validity due to the possible influence of peak concentrations. The authors concluded that, on the basis of the above-mentioned qualitative criteria, it is necessary to examine in each individual case whether a route-to-route extrapolation appears justified.

According to Vermeire et al. (1999), data on absorption or acute toxicity are often used to account for differences between routes of exposure, but these methodologies are not validated and are based on broad assumptions.

A Dutch study (Wilschut et al. 1998, as reviewed in Vermeire et al. 1999) has evaluated route-to-route extrapolation on the basis of absorption or acute toxicity data. Data were collected primarily on dermal and inhalation repeated dose toxicity. An extrapolation factor, defined as the factor that is applied in route-to-route extrapolation to account for differences in the expression of systemic toxicity between exposure routes, was determined for each substance by using data on absorption and acute toxicity data. As experimental data on absorption often were not available, default values for absorption were also used to determine an extrapolation factor. Despite a rather large overall database, relatively few data could be used for the evaluation and the selection criteria were modified in order to include data that initially were considered less suitable for data analysis: interspecies extrapolation based on caloric demands was introduced, and a factor of 3 was applied in case a LOAEL instead of a NOAEL was available. The choice of NOAELs for different exposure routes known for a substance suitable for analysis was based primarily on the same effect, but this criterion could not be maintained.

For oral to inhalation route-to-route extrapolation (28 substances), the Predicted Inhalation No-Adverse-Effect Level (NAEL) was often higher than the observed NOAEL (for inhalation) implicating that the substance was considered less toxic after extrapolation when compared with the experimental observations. Based on the 95th percentile of the lognormal distribution of the ratios between the predicted NAEL and the observed NOAEL, UFs ranging from 75 to 201 for the different extrapolation methodologies were found.

For oral to dermal route-to-route extrapolation (25 substances), the predicted dermal NAEL (for inhalation) was often lower than the observed NOAEL (dermal) implicating that the substance was considered more toxic after extrapolation when compared with the experimental observations. Based on the 95th percentile of the lognormal distribution of the ratios between the predicted NAEL and the observed NOAEL, UFs ranging from 2.7 to 35 for the different extrapolation methodologies were found.

The authors noted that the reliability of the data is questionable, because the influence of the several assumptions made in order to derive comparable data on the ratio of the predicted NAEL and the NOAEL is unknown. They concluded that the development of scientifically based principles and procedures for route-to-route extrapolation appears to be a difficult task without the availability of adequate experimental data.

Vermeire et al. (1999) noted that for both extrapolations, the results were hardly influenced by the assumptions made on absorption, indicating that other factors may be important in route-to-route extrapolation and/or the reliability of the estimates of absorption used in the study was poor. Vermeire et al. (1999) also concluded that scientific justification for the application of route-to-route extrapolation was not derived in this study and heavily depends on expert judgment.

ECETOC (1995) evaluated route-to-route extrapolation using acute study data or chronic study data.

For acute study data (oral LD_{50} values compared with inhalation LC_{50} values), a wide variation was observed for extrapolation from the oral route to the inhalation route suggesting that such an extrapolation is extremely difficult and should not be undertaken. ECETOC also noted that, in most cases, it can be assumed that the dermal route leads to less absorption of a substance than the oral route, because of the skin barrier, and recommended that this should be assessed on a case-by-case basis.

For chronic study data, it was concluded that whilst it may be possible to undertake route-to-route extrapolation, caution is advised when doing so. Default values were therefore not recommended and conversion factors (CFs) should be calculated for each individual situation, making appropriate assumptions about body weight, minute volume, and percentage absorption.

In a more recent publication, ECETOC (2003) noted that route-to-route extrapolation is only feasible for substances with a systemic mode of action, and should take dose rate and toxicokinetic data into account. It was noted that the following points need to be taken into consideration when conduction a route-to-route extrapolation with systemic toxicity data:

- Absorption efficiency is known for both routes, or can be quantified.
- Elimination half-life of the chemical is relatively long compared to the absorption half-life.
- First-pass metabolism is minimal.
- Critical target organ is not the port of entry.
- Chemical undergoes no significant metabolism by intestinal microflora or pulmonary macrophages.
- Chemical is relatively soluble in body fluids.
- Adequate systemic toxicity data are available for the route used as a basis for extrapolation.

If route-to-route extrapolation implies a lower rate of dosing, this can be considered to provide a built-in safety margin and in such cases no assessment factor is needed, i.e., an assessment factor of 1 is appropriate. It is not appropriate to define a default assessment factor (ECETOC 2003).

The WHO/IPCS (1994, 1996, 1999) did not consider route-to-route extrapolation specifically, but the uncertainty related to this element is probably included in a broader defined "additional factor" addressing the adequacy of the overall database (Section 5.9).

According to US-EPA (1993), its position, in general is that the potential for toxicity manifested via one route of exposure is relevant to considerations of any other route of exposure, unless convincing evidence exists to the contrary. Consideration is given to potential differences in absorption or metabolism resulting from different routes of exposure, and whenever appropriate data (e.g., comparative metabolism studies) are available, the quantitative impacts of these differences on the risk assessment are delineated.

It was noted in the 2002 review of the RfD and RfC processes (US-EPA 2002) that the most appropriate route of exposure is the route for which an evaluation is to be made. The toxicity of the chemical may differ with route of exposure because of differences in mechanism of action or toxicokinetics. Development of data to establish dosimetry for the purpose of route-to-route extrapolation is encouraged; however, route-to-route extrapolation is inappropriate when based exclusively upon default assumptions regarding exposure and toxicokinetics. Even within the same route of exposure, responses may differ due to alterations in toxicokinetics, e.g., dietary or water exposure versus oral gavage.

The EU TGD (EC 2003) mentions that when no reliable or adequate toxicity data are available for a relevant route of human exposure, but are available for another route, the possibility of using route-to-route extrapolation may be considered. Route-to-route extrapolation is defined as the prediction of an equivalent dose and dosing regime that produces the same toxic endpoint or response as that obtained for a given dose and dosing regime by another route. In general, route-to-route extrapolation is thought to be a poor substitute for toxicity data obtained using the appropriate route of exposure. However, a procedure for route-to-route extrapolation is described.

When route-to-route extrapolation is to be used, the following aspects should be carefully considered:

- Nature of effect: Route-to-route extrapolation is only applicable for the evaluation of systemic effects. For the evaluation of local effects after repeated exposure, only results from toxicity studies performed with the route under consideration can be used.

- Toxicokinetic data: The major factors responsible for differences in toxicity due to route of exposure include (1) differences in bioavailability (absorption), (2) differences in metabolism (e.g., first-pass effects), and (3) differences in internal exposure pattern (kinetics).

The TGD has noted that in practice, relevant data on kinetics and metabolism, especially after dermal and inhalation exposure, are frequently missing. As a consequence, corrections can only be made for differences in bioavailability. There are some pragmatic approaches in order to calculate a NAEL (or LAEL) by extrapolation, when specific data are not available. The methods described are for extrapolating from oral toxicity data since this is the route most often used for repeated dose toxicity studies in animals. The TGD emphasized that it should be noted that insight into the reliability of the current methodologies for route-to-route extrapolation has not been obtained yet, with a reference to the study performed by Wilschut et al. (1998), see above.

Regarding an approximate inhalation NAEL from an oral NOAEL, the TGD suggests that from the inhalation LC_{50} value (concentration inhaled) the equivalent inhalation LD_{50} (dose absorbed) value can be calculated by assuming a percentage value for absorption via the lungs (values of 75% to 100% are commonly used) and taking into account the respiration rate and body weight. If the inhalation absorption value is known, this should be used. The ratio of the calculated inhalation LC_{50} value to the measured oral LD_{50} value can then be used to estimate the inhalation NAEL from the oral NOAEL. An alternative approach, which could be used in the absence of an LC_{50} value is to convert an oral repeated NOAEL to an approximate inhalation NAEL taking into account the respiration rate and body weight.

Regarding an approximate dermal NAEL from an oral NOAEL, the TGD suggests that unless there are data that contraindicate route-to-route extrapolation (e.g., the oral LD_{50} is much greater than the dermal LD_{50}), it can be assumed that the NOAEL for repeated dose toxicity studies is the same for both routes on a mg/kg body weight per day basis. Dermal absorption is mostly less than, or no more than equal to, oral absorption and generally dermal absorption is slower than oral absorption (especially after gavage application). Extrapolation therefore errs on the side of caution. In case data on dermal absorption are available and/or in case data from dermal absorption studies exist, the available information should be used.

TNO has not suggested a default value for route-to-route extrapolation (Hakkert et al. 1996), see also Section 5.2.1.7.

KEMI (2003) noted that route-to-route extrapolation can only be performed in the case of systemic toxicity and that possible local toxicity in the airways cannot be detected. Not only the degree of absorption but also metabolism should be considered, as, e.g., compounds may be highly metabolized in the liver due to first-pass effect in case of oral exposure but much less metabolized in the case of other routes of exposure. KEMI also noted that there are databased distributions of NOAEL ratios from different routes of exposure, but the size and reliability of this database are limited and they therefore suggested that these distributions should not be used. KEMI suggested that kinetic data are required if possible and that route-to-route extrapolation should be performed in a case-by-case manner based on expert judgment of scientific information. In case no data are available to base the extrapolation upon, 100% should be used as the default degree of absorption. It was emphasized that this default level is generally very conservative in the case of dermal exposure, while in case of inhalation exposure, this default level may not be conservative at all and may even be the opposite.

5.5.1 ROUTE-TO-ROUTE EXTRAPOLATION: SUMMARY AND RECOMMENDATIONS

In case that relevant data are lacking on the exposure route of interest for the derivation of a tolerable intake, a route-to-route extrapolation might be considered. There is no simple and generally applicable way in which toxicity data derived from one route of exposure can be used to evaluate the effects of another exposure route. It should be noted that, in general, route-to-route extrapolation is a poor substitute for toxicity data obtained using the appropriate route of exposure, and in case of a total absence of information on the other route, a route-to-route extrapolation should generally not be undertaken.

A route-to-route extrapolation includes an examination of whether knowledge resulting from studies involving one exposure route can be applied to another exposure route. It is only applicable for the evaluation of systemic effects, especially if the substance has a relatively long half-life; for the evaluation of local effects, only results from toxicity studies performed with the route under consideration can be used. Not only the degree of absorption but also the metabolism should be considered.

Data on absorption or acute toxicity are often used to account for differences between routes of exposure, but these methodologies are not validated and are based on broad assumptions.

A Dutch study (Wilschut et al. 1998, as reviewed in Vermeire et al. 1999) has evaluated route-to-route extrapolation on the basis of absorption or acute toxicity data. For oral to inhalation route-to-route extrapolation, the predicted inhalation NAEL was often higher than the observed NOAEL (for inhalation) implicating that the substance was considered less toxic after extrapolation when compared with the experimental observations. For oral to dermal route-to-route extrapolation, the predicted dermal NAEL (for inhalation) was often lower than the observed NOAEL (dermal) implicating that the substance was considered more toxic after extrapolation when compared with the experimental observations.

No default factor has been suggested for route-to-route extrapolation. It is generally recommended that whether an extrapolation of data is justified in a given situation should be decided on a case-by-case basis, based on expert judgment of scientific information.

In conclusion, the assessment factor for route-to-route extrapolation should preferentially be described probabilistically; however, at present there is no valid and reliable database-derived distribution. No default factor can be suggested for route-to-route extrapolation. Whether a route-to-route extrapolation is justified in a given situation should be decided on a case-by-case basis, based on expert judgment of scientific information, and should generally not be undertaken in cases where no information is available on the other route.

5.6 DURATION OF EXPOSURE EXTRAPOLATION

The most relevant study to base a hazard assessment and derivation of a tolerable intake upon is a study that reflects the human exposure situation as well as possible. In many cases a lifelong exposure is the most relevant exposure scenario for humans and a lifetime animal study (in practice a chronic study) is the most relevant study on which to base the assessment. In other situations where the expected human exposure is of limited duration, for example in seasonal work with plant protection products lasting 2 or 3 months per year, or occasionally, for example use of certain consumer products, the assessment should preferably be based on studies of shorter duration.

It is generally assumed that a longer duration of exposure leads to a lower NOAEL (Vermeire et al. 1999, KEMI 2003, EC 2003). It might be due to accumulation of a substance in the body and consequently, a critical threshold in the body, at which the adverse effect(s) occur, may only be reached after a longer period of time. It might also be that there is a long period of latency before effects become apparent or that a study of longer duration may reveal a target tissue/organ that was not affected in a study of shorter duration. Moreover, fewer animals are generally used per dose group in subacute/subchronic studies than in chronic studies and thus it could be expected, for statistical reasons, that NOAELs from subacute/subchronic studies tend to be higher than NOAELs from chronic studies, even if the dose–relationships in both studies are identical.

For numerous chemicals, a lifetime or chronic study may not be available for the assessment. In such cases it may be necessary to base the assessment on data from a shorter duration study, e.g., a 90-day study and then the lack of data from a long-term study needs to be accounted for in the assessment. Assessment factors of 1 to 10 have been suggested or applied by various national and international bodies. A number of analyses have been performed in order to evaluate the magnitude of an appropriate assessment factor, based on evaluations of the ratios of NOAELs and LOAELs for studies of different durations. This section gives an overview of a number of these analyses. Then, the key issues are summarized and our recommendations are presented.

Weil and McCollister (1963) evaluated 33 oral studies performed with rats and dogs. In each study, a NOAEL was available. For the short-term studies, the exposure duration ranged from 29 to 210 days, with approximately 90 days for 21 of the studies; the long-term studies were all of 2 years duration. Different classes of chemicals were evaluated including agrochemicals, stabilizers, food additives, water-treatment agents, and food packaging materials. The ratio $NOAEL_{short-term}/NOAEL_{long-term}$ was 2 or less for about 50% of the substances, larger than 3 for 21% (6/28) of the substances where the duration of the short-term study was 130 days or less, and less than 10 for 97% (32/33) of the substances. For one substance, a factor of 10 was clearly exceeded.

McNamara (1976, as cited in Kalberlah and Schneider 1998; ECETOC 1995, 2003) analyzed the NOAEL ratios for 41 chemicals (pesticides, food additives, pharmaceuticals) from "short-term" and "long-term" studies; the exposure duration was not stated explicitly. The major part of the studies was performed with rats, three studies involved dogs and one study involved monkeys. Most studies used oral administration, three studies involved inhalation, and one study involved subcutaneous injection. The studies were performed by numerous investigators using diverse study durations and techniques. For the majority of the substances (28), the ratio $NOAEL_{short-term}/NOAEL_{long-term}$ was about 1, for 6 substances the long-term NOAEL was higher than the short-term NOAEL (i.e., the ratio was below 1), and for 7 substances, the long-term NOAEL was lower than the short-term NOAEL (i.e., the ratio was above 1, maximum factor 6.25). According to ECETOC (1995, 2003), ratios of less than 3 were reported for all cases, with a mean ratio of about 1.

Dourson and Stara (1983) evaluated ratios of subchronic to chronic exposure for either NOAELs (30 ratios), LOAELs (22 ratios), or their combination (52 ratios) derived from the toxicity studies compiled by Weil and McCollister (1963), see above. For more than half of the observed chemicals, ratios were 2 or less, and approximately 96% of the ratios were below a value of 10. According to the authors, this supports a 10-fold UF to account for estimating an ADI from a subchronic effect level for a chemical if a chronic level is unavailable.

Woutersen et al. (1985, as cited in Kalberlah and Schneider 1998; ECETOC 1995, 2003) evaluated toxicity data relating to 82 substances including stabilizers, plasticizers, antioxidants, disinfectants, food additives, pesticides, other agrochemicals, and industrial chemicals. The substances were each tested (oral administration to rats) for a subacute (2–4 weeks) and a subchronic (13–18 weeks) duration of exposure. Both the NOAEL and the LOAEL were included in the comparison. For 56% of the substances (46), the ratio $NOAEL_{subacute}/NOAEL_{subchronic}$ was about 1. For 44% of the substances (36), the subchronic NOAEL was lower than the subacute NOAEL (i.e., the ratio was above 1), and for 3/82 substances, the ratio was above 100. The 95th percentile was about 10. A factor of 4 covered 70%–80% of the substances.

Rulis and Hattan (1985, as cited in Kalberlah and Schneider 1998) evaluated 20 food additives by comparing NOAELs or LOAELs and differentiating between long-term (>200 days) and short-term studies (<200 days). The mean value (arithmetic mean) of the time factor ($N/LOAEL_{short-term}/N/LOAEL_{long-term}$) differed slightly depending on the properties of the substance (chemical structure). For the NOAELs and LOAELs, the mean value of the time factor was between 1.5 and 5.4 (structure group A), and 3.1 and 2.8 (structure group B). From a distribution of the ratios $LOAEL_{short-term}/LOAEL_{long-term}$, (independent of the structure group), the 50th percentile corresponded to a factor of 2 and the 70th percentile to a factor of 5; the 95th percentile was covered by a factor of 10.

Lewis (1993, as cited in Dourson et al. 1996) performed an analysis of subchronic-to-chronic NOAEL ratios based on peer-reviewed literature or information from NTP (U.S. National Toxicology Program). Criteria for inclusion in their analysis were rigorous and of 54 chemicals considered, only 18 chemicals were analyzed. Of these, 78% (14 chemicals) had ratios of 3.5 or less, and all but one of these chemicals had ratios of 10-fold or less.

Nessel et al. (1995, as cited in ECETOC 2003) examined data for subchronic (90 days) to chronic (2 years) extrapolation. The median of the ratios $NOAEL_{subchronic}/NOAEL_{chronic}$ was approximately 2 for 23 oral studies, and 4 for 9 inhalation studies in rodents. Using these data and those of Weil and McCollister (1963) and of McNamara (1976), the authors recommended the following for an appropriate assessment factor for extrapolation from subchronic to chronic data:

- Where there is no evidence for bioaccumulation and/or cumulative toxicity, no downward adjustment is necessary, i.e., a factor of 1 should be used.
- For extrapolation of typical subchronic toxicity data (e.g., NOAELs), a value of 2 is most plausible, and a value of 3 or less should be employed.
- When there is evidence of significant potential for bioaccumulation and/or cumulative injury with prolonged or repeated exposure, a larger adjustment factor is required.

Dourson et al. (1996) referred to a number of examinations of subchronic-to chronic NOAEL ratios, which showed that the average difference between subchronic and chronic values was only 2–3. Based on these examinations as well as on the analysis by Lewis (1993) and unpublished work in US-EPA, the authors concluded that the routine use of a 10-fold default factor for this area of uncertainty should be examined closely. They noted that short-term (2 weeks) and subchronic (90 days) NOAELs are often available and can give an indication of the possible differences in the subchronic NOAEL and the expected chronic NOAEL. When such data are not available, a 10-fold UF may not be unreasonable, but should be considered as a loose upper-bound estimate to the overall uncertainty.

The Federal Institute for Occupational Safety and Health (BAuA 1996, internal working paper as cited in Kalberlah and Schneider 1998) has made an evaluation based on 17 NTP reports on inhalation studies performed with rats and mice, for both the LOAEL and the NOAEL if provided in the NTP report. It was found that, in a great many cases, there is no correspondence in the target organ when the critical endpoints are compared for subacute and chronic exposure: in the case of systemic effects, the authors found the same target organs for chronic as compared with subacute exposure in only 9% of the cases. They also calculated the ratios $N(L)OAEL/N(L)OAEL$ for various exposure durations and examined the ratio distributions. For the ratio $N(L)OAEL_{subacute}/N(L)OAEL_{subchronic}$, the GM was 2.0 and 2.2 for rats (10 ratios) and mice (9 ratios), respectively, and the 90th percentile was 11 and 8, respectively. For the ratio $N(L)OAEL_{subacute}/N(L)OAEL_{chronic}$, the GM was 3.2 and 7.0 for rats (13 ratios) and mice (10 ratios), respectively, and the 90th percentile was 12 and 34.5, respectively. For the ratio $N(L)OAEL_{subchronic}/N(L)OAEL_{chronic}$, the GM was 2.8 and 3.3 for rats (12 ratios) and mice (16 ratios), respectively, and the 90th percentile was 11 and 22.2, respectively.

The Research and Advisory Institute for Hazardous Substances (FoBiG 1996a, internal working paper as cited in Kalberlah and Schneider 1998) examined NTP studies with oral exposure (gavage) performed with rats (30 substances) and mice (27 substances), for both the LOAEL and the NOAEL if provided in the NTP report. The ratios $N(L)OAEL/N(L)OAEL$ for various exposure durations were calculated and the ratio distributions were examined. For the ratio $N(L)OAEL_{subacute}/N(L)OAEL_{subchronic}$, the GM was 3.3 and 2.6 for rats (87 ratios) and mice (78 ratios), respectively, and the 90th percentile was 10 for both rats and mice. For the ratio $N(L)OAEL_{subacute}/N(L)OAEL_{chronic}$, the GM was 5.1 and 4.2 for rats (76 ratios) and mice (51 ratios), respectively, and the 90th percentile was 14.1 and 10.6, respectively. For the ratio $N(L)OAEL_{subchronic}/N(L)OAEL_{chronic}$, the GM was 2.9 and 2.5 for rats (71 ratios) and mice (55 ratios), respectively, and the 90th percentile was 8.6 and 6.0, respectively.

TABLE 5.6
Duration-of-Exposure Extrapolation

	Subacute-to-subchronic			Subacute-to-chronic			Subchronic-to-chronic		
	Rat	Mouse	All	Rat	Mouse	All	Rat	Mouse	All
N	21	5	26	21	11	32	24	18	42
AM	3.0	4.8	3.3	5.3	15.4	8.8	2.1	3.6	2.7
GM	2.1	2.6	2.2	3.2	12	5.0	1.7	2.0	1.9
GSD	2.2	3.2	2.3	2.8	2.2	3.1	1.9	2.4	2.1
Median	1.7	2.0	1.7	1.8	16.0	5.0	1.4	1.8	1.5
P_{75}	3.8	nd	4.0	7.0	20.0	14.3	2.5	4.0	3.0
P_{90}	8.1	nd	8.3	14.7	29.7	20.0	5.0	5.0	5.0

Source: Modified from Kalberlah, F. and Schneider, K., *Quantification of extrapolation factors.* Final report of the research project No. 11606113 of the Federal Environmental Agency, Bremerhaven, 1998.

N: number of ratios.
AM: arithmetic mean.
GM: geometric mean.
GSD: geometric standard deviation.
P_{75}: 75th percentile.
P_{90}: 90th percentile.
nd: not determined, too few data.

The Research and Advisory Institute for Hazardous Substances (FoBiG 1996b, internal working paper as cited in Kalberlah and Schneider 1998) also performed an evaluation based on toxicological data on 10 agrochemicals and a further 3 chemicals (anonymous), by 2 manufacturers. In most cases, the LOAEL and the NOAEL for short-term (28 days), subchronic (90 days), and chronic (12–24 months) were reported. In most of the cases, it appeared immaterial whether LOAEL-to-LOAEL or NOAEL-to-NOAEL ratios for the various exposure durations were compared and consequently, overall ratios N(L)OAEL/N(L)OAEL for the various exposure durations were calculated and the ratio distributions were examined, see Table 5.6.

For the ratio N(L)OAEL$_{subacute}$/N(L)OAEL$_{subchronic}$, the GM was 2.1 and 2.6 for rats (21 ratios) and mice (5 ratios), respectively, and the 90th percentile was 8.1 for rats (too few data for mice). For the ratio N(L)OAEL$_{subacute}$/N(L)OAEL$_{chronic}$, the GM was 3.2 and 12 for rats (21 ratios) and mice (11 ratios), respectively, and the 90th percentile was 14.7 and 29.7, respectively. For the ratio N(L)OAEL$_{subchronic}$/N(L)OAEL$_{chronic}$, the GM was 1.7 and 2.0 for rats (24 ratios) and mice (18 ratios), respectively, and the 90th percentile was 5.0 for both rats and mice.

Kramer et al. (1996) assessed CFs to estimate a chronic NOAEL from short-term toxicity data by evaluating distributions of ratios between (sub)acute (3–6 weeks) and chronic (1–2 years) toxicity data. The database used was composed of toxicological data evaluated by international agencies. The final database contained 425 records comprising 332 different compounds (pesticides (50%), solvents (25%), metal-containing compounds (5%), phthalates (3%), other compounds (17%)) for several test species; most data were derived from rat studies, to a lesser extent, from mouse studies, and a few studies were available involving dogs, rabbits, or humans. By defining the CF as the upper 95% confidence limit of the 95th percentile for the relevant ratio distribution, both the variation between compounds (95th percentile) and the estimation error (upper 95% confidence limit) could be taken into account. They found a CF of 17,000 for an LD_{50} based on 244 ratios, and a CF of 87 for a NOAEL$_{subacute}$ based on 71 ratios. The authors found the NOAEL$_{subacute}$ to be a better predictor of the NOAEL$_{chronic}$ than the LD_{50}, and that the beneficial value of an LD_{50} for an estimation of a NOAEL$_{chronic}$ appeared to be limited when a NOAEL$_{subacute}$ was available.

Pieters et al. (1998) used the same database as Kramer et al. (1996) for an evaluation of subchronic (10–26 weeks) to chronic (1–2 years) extrapolation. They calculated the ratio $NOAEL_{subchronic}/NOAEL_{chronic}$ and examined the ratio distribution. The GM was 1.7 and the GSD was 5.6 (149 compounds). The point estimate of the 95th percentile of the ratio distribution was 29 and the 95% confidence interval (CI) ranged from 20 to 46, i.e., the CF was 46. To gain insight in the predictive value of the $NOAEL_{subchronic}$ for the $NOAEL_{chronic}$, the authors performed a linear regression analysis, which revealed that the $NOAEL_{subchronic}$ explained 55% of the variance of the $NOAEL_{chronic}$ and therefore the predictive value of the $NOAEL_{subchronic}$ was considered as being rather poor. It was expected that the use of NOAELs from various animal species introduced a large variation in the results and therefore the same analysis was performed on rat/rat ratios only (70 ratios). The GM, GSD, and the 95th percentile were 1.5, 6.3, and 31, respectively, i.e., the variance increased rather than decreased. The authors concluded that these data do not support a lowering of the UF of 10 generally used for extrapolating a subchronic NOAEL to a chronic NOAEL.

Kalberlah and Schneider (1998) reviewed and critically assessed the statistical evaluations reported in the literature regarding the time extrapolation factor (Weil and McCollister 1963, McNamara 1976, Woutersen et al. 1984, Rulis and Hattan 1985, BAuA 1996, Kramer et al. 1996) as well as their own evaluations (FoBiG 1996a,b), see Table 5.7. They concluded that these evaluations revealed a number of problems with the time extrapolation including (1) spacing of the doses in the various studies, (2) variation in the depth of investigations in the various studies as some studies included a thorough investigation while other studies only included some parameters, (3) number of animals examined, (4) nonexamined endpoints such as, e.g., immunotoxicity, (5) specificity of the observed effects, and (6) adaptation.

For subchronic-to-chronic time extrapolation, the authors concluded that the factor of 10, which is traditionally used, presumably is sufficient for approximately 75% of the examined substances. A factor of 2–3 should be suitable if a lower probability (e.g., 50%) is considered sufficient. For subacute-to-chronic time extrapolation, the authors concluded that a factor of 6 appeared to be justified when the GM (50th percentile) of the distribution is taken as the basis for calculations. It should be noted, according to the authors, that the resultant factor does not directly represent a measure for the actual dose/time relationship but, instead, only for the usually observed dose/time relationship since the study design has a strong influence on the result, i.e., it would presumably be possible to find more effects in a well-conducted subchronic study if a more in-depth investigation was performed and a higher number of animals were selected. The authors noted considerable differences between the evaluations from the literature and their own evaluations and therefore concluded that additional discussion is required before the 90th or 95th percentile is finally stated.

The authors noted that there are fundamental difficulties in the prediction of a chronic effect when only subacute data are available and thus the overall uncertainty is very high and is possibly increased further by the assignment of an extrapolation factor; therefore, such extrapolations should be avoided, at least in cases where the uncertainty is extended by other extrapolation steps such as, e.g., LOAEL-to-NAEL, interspecies, interindividual, and route-to-route. They also noted that, due to great uncertainties, the extrapolation from single short-term exposure in the high-dose range to lifetime exposure is rejected. However, they stated that, currently, no justified "cutoff criterion" has been identified to determine the minimum exposure for which an extrapolation to lifetime exposure may be permitted or should not be carried out.

The authors also pointed out the connection between the duration of exposure extrapolation factor and the interindividual variance. If, e.g., young adult animals were exposed (normal case) and observed over 90 days, this might have to be assessed differently from a study on neonates or older animals. Therefore, it is necessary to examine which elements have already been included in the interindividual factor. If the time factor (children, elderly people) was considered as a separate subfactor in the interindividual factor, then the specific data on the exposure period must be considered in relation to the duration of exposure extrapolation factor in order to avoid double assessment.

TABLE 5.7

Duration-of-Exposure Extrapolation, Systemic Effects, Overall Evaluation

Subacute-to-subchronic			Subacute-to-chronic			Subchronic-to-chronic			Comments	Reference
N	GM	P_{90}	N	GM	P_{90}	N	GM	P_{90}		
						33	2.2	5.8	Oral, various species	Weil and McCollister (1963)
82	2.0	6.6							Oral, various species	Woutersen et al. (1984)
						20	2.0	9.0	Oral, various species	Rulis and Hattan (1985)
91	2.2	62*	57	6.5	53*	22	1.7	16*	Oral, various species	Kramer et al. (1995)
			37	3.4					Oral, rat	Kramer et al. (1995)
87	3.3	10.0	76	5.4	14.1	71	2.9	8.6	Gavage, rat, NTP	FoBiG (1996a)
78	2.6	10.0	51	4.2	10.6	55	2.5	6.0	Gavage, mouse, NTP	FoBiG (1996a)
21	2.1	8.1	21	3.2	14.7	24	1.7	5.0	Oral, rat, IND	FoBiG (1996b)
5	2.6	nd	11	12	29.7	18	2.0	5.0	Oral, mouse, IND	FoBiG (1996b)
22	1.7	16*	10	10	1400*	10	1.9	142*	Inhalation, various species	Kramer et al. (1995)
10	2.0	11.0	13	3.2	12	12	2.8	11	Inhalation, rat, NTP	BAuA (1996)
9	2.2	8.0	10	7.0	34.5	16	3.3	22.2	Inhalation, mouse, NTP	BAuA (1996)
	2.3	16.5		6.1	196		2.3	23	Mean	
	3.3	62*		12	1400*		3.3	142*	Upper limit	

Source: Modified from Kalberlah, F. and Schneider, K., *Quantification of extrapolation factors.* Final report of the research project No. 11606113 of the Federal Environmental Agency, Bremerhaven, 1998.

N: number of ratios.
GM: geometric mean.
P_{90}: 90th percentile.
*: 95th percentile.
nd: not determined, too few data.

Vermeire et al. (1999) reviewed several studies comparing NOAELs from chronic and subacute/subchronic studies in order to evaluate the distribution of the extrapolation factor for duration of exposure. The ratios of observed NOAELs from oral studies using historical data for various compounds were calculated. The most likely distribution of the ratios was considered to be lognormal and the parameters of the distributions of the ratios were estimated, see Table 5.8.

The GMs of the ratios NOAEL$_{subacute}$/NOAEL$_{chronic}$ ranged from 3.1 to 4.1 in the three studies, and the GSDs ranged from 1.9 to 4.4 (see Table 5.8). Based on these data, Vermeire et al. considered it reasonable to approximate their real distribution with a GM of 4 and a GSD of 4. Based on this proposed distribution, default values for 90th, 95th, and 99th percentiles could be calculated to be 24, 39, and 101, respectively.

TABLE 5.8

Subacute/Subchronic-to-Chronic Oral NOAEL Ratios

N	GM	GSD	P_{90}	P_{95}	Exposure sa/sc	Exposure c	Species	Reference
Subacute-to-chronic								
71	4.1	4.4	27	46	3–6 weeks	1–2 years	Various	Kramer et al. (1996)
20	3.1	1.9	7.0	8.9	14 days	2 years	Mice	Kalberlah and Schneider (1998)
26	3.9	2.2	10.7	14.3	14 days	2 years	Rats	Kalberlah and Schneider (1998)
Subchronic-to-chronic								
33	2.2	2.3	6.4	8.7	30–210 days	2 years	Rats	Weil and McCollister (1963)
41	2.0	1.7	2.0	2.5	Not specified	Not specified	Rats, dogs[a]	McNamara (1976)
20	1.9	3.0	8.0	12	<200 days	>200 days	Various	Rulis and Hattan (1985)
149	1.7	5.6	15.4	29	10–26 weeks	1–2 years	Various	Pieters et al. (1998)
23	2.0	1.8	4.2	5.1	90 days	2 years	Rodents[b]	Nessel et al. (1995)
9	2.4	1.3	3.4	3.7	90 days	1–2 years	Mice	Kalberlah and Schneider (1998)[c]
11	1.7	1.8	3.6	4.5	90 days	1–2 years	Rats	Kalberlah and Schneider (1998)[c]
20	2.0	2.4	6.1	8.4	90 days	1–2 years	Mice, Rats	Kalberlah and Schneider (1998)[d]
21	1.7	1.7	3.3	4.1	90 days	2 years	Mice	Kalberlah and Schneider (1998)[d]
22	2.5	1.9	5.7	7.2	90 days	2 years	Rats	Kalberlah and Schneider (1998)[d]

Source: Modified from Vermeire, T., Stevenson, H., Pieters, M.W., Rennen, M., Slob, W., and Hakkert, B.C., *Crit. Rev. Toxicol.* 29, 439, 1999.

N: number of ratios.

GM: geometric mean.

GSD: geometric standard deviation.

P_{90}: 90th percentile.

P_{95}: 95th percentile.

sa: subacute.

sc: subchronic.

c: chronic.

[a] 39 rat pairs, 2 dog pairs.

[b] Matched pairs.

[c] Industry data from 13 agrochemicals.

[d] Data from the U.S. National Toxicology Program.

The GMs of the ratios NOAEL$_{subchronic}$/NOAEL$_{chronic}$ were similar for all 10 distributions, approximately 2, i.e., the NOAELs were on average twice as high in the subchronic compared to the chronic studies for the chemicals evaluated, whereas the GSDs ranged from 1.3 to 5.6 (see Table 5.8). Based on these data, Vermeire et al. considered that a GSD of 4 was a reasonable approximation of the real standard deviation for the distribution of the ratios NOAEL$_{subchronic}$/NOAEL$_{chronic}$ and, based on this proposed distribution, default values for 90th, 95th, and 99th percentiles could be calculated to be 12, 20, and 50, respectively. The often-applied default factor of 10 coincided with the 88th percentile.

The authors noted that it may be expected that the NOAELs from subacute/subchronic studies tend to be larger than NOAELs from chronic studies and it was considered that the GM ratios for the NOAELs assessed in the studies most likely overestimated the median of the distribution of the extrapolation factor for duration of exposure. The authors also noted that it is very likely that the databases used in these studies overlap each other significantly. It was also pointed out that the distributions presented in Table 5.8 were based on rather variable exposure periods for the subchronic NOAELs, included interspecies variation (no matching for species) for which no correction was made, differences in endpoints were not considered, and in several cases used rather old data. The authors considered that the distributions obtained from NOAELs of various species would probably be too wide, whereas the distributions obtained from NOAELs of one species and more strict criteria with respect to exposure period and overall study design might be too narrow.

They concluded that it did not seem appropriate to rely on one particular metastudy because the selection criteria used for different distributions all have advantages and disadvantages. In addition, they also concluded that the most relevant NOAEL ratios were those based on the same species, and that the most relevant distributions of NOAEL ratios were those that included a sufficient number of matched pairs of NOAELs from various species. Whether the distributions also apply to inhalation and dermal subchronic to chronic ratios was considered to be questionable as it might be possible that the influence of the exposure period on the toxicological effect depends on the route of exposure.

Vermeire et al. (2001) adjusted the GSD for the distribution of the ratios $NOAEL_{subchronic}/NOAEL_{chronic}$ to 3.5, based on the review by Vermeire et al. (1999) and taking into account another detailed study with 70 pairs of NOAELs (Groeneveld et al. 1998, as cited in Vermeire et al. 2001); the GM remained 2. KEMI (2003) calculated the 90th, 95th, and 99th percentiles for this distribution to 10, 16, and 37, respectively; the traditional 10-fold factor thus coincided with 90% of the variation among different substances. For subacute to chronic exposure, the GM was adjusted to 5 and the GSD to 3.5, based on the review by Vermeire et al. (1999) and taking into account another, detailed study with 35 pairs of NOAELs (Groeneveld et al. 1998, as cited in Vermeire et al. 2001). KEMI (2003) calculated the 90th, 95th, and 99th percentiles for this distribution to 25, 39, and 92, respectively. For subacute to subchronic exposure, a default lognormal distribution with a GM of 2 and a GSD of 4 was concluded based on one study with a data set of 35 pairs of NOAELs (Groeneveld et al. 1998, as cited in Vermeire et al. 2001). In contrast to Vermeire et al. (1999), Vermeire et al. (2001) assumed that although the distributions were derived from oral data, they could also be applied to systemic effects caused by inhalation or dermal exposure, after estimation of the systemic dose.

ECETOC (1995) stated that the subchronic to chronic extrapolation factor depends on the particular substance involved and may be as low as 1 for chemicals that neither produce cumulative effects nor accumulate in the body. A higher factor may be applied in cases of cumulating effects or accumulating compounds, but double correction of this fact, e.g., as an interspecies kinetic factor, should be avoided. An additional factor may be needed if the exposure duration in the short-term test is 14–28 days, because the NOAELs of subacute studies cannot simply replace NOAELs of subchronic studies. A further factor may be needed where the route of concern is inhalation, the human predicted NAEL required is for continuous exposure, and the animal studies involve a discontinuous exposure (i.e., 6–8 h a day for 5 days per week). ECETOC suggested the simplest way of accommodating this situation by using a further factor of 4 to allow arithmetically for the difference in total hours of exposure. This factor was considered to be justifiable for substances for which toxicokinetic data are available. Those substances with a short half-life (<1 h) will reach equilibrium in the body well within 6–8 h and the total body burden will not be increased by extending the exposure over the full 24 h. Those with a long half-life (>1 week) will accumulate in the body. Where the half-life is between these two extremes, a further factor, in addition to the above fourfold factor, may be required. In the absence of such information, the factor of 4 was recommended and the increased uncertainty involved in the extrapolation to be accounted for in the overall review of the total UF.

ECETOC emphasized that the scientific basis for establishing meaningful extrapolation factors in this area is still weak. Nevertheless, a provisional default value of 2–3 was considered to be consistent with the available scientific data when extrapolating from subchronic to chronic exposure. For extrapolation from short-term repeated to subchronic exposure, a factor of 3 was recommended as a provisional default value.

ECETOC (2003) considered that application of the ratio approach ignores several sources of error, the main contribution being the imprecision of the NOAEL, and noted that the ratio approach could be potentially improved by replacing the NOAEL by the point estimate of the BMD (Section 4.2.5). The problem of extrapolating the change in effect concentrations over time, is that the available data comparing NOAELs after different durations of exposure indicate that although the GM value for subacute to chronic oral NOAEL rations may be a factor of 4, there is wide variability with the 95th percentile being up to 46 (i.e., an extrapolation factor of 46). Similarly, the GM value for the subchronic to chronic NOAEL ratios is 1.7, while the 95th percentile may be up to 29. They also noted that a further concept relevant to extrapolation for study duration is emerging from investigations, which aim to identify the optimum duration of a study. Using the 95th percentile ratios as assessment factors for exposure duration would be in conflict with the results of these studies for an optimum study duration (6 months).

For systemic effects, ECETOC (2003) recommended a default assessment factor of 6 for extrapolation from subacute (28 days) to chronic exposure, and a factor of 2 from subchronic (90 days) to chronic exposure. For local effects, no additional assessment factor is needed for duration of exposure extrapolation for substances with a local effect below the threshold of cytotoxicity.

WHO/IPCS (1994, 1996, 1999) did not consider an extrapolation factor for duration of exposure specifically, but the uncertainty related to this element is included in a broader defined "additional factor" addressing the adequacy of the overall database (Section 5.9).

According to US-EPA (1993), a 10-fold factor is generally used, in addition to the assessments factors for interspecies and interindividual differences, when extrapolating data from less than chronic results on experimental animals when there are no useful long-term human data. This factor is intended to account for the uncertainty involved in extrapolating from less than chronic NOAELs to chronic NOAELs and is referenced as "10S."

It was noted in the 2002 review of the RfD and RfC processes (US-EPA 2002) that a duration adjustment currently in use is the application of a UF when only a subchronic duration study is available to develop a chronic reference value. A default value of 10 for this UF is applied to the NOAEL/LOAEL or benchmark dose (BMDL/BMCL) from the subchronic study on the assumption that effects from a given compound in a subchronic study occur at a 10-fold higher concentration than in a corresponding (but absent) chronic study; this factor would be applied subsequent to the adjustment of the exposures from intermittent to continuous. The specific use of a UF applied to a subchronic study in the derivation of a chronic reference value was stated as being reasonable. No chronic reference value is derived if neither a subchronic nor a chronic study is available. The application of a UF to less-than-subchronic studies is not part of the current practice, but further exploration of this issue may be appropriate. For short- and longer-term reference values, the application of a UF analogous to the subchronic-to-chronic duration UF also needs to be explored, as there may be situations in which data are available and applicable but they are from studies in which the dosing period is considerably shorter than that for the reference value being derived. Guidance for replacement of the default factor of 10 by a CSAF (chemical-specific adjustment factor, Section 5.2.1.12) may be forthcoming.

The EU TGD (EC 2003) mentions that differences in the duration of exposure between the exposed humans and the studies from which toxicological data are available may, in part, be addressed when considering the acceptability (for the situation of interest) of the exposure/NOAEL ratio. It is also mentioned that it is assumed that, often, the NOAEL will decrease as the duration of exposure in the toxicity study increases. In addition, it is necessary to take account of the possibility that a study of longer duration may reveal a target tissue/organ/system which was not affected in a relatively short-term study.

TNO has, for differences between experimental conditions and exposure patterns for workers, suggested a number of default values (Hakkert et al. 1996). For extrapolation of data from subacute-to-subchronic exposure and from subchronic-to-chronic exposure, the factor ranges generally between 1 and 5; however, in cases when no conclusions can be drawn as to the effect of exposure time on the NOAEL, a default value of 10 should be used for the extrapolation. In other cases, a default value of 1 has been suggested, see also Section 5.2.1.7.

KEMI (2003) suggested that, if necessary, extrapolation can be performed from subchronic to chronic exposure and that such an extrapolation should be based on the distribution of NOAEL ratios reported by Vermeire et al. (2001). If the 95th percentile is chosen, i.e., covering 95% of the substances compared, the corresponding assessment factor is 16. Extrapolation from subacute to chronic exposure should preferably not be performed, but if it is necessary a similar approach is suggested; for this extrapolation, an assessment factor of 39 corresponds to the 95th percentile based on the lognormal distribution of NOAEL ratios from subacute and chronic exposure studies in Vermeire et al. (2001).

5.6.1 Duration of Exposure Extrapolation: Summary and Recommendations

The most relevant study to base a hazard assessment and derivation of a tolerable intake upon is a study that reflects the human exposure situation as well as possible. For numerous substances, data are only available from acute (single exposure), subacute (14–28 days), or subchronic (90 days) animal studies. In order to derive, e.g., a TDI or RfD for such a substance, it may be necessary to base the assessment on data from a shorter duration study. An assessment factor allowing for differences in the experimental exposure duration and the duration of exposure for the population and scenario under consideration needs to be considered taking into account that, in general, the experimental NOAEL will decrease with increasing exposure duration as well as other and more serious adverse effects may appear with increasing exposure duration.

Also bioaccumulating substances may in some situations call for a higher assessment factor. If accumulation is likely, the toxicity studies need to be of sufficient length to cover the accumulation period (e.g., the time to reach a steady-state concentration). If there is limited information on these aspects, it has to be considered to which extent this lack of information should affect the assessment factor.

A lower assessment factor may be applied if there is evidence that the exposure duration is of no or low importance. If it is assumed that the effect is concentration-dependent rather than dose-dependent, which might be the case for certain local effects, no assessment factor for duration of exposure may be considered necessary.

A default factor of 10 has been applied by some national and international bodies. A number of analyses for an appropriate assessment factor have been based on evaluations of the ratios of NOAELs and LOAELs for studies of different duration performed in various animal species exposed orally or by inhalation. The most extensive analyses are those performed by Kalberlah and Schneider (1998) and Vermeire et al. (1999, 2001).

Kalberlah and Schneider (1998) reviewed and critically assessed the statistical evaluations reported in the literature regarding the time extrapolation factor. They noted that the difference between a factor of 10 and a factor of 2–3 was mainly due to the variations in the percentiles (degree of statistical certainty) used in each case; if 50% of the examined substances were covered, a lower factor resulted than if approximately 90% of the examined substances should be covered. Based on the evaluations from the literature as well as their own evaluations (see Table 5.7), they concluded that, for subchronic-to-chronic extrapolation, the traditionally used factor of 10 presumably is sufficient for approximately 75% of the examined substances, and a factor of 2–3 to be suitable if a lower probability (e.g., 50%) is considered sufficient. For subacute-to-chronic time extrapolation, the authors concluded that a factor of 6 appeared to be justified for 50% of the substances. The authors noted considerable differences between the evaluations from the literature and their own evaluations and therefore concluded that additional discussion was required before the 90th or 95th percentile is finally stated. They also noted that the evaluations revealed a number of problems with the time extrapolation including (1) spacing of the doses, (2) variation in the depth of investigations,

(3) number of animals examined, (4) nonexamined endpoints, (5) specificity of the observed effects, and (6) adaptation to some toxicological effects. Generally, it was recommended to avoid subacute-to-chronic extrapolations, at least in cases where the uncertainty is extended by other extrapolation steps (interspecies, interindividual, LOAEL-to-NAEL, and route-to-route). The extrapolation from single acute exposure in the high-dose range to lifetime exposure should be rejected.

Vermeire et al. (1999) reviewed several studies comparing oral NOAELs from chronic and subacute/subchronic studies in order to evaluate the distribution of the extrapolation factor for duration of exposure. The studies were partly the same literature studies as evaluated by Kalberlah and Schneider as well as the analyses performed by FoBiG (1996a,b, internal working papers as cited in Kalberlah and Schneider 1998). The most likely distribution of the NOAEL ratios was considered by the authors to be lognormal, and the parameters of the distributions of the ratios were estimated, see Table 5.8. Taken together, the GM for the 10 distributions of oral subchronic-to-chronic NOAEL ratios was approximately 2, and a GSD of 4 was considered to be a reasonable approximation of the real standard deviation for the distribution of the ratios; default values for 90th, 95th, and 99th percentiles were estimated at 12, 20, and 50, respectively; the traditional default factor of 10 coincided with the 88th percentile. For subacute-to-chronic NOAEL ratios (three distributions), the authors considered it reasonable to approximate their real distribution with a GM of 4 and a GSD of 4; default values for 90th, 95th, and 99th percentiles were estimated at 24, 39, and 101, respectively.

Vermeire et al. (2001) adjusted the GSD for the distribution of the subchronic-to-chronic NOAEL ratios to 3.5 and the GM remained 2; default values for 90th, 95th and 99th percentiles were estimated at 10, 16, and 37, respectively, and the 10-fold factor thus coincided with 90% of the variation among different substances. For subacute-to-chronic exposure, the GM was adjusted to 5 and the GSD to 3.5; default values for 90th, 95th, and 99th percentiles were estimated at 25, 39, and 92, respectively. For subacute to subchronic exposure, a default lognormal distribution with a GM of 2 and a GSD of 4 was concluded based on one study with a data set of 35 pairs of NOAELs.

WHO/IPCS (1994, 1996, 1999) did not consider an extrapolation factor for duration of exposure specifically, but the uncertainty related to this element is included in a broader defined "additional factor" addressing the adequacy of the overall database (Section 5.9). The US-EPA (1993) has adopted the 10-fold factor to account for the uncertainty involved in extrapolating from less than chronic NOAELs to chronic NOAELs. This default value has later on been reconfirmed (US-EPA 2002) when only a subchronic duration study is available to develop a chronic reference value; no chronic reference value is derived if neither a subchronic nor a chronic study is available. For systemic effects, ECETOC (2001) recommended a default assessment factor of 6 for extrapolation from subacute (28 days) to chronic exposure, and a factor of 2 from subchronic (90 days) to chronic exposure. For local effects, no additional assessment factor is needed for duration of exposure extrapolation for substances with a local effect below the threshold of cytotoxicity. KEMI (2003) suggested that extrapolation from subchronic to chronic exposure should be based on the distribution of NOAEL ratios reported by Vermeire et al. (2001) with an assessment factor of 16 covering 95% of the substances compared; and for extrapolation from subacute to chronic exposure, with an assessment factor of 39 covering 95% of the substances.

In conclusion, the most relevant study duration reflecting the relevant human exposure situation as well as possible should be selected for the hazard assessment. If a chronic study is not available in order to derive, e.g., a TDI or RfD for such a substance, extrapolation can be performed from a subchronic study by using an assessment factor. The assessment factor for subchronic-to-chronic extrapolation should be derived probabilistically as suggested by KEMI (2003) and based on the distribution proposed by Vermeire et al. (2001); an exact value of the assessment factor is not recommended as it depends on the percentile (degree of statistical certainty) decided to be used. It should be recognized that the higher percentage of substances to be covered the higher the default assessment factor is. Extrapolation from subacute or single short-term exposure to chronic exposure should not be performed due to the high uncertainty related to such an extrapolation.

5.7 DOSE–RESPONSE CURVE (LOAEL-TO-NOAEL EXTRAPOLATION)

One of the most evident limitations in the NOAEL approach in the derivation of tolerable intakes is that it does not take into account the slope of the dose–response curve for the particular response of interest (Section 4.2.4). The NOAEL is by definition one of the doses tested, and apart from ensuring that the number and spacing of data points are adequate to provide a reasonable estimate of the NOAEL, all other data points are ignored. Although the NOAEL could be considered an estimate of the true NAEL, the quality of the estimate cannot be assessed. For the dose–response relationship and precision in the NOAEL, consideration should therefore be given to the uncertainties in the NOAEL as the surrogate for the NAEL.

In case a NOAEL cannot be set for the critical effect, a LOAEL is then set and extrapolated to a NOAEL. The extrapolation from a LOAEL to a NOAEL can be regarded as part of the dose–response analysis. Consideration should therefore also be given to the uncertainties in the extrapolation of the LOAEL to the NAEL in cases where only a LOAEL is available as the starting point for the assessment.

Assessment factors between 1 and 10 have been suggested or applied by various national and international bodies for the extrapolation from a LOAEL to a NOAEL. Some analyses have been performed in order to evaluate the magnitude of an appropriate assessment factor for the LOAEL-to-NOAEL extrapolation, based on evaluations of LOAEL/NOAEL ratios. This section gives an overview of such analyses and evaluations. Then, the key issues are summarized and our recommendations are presented.

Dourson and Stara (1983) made a plot of frequency versus ratios of LOAEL to NOAEL for either subchronic (27 comparisons) or chronic exposure (25 comparisons), or their combination (52 comparisons) based on data adapted from Weil and McCollister, see Section 5.6. They stated that these experimentally determined ratios can be thought of as reductions in a LOAEL found after subchronic or chronic exposure in order to yield the corresponding NOAEL. For example, a ratio of 3 indicates that the NOAEL is threefold less than the corresponding LOAEL for a particular chemical. All chemicals had ratios of 10 or less and of these ratios, 96% had values of 5 or less. The authors concluded that this analysis supported a UF between 1 and 10 to account for estimating an ADI from a LOAEL if a NOAEL is not available.

Kadry et al. (1995) evaluated the UF for LOAEL to NOAEL extrapolation by comparing subchronic (23) and chronic (23) LOAEL/NOAEL ratios for six chlorinated compounds. All of the ratios were 10 or less; for subchronic exposure, 91.3% of the ratios were sixfold or less and for chronic exposure, 87% of the ratios were 5 of less. The authors concluded that automatic safety factors of 10-fold are not scientifically supportable and are overly conservative for the chlorinated compounds included in their analysis.

For LOAEL to NOAEL extrapolation, Dourson et al. (1996) noted that analysis of several databases suggest that a factor of 10 or lower is adequate, and that use of data does support a lower factor with certain chemicals. The authors noted that the results of the research on LOAEL to NOAEL extrapolation are not unexpected as experimental studies are seldom designed with doses in excess of 10-fold apart, leading to the common statement that these ratios depend more on dose spacing than inherent toxicity. They also noted that the choice of dose spacing often reflects the judgment on the likely steepness of the dose–response slope, with steeper slopes resulting in tighter dose spacing. According to the authors, the data indicate that when a NOAEL is not available, the choice of UF should generally depend on the severity of the effect seen at the LOAEL. More severe effects should be judged to need a larger UF because the expected NOAEL is further away from the LOAEL. Less severe effects would not require a large factor, because, presumably, the LOAEL is closer to the unknown NOAEL.

Pieters et al. (1998) have performed a statistical analysis on LOAEL/NOAEL ratios derived from oral subacute, subchronic, and chronic studies. For the subacute studies (95 ratios), the GM was 3.5, the GSD was 1.8, the 95th percentile was 9, and the 95% CI was 8–11. For the subchronic studies (226 ratios), the GM was 4.3, the GSD was 2.2, the 95th percentile was 16, and the CI was 14–19. For the chronic studies (175 ratios), the GM was 4.5, the GSD was 1.7, the 95th percentile was 11, and the CI was 10–12. Although the results may be considered supportive of a UF 10, the authors believed that there is no justification for the use of such a factor. They considered the observed LOAEL/NOAEL ratios as irrelevant since they only give information on applied intervals between dose levels in the studies, and stated that they do not support the idea that observed LOAEL/NOAEL ratios have any supportive value for the UF for LOAEL-to-NOAEL extrapolation. Furthermore, if a particular study results in a LOAEL, there is no guarantee whatsoever that at one dose interval lower, the effect would be statistically nonsignificant. Instead, the authors recommended the use of dose–response modeling, which would make the use of LOAEC-to-NOAEL extrapolation redundant.

Kalberlah and Schneider (1998) noted that the traditional factor of 10 for the estimation of a NAEL on the basis of a LOAEL was first used without any toxicological basis, and later justified, in particular as a result of the evaluation undertaken by Dourson and Stara (1983); however, the authors concluded, based on evaluations of available literature data, that the justification for the selection of an extrapolation factor of 10, as given by Dourson and Stara, does not represent a suitable basis. They noted, however, that the data reported in the literature reflect more the size of the usual dose graduations (spacing), which are dependent on the study design, than the actual steepness of the dose–response relationship. They considered the use of the BMD concept to be preferred to the selection of a general factor for LOAEL to NAEL extrapolation. In case the BMD concept is associated with a high degree of uncertainty, the following alternatives are available although the authors stated that currently it is not possible to propose any well-justified method:

- Dispensing with an extrapolation. The data quality in these cases are so poor that consideration should be given to dispensing with a quantitative risk estimation altogether.
- If, in the case of poor data, dispensing with an extrapolation is not decided, a (poorly validated) benchmark extrapolation is to be preferred to a (likewise poorly validated) extrapolation using a standard extrapolation factor.
- If, in the case of poor data, extrapolation using a standard extrapolation factor is decided, the retention of the traditional factor 10 (upper boundary of the mean values in the literature data), which would take into account the uncertainty to a certain extent, would serve as a suitable convention.
- Decision about whether to use the benchmark method or an extrapolation with an extrapolation factor must be taken in consideration of the available data.
- If a standard extrapolation factor is taken as the starting point, additional substance-specific information may result in a modification of the standard value, i.e., a reduction or an increase in relation to 10.

Vermeire et al. (1999) pointed out that there is no scientific basis for any value of a default factor to account for uncertainty in the NOAEL, nor any distribution. The authors considered the use of LOAEL/NOAEL ratios to estimate a NOAEL from a LOAEL as questionable, since doses in toxicological tests are usually spaced at fixed intervals and the observed distribution of LOAEL/NOAEL ratios primarily reflects the historical frequency of use of various dose spacing. There is no guarantee whatsoever that extrapolation of a LOAEL with any factor will yield an estimate of the NOAEL. Therefore, this factor can only be assigned using expert judgment in which the shape of the dose–response curve and the magnitude of the effect at the LOAEL are taken into account.

According to ECETOC (1995), a survey of the literature showed that the magnitude of the assessment factor for LOAEL-to-NOAEL extrapolation can vary between 1 and 10. They noted that most of the experimentally derived arguments for the size of this factor rely on data adapted from results published by Weil and McCollister (1963). With a reference to Dourson and Stara (1983), ECETOC noted that LOAEL/NOAEL ratios for subchronic and chronic exposure revealed that the NOAEL values were, in most cases, maximally fivefold less than the corresponding LOAEL with the majority of the values ranging from 2 to 3 (mean 3.02) for subchronic exposure, and from 2 to 4 (mean 3.8) for chronic exposure. ECETOC also noted that the magnitude of the LOAEL/NOAEL ratio in any individual case will depend on the slope of the dose–response curve, the group size, and the interval between doses. Thus the study design may affect not only the ratio, but may also result in an observed NOAEL being considerably lower than the NAEL where large dose intervals have been used. If the dose interval is large (e.g., 10) and if the effects seen at the LOAEL are minimal (indicating that the NAEL is close to the LOAEL), then it may be more appropriate to base the assessment on the LOAEL, rather than on the observed NOAEL, with a default extrapolation factor of 10.

ECETOC (1995) recommended a factor of 2 to be used in case the extent of the relevant effect is of minor importance, and the slope of the dose–response curve reasonably justifies the assumption that a halving of the LOAEL would be likely to arrive at the no-effect dose. A factor of 3 was recommended as a default value, which would be used in the majority of cases. Extent and severity of the effect at the LOAEL and/or a very flat dose–response curve may justify the use of a higher factor.

ECETOC (2003) noted that the generally used default assessment factor of 10, in cases where only a LOAEL is available for the critical effect, is overly conservative with a reference to published studies in which the LOAEL/NOAEL ratios were compared for a range of different chemicals and different study duration (subacute, subchronic, and chronic). According to ECETOC, these studies indicated that the LOAEL rarely exceeded the NOAEL by more than about 5–6-fold and was typically closer to a value of 3. They noted that the LOAEL/NOAEL ratio is highly dependent on the spacing between the doses, and since recent study design generally uses a dose spacing of 2–4, it is logical to conclude that the ratio data support a value of 3 as a default. ECETOC therefore considered that the size of the assessment factor for LOAEL/NOAEL extrapolation should take into account not only the interval of the doses in the experimental study, but also the shape of the dose–response curve, including its slope, and the extent and severity of the effect seen at the LOAEL.

ECETOC (2003) recommended that if an appropriate NOAEL is available, then no extrapolation and hence, no assessment factor is necessary. Where it is considered more appropriate to use the LOAEL, a default assessment factor of 3 was recommended; however, the factor may need to be adjusted depending on the effects observed at the LOAEL and the slope of the dose–response curve. The BMD could be an alternative approach for defining or confirming a NOAEL depending on the data quality and dose spacing.

WHO/IPCS (1994, 1996, 1999) have adopted the approach that in situations where a NOAEL has not been achieved but the data on effects are of sufficient quality to be the basis of the risk assessment, a NAEL should be developed by the application of an appropriate UF to the LOAEL. According to WHO/IPCS (1994), UFs of 3, 5, or 10 have been used previously to extrapolate from a LOAEL to a NOAEL depending on the nature of the effect(s) and the dose-response relationship. A BMD may be developed as an alternative to the UF in extrapolating to the NOAEL.

According to US-EPA (1993), a 10-fold factor is generally used, in addition to the assessments factors for interspecies and interindividual differences, when deriving an RfD from a LOAEL, instead of a NOAEL. This factor is intended to account for the uncertainty involved in extrapolating from LOAELs to NOAELs and is referenced as "10L."

It was noted in the 2002 review of the RfD and RfC processes (US-EPA 2002) that a UF (default 10) is typically applied to the LOAEL when a NOAEL is not available. The size of the LOAEL-to-NOAEL UF may be altered, depending on the magnitude and nature of the response at the LOAEL. It is also noted to be important to consider the slope of the dose–response curve in the range of the LOAEL in making the determination to reduce the size of the LOAEL-to-NOAEL UF. With reference to several papers describing the magnitude of the difference between the dose at the LOAEL and at the NOAEL, it was noted that the ratio of the LOAEL-to-NOAEL in many cases was approximately threefold, but in a few cases the difference was as much as 10-fold. In general, the ratio of the doses at the LOAEL and the NOAEL is likely to vary considerably among studies and may not be informative. This is because the lowest dose in a study is often selected to ensure that no statistically significant response above control is observed and the next higher dose is selected to ensure that some significant response is observed, rather than selecting doses that will give a maximum NOAEL and a minimum LOAEL. Data should be carefully evaluated, taking into consideration the level of response at the LOAEL and the NOAEL and the slope of the dose–response curve before reducing the size of the UF applied to the LOAEL.

The EU TGD (EC 2003) recognized that the NOAEL is not very accurate with respect to the degree to which it corresponds with the (unknown) true NAEL. In the case of a steep curve the derived NOAEL can be considered as more reliable (the greater the slope, the greater the reduction in response to reduced doses); in the case of a shallow curve, the uncertainty in the derived NOAEL may be higher and this has to be taken into account in the assessment. If a LOAEL has to be used in the assessment, then this value can only be considered reliable in the case of a very steep curve. According to KEMI (2003), extrapolation factors of between 3–5 are used for LOAEL-to-NOAEL extrapolation without any scientific basis in risk assessment reports of existing substances within the European Union.

TNO has stated that when a reliable dose–response curve for the relevant adverse effect has been established, the slope of this curve should be taken into account (Hakkert et al. 1996). The steeper the dose–response curve, the smaller the assessment factor. The assessment factor depends on expert judgment, the default value is 1.

In relation to the dose–response curve, KEMI (2003) stated that the slope always has to be considered. A moderate assessment factor (not further specified) may provide an adequate MOS if the dose–response relationship is relatively steep, but may not be sufficiently conservative if the dose–response curve is relatively shallow, see Figure 5.6. In relation to extrapolation from LOAEL to NOAEL, KEMI considered that analysis of several databases does support the statement that a

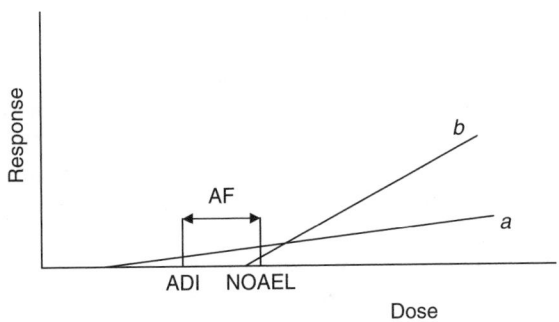

FIGURE 5.6 Schematic illustration of the traditional setting of an acceptable level of exposure (ADI) by dividing the NOAEL from an animal study by an assessment factor (AF). The two dose–response relationships have identical NOAEL. If a uniform assessment factor is applied, there will be an adequate MOS at the ADI for effect "b" but not for effect "a." (Modified from KEMI, *Human health risk assessment. Proposals for the use of assessment (uncertainty) factors. Application to risk assessment for plant protection products, industrial chemicals and biocidal products within the European Union.* Report No. 1/03, Solna, Sweden, 2003.

factor of 10 or lower is adequate for the extrapolation from LOAEL to NOAEL. They noted that the suggestion of using a default factor of up to 10 is supported by the fact that, according to OECD guidelines for a 90-day repeated dose toxicity study, the interval between the doses should be 2–4, with a maximum of 10.

KEMI recommended that extrapolation using the historical LOAEL/NOAEL ratio should not be undertaken in order to arrive at the dose without adverse effects from the LOAEL. The BMD approach should be used if data are adequate. If it is not possible to use the BMD approach, or to set a NOAEL, an extrapolation factor of 3–10, depending on the shape of the curve, is suggested for extrapolation from LOAEL to NOAEL. A LOAEL should preferably only be used in the case of a steep dose–response curve, and there is no guarantee that extrapolation of a LOAEL with any factor will yield an estimate of the NOAEL.

5.7.1 Dose–Response Curve (LOAEL-to-NOAEL Extrapolation): Summary and Recommendations

The NOAEL is not very accurate with respect to the degree to which it corresponds with the (unknown) true NAEL. One of the most evident limitations in the NOAEL setting is that it does not take into account the shape of the dose–response curve, including its slope, for the effect as the NOAEL by definition is one of the doses tested in the specific experimental study, and all other data points are ignored. In case a NOAEL cannot be set for the critical effect, a LOAEL is then set and extrapolated to a NOAEL; this extrapolation can also be regarded as part of the dose–response analysis.

If sufficient data are available, the shape of the dose–response curve should be taken into account. In the case of a steep curve the derived NOAEL can be considered as more reliable (the greater the slope, the greater the reduction in response to reduced doses); in the case of a shallow curve, the uncertainty in the derived NOAEL may be higher and this has to be taken into account in the assessment. If a LOAEL has to be used in the assessment, then this value can only be considered reliable in the case of a very steep curve.

In theory, the steeper the slope of the dose–response curve, the smaller the assessment factor and vice versa. However, there is no scientific basis for any value of a default factor to account for uncertainty in the NOAEL, nor any distribution.

A default assessment factor of 10 has been used traditionally for the extrapolation from a LOAEL to a NOAEL. Some analyses have been performed in order to evaluate the size of an appropriate assessment factor for the LOAEL-to-NOAEL extrapolation, based on evaluations of LOAEL/NOAEL ratios. A number of evaluations have been based on data from Weil and McCollister (1963) and adapted by Dourson and Stara (1983). Some evaluations found that this analysis supports that a factor of 10 or lower is adequate while others found the factor of 10 to be overly conservative as the LOAEL rarely exceeded the NOAEL by more than about 5–6 fold and was typically closer to a value of 3. The analysis on LOAEL/NOAEL ratios performed on LOAEL/NOAEL ratios showed 95th percentiles of 9, 16, and 11 for subacute, subchronic, and chronic exposure durations, respectively, supporting the 10-fold factor to account for about 95% of the chemicals evaluated.

In general, the use of LOAEL/NOAEL ratios to estimate a NOAEL from a LOAEL is questionable as these ratios reflect more the applied intervals between the dose levels in the studies (dose spacing, which is dependent on the study design), rather than the steepness of the dose–response relationship, i.e., the inherent toxicity. It has also been pointed out that there is no guarantee whatsoever that at one dose interval lower (extrapolation from a LOAEL to a NOAEL), the effect would be statistically nonsignificant.

WHO/IPCS (1994, 1996, 1999) have used UFs of 3, 5, or 10 to extrapolate from a LOAEL to a NOAEL depending on the nature of the effect(s) and the dose–response relationship; a BMD may be developed as an alternative to the UF in extrapolating to the NOAEL. The US-EPA (1993) has adopted the 10-fold factor to account for the uncertainty involved in extrapolating from a LOAEL to a NOAEL. This default value has later on been reconfirmed (US-EPA 2002), but it was stated that the size of the UF may be altered, depending on the magnitude and nature of the response at the LOAEL. TNO (Hakkert et al. 1996) stated that the default value is 1, but the factor depends on expert judgment. KEMI (2003) recommended that extrapolation using the historical LOAEL/NOAEL ratio should not be undertaken in order to arrive at the dose without adverse effects from the LOAEL, and the BMD approach should be used if data are adequate. If it is not possible to use the BMD approach, or to set a NOAEL, an extrapolation factor of 3–10, depending on the shape of the curve, is suggested for extrapolation from LOAEL to NOAEL. ECETOC (2003) recommended that if an appropriate NOAEL is available, then no assessment factor is necessary; if a LOAEL is used, a default assessment factor of 3 was recommended, but the factor may need to be adjusted depending on the effects observed at the LOAEL and the slope of the dose–response curve. The BMD could be an alternative approach for defining or confirming a NOAEL depending on the data quality and dose spacing. Kalberlah and Schneider (1998) considered the use of the BMD concept to be preferred to a general factor for LOAEL-to-NAEL extrapolation; if extrapolation using a standard factor is decided, the traditional factor 10 would serve as a suitable convention. Vermeire et al. (1999) did not recommend a specific size of a default factor to account for uncertainty in the LOAEL-to-NOAEL extrapolation, but pointed out that this factor can only be assigned using expert judgment in which the shape of the dose–response curve and the magnitude of the effect at the LOAEL are taken into account.

In conclusion, consideration should be given to the uncertainties in the NOAEL as the surrogate for the NAEL, as well as to the extrapolation of the LOAEL to the NAEL in cases where only a LOAEL is available as the starting point for the assessment. However, the BMD approach should be preferred instead of the NOAEL approach provided that the data are sufficient for a valid estimation of a BMD. If the NOAEL approach is selected and the starting point for the assessment is a LOAEL, it is suggested to use an assessment factor between 3 and 10; the size of the factor should take into account the dose spacing in the experimental study (in recent study designs generally spacing of 2–4-fold), the shape and slope of the dose–response curve, and the extent and severity of the effect seen at the LOAEL. When the starting point for the assessment is a NOAEL, the default assessment factor is 1; however, a larger factor may be considered in case of uncertainty about the statistical derivation of the NOAEL, e.g., a shallow dose–response curve, poor quality of the study from which the NOAEL is derived (e.g., few animals and inconsistent spacing between doses), other concerns related to the identified NOAEL, e.g., whether the mode of action for a threshold carcinogen is completely revealed. In general, the size of the assessment factor accounting for the dose–response curve, including LOAEL-to-NOAEL extrapolation, should be determined on a case-by-case basis using expert judgment.

5.8 NATURE AND SEVERITY OF EFFECTS

The nature of an effect includes the adversity of the toxicity expressed as the level of and the basis for NOAEL and LOAEL values (Section 4.2.2), and the severity of the specific endpoint or key event (Section 4.2.3), for example judging skin irritation less severe than teratogenicity. An additional assessment factor for severe and/or irreversible effects has been proposed by various groups. This section gives an overview of such proposals and evaluations. Then, the key issues are summarized and our recommendations are presented.

Renwick (1995) discussed the possible rationale for the use of an additional safety factor for nature of toxicity in the estimation of Acceptable/Tolerable Daily Intake (ADI/TDI) values, including a survey of the historical use of such a factor. He concluded that the principal rationale for an additional factor for nature of toxicity is to provide a greater degree of separation of the ADI/TDI from the NOAEL observed in animal studies. In scientific terms, it was recognized that there are a number of uncertainties in extrapolating from high-dose animal studies to sensitive subgroups of the human population. Because of the possibility that the total UF may be only barely adequate, the possible production of an irreversible effect could be minimized by the use of an extra factor. Renwick noted that the main use of a factor for nature of toxicity was in the WHO 1993 assessment of water quality guidelines, the JMPR had occasionally applied a factor specifically for nature of toxicity, whereas the JECFA had rarely applied an additional factor. The reasons for giving an additional factor include:

- Carcinogenicity, usually but not exclusively for non-genotoxic chemicals, the most common reason
- Teratogenicity, usually linked to an expression of concern about the nature of the effect
- Steep dose–response, when there is serious toxicity detected at doses just above the NOAEL, used only very rarely and usually in connection with relatively weak databases

According to Renwick, the nature of toxicity factor was not always applied to the NOAEL for the toxicity of concern, for example:

- Extra factor applied to the NOAEL for a related but less severe lesion in the same target organ and same species, e.g., application of a factor for carcinogenic potential to hyperplasia or changes in organ weight at doses below the NOAEL for carcinogenicity in the organ
- Extra factor applied to the NOAEL for a different target organ in the same species, e.g., a factor for carcinogenicity based on observations in one organ is applied to the NOAEL based on toxicity in a different organ
- Extra factor applied to the NOAEL for a different endpoint in a different species, e.g., detection of cancer in experimental animals can trigger an extra factor, which is then applied irrespective of the NOAEL, which is the basis for the ADI/TDI
- Extra factor for an effect via one route applied to the NOAEL based on an effect by a different route in the same species, e.g., an extra factor from inhalation to derive a guideline for drinking-water quality

Renwick considered that in relation to carcinogenicity for non-genotoxic chemicals and teratogenicity, the application of an extra factor for nature of toxicity is difficult to justify scientifically. He concluded that if a safety factor for nature of toxicity is to be used then logically it should be applied to the NOAEL for the toxicity, which resulted in its use. For example, in relation to teratogenicity, a factor for nature of toxicity should be applied to the NOAEL for teratogenicity and not for maternal toxicity or some other endpoint. For carcinogenicity, the extra factor should be applied only to the NOAEL for the detection of tumors in those studies where this effect was the rationale for the use of an extra factor. In relation to a steep dose–response, it was concluded that this, in reality, concerns the precision of the NOAEL and therefore relates to the adequacy of the database rather than nature of toxicity.

Dourson et al. (1996) considered that when a NOAEL is not available, the choice of a UF for LOAEL-to-NOAEL extrapolation should generally depend on the severity of the effect seen at the LOAEL, see Section 5.7.

Vermeire et al. (1999) considered that the type of critical effect should be taken into account. Assessment factors may be applied by expert judgment depending on each individual case; by default, it can be assumed that no extra correction is necessary.

ECETOC (2003) considered that the size of the assessment factor for LOAEL-to-NOAEL extrapolation should take into account, among others, the extent and severity of the effect seen at the LOAEL, see Section 5.7.

WHO/IPCS (1994, 1996, 1999) stated that the nature of toxicity, i.e., whether the effect is adverse or not, is considered in the determination of NOAEL and LOAEL, see Section 4.2.4. An additional safety factor of up to 10 has been incorporated in cases where the NOAEL is derived for a critical effect, which is a severe and irreversible phenomenon, such as teratogenicity or non-genotoxic carcinogenicity, especially if associated with a shallow dose–response relationship. This additional factor has been applied in such cases to provide a greater margin between the exposure of any particularly susceptible humans and the dose–response curve for such toxicity in experimental animals. However, for other types of toxic effects, e.g., changes in organ weight or histopathology, a value of 1, i.e., no further correction would be appropriate.

According to WHO/IPCS (1994), UFs of 3, 5, or 10 have been used previously to extrapolate from a LOAEL to a NOAEL depending on the nature of the effect(s) and the dose–response relationship, see Section 5.7.

The EU TGD (EC 2003) pointed out that the nature and severity of the effect needs to be considered in the evaluation of the MOS (can be interpreted as an overall assessment factor).

TNO considered that the biological significance of the critical adverse effect in terms of its presumable health consequence should be considered in the selection of assessment factor (Hakkert et al. 1996). For example, a reversible change in a biochemical parameter of doubtful toxicological significance may warrant the use of an additional factor smaller than 1, whereas, e.g., microscopically visible tissue damage may indicate application of a factor higher than 1. The default value is 1.

KEMI (2003) recommended that the nature of effect should be taken into account and that the adversity, potency, and severity of a toxic substance should be considered, although realizing that it is extremely difficult to set forth other than very rough guidelines on how to assess the nature of effect. Depending on the complexity of the nature of effect, a case-by-case expert judgment appears at present to be the most appropriate method to use and, in comparison with other organizations, a factor of 1 to 10 was suggested when judging the nature of effect.

5.8.1 NATURE AND SEVERITY OF EFFECTS: SUMMARY AND RECOMMENDATIONS

An additional assessment factor, of up to 10, has been applied in some cases where the NOAEL has been derived for a critical effect, which is considered as a severe and irreversible effect, such as teratogenicity or non-genotoxic carcinogenicity, especially if associated with a shallow dose–response relationship. The principal rationale for an additional factor for nature of toxicity has been to provide a greater margin between the exposure of any particularly susceptible humans and the dose–response curve for such toxicity in experimental animals.

The application of an extra factor for nature of toxicity is difficult to justify scientifically. Renwick (1995) concluded that if a factor for nature of toxicity is to be used then it should be applied to the NOAEL for the toxicity which resulted in its use. Vermeire et al. (1999) considered that the type of critical effect should be taken into account but, by default, no extra correction is necessary. WHO/IPCS (1994, 1996, 1999) stated that the nature of toxicity, i.e., whether the effect is adverse or not, is considered in the determination of the NOAEL and LOAEL; however, an additional factor of up to 10 could be incorporated in cases where the NOAEL is derived for a critical effect which is severe and irreversible. The EU TGD (EC 2003) pointed out that the nature and severity of the effect need to be considered in the evaluation of the MOS. TNO (Hakkert et al. 1996) considered that the biological significance of the critical adverse effect in terms of its presumable health consequence should be considered in the selection of assessment factor; the default value is 1. KEMI (2003) recommended that the nature of effect should be taken into account case-by-case on expert judgment and, in comparison with other organizations, a factor of 1 to 10 was suggested.

In conclusion, consideration should be given to the nature and severity of effects in the selection of assessment factor. However, in reality, this should be considered in the determination of the NOAEL and LOAEL and therefore relates to the dose–response curve, see Section 5.7, rather than to the nature and severity of effects. The recommended default factor for nature and severity of effects is therefore 1.

5.9   CONFIDENCE IN THE DATABASE

In the hazard assessment, it is important to evaluate the toxicological database with regard to its adequacy. The adequacy of a study includes its validity and its relevance. The relevance refers to what has been studied in relation to what is needed for the hazard and risk assessment, and the validity refers to how the study was performed, e.g., conforming with a particular test guideline. The validity and the relevance of a study, or a whole database, has to be considered in relation to the reliability and thus the confidence. The data for hazard assessment are described in detail in Chapter 3.

It has been suggested to apply an assessment factor for the confidence in the database in case there are limitations in the database, including lack of data for children, which are important in relation to the purpose of the assessment. This section gives an overview of such proposals and evaluations. Then, the key issues are summarized and our recommendations are presented. The question of an extra assessment factor in the hazard and risk assessment for chemicals of concern for children is specifically addressed in Section 5.2.1.13.

Dourson et al. (1992) examined the use of an assessment factor for database insufficiencies in relation to the estimation of oral RfDs through an analysis of frequency histograms of NOAEL ratios for various types of studies in various species. The toxicity data for 69 pesticides were obtained from the US-EPA IRIS database, and the chemicals selected had the most complete toxicity database possible in 1–2-year rat, 1–2-year dog, and 1–2-year mouse studies, and rat reproductive and rat developmental toxicity studies. A total of 296 studies were included in the analysis. After quantitative and statistical analysis of these data, it was concluded that, on average, a 2-year dog study detected toxic responses at similar doses as a 2-year rat study, and that both of these studies detected toxic responses at lower doses than either a rat two-generation study, a rat developmental toxicity study, or a 2-year mouse study. The authors pointed out that, although these chronic dog and rat studies were found to detect toxic responses at lower doses than the other studies listed, this analysis does not reflect the seriousness of the effects that were compared. They concluded that, within the confines of this analysis, it appears that a 2-year dog and rat study, and reproductive and developmental studies are a sufficient database on which to estimate high confidence RfDs, i.e., that more than one of the listed studies are needed to develop a high confidence estimate of an RfD. Furthermore, if one or more of the listed studies are missing, then an additional UF is needed to estimate RfDs to account for this interbioassay variability.

Dourson et al. (1996) noted that if data are only available from one chronic study on which to base the estimation of a "sub-threshold" dose, the question could be asked whether data from chronic studies in other species or data from different types of bioassays (e.g., reproductive or developmental toxicity) would yield lower NOAELs. The uncertainty related to this issue must therefore be addressed and, according to the authors, the default approach to address this uncertainty is to apply a 3- or 10-fold UF, based on the assumption that the critical effect can be discovered in a reasonably small selection of toxicity studies. With a reference to some analyses performed within this area, the authors suggested the use of a UF to account for missing bioassays; however, the quantification of this UF was considered to require additional work.

Vermeire et al. (1999) noted that the size, quality, completeness, and consistency of the database should be considered. They also noted that the assessment factor should be higher than unity if one is less confident about the database, and probably up to 100. The assessment factor can only be assigned on the basis of expert judgment, preferably made transparent through the application of a set of criteria. The authors pointed out that it may be argued that a database necessitating very high assessment factors is probably inadequate for the risk assessment altogether.

ECETOC (1995) has suggested an approach where the risk assessor should formulate a statement about the degree of confidence he or she has in the predicted NAELs derived. This statement should guide the risk manager in selecting between risk management alternatives. The degree of confidence may be high, medium, or low. Assessment factors have been suggested as a guide:

- High degree of confidence: The database contains high quality human or animal studies, i.e., two or more studies with the same endpoint. The database should be sufficiently extensive to give confidence that the correct critical effect has been selected, and that there are no major uncertainties in this respect. No additional numerical UF required, i.e., the default factor is 1.
- Medium degree of confidence: The database falls short of the quality described above in some significant respect, which limits the overall confidence to "medium". Assess on a case-by-case basis, perhaps consider to use a low numerical UF, in the range of 1–2.
- Low degree of confidence: The human or animal studies fall short of the highest standards in some important respects. Consider the need to generate more data, either on effects or on exposure, to increase the degree of confidence or, alternatively, use a larger UF (value not further specified).

WHO/IPCS (1994, 1999) stated that a minimum data set considered adequate for an assessment will vary according to the purpose of the assessment. The major deficiencies in a toxicity database, other than those related to the pivotal study, which increase the uncertainty of the extrapolation should be recognized by the use of an additional UF. Since the quality and/or completeness of different databases vary, the additional UF will also vary. For example, a value of 1 would be applied to a database that was considered complete for the evaluation of the compound under consideration, but a factor of 1–100 might be necessary for limited databases. If minor deficiencies in the data exist with respect to quality, quantity, or omission, then an extra factor of 3 or 5 would be appropriate. An extra factor of 10 would be appropriate where major deficiencies in the data exist, e.g., a lack of chronic toxicity studies and reproductive toxicity studies. It was pointed out that inadequacies of the pivotal study could also be considered as a subset of inadequacies of the database and that the total factor for limitations of the pivotal study plus adequacy of the overall database should not exceed 100 since such a database is generally not acceptable for development of a TDI.

US-EPA (1993) stated that in addition to the standard factors (for inter- and intraspecies differences, less than chronic duration studies, and LOAEL-to-NOAEL extrapolation), an extra factor should be included if the total toxicological database is incomplete, i.e., the so-called modifying factor (MF). It was stated that the magnitude of the MF depends upon a professional assessment of scientific uncertainties of the study and database not explicitly accounted for by the standard factors, e.g., the completeness of the overall database and the number of species tested. The default value for the MF is 1.

It was noted in the 2002 review of the RfD and RfC processes (US-EPA 2002) that the database UF, generally 10-fold, is intended to account for the potential for deriving an underprotective RfD/RfC as a result of an incomplete characterization of the chemical's toxicity, i.e., is intended to account for the uncertainty associated with extrapolation from animal and human data when the database is incomplete. In addition to identifying toxicity information that is lacking, review of existing data may also suggest that a lower reference value might result if additional data were available. Consequently, in deciding to apply this factor to account for deficiencies in the available data set and in identifying its magnitude, the assessor should consider both the data lacking and the data available for particular organ systems as well as life stages. For example, depending on the database and what is known about the chemical, the lack of a two-generation animal reproductive toxicity study might be considered a deficiency. In any case, the size of the database UF will depend on other information in the database and on how much impact the missing data may have on determining the toxicity of a chemical and, consequently, the point of departure (POD) for the setting of RfDs/RfCs. If the RfD/RfC is based on animal data, a factor of 3 is often applied if either a prenatal toxicity study or a two-generation reproduction study is missing, or a factor of 10 may be applied if both are missing. If data from the available toxicology studies raise suspicions of developmental toxicity and signal the need for developmental data on specific organ systems (e.g., detailed nervous system, immune system, carcinogenesis, or endocrine system), then the database factor should take into account whether or not these data are available and used in the assessment and their potential to affect the POD for the particular duration RfD/RfC under development.

It was also noted in the 2002 review that the description of the database UF shows substantial similarity to that of the MF and that the purpose of the MF is to be sufficiently covered by the general database UF; therefore, it was recommended that use of the MF be discontinued.

Furthermore it was noted that, in many respects, the additional 10-fold factor for infants and children called for in the 1996 FQPA is similar to the database UF, see Section 5.2.1.13. It was concluded that an additional UF, i.e., a children-specific factor, is not needed because the traditional factors (interspecies, intraspecies, LOAEL-to-NOAEL, subchronic-to-chronic, and database-deficiency) were considered sufficient to account for uncertainties in the database from which the reference values are derived, including lack of data on children.

The EU TGD (EC 2003) pointed out that the overall confidence in the database needs to be considered in the evaluation of the MOS (can be interpreted as an overall assessment factor).

TNO has stated that the size, quality, completeness, and consistency of the database should be considered (Hakkert et al. 1996). Major aspects for the evaluation of the quality of the data supporting the NOAEL are (1) deviations from official guidelines, which are not properly substantiated, (2) number of animals used, (3) number of dose levels tested, and (4) adequacy of hematological, biochemical, and pathological examinations. Indications for doubts on the confidence in the database are (1) the absence of certain types of studies, (2) conflicting results between studies, and (3) doubts on the reliability of the route-to-route extrapolation. However, consistency of results from different studies, consistency of animal and human data, and reliable mechanistic data are indicative for a high-confidence database. The default assessment factor for confidence of the database is 1.

KEMI (2003) noted that it is important in the effect assessment to evaluate the toxicological database with regard to its adequacy. They also noted that no systematic approach to the basis of the use of an assessment factor for the confidence in the database has been found. Therefore, transparent expert judgment of the adequacy of the database in a case-by-case manner was recommended at present as the most useful tool in considering the database. They also recommended to be extra prudent in the risk characterization when dealing with a poor database. Based on other organizations' default values, for practical reasons, a factor of 1–5 could be chosen when judging the quality of the database. In the special case of children as a target group for the risk assessment, and if there is a shortage of data for children, a larger factor of 1–10 was proposed to compensate for a poor database in this respect. The assessment factor is based on the severity of the effect (Section 5.8) and uncertainty of the difference in the sensitivity between young and adult individuals for the particular effect, see also Section 5.4.

5.9.1 Confidence in the Database: Summary and Recommendations

In the hazard assessment, it is important to evaluate the toxicological database with regard to its adequacy, i.e., the overall confidence regarding the quality, completeness, and consistency of the database should be considered.

Major aspects for the evaluation of the quality of the data supporting the NOAEL or LOAEL are essential deviations from test guidelines that are not properly substantiated, number of animals used, number of dose levels tested, and adequacy of the examined endpoints.

The minimum database considered to be adequate for an assessment, i.e., the completeness of the database, will vary according to the purpose of the assessment. Both the data lacking and the data available for particular target organs and systems as well as life stages should be considered. The lack of certain types of studies is an indication for a lower confidence in the database; for example, depending on the database and what is known about the chemical, the lack of a two-generation animal reproductive toxicity study or a chronic toxicity study, or both, might be considered a deficiency.

Regarding consistency of the database, conflicting results between studies are an indication of a lower confidence in the database, whereas consistency of results from different studies, consistency of animal and human data, and reliable mechanistic data are indicative of a high-confidence database.

It has been suggested to apply an assessment factor for the confidence in the database in case there are limitations in the database, including lack of data for children, which are important in relation to the purpose of the assessment. WHO/IPCS (1994, 1999) stated that a UF of 1 would be applied to a database that was considered complete. If minor deficiencies in the data exist with respect to quality, quantity, or omission, then an extra factor of 3 or 5 would be appropriate. An extra factor of 10 would be appropriate where major deficiencies in the data exist, e.g., a lack of chronic toxicity studies and reproductive toxicity studies. US-EPA (2002) noted that the database UF generally has a value of 10. If the RfD/RfC is based on animal data, a factor of 3 is often applied if either a prenatal toxicity study or a two-generation reproduction study is missing, or a factor of 10 may be applied if both are missing. TNO (Hakkert et al. 1996) stated that the default assessment factor for confidence of the database is 1. KEMI (2003) recommended, based on other organizations' default values, for practical reasons, that a factor of 1–5 could be chosen when judging the quality of the database. In the special case of children as a target group for the risk assessment, and if there is a shortage of data for children, a larger factor of 1–10 was proposed to compensate for a poor database in this respect. ECETOC (1995) proposed to distinguish between a high, medium, and low degree of confidence and provided criteria for evaluating the degree of confidence. For a high degree of confidence in the database, the proposed default factor was 1; for a medium degree of confidence, a low numerical UF (in the range of 1–2) was proposed; and for a low degree of confidence, a larger UF (value not further specified) should be applied. Dourson et al. (1996) noted that the default approach to address the uncertainty related to the lack of data from chronic studies or from reproductive or developmental toxicity studies is to apply a 3- or 10-fold UF, based on the assumption that the critical effect can be discovered in a reasonably small selection of toxicity studies. Vermeire et al. (1999) noted that the assessment factor should be higher than unity if one is less confident about the database, and probably up to 100.

The question of an extra assessment factor in the hazard and risk assessment for chemicals of concern for children is specifically addressed in Section 5.2.1.13. The U.S. Food Quality Protection Act (FQPA) (US-EPA 1996) directed the US-EPA to apply an extra safety factor of 10 in assessing the risks of pesticides to infants and children. The US-EPA (2002) noted the overlap of areas covered by the FQPA factor and those addressed by the traditional UFs, and it was concluded that an additional UF (children-specific) is not needed in the setting of reference values because the currently available UFs (interspecies, intraspecies, LOAEL-to-NOAEL, subchronic-to-chronic, and database-deficiency) were considered sufficient to account for uncertainties in the database from which the reference values are derived. Renwick et al. (2000) concluded that the available data did not provide a scientific rationale for an additional 10-fold UF for infants and children and pointed out that when adequate reproduction, multigeneration, or developmental studies are conducted, there will be no need for an additional 10-fold factor.

In conclusion, the uncertainty related to the confidence in the database should be taken into account by the use of an assessment factor. Since the quality, completeness, and/or consistency of different databases vary, the assessment factor will also vary and can only be assigned on the basis of expert judgment, preferably made transparent through the application of a set of criteria. In any case, the size of the factor should be considered in terms of other information in the database. The default value should be 1 in case of a high-confidence database, and a factor of 10 would be appropriate where major deficiencies in the data exist, e.g., a lack of chronic and reproductive toxicity studies when setting a tolerable intake.

No children-specific assessment factor is warranted when setting a tolerable intake as the traditional assessment factors (interspecies, intraspecies, subchronic-to-chronic, LOAEL-to-NOAEL, and database-deficiency) are considered to cover the concerns and uncertainties for children adequately.

5.10 OVERALL ASSESSMENT FACTOR

One of the crucial assumptions affecting how the assessment factors are implemented in the derivation of tolerable intakes is that they are independent of each other. This assumption has led to the conclusion that the overall assessment factor is obtained by multiplication of the individual assessment factors discussed in the previous Sections 5.3 through 5.9. This section gives an overview of the validity of this approach. Then, the key issues are summarized and our recommendations are presented.

Calabrese and Gilbert (1993) reassessed the use of multiple UFs in the risk assessment process. Their intention was to establish that the UFs display a mixture of independence and interdependence based on their inherent properties and the nature of specific toxicological and epidemiological studies. The authors demonstrated the lack of independence of the interspecies and intraspecies UFs, as well as of the intraspecies and the less-than-lifetime UFs, and proposed revised UF values based on the concept of the relationship of independent and interdependent UFs, see Section 5.2.1.2.

Kalberlah and Schneider (1998) stated that the overall assessment factor is the resultant of the sub-factors multiplicatively, but that this is only statistically correct if the individual factors are independent of each other, which is not always the case - with a reference to Calabrese and Gilbert (1993). They also stated that interdependence between the individual extrapolation factors should be kept low through the definition and derivation of the factor, but overlap may, however, occur for a specific substance; in such a case, the standard values for the factors should be modified. They also noted that if several extrapolation factors are multiplied by each other, the total degree of statistical certainty is altered, even in the case of independent sub-factors. In general, the degree of statistical certainty increases if the individual factors are combined. It would therefore be justifiable to keep the overall level of certainty constant by reducing the degree of statistical certainty of an individual factor. However, a different combination of sub-factors may be applied for various chemicals. A more precise estimation of the degree of statistical certainty could be attempted by using probability distributions (Section 5.11).

Vermeire et al. (1999) stated that in the standard procedure for deriving acceptable limit values, various assessment factors are multiplied to obtain an overall assessment factor. However, multiplication of assessment factors implies a piling up of worst-case assumptions: the probability of simultaneous occurrence of worst-case situations for the same chemical is smaller than the occurrence of a single worst-case situation. Therefore, the higher the number of extrapolation steps, the higher the level of conservatism. The piling up of worst-case assumptions can be avoided by using probability distributions (Section 5.11).

Gaylor et al. (1999) have proposed a total default UF in the order of 10,000 for severe, irreversible adverse health effects when establishing human exposure guidelines based on a benchmark dose ($BMDL_{10}$) derived from animal data. For reversible biological effects, a smaller default UF in the order of 1000 may be employed.

WHO/IPCS (1994, 1996) stated the importance of a final review of the total UF applied, particularly in cases where a low value has been used, based on toxicokinetic or toxicodynamic data, to replace one of the default values. Under such circumstances, a TDI derived on the basis of the appropriate overall UF for that toxic effect might be greater than that which would be produced by an alternative, well-defined toxic endpoint observed at slightly higher intakes or exposures. It was also pointed out that the precision of the TDI largely depends on the magnitude of the overall UF. Because of the imprecision of the default factors, the total default UF should not exceed 10,000. If the assessment would lead to a higher factor, then the resulting TDI would be so imprecise as to lack meaning. For substances for which UFs were greater than 1000, the guideline values for drinking-water quality are designated as provisional in order to emphasize the high level of uncertainty inherent in these values.

In the 2002 review (US-EPA 2002) of the RfD and RfC processes the US-EPA stated that the exact value of the UFs chosen should depend on the quality of the studies available, the extent of the database, and scientific judgment. It was recognized that there is overlap in the individual UFs (interspecies, intraspecies, LOAEL-to-NOAEL, subchronic-to-chronic, and database-deficiency) and that the application of these five UFs with a default value of 10 for the chronic reference value (yielding a total UF of 100,000) is inappropriate. It was recommended to limit the total UF applied to a chronic reference value for any particular chemical to no more than 3000, and to avoid the derivation of a reference value that involves application of the full 10-fold UF in four or more areas of extrapolation.

TNO has stated that principally the overall factor is established by multiplication of the separate factors unless the data indicate another method to be used (Hakkert et al. 1996). They also noted that in practice it is not possible to distinguish all the separate assessment factors and some factors are not independent of each other. Therefore, straightforward multiplication may lead to unreasonably high factors, and discussion and weighing of individual factors are essential to establish a reliable and justifiable overall assessment factor.

KEMI (2003) noted the high level of conservatism in the standard procedure to obtain an overall assessment factor by multiplication of the individual assessment factors. It was therefore recommended that distributions of the assessment factors should be used in the calculation of the overall assessment factor, if available (Section 5.11). However, it was recognized that such data are only available for the interspecies factor and the factor for duration of exposure.

5.10.1 OVERALL ASSESSMENT FACTOR: SUMMARY AND RECOMMENDATIONS

The overall assessment factor is the product of a number of assessment factors accounting for uncertainties related to various extrapolation steps (inter- and intraspecies, route-to-route, subchronic-to-chronic, LOAEL-to-NOAEL, nature and severity of effect, and database-deficiency). However, the higher the number of extrapolation steps, the higher the level of conservatism. Since the different assessment factors are not always independent of each other, straightforward multiplication may lead to unreasonably high factors, and discussion are and weighing of individual factors are essential to establish a reliable and justifiable overall assessment factor. Some aspects to consider in the final qualitative discussion are:

- How extensive is the database?
- What is the quality of the studies, e.g., study design, performance, and reporting?
- What are the data gaps?
- Are both human and animal data available, and are the results consistent?
- Are there data on more than one species, and are the results consistent?
- How adverse/severe is the critical effect?
- Are data available for the relevant route of exposure?
- What are the scientific uncertainties?
- What is the overall confidence in the database?

It should be recognized that these aspects might partly have been covered by some of the above-mentioned assessment factors. No further assessment factor should be introduced in this final qualitative step.

WHO/IPCS (1994, 1996) stated that the total UF applied when setting a TDI should not exceed 10,000. A higher UF than 10,000 indicates that the resulting TDI would be so imprecise as to lack meaning. US-EPA (2002) noted that the application of their five standard UFs with a default value of 10 for the chronic reference value (yielding a total UF of 100,000) is inappropriate, and it was recommended to limit the total UF for any particular chemical to no more than 3,000, and to avoid the derivation of a reference value that involves application of the full 10-fold UF in four or more areas of extrapolation. KEMI (2003) recommended that probability distributions of the assessment factors should be used in the calculation of the overall assessment factor, if available. Kalberlah and

Schneider (1998) noted that if several extrapolation factors are multiplied by each other, the total degree of statistical certainty is altered, even in the case of independent sub-factors, and that a more precise estimation of the degree of statistical certainty could be attempted by using probability distributions. Vermeire et al. (1999) also noted that the piling up of worst-case assumptions could be avoided by using probability distributions. Gaylor et al. (1999) have proposed a total default UF of 10,000 for severe, irreversible adverse health effects when establishing human exposure guidelines based on a benchmark dose ($BMDL_{10}$) derived from animal data; for reversible biological effects, a factor of 1000 may be employed.

In conclusion, discussion and weighing of the individual assessment factors are essential in order to establish a reliable and justifiable overall assessment factor, and the possible overlap in the individual assessment factors should be recognized in the justification for the overall factor. If an unreasonable high total factor (in the order of 10,000) is established, then the resulting tolerable intake is considered to be too imprecise, and it should be realized that the database is too limited in order to derive a tolerable intake.

5.11 ASSESSMENT FACTORS: PROBABILISTIC APPROACH

It has been suggested by, e.g., Slob and Pieters (1998), Kalberlah and Schneider (1998), Vermeire et al. (1999), and KEMI (2003) to use probability distributions for the various types of assessment factors in order to achieve a more precise estimation of the degree of statistical certainty and to avoid the piling up of worst-case assumptions in the overall assessment factor.

The probabilistic approach allows for a closer link with specific knowledge or lack of knowledge in specific assessments. For example, one may be more confident in the magnitude of the possible interspecies difference in one case than another. This may be expressed in the width of the relevant distribution for the assessment factor. However, in many cases, even the range of uncertainty is uncertain, and for those situations default distributions are called for.

In this method, each assessment factor is considered uncertain and characterized as a random variable with a lognormal distribution with a GM and a GSD. Propagation of the uncertainty can then be evaluated using Monte Carlo simulation (a repeated random sampling from the distribution of values for each of the parameters in a calculation to derive a distribution of estimates in the population), yielding a distribution of the overall assessment factor. This method requires characterization of the distribution of each assessment factor and of possible correlations between them. As a first approach, it can be assumed that all factors are independent, which in fact is not correct.

Baird et al. (1996) suggested a probabilistic alternative to the practice used by the US-EPA to derive RfDs from a NOAEL and application of UFs. The probabilistic approach expresses the human population threshold for a given substance as a probability distribution of values, rather than a single RfD value, taking into account the major sources of scientific uncertainty in such estimates. The approach was illustrated by using much of the same data that US-EPA used to justify their RfD procedure. For the four key extrapolations that were considered necessary to define the human population threshold based on animal data (interspecies, interindividual, LOAEL-to-NOAEL, and subchronic-to-chronic), the proposed approach used available data to define a probability distribution of each adjustment factor, rather than using available data to define point estimates of UFs.

Slob and Pieters (1998) have proposed a probabilistic approach for deriving acceptable human intake limits and human health risks from toxicological studies in which it is acknowledged that both the effect parameter (e.g., NOAEL, BMD) and the assessment factors are uncertain and can best be described by lognormal distributions.

They suggested the effect parameter the Critical Effect Dose (CED, a benchmark dose, Section 4.2.5) derived from the dose–response data by regression analysis. This CED was defined as the dose at which the average animal shows the Critical Effect Size (CES) for a particular toxicological endpoint, below which there is no reason for concern. The distribution of the CED can probabilistically be combined with probabilistic distributions of assessment factors for deriving standards such as, e.g., ADI/TDI and RfD by using uncertainty distributions instead of extrapolation factors. It was assumed that each assessment factor is lognormally distributed. Theoretical distributions were proposed for the interspecies, the intraspecies, and the duration-of-exposure factors. Furthermore, the general default UF of 10 was regarded as worst-case assumptions, and therefore was taken to correspond with the 99th percentile of the distributions for the extrapolation factors.

The distributions of two individual assessment factors can be combined forming a new distribution (with a new GM and GSD) characterizing the combination of the two individual factors; a certain percentile can be chosen from this new distribution. Which percentile of a distribution to be chosen is a policy issue. When distributions of the assessment factors are not available, the point estimate of a particular factor could be used. Distributions and point estimates can be used in parallel and combined when necessary; for example the chosen percentile from a distribution of one factor can be combined with the point estimate of another factor by multiplication.

Characterization of a databased distribution is only available at present for the interspecies factor (Section 5.3.3) and for the factor for duration of exposure (Section 5.6). A probabilistic distribution has also been proposed for the interindividual factor (Section 5.4.2).

5.12 TOLERABLE INTAKE

The tolerable intake (TI) is calculated by dividing the NOAEL (or LOAEL) for the critical effect(s) by the derived overall assessment factor (AF):

$$TI = NOAEL \text{ (or LOAEL)}/AF$$

It should be remembered that the tolerable intake covers, e.g., the ADI, TDI, PTWI, RfD, and RfC. The term "acceptable" is used widely to describe "safe" levels of intake and is applied for chemicals to be used in food production such as, e.g., food additives, pesticides, and veterinary drugs, and the term "tolerable" is applied for chemicals unavoidably present in a media such as contaminants in, e.g., drinking water and food. The term "PTWI" is generally used for contaminants that may accumulate in the body, and the weekly designation is used to stress the importance of limiting intake over a period of time for such substances. The terms RfD/RfC are widely used by, e.g., the US-EPA.

The precision of the tolerable intake depends therefore largely on the magnitude of the overall assessment factor. The precision is probably to one significant figure at best, and more usually to one order of magnitude; for assessment factors of 1000 or more, the precision becomes even less.

Implicit from the definition of the tolerable intake, i.e., "an estimate of the intake of a substance over a lifetime that is considered to be without appreciable health risk," arises the question: What are the health implications of exceeding the tolerable intake? This issue has been discussed at an ILSI (International Life Science Institute) Europe Workshop on the Significance of Excursions of Intake above the Acceptable Daily Intake (ADI) in 1999. The following questions were asked (Larsen and Richold 1999, Larsen 2006):

- By how much can the ADI be exceeded?
- For how long can excursions above the ADI be tolerated with respect to chronic toxicity, accumulation, and mechanisms of toxicity?
- What methods should be used to estimate intakes so that the estimates are relevant to the ADI?
- Do the same principles apply to contaminants that have TDI or PTWI values?

From the workshop discussions, a number of conclusions and recommendations could be outlined to provide guidance on the assessment of the significance of the excursions of intake above the ADI (Larsen and Richold 1999).

General discussion:

For most food additives, veterinary drug residues, and pesticide residues, excursions above the ADI are probably not frequent, whereas they may occur more frequently for some food contaminants. There was general agreement that the ADI/TDI and its derivation are an appropriate and scientifically credible basis for the safety assurance of food additives, pesticide residues, veterinary drug residues, and contaminants (in this case expressed as TDI), which show thresholded toxicity. By experience, toxicologists had no serious health concerns about occasional excursions of intake above the ADI as a large MOS is built into its derivation and proved that the intake over a longer period is averaged at or below the ADI. It was pointed out, however, that excursions of intake above the ADI are generally undesirable, in particular for a prolonged period. But it was also stated that despite the regulatory efforts, it is not always possible to prevent that the estimated intake of some substances, by some individuals, may exceed the ADI. There was also general agreement that the assessment of the significance of exceeding the ADI should be performed on a case-by-case basis.

What methods should be used to estimate intakes?

It was recommended that the first step is to use a conservative, theoretical/hypothetical approach (such as the budget method); if no problems are encountered, then there is no need for further estimation. The next step would be a refinement of the intake estimate by undertaking a 3-day dietary study supplemented with a food frequency questionnaire to estimate percentage of consumers; a minimum study population size would be 200 persons. If the intake estimate is still above the ADI, it would be necessary to carry out a risk assessment.

For how long can excursions above the ADI be tolerated and by how much can it be exceeded?

It was concluded that no general guidance could be given on how long and by how much excursions above the ADI could be tolerated, and that a case-by-case evaluation is required. However, consideration of typical situations was presented, which may help in evaluating the significance of the duration and the magnitude of exceeding the ADI: when the ADI is based on a chronic study, an intake above the ADI should not be acceptable if it occurred throughout the major part of human life because it reduces the overall safety margin. When the excess intake is for a period shorter than the pivotal study on which the ADI was based, consideration should be given to using a NOAEL from a study of shorter duration to determine whether the excess intake is indeed of concern. For food additives and most pesticides and veterinary drugs, the half-life was considered normally to be short and toxicity therefore might be due to chronic stress rather than cumulative toxicity. When the effects are not fully reversible, or are even progressive, the consequences of short-term peaks of intake above the ADI would require careful evaluation against the NOAEL or LOAEL in subacute or subchronic studies.

Do the same principles apply to contaminants that have TDI or PTWI values?

It was recommended that excursions above the TDI/PTWI should be considered in the way as for substances where an ADI has been set. It was emphasized that contaminants having very long half-lives accumulate in the body and the chronic toxicity is most often manifested when critical concentrations are achieved in target tissues. Furthermore, there are usually large differences between the acute or shorter-term toxic doses and the chronic LOAELs. In such cases, peak excursions of several times the PTWIs for short periods (days, weeks, or even months) or lower peak intakes for even longer periods (months to years) may be of no consequence provided that the integrated exposure over longer periods does not lead to critical steady-state tissue concentrations.

REFERENCES

Baird, S.J.S., J.T. Cohen, J.D. Graham, A.I. Shylakter, and J.S. Evans. 1996. Noncancer risk assessment: Alternatives to current Practice. *Hum. Ecol. Risk Assess.* 2:79–102.

Beltoft, V., E. Nielsen, O. Meyer, and O. Ladefoged. 2001. *Individual variations in biological susceptibility to xenobiotics: A review of the current knowledge.* Report prepared for the Danish Environmental Protection Agency, March 2001.

Burin, G.J. and D.R. Saunders. 1999. Addressing human variability in risk assessment - The robustness of the intraspecies uncertainty factor. *Regul. Toxicol. Pharmacol.* 30:209–216.

Calabrese, E.J. 1985. Uncertainty factors and interindividual variation. *Regul. Toxicol. Pharmacol.* 5:190–196.

Calabrese, E.J. and C.E. Gilbert. 1993. Lack of total independence of uncertainty factors (UFs): Implications for the size of the total uncertainty factor. *Regul. Toxicol. Pharmacol.* 17:44–51.

Calabrese, E.J., B.D. Beck, and W.R. Chappell. 1992. Does the animal-to-human uncertainty factor incorporate interspecies differences in surface areas? *Regul. Toxicol. Pharmacol.* 15:172–179.

Clewell, H.J., M.E. Andersen, and H.A. Barton. 2002a. A consistent approach for the application of pharmacokinetic modeling in cancer and noncancer risk assessment. *Environ. Health Perspect.* 110:85–93.

Clewell, H.J., J. Teeguarden, T. McDonald, et al. 2002b. Review and evaluation of the potential impact of age- and gender-specific pharmacokinetic differences on tissue dosimetry. *Crit. Rev. Toxicol.* 32:329–389.

Davidson, I.W.F., J.C. Parker, and R.P. Beliles, 1986. Biological basis for extrapolation across mammalian species. *Regul. Toxicol. Pharmacol.* 6:211–237.

D-EPA. 2006. *Principles for establishment of health based quality criteria for ambient air, soil and drinking water.* Vejledning fra Miljøstyrelsen Nr. 5 2006. Copenhagen: Danish Environmental Protection Agency, Danish Ministry of the Environment (in Danish). http://www2.mst.dk/Udgiv/publikationer/2006/87-7052-182-4/pdf/87-7052-182-4.pdf

Dorne, J.L. and A.G. Renwick. 2005. The refinement of uncertainty/safety factors in risk assessment by the incorporation of data on toxicokinetic variability in humans. *Toxicol. Sci.* 86:20–26.

Dorne, J.L., K. Walton, and A.G. Renwick. 2005. Human variability in xenobiotic metabolism and pathway-related uncertainty factors for chemical risk assessment: A review. *Food Chem. Toxicol.* 43:203–216.

Dourson, M.L. and J.F. Stara. 1983. Regulatory history and experimental support of uncertainty (safety) factors. *Regul. Toxicol. Pharmacol.* 3:238–244.

Dourson, M.L., L.A. Knauf, and J.C. Swartout. 1992. On reference dose (Rfd) and its underlying toxicity data base. *Toxicol. Ind. Health* 8:171–183.

Dourson, M.L., S.P. Pelter, and D. Robinson. 1996. Evolution of science-based uncertainty factors in noncancer risk assessment. *Regul. Toxicol. Pharmacol.* 24:108–120.

Dybing, E. and E.J. Søderlund. 1999. Situations with enhanced chemical risks due to toxicokinetic and toxicodynamic factors. *Regul. Toxicol. Pharmacol.* 30:S27–S30.

EC. 1996. Technical Guidance Document in support of Commission Directive 93/67/EEC on risk assessment for new notified substances and Commission Regulation (EC) No 1488/94 on risk assessment for existing substances. http://ecb.jrc.it/tgd

EC. 2003. Technical Guidance Document on risk assessment in support of Commission Directive 93/67/EEC on risk assessment for new notified substances, Commission Regulation (EC) No 1488/94 on risk assessment for existing substances and Directive 98/8/EC of the European Parliament and of the Council concerning the placing of biocidal products on the market. http://ecb.jrc.it/tgd

ECETOC. 1995. *Assessment factors in human health risk assessment.* Technical Report No. 66. Brussels: ECETOC.

ECETOC. 2003. *Derivation of assessment factors for human health risk assessment.* Technical Report No. 86. Brussels: ECETOC.

EU. 2006. The DG Environment REACH website. http://ec.europa.eu/environment/chemicals/reach/reach_intro.htm

Feron, V.J., P.J. van Bladeren, and R.J.J. Hermus. 1990. A viewpoint on the extrapolation of toxicological data from animals to man. *Food Chem. Toxicol.* 28:783–788.

Gaylor, D.W., R.L. Kodell, J.J. Chen, and D. Krewski. 1999. A unified approach to risk assessment for cancer and noncancer endpoints based on benchmark doses and uncertainty/safety factors. *Regul. Toxicol. Pharmacol.* 29:151–157.

Ginsberg, G., D. Hattis, B. Sonawane, et al. 2002. Evaluation of child/adult pharmacokinetic differences from a database derived from the therapeutic drug literature. *Toxicol. Sci.* 66:185–200.

Grönlund, M.H. 1992. *Kvantitativ riskbedömning av icke-genotoxiska substanser - en metodikstudie. IMM-rapport 4/92.* Stockholm: Institutet för miljömedicin. Karolinska institutet (in Swedish).

Hakkert, B.C., H. Stevenson, P.M.J. Bos, and J.J. van Hemmen. 1996. *Methods for the establishment of health-based recommended occupational exposure limits for existing substances.* TNO Report V96.463. Zeist: TNO Nutrition and Food Research, Zeist, The Netherlands.

Hattis, D., L. Erdreich, and M. Ballew. 1987. Human variability in susceptibility to toxic chemicals - A preliminary analysis of pharmacokinetic data from normal volunteers. *Risk Anal.* 7:415–426.

IGHRC. 2003. The Interdepartmental Group on Health Risks from Chemicals. *Uncertainty factors: Their use in human health risk assessment by UK Government.* Leicester: Institute for Environment and Health, University of Leicester. http://www.silsoe.cranfield.ac.uk/ieh/pdf/cr9.pdf

Kadry, A.M., G.A. Skowronski, M.S. Abdel-Rahman. 1995. Evaluation of the use of uncertainty factors in deriving RfDs for some chlorinated compounds. *J. Toxicol. Environ. Health* 45:83–95.

Kalberlah, F. and K. Schneider. 1998. *Quantification of extrapolation factors.* Final report of the research project No. 116 06 113 of the Federal Environmental Agency, Schriftenreihe der Bundesanstalt für Arbeitsschutz und Arbeitsmedizin - Forschung - FB 797. Bremerhaven: Wirtschaftsverlag NW Verlag für neue Wissenschaft GmbH 1998.

KEMI. 2003. *Human health risk assessment. Proposals for the use of assessment (uncertainty) factors. Application to risk assessment for plant protection products, industrial chemicals and biocidal products within the European Union.* Report No. 1/03. Solna, Sweden: Body for Competence and Methodology Development, the Swedish National Chemicals Inspectorate and Institute of Environmental Medicine Karolinska Institutet, Solna, Sweden.

Kramer, H.J., W.A. van den Ham, W. Slob, and M.N. Pieters. 1996. Conversion factors estimating indicative chronic no-observed-adverse-effect levels from short-term toxicity data. *Regul. Toxicol. Pharmacol.* 23:249–255.

Larsen, J.C. 2006. Risk assessment of chemicals in European traditional foods. *Trends Food Sci. Technol.* 17:471–481.

Larsen, J.C. and M. Richold. 1999. Report of workshop on the significance of excursions of intake above the ADI. *Regul. Toxicol. Pharmacol.* 20:S2–S12.

Lehman, A.J. and O.G. Fitzhugh. 1954. 100-fold margin of safety. *Assoc. Food Drug Off. U.S. Bull.* 18:33–35.

Lewis, S.C., J.R. Lynch, and A.L. Nikiforov. 1990. A new approach to deriving community exposure guidelines from "no-observed-adverse-effect levels." *Regul. Toxicol. Pharmacol.* 11:314–330.

Nielsen, E., I. Thorup, A. Schnipper, et al. 2001. *Children and the unborn child. Exposure and susceptibility to chemical substances - An evaluation.* Environmental Project No. 589. Copenhagen: Danish Environmental Protection Agency, Ministry of the Environment and Energy. http://www2.mst.dk/Udgiv/publications/2001/87-7909-574-7/html/default_eng.htm

Nielsen, E., G. Østergaard, J.C. Larsen, and O. Ladefoged. 2005. *Principles for human health assessments of chemical substances in relation to the establishment of health based quality criteria for ambient air, soil and drinking water.* Environmental Project No. 974/2004. Copenhagen: Danish Environmental Protection Agency, Danish Ministry of the Environment (in Danish with a summary in English).

OECD. 2003. *Descriptions of Selected Key Generic Terms Used in Chemical Hazard/Risk Assessment.* Joint Project with IPCS on the Harmonisation of Hazard/Risk Assessment Terminology. OECD Series on Testing and Assessment No. 44. Environment Directorate, Joint Meeting of the Chemicals Committee and the Working Party on Chemicals, Pesticides and Biotechnology. ENV/JM/MONO(2003)15. Paris: OECD.

Olson, H., G. Betton, D. Robinson, et al. 2000. Concordance of the toxicity of pharmaceuticals in humans and in animals. *Regul. Toxicol. Pharmacol.* 32:56–67.

Pieters, M.N., H.J. Kramer, and W. Slob. 1998. Evaluation of the uncertainty factor for subchronic-to-chronic extrapolation: Statistical analysis of toxicity data. *Regul. Toxicol. Pharmacol.* 27:108–111.

Rennen, M.A.J., B.C. Hakkert, H. Stevenson, and P.M.J. Bos. 2001. Data-base derived values for the interspecies extrapolation. *Comm. Toxicol.* 7:423–426.

Renwick, A.G. 1991. Safety factors and establishment of acceptable daily intakes. *Food Addit. Contam.* 8:135–150.

Renwick, A.G. 1993. Data-derived safety factors for evaluation of food additives and environmental contaminants. *Food Addit. Contam.* 10:275–305.

Renwick, A.G. 1995. The use of an additional safety or uncertainty factor for nature of toxicity in the estimation of acceptable daily intake and tolerable daily intake values. *Regul. Toxicol. Pharmacol.* 27:3–20.

Renwick, A.G. 1999. Subdivision of uncertainty factors to allow for toxicokinetics and toxicodynamics. *Hum. Ecol. Risk Assess.* 5:1035–1050.

Renwick, A.G. and N.R. Lazarus. 1998. Human variability and noncancer risk assessment—An analysis of the default uncertainty factor. *Regul. Toxicol. Pharmacol.* 27:3–20.

Renwick, A.G., J.L. Dorne, and K. Walton. 2000. An analysis of the need for an additional uncertainty factor for infants and children. *Regul. Toxicol. Pharmacol.* 31:286–296.

RIVM. 2007. RIVM Interspecies info website. http://www.rivm.nl/interspeciesinfo/about

Sharratt, M. 1988. Assessing risks from data on other exposure routes. Possibilities and limitations of using toxicity data derived from one exposure route in assessing risks from other exposure routes. *Regul. Toxicol. Pharmacol.* 8:399–407.

Slob, W. and M.N. Pieters. 1998. A probabilistic approach for deriving acceptable human intake limits and human risks from toxicological studies: General framework. *Risk Anal.* 18:787–798.

US-EPA. 1988. Recommendations for and documentation of biological values for use in risk assessment. EPA 600/6-87/008, NTIS PB88-179874/AS, February 1988. http://cfpub.epa.gov/ncea/cfm/recordisplay. cfm?deid = 34855

US-EPA. 1993. IRIS Reference dose (RfD): Description and use in health risk assessment. http://www.epa. gov/iris/rfd.htm

US-EPA. 1994. Methods for derivation of inhalation reference concentrations and application of inhalation dosimetry. EPA/600/8-90/066F. http://cfpub.epa.gov/ncea/raf/recordisplay.cfm?deid = 71993

US-EPA. 1996. Food Quality Protection Act of 1996. Public Law 104–170 104th Congress. http://www.epa. gov/pesticides/regulating/laws/fqpa/gpogate.pdf

US-EPA. 2002. *A review of the reference dose and reference concentration processes.* EPA/630/P-02/002F, December 2002, Final Report. Washington, DC: Risk Assessment Forum, U.S. Environmental Protection Agency. http://www.epa.gov/iris/RFD_FINAL[1].pdf

US-EPA. 2003. Glossary of Frequently Used Modeling Terms. http://www.epa.gov/ord/crem/library/CREM %20Modeling%20Glossary%2012_03.pdf

US-EPA. 2004. An examination of EPA risk assessment principles and practices. Staff paper prepared for the U.S. Environmental Protection Agency by members of the risk assessment Task Force. EPA/100/B-04/001 March 2004. http://www.epa.gov/OSA/ratf.htm

US-EPA. 2005. *Guidelines for Carcinogen Risk Assessment.* EPA/630/P-03/001F, March 2005. Washington, DC: Risk Assessment Forum, U.S. Environmental Protection Agency. http://cfpub.epa.gov/ ncea/cfm/recordisplay.cfm?deid = 116283

van Genderen, H. 1988. General conclusions of the chairman (presented at the workshop May 2–3, 1988 at the National Institute of Public Health and Environmental Protection, BA Bilthoven, The Netherlands). *Regul. Toxicol. Pharmacol.* 8:431–436.

Vermeire, T., H. Stevenson, M.N. Pieters, M. Rennen, W. Slob, and B.C. Hakkert. 1999. Assessment factors for human health risk assessment: A discussion paper. *Crit. Rev. Toxicol.* 29:439–490.

Vermeire, T., M. Pieters, M. Rennen, and P. Bos. 2001. *Probabilistic assessment factors for human health risk assessment: A practical guide.* RIVM report 601516 005, TNO report V3489. Bilthoven: RIVM. http:// www.rivm.nl/bibliotheek/rapporten/601516005.pdf

Vocci, F. and T. Farber. 1988. Extrapolation of animal toxicity data to man. *Regul. Toxicol. Pharmacol.* 8:389–398.

Voisin, E.M., M. Ruthsatz, J.M. Collins, and P.C. Hoyle. 1990. Extrapolation of animal toxicity to humans: Interspecies comparisons in drug development. *Regul. Toxicol. Pharmacol.* 12:107–116.

Walton, K., J.L. Dorne, and A.G. Renwick. 2001a. Uncertainty factors for chemical risk assessment: Inter-species differences in the in vivo pharmacokinetics and metabolism of human CYP1A2 substrates. *Food Chem. Toxicol.* 39:667–680.

Walton, K., J.L. Dorne, and A.G. Renwick. 2001b. Uncertainty factors for chemical risk assessment: Inter-species differences in glucuronidation. *Food Chem. Toxicol.* 39:1175–1190.

Walton, K., J.L. Dorne, and A.G. Renwick. 2001c. Categorical default factors for interspecies differences in the major routes of xenobiotic elimination. *Hum. Ecol. Risk Assess.* 7:181–201.

Walton, K., J.L. Dorne, and A.G. Renwick. 2004. Species-specific uncertainty factors for compounds elimin-ated principally by renal excretion in humans. *Food Chem. Toxicol.* 42:267–280.

Weil, C.S. and D.D. McCollister. 1963. Relationship between short- and long-term feeding studies in designing an effective toxicity test. *Agric. Food Chem.* 11:486–491.

WHO/IPCS. 1987. *Principles for the safety assessment of food additives and contaminants in food.* Environ-mental Health Criteria 70. Geneva: WHO. http://www.inchem.org/documents/ehc/ehc/ehc70.htm

WHO/IPCS. 1993. *Principles of evaluating chemical effect on the aged population*. Environmental Health Criteria 144, Geneva: WHO. http://www.inchem.org/documents/ehc/ehc/ehc144.htm

WHO/IPCS. 1994. *Assessing human health risks of chemicals: Derivation of guidance values for health-based exposure limits*. Environmental Health Criteria 170. Geneva: WHO. http://www.inchem.org/documents/ehc/ehc/ehc170.htm

WHO/IPCS. 1996. 12. *Chemical and physical aspects: Introduction*. In: *Guidelines for Drinking-Water Quality, Second Edition*. Volume 2: Health Criteria and other Supporting Information. Geneva: WHO.

WHO/IPCS. 1999. *Principles for the assessment of risks to human health from exposure to chemicals*. Environmental Health Criteria 210. Geneva: WHO. http://www.inchem.org/documents/ehc/ehc/ehc210.htm

WHO/IPCS. 2005. Chemical-specific adjustment factors for interspecies differences and human variability: Guidance document for use of data in dose/concentration–response assessment. Harmonization Project Document No. 2. Geneva: WHO. http://whqlibdoc.who.int/publications/2005/9241546786_eng.pdf

6 Standard Setting: Non-Threshold Effects (Carcinogenicity)

Adverse health effect(s) can be considered to be of two types (see Section 4.2): those considered to have a threshold, known as "threshold effects" (effects such as, e.g., organ-specific, neurological, immunological, non-genotoxic carcinogenicity, reproductive, developmental), and those for which there is considered to be some risk at any exposure level, known as "non-threshold effects" (effects such as, e.g., mutagenicity, genotoxicity, genotoxic carcinogenicity). Though it is not possible to demonstrate experimentally the presence or absence of a threshold, differences in the approach to the hazard assessment of threshold effects versus non-threshold effects have been adopted widely. The distinction in approaches is based primarily on the premise that simple events such as *in vitro* activation and covalent binding may be linear over many orders of magnitude, i.e., that these events occur even at very low exposure levels. However, a simple pragmatic distinction on this basis is increasingly problematic as it is likely that there is a threshold for a number of genotoxic effects.

In the hazard assessment process, described in detail in Chapter 4, all effects observed are evaluated in terms of the type and severity (adverse or non-adverse), their dose–response relationship, and the relevance for humans of the effects observed in experimental animals.

At present, there is no clear consensus on an appropriate methodology for the hazard characterization of non-threshold effects. Cancer risks have traditionally been assessed by a quantitative extrapolation by mathematical modeling of the experimental dose–response data to estimate the risk at much lower human exposure levels, i.e., low-dose risk extrapolation. The outcome of low-dose extrapolation is the resulting lifetime cancer risk associated with estimated exposure for a particular population. It should be recognized that such low-dose risk estimation is uncertain. Owing to this uncertainty, alternative measures of dose–response are increasingly being developed and adopted; for example, there is an increasing reliance on specification of the margin between potency in the experimental range and exposure as the measure of risk for carcinogens.

It should be recognized that the derivation of a "tolerable intake" for substances exerting non-threshold effects is generally considered as being inappropriate, as it has generally been assumed that there is no exposure level without a potential effect for such substances, i.e., there is considered to be some risk at any exposure level. However, there is an increasing and ongoing scientific debate regarding the sound rationale underlying the assumption of non-threshold effects. The term "tolerable intake" has therefore been kept for the outcome of the quantitative assessment in this chapter in order to acknowledge the emerging new approaches in the assessment of compounds that are both genotoxic and carcinogenic as well as to seek consistency in terminology. However, acknowledging the possible inappropriateness of the term "tolerable intake" in terms of the assessment of such substances, the term will be kept in quotation marks (i.e., "tolerable intake") throughout this chapter unless otherwise strictly stated.

The approach of deriving a "tolerable intake" for non-threshold effects has been described and discussed extensively in the scientific literature. It is beyond the scope of this book to review all these references. This chapter presents an overview of approaches for the derivation of a "tolerable intake" for the non-threshold carcinogenic endpoint, i.e., limited to address compounds with both direct DNA-acting genotoxic properties and carcinogenic properties, and is thus not meant to be exhaustive. The main focus is on the quantitative dose–response assessment, i.e., low-dose extrapolation approaches. Pertinent guidance documents and reviews for the issues addressed in this chapter include WHO/IPCS (1994, 1999), WHO (1996, 2000), US-EPA (2005), EU TGD (EC 2003), and Nielsen et al. (2005).

The approach of deriving a tolerable intake for threshold effects is addressed in Chapter 5.

The development of health-based guidance values, which are derived from the tolerable intake, is addressed in Chapter 9.

6.1 INTRODUCTION

According to the OECD/IPCS definitions listed in Annex 1 (OECD 2003):

Threshold is "Dose or exposure concentration of a substance below that a stated effect is not observed or expected to occur."

Unless a threshold mechanism of action is clearly demonstrated, mutagenicity and genotoxicity are considered as being "non-threshold effects". For compounds with both non-threshold genotoxic properties and carcinogenic properties (short form "genotoxic carcinogens" in this chapter), it is generally assumed that there is no level of exposure without a potential effect and, in theory, there will always be a risk associated with exposure even to the lowest exposure levels. It should be kept in mind, however, that both direct and indirect mechanisms of genotoxicity can be nonlinear (i.e., thresholded) and thus it is also recognized that for certain genotoxic carcinogens, a threshold may exist for the underlying genotoxic effect.

Cancer risk assessment is basically a two-step procedure, involving a qualitative assessment of how likely it is that a compound is a human carcinogen, and a quantitative assessment of the cancer risk that is likely to occur at given exposure levels and duration of exposure.

The qualitative assessment is based on all available information on carcinogenicity and is described in detail in Section 4.9.

Valid epidemiological studies are preferable for the quantitative risk assessment of genotoxic carcinogens for the purpose of deriving a "tolerable intake." If such data are available, for example in the working environment, they can be used quantitatively to convert work exposure to lifetime exposure, i.e., to convert intermittent exposure to continuous exposure (see Section 5.1 for adjustment of concentrations). However, as addressed in Chapter 3, valid human data are seldom available.

The quantitative risk assessment is therefore generally based on long-term studies with experimental animals. In these studies, the animals are exposed to a substance for the major part of their lifetime at high dose levels, so that a detectable and statistically significant tumor incidence can be produced. The aim of the quantitative risk assessment is to use such data to predict the risk to the general population posed by exposure to ambient levels of carcinogens. In general, therefore, quantitative risk assessment includes the extrapolation of risk from relatively high exposure levels (characteristic of animal studies) where cancer responses can be measured, to the usually much lower human exposure levels (characteristic for the environment of the general population) where cancer responses and thus risks are too small to be measured directly by experimental studies.

The quantitative risk assessment can be divided into at least three types of extrapolation:

- From animal to human (species-to-species)
- From effect to mechanism of action (model)
- From high dose to low dose (dose–response)

The first step, extrapolation of data from experimental animals to the human situation, is similar to the interspecies extrapolation described in detail for threshold effects (Section 5.3). The second step, evaluation of a carcinogen's mechanism(s) or mode of action(s), is very important for the choice of model for the risk assessment, i.e., non-threshold or threshold; this issue is addressed in Section 4.9. The third step, quantitative dose–response assessment, is the main focus of this chapter and is addressed in more detail in the following text.

6.2 QUANTITATIVE DOSE–RESPONSE ASSESSMENT: GENERAL ASPECTS

The quantitative dose–response assessment involves two different challenges, namely to determine the relationship between doses and the frequency of cases of cancer (i.e., potency evaluation), and to determine what statistical risk is tolerable or acceptable. This section gives a very short overview of some general aspects related to the quantitative dose–response assessment. The currently used approach by the WHO, the US-EPA, and the EU, as well as new approaches for the risk assessment of compounds that are both genotoxic and carcinogenic, are presented in Sections 6.3 and 6.4, respectively.

For most of the toxic effects that might be exerted by a chemical substance, including at least certain types of genotoxic carcinogenicity, the dose–response curve is S-shaped as illustrated in Figure 4.1 where the carcinogenic response is the tumor incidence. This means that no or only a few tumors occur at the lower dose levels, but the tumor incidence increases as the dose level increases, in many cases until a plateau is reached.

However, for non-threshold carcinogens it is generally assumed that there is a dose–dependent response at all doses above zero. The term linear, respectively nonlinear, is used in this context essentially for describing the dose–response characteristics for the low-dose region. The assumption of linearity (a straight line through origin, i.e., zero dose/zero response) at low doses does not necessarily imply linearity at higher doses near the experimental results. The term nonlinear for a dose–response relation is anticipated to display some curvature at low doses (e.g., sublinearity or supralinearity) instead of a straight line but still shows some dose-dependent response at all doses above zero. The term nonlinearity thus does not automatically imply a biological threshold dose. For a specified risk level, the assumption of low-dose sublinearity compared to the assumption of linearity implies a higher effective exposure level and thus a lower carcinogenic potency while supralinearity implies a lower effective exposure level and thus a higher carcinogenic potency. These various types of dose–response relationships at low doses are illustrated in Figure 6.1.

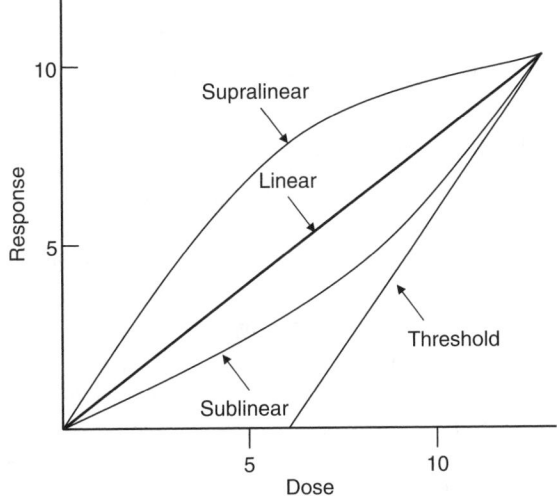

FIGURE 6.1 Various types of dose–response relationships at low doses.

The default approach for non-threshold carcinogens is to assume low-dose linearity. This default approach applies when the information on the mode of action is supportive of linearity or the mode of action is not sufficiently understood.

The concept of categorizing carcinogens into threshold carcinogens and non-threshold carcinogens is a pragmatic approach that simplifies the reality of dose–response relationships. The observed dose–response curve for tumor formation in some cases represents a single rate-determining step; however, in many cases it may be more complex and represent a superposition of a number of dose–response curves for the various steps involved in the tumor formation. It is therefore more realistic to assume that there is a continuum of shapes of dose–response relationships which cannot be easily differentiated by data and information usually available.

At present, there is no clear consensus on an appropriate methodology for the risk assessment of genotoxic carcinogens. A number of approaches based largely on characterization of dose–response have been adopted for the assessment of genotoxic carcinogens:

- Quantitative extrapolation by mathematical modeling of the dose–response curve to estimate the risk at likely human exposures, i.e., low-dose risk extrapolation
- Relative ranking of potencies in the experimental range
- Division of effect levels by an uncertainty factor (UF)

Dose–response assessment today is generally performed in two steps: (1) assessment of observed data to derive a dose descriptor as a point of departure and (2) extrapolation to lower dose levels for the tumor type under consideration. The extrapolation is based on extension of a biologically based model (see Section 6.2.1) if supported by substantial data. Otherwise, default approaches that are consistent with current understanding of mode of action of the agent can be applied, including approaches that assume linearity or nonlinearity of the dose–response relationship, or both. The default approach is to extend a straight line to the human exposure doses.

As genotoxic carcinogenicity generally is assumed to be a non-threshold effect, implying that there is a probability of harm at any level of exposure to a genotoxic carcinogen, the derivation of a "tolerable intake" is generally considered as being inappropriate. With regard to the quantitative risk assessment approach, two approaches of presenting risk-based considerations have been set forth:

- Lifetime cancer risk approach essentially results in risk estimates for specific scenarios presented as the lifetime cancer risk.
- MOE (Margin of Exposure) approach is formally similar to the MOS (Margin of Safety) approach for threshold effects (see Section 8.3.3 for MOS approach).

6.2.1 Low-Dose Risk Extrapolation

The outcome of low-dose extrapolation is the resulting lifetime cancer risk associated with estimated exposure for a particular population. A wide range of models have been developed for low-dose extrapolation of animal data to calculate a "tolerable intake" for an acceptable risk, often set at one extra cancer per million exposed persons (see Section 6.2.4 for acceptable risk).

Models have been developed that were suitable for extrapolating the effect of substances that act through one or more "hits" i.e., critical, genotoxic damage (DK NFA 1990). This group of models included the one-hit model (one-stage), two-hit (two-stage)...up to the multi-hit (multistage) models. The multi-hit models were generally believed to make best use of existing knowledge concerning the cancer process; however, in order for these models to be used properly, more experimental data than are normally available are required, so they had to be simplified by making use of any knowledge concerning the mechanism of action.

The multi-hit models are most suitable for extrapolating the effect of genotoxic substances. It is implicit in these models that all hits occur in one specific cell that only begins to divide and develop into a tumor when it has received the necessary number of hits. However, this is in poor agreement with experimental data, which show that proliferation of the cells that have had their first hit (the initiated cells) into pre-neoplastic lesions considerably increases the risk of a second hit in an initiated cell. While the one-hit model often oversimplifies the process, the multi-hit models impose an unreasonable tight restriction of the possibility of more than one critical hit affecting the same cell.

To take account of these problems, the so-called multistage models were developed. The original Armitage and Doll multistage mathematical model (published in 1954) assumed that the carcinogenic mechanism could be considered as a series of somatic mutations. After a cell has gone through these series of mutational stages, it became malignant and proceeded to develop into a tumor, see Figure 4.9. It was also assumed that several random hits or biological events were required in a specific sequence before a tumor developed. The model assumed mathematically that a carcinogen would affect at least one of the transitions between the different mutational stages. The original Armitage and Doll model was modified by making the assumption that the rates of occurrence of the different changes were all directly proportional to the dose of a carcinogen. This allowed the cumulative tumor incidence to be approximated by a relatively simple equation. This equation could then be further modified to take into account a background incidence of the changes in the absence of any dose of the chemical, i.e., to take into account spontaneous occurring tumors. The most widely used multistage model is the so-called Linearized Multistage (LMS) model (Section 6.2.1.2).

The choice of the extrapolation model depends on the understanding of the mechanisms of carcinogenic effects of the chemical under evaluation as the different models to various extents take into account the different stages in the process of carcinogenesis as well as essential biological parameters. The various models will not predict exactly the same lifetime cancer risk for a specific data set, as the models generally show different dose–response relationships in the low-dose area. One major drawback in low-dose extrapolation is that, because of the small number of doses in experimental studies (usually only 2 or 3 treated group and a control group), almost all data sets fit equally well various mathematical functions, and it is generally not possible to determine valid dose–response curves on the basis of mathematical modeling, see Section 4.2.5. Another drawback is that the slope of the dose–response curve and the mathematical model used are the decisive factors that determine the risk estimates when results are extrapolated over several orders of magnitude. No single extrapolation model can therefore be regarded as fully appropriate for low-dose extrapolation.

According to the WHO (WHO/IPCS 1994, 1999; WHO 1996, 2000), it should be noted that crude expression of risk in terms of excess incidence or numbers of cancers per unit of the population at doses or concentrations much less than those on which the estimates are based may be inappropriate, owing to the uncertainties of the quantitative extrapolation over several orders of magnitude. Estimated risks are therefore considered to represent only the plausible upper bounds and vary depending upon the assumptions on which they are based.

In view of the considerable uncertainties in the extrapolation of results over several orders of magnitude, specification of risks in terms of predicted incidence or numbers of excess cancers per unit of the population implies a degree of precision that is considered misleading by some. Larsen (2006), e.g., noted that the model most often used in low-dose extrapolation is a linear extrapolation from the observable range, and the apparent precision of the calculations does not reflect the uncertainty in the risk estimate; the results are therefore open to misinterpretation because the numerical estimates may be regarded as quantification of the actual risk.

The risk estimates should therefore not be regarded as being equivalent to the true cancer risk.

6.2.1.1 Linear Extrapolation

The dose–response relationship is by default assumed to be linear in the absence of mechanistic evidence to the contrary, at least in the observable range of the response, and risks are often linearly extrapolated into the low-dose range.

Models can predict low-dose linearity provided only that the response increases smoothly with dose. However, it is difficult to prove or disprove low-dose linearity experimentally even in bioassays involving extremely large numbers of animals. Often, linear extrapolation is criticized as being too conservative.

The concept of a linear relationship down to zero dose originates from studies on covalent binding to DNA of compounds that are both genotoxic and carcinogenic (Larsen 2006). DNA binding normally shows a linear dose–response relationship in the low-dose range, with no indication of a threshold. This has therefore been the argument to suggest a linear decrease of mutagenicity, and eventually of cancer risk, at low doses. However, a DNA adduct does not in itself have genetic consequences, but needs to be fixed into a mutation through DNA replication. The probability for a DNA adduct to be fixed is dependent on the rates of DNA repair and cell proliferation, which are influenced by dose. There is now an emerging scientific database on cell "household" mechanisms, like DNA repair, that indicates nonlinear dose–response relationships. In addition, endogenous physiological processes normally produce a relatively high level of DNA damage that is efficiently repaired. This suggests that the contribution of very low doses of compounds that are both genotoxic and carcinogenic to background damage may be negligible. As the high doses applied in carcinogenicity bioassays usually elicit significant toxicity with regenerative cell proliferation in target organs, simple linear extrapolation from experimental data to effects at low doses may lead to a considerable overestimation of the true incidence.

Three different approaches have been used or proposed by regulatory authorities in Europe and the United States as well as by the WHO. Previously the LMS model was the most widely adopted approach for low-dose extrapolations (Section 6.2.1.2), and has been used by the US-EPA (1986) for many years, as well as by the WHO in relation to derivation of drinking water guideline values for potential carcinogens (WHO 1996) (Section 6.3.1). More recently, an MOE approach has been recommended by the US-EPA (2005) based on curve fitting within the range of observation with extrapolation from a LED (the 95% lower confidence limit on a dose associated with an extra tumor risk) chosen to be representative of the lower end of the observed range (see Section 6.3.2). Within the EU chemicals regulation, a more simple approach based on the dose descriptor T25 has been proposed (see Section 6.3.3).

Similarly, in order to avoid any quantitative estimate, an MOE approach has been recommended by, e.g., JECFA (the Joint FAO/WHO Expert Committee on Food Additives) and EFSA (the European Food Safety Authority) in the assessment of compounds that are both genotoxic and carcinogenic by using a benchmark dose (BMD) approach to estimate the $BMDL_{10}$ (benchmark dose lower limit) representing the lower bound of a 95% confidence interval on the BMD corresponding to a 10% tumor incidence (see Section 6.4).

6.2.1.2 Linearized Multistage Model

The most widely used of the many mathematical models proposed for extrapolation of carcinogenicity data from animal studies to low-dose human exposures (i.e., low-dose extrapolation) is the LMS model. This has, in effect, become the default approach for quantitative risk assessment and has been used by, e.g., the US-EPA for many years as well as by the WHO in relation to derivation of drinking-water guideline values for potential carcinogens (WHO 1996) (see Section 9.2.1.2 for drinking-water guideline values).

The following overview of the LMS model is based primarily on the paper by Lovell and Thomas (1996).

In its 1986 "*Guidelines for Carcinogen Risk Assessment*," the US-EPA introduced the LMS model into the U.S. regulatory framework. The multistage model was chosen for regulatory purposes, because the model appeared to have parallels with biological explanations of cancer as a cell passing through a series of stages as initiation, promotion, and progression, see Figure 4.9). The LMS model introduced by the US-EPA was based on the multistage model of cancer originally developed by Armitage and Doll (in 1954) and modified by Crump and coworkers by a reformulation of the multistage model as a polynomial with respect to dose. This formulation assumed that all carcinogenesis operated by a common mechanism, and any carcinogen increased that part of the ongoing process. This formulation has been included in a number of software packages. The dose–response relationship is essentially linear at low doses.

This model assumes that any "dosage effect" has the same mechanism as that which causes the background incidence. Low-dose linearity follows directly from this additive assumption, provided that any fraction of the background effect is additive no matter how small. A "best fit" curve is fitted to the data obtained from a long-term rodent cancer bioassay using computer programs. The estimates of the parameters in the polynomial are called Maximum Likelihood Estimates (MLE), based upon the statistical procedure used for fitting the curve, and can be considered as "best fit" estimates. Provided the fit of the model is satisfactory, the estimates of these parameters are used to extrapolate to low-dose exposures.

The linear component of the LMS model, q_1 (i.e., one of the parameters of the polynomial), is approximately equivalent to the slope at low doses of the dose–response relationship between the tumor incidence and the dose. This linearity at low dose is a property of the formulation developed for the multistage model and is considered by proponents to be one of its important properties. This linear component of the polynomial, q_1, is used to carry out low-dose extrapolation. The linear response at low doses is considered to be conservative with regard to risk, as the dose–response relationship at low doses may well be sublinear. Although supralinearity at low doses cannot be excluded, it is usually considered to be unlikely.

The 95% confidence limits of the estimate of the linear component of the LMS model, q_1, can also be calculated. The 95% upper confidence limit is termed q_1^* and is central to the US-EPA's use of the LMS model in quantitative risk assessment, as q_1^* represents an upper bound or "worst-case" estimate of the dose–response relationship at low doses. It is considered a plausible upper bound, because it is unlikely that the true dose–response relationship will have a slope higher than q_1^*, and it is probably considerably lower and may even be zero (as would be the case if there was a threshold). Use of the q_1^* as the default, therefore, may have considerable conservatism incorporated into it. The values of q_1^* have been considered as estimates of carcinogenic potency and have been called the unit carcinogenic risk or the Carcinogen Potency Factor (CPF).

The estimates of the parameters, q_1 and q_1^*, are used to provide estimates either of the risks associated with specific doses, or conversely the dose associated with a specific increase in risk. The risk associated with a one in a million, 10^{-6}, extra lifetime incidence of cancer in the experimental species can be related to the dose; this dose is often referred to as the Virtually Safe Dose (VSD).

Some of the assumptions implicit in the LMS version of the multistage model used for regulatory purposes can clearly no longer be considered biologically realistic. These assumptions include acceptance that the order of the progression of the cell through the stages is fixed and irreversible, that the "waiting time" in the various stages are statistically independent and follow the exponential distribution when the exposure is constant, and that cells go through this progression independently of one another so that the effect of cell divisions is missed.

Analysis of simulated data using the LMS model performed by Lovell and Thomas (1996) showed

- That the MLE of the low-dose slope, q_1, was unstable and extremely sensitive to small changes in the data
- That the 95% upper confidence limit estimate, q_1^*, was insensitive with only small changes in values being obtained for large changes in the data
- Data sets where there was no statistical significance could give risk estimates similar to those obtained from data sets with clear dose-related effects

- Size of the values of the VSD obtained did not necessarily relate to the biological interpretation of the data sets
- Value of $q_1{}^*$ obtained was closely related to the top dose used in the study

6.2.2 Relative Ranking of Potencies

Animal studies performed with carcinogenic substances have shown that there can even be extremely wide differences in the exposure levels of various substances necessary to induce tumors in the animals, i.e., the potency in terms of inducing carcinogenicity can vary widely from substance to substance. There are no reasons to believe that humans should be different from experimental animals in this respect.

In order to express the carcinogenic response or potency, a dose descriptor is used, for example the Tumorigenic Dose (TD). The TD is often set at a defined incidence, for example 5%, the TD_5, defined as the dose (or concentration) associated with a 5% incidence of tumors. The dose descriptor can serve as the basis for development of an Exposure/Potency Index (EPI), which is the estimated daily human exposure divided by the TD. A calculated EPI of 10^{-6} for the TD_5 represents a one million-fold difference between the human exposure and that at the lower end of the dose–response curve, on which the estimate of potency is based.

The Carcinogenic Potency Database (CPDB) is a unique and widely used international database containing results from 6153 chronic, long-term animal cancer studies on 1485 chemicals (CPDB 2007), see also Section 4.9.8.

According to WHO/IPCS (1999), any model that fits the empirical data well is likely to provide a reasonable estimation of the TD_5. Choice of the model may not be critical since estimation is within the observed dose range, thereby avoiding the numerous uncertainties associated with low-dose extrapolation.

The value of 5% is arbitrary and selection of another value would not affect the relative potencies for each of a range of compounds. The TD_5 has been adopted as the measure of dose–response for assessment of genotoxic carcinogens under the Canadian Environmental Protection Act (Health Canada 1994, as cited in WHO/IPCS 1999). In the literature, others have proposed the TD_{25} and the TD_{50}. The Committee on Carcinogenicity of Chemicals in Food, Consumer Products and the Environment in the United Kingdom has concluded that the TD_{50} is the most practical quantitative estimate of carcinogenic potency for the ranking of genotoxic carcinogens (UK DOH 1995, as cited in WHO/IPCS 1999).

The US-EPA has in its 1996 "*Proposed Guidelines for Carcinogen Risk Assessment*" (US-EPA 1996) adopted the dose descriptor LED_{10} (the 95% lower confidence limit on a dose associated with a 10% extra tumor risk) whereas in its 2005 "*Guidelines for Carcinogen Risk Assessment*" (US-EPA 2005), no defined incidence has been recommended (see Section 6.3.2). Within the EU chemical's regulation, the dose descriptor T25 has been proposed (see Section 6.3.3). In the newly proposed MOE approach, the JECFA and the EFSA have recommended the dose descriptor $BMDL_{10}$ (see Section 6.4).

6.2.3 Division of Effect Levels by an Uncertainty Factor

An alternative approach to a quantitative assessment is to divide the highest dose at which there is no observed increase in tumor incidence in comparison with controls by a large composite UF, for example 5000 as suggested by Weil (1972). The magnitude of the factor could be a function of the weight of evidence, e.g., numbers of species in which the tumors have been observed or nature of the tumors (WHO/IPCS 1994). The adequacy of this approach, which is sometimes used when data on dose–response are limited, must be judged by criteria similar to those used in developing a tolerable intake for threshold effects; this is addressed in detail in Chapter 5.

It should be noted that this alternative approach in some aspects is comparable to the more simple T25 approach more recently proposed within the EU chemical's regulation (see Section

6.3.3) as well as to the MOE approach more recently recommended by the US-EPA (see Section 6.3.2), as well as by the JECFA and the EFSA (see Section 6.4).

6.2.4 Acceptable/Tolerable Lifetime Cancer Risk

The outcome of low-dose extrapolation is the resulting lifetime cancer risk associated with estimated exposure for a particular population. An acceptable/tolerable lifetime cancer risk is often used as a reference value to compare with the estimated lifetime cancer risk. The important question is thus: "Which lifetime cancer risk is acceptable/tolerable?"

According to the OECD/IPCS definitions listed in Annex 1 (OECD 2003):

Acceptable risk is "A risk management term. The acceptability of the risk depends on scientific data, social, economic, and political factors, and on the perceived benefits arising from exposure to an agent."

There are no regulations, neither nationally nor internationally, governing acceptable/tolerable lifetime cancer risks as this decision is a policy issue. The acceptable/tolerable lifetime cancer risk may therefore vary from one authority to another and might be dependent on the target population as well as on policy issues such as social, economic, and political factors. As an administrative practice, an acceptable/tolerable lifetime cancer risk has often been set as 10^{-6}, i.e., at one additional cancer case per million exposed persons.

The WHO considers in its drinking-water quality guidelines, in relation to genotoxic carcinogens, that a lifetime cancer risk for consumers of less than 10^{-5} represents a tolerable risk (WHO 1996). Guideline values associated with excess lifetime cancer risks of 10^{-4} and 10^{-6} are also presented for the genotoxic carcinogens to emphasize the fact that each country should select its own appropriate risk level.

In relation to their air quality guidelines, the WHO has estimated the risk associated with lifetime exposure to a certain concentration of a genotoxic carcinogen by linear extrapolation and the carcinogenic potency expressed as the incremental unit risk estimate. However, the guideline sections for carcinogenic pollutants also provide the concentration in air associated with an excess cancer risk of 10^{-4}, 10^{-5}, and 10^{-6}, calculated from the unit risk, in order to support authorities in the decision-making process.

In the development of national primary drinking water regulations under the Safe Drinking Water Act (SDWA) (see also Section 9.2.2.2), the US-EPA policy has been to set the Maximum Contaminant Level Goal (MCLG) at zero for chemicals with strong evidence of carcinogenicity associated with exposure from water (US-EPA 1998). For chemicals with limited evidence of carcinogenicity, including many Group C carcinogens (see Section 4.9.7.2), the MCLG is usually obtained using the RfD (Reference Dose) for that chemical based on its noncancer effects with the application of an additional UF of 1–10 to account for its possible carcinogenicity. If valid noncancer data for a Group C carcinogen are not available to establish an RfD but adequate data are available to quantify the cancer risk, then the MCLG is based upon a nominal lifetime excess cancer risk calculation in the range of 10^{-5}–10^{-6} (ranging from one case in a population of 100,000 to one case in a population of one million).

Some EU Member States have also applied similar lifetime cancer risk estimates in judging tolerable risk levels. There is as yet no EU harmonized view on such default risk estimates at a policy level, although the starting point for the derivation of limit values for the general population in relation to the EU directives on ambient air and drinking water quality is the 10^{-6} lifetime risk for genotoxic carcinogens.

6.3 QUANTITATIVE DOSE–RESPONSE ASSESSMENT: CURRENTLY USED APPROACHES

Previously the LMS model (Section 6.2.1.2) was the most widely adopted approach for low-dose extrapolations for data from studies in experimental animals. More recently, an MOE approach has been recommended by the US-EPA (Section 6.3.2), and the JECFA and the EFSA (Section 6.4). Within the EU chemicals regulation, a more simple approach based on the dose descriptor T25 has been adopted. This section gives a short overview of the currently used approaches in the WHO, the US-EPA, and the EU, as well as the new approach for the risk assessment of compounds that are both genotoxic and carcinogenic. The overview comprises a few selected, essential key references, and is not meant to be exhaustive. For more detailed information, the reader is recommended to read the key references.

6.3.1 WHO APPROACH: DRINKING WATER AND AIR QUALITY GUIDELINES

6.3.1.1 Drinking-Water Guidelines

In the derivation of drinking-water guideline values for potential carcinogens, a mathematical low-dose extrapolation is applied for the quantitative risk assessment (WHO 1996). The LMS model was generally adopted in the development of the drinking-water guideline values; however, other models were considered more appropriate in a few cases. The guideline values presented are the concentrations in drinking water associated with an estimated upper bound excess lifetime cancer risk of 10^{-5}, i.e., one additional cancer case per 100,000 of the population ingesting drinking water containing the substance at the guideline value for 70 years. Concentrations associated with estimated excess lifetime cancer risks of 10^{-4} and 10^{-6} can be calculated by multiplying or dividing the guideline value by 10, respectively; these values are also presented. See Section 9.2.1.2 for WHO drinking-water guideline values in general.

The WHO emphasized that guideline values for carcinogenic substances computed using mathematical models must be considered at best as a rough estimate of the cancer risk, as these models do not usually take into account a number of biologically important considerations, such as toxicokinetics, DNA repair, or immunological protection mechanisms. However, the models used are conservative and probably err on the side of caution.

In order to account for differences in metabolic rates between experimental animals and humans, a surface area to body weight correction (Section 5.3.2.2) is sometimes applied to quantitative estimates of cancer risk derived by low-dose extrapolation. The WHO stated that incorporation of this factor increases the risk by approximately one order of magnitude, depending on the species upon which the estimate is based, and increases the risk estimated on the basis of studies in mice relative to that in rats. The WHO considered incorporation of this factor to be overly conservative, particularly in view of the fact that linear extrapolation more likely overestimates risk at low doses. Therefore, the guideline values for carcinogens were developed on the basis of quantitative estimates of risk that were not corrected for the ratio of surface area to body weight.

6.3.1.2 Air Quality Guidelines

In the derivation of air quality guidelines for potential carcinogens, the WHO (2000), as a general rule, used quantitative assessment with low-dose risk extrapolation for compounds in IARC Groups 1 and 2A (see Section 4.9.7.1 for the IARC Groups). The risk associated with lifetime exposure to a certain concentration of a genotoxic carcinogen in the air was estimated by linear extrapolation and the carcinogenic potency expressed as the incremental unit risk estimate.

The incremental unit risk estimate for an air pollutant is defined as "the additional lifetime cancer risk occurring in a hypothetical population in which all individuals are exposed continuously from birth throughout their lifetime to a concentration of 1 $\mu g/m^3$ of the agent in the air they breathe." The unit risk estimates provide the opportunity to compare the carcinogenic potency of different compounds. It is stated that, by using unit risk estimates, any reference to the "acceptability" of risk is avoided, and that the decision on the acceptability of a risk should be made by national authorities within the framework of risk management. However, the guideline sections for carcinogenic pollutants also provide the concentration in air associated with an excess cancer risk of one in a population of 10,000, 100,000, or 1,000,000 (10^{-4}, 10^{-5}, and 10^{-6}), calculated from the unit risk, in order to support authorities in the decision-making process.

For those substances for which appropriate human studies are available, the so-called "average relative risk model" has been used. Quantitative assessments using this model comprises four steps: (1) selection of studies; (2) standardized description of study results in terms of relative risk, exposure level, and duration of exposure; (3) extrapolation towards zero dose; and (4) application to a general (hypothetical) population.

The relative risk is calculated as a measure of response and is then used to calculate the excess lifetime cancer risk expressed as unit risk (associated with a lifetime exposure to 1 $\mu g/m^3$).

The WHO has cautioned that the risk estimates presented should not be regarded as being equivalent to the true cancer risk, and that the crude expression of risk in terms of excess incidence or numbers of cancers per unit of the population at doses or concentrations much less than those on which the estimates are based may be inappropriate. Estimated risks are believed to represent only the plausible upper bounds, and may vary widely depending on the assumptions on which they are based.

6.3.2   US-EPA: GENERAL APPROACH

In 1996, the US-EPA published their "*Proposed Guidelines for Carcinogen Risk Assessment*" (US-EPA 1996). These Proposed Guidelines were a revision of the 1986 "*Guidelines for Carcinogen Risk Assessment*" (US-EPA 1986) and introduced, among others, a new approach for the quantitative risk assessment. A revised draft "*Guidelines*" was launched in 1999 (US-EPA 1999) and the final version was published in 2005 (US-EPA 2005).

The major change from the previous guidelines in terms of the quantitative risk assessment is that the LMS model no longer is the recommended default approach for low-dose extrapolation. Instead, an MOE approach is recommended based on curve fitting within the range of observation with extrapolation from a LED (the 95% lower confidence limit on a dose associated with an extra tumor risk) chosen to be representative of the lower end of the observed range.

The following overview of the US-EPA revised quantitative approach for cancer risk assessment is based on the final version of the "*Guidelines for Carcinogen Risk Assessment*" (US-EPA 2005).

In the absence of sufficient data to develop a robust, biologically based model for quantitative risk assessment, a single curve-fitting model for each type of data set is preferred. It is noted that many different curve-fitting models have been developed, and those that fit the observed data reasonably well may lead to several-fold differences in estimated risk at the lower end of the observed range. Therefore, the US-EPA uses a standard curve-fitting procedure for tumor incidence data.

The use of information on the mode of action in the assessment of potential carcinogens is a main focus of the revised cancer guidelines because of the significant scientific advances that have developed concerning the causes of cancer induction.

Dose–response assessment evaluates potential risks to humans at particular exposure levels. The approach to dose–response assessment for a particular agent is based on the conclusion reached as to its potential mode(s) of action for each tumor type. If the mode of action for known carcinogens is anticipated to be a DNA-reactive and direct mutagenic activity, such substances are assessed with a linear approach. Other modes of action may be modeled with either linear or nonlinear approaches after a rigorous analysis of available data.

Because an agent may induce multiple tumor types, the dose–response assessment includes an analysis of all tumor types, followed by an overall evaluation that includes a characterization of the risk estimates across tumor types, the strength of the mode of action information of each tumor type, and the anticipated relevance of each tumor type to humans, including susceptible populations and life stages (e.g., childhood).

Data from epidemiological studies, of sufficient quality, are generally preferred for estimating risks. When the evaluation is based on animal studies, the estimation of a human-equivalent dose should utilize toxicokinetic data for cross-species dose scaling if adequate data are available. Otherwise, a default procedure should be applied. The aim of the cross-species dose scaling is to define exposure levels for humans and animals that are expected to produce the same degree of effect, taking into account differences in scale between test animals and humans, such as size and life span.

For oral exposures, administered doses should be scaled from animals to humans on the basis of the caloric requirement approach (Section 5.3.2.3), i.e., body weight normalized by the 3/4 power. It is noted that the 3/4 power is consistent with current science, including empirical data that allow comparison of potencies in humans and animals, and it is also supported by analysis of the allometric variation of key physiological parameters across mammalian species. It is also noted that it is generally more appropriate at low doses, where sources of nonlinearity such as saturation of enzyme activity are less likely to occur.

For inhalation exposure, an appropriate default methodology estimates respiratory deposition of particles and gases, and estimates internal doses of gases with different absorption characteristics.

The principle underlying the cancer guidelines is to use approaches that include as much information as possible. One of the new principles in the revised cancer guidelines is to use quantitative information about key precursor events to develop a toxicodynamic model. When data for development of a toxicodynamic model are not available, empirical modeling (sometimes called "curve fitting") should be used in the range of observation. A model can be fitted to data on either tumor incidence or a key precursor event, and the analyses of data on tumor incidence and on precursor effects may be used in combination. To the extent the relationship between precursor effects and tumor incidence are known, precursor data may be used to estimate a dose–response function below the observable tumor data. Study of the dose–response function for effects believed to be part of the carcinogenic process influenced by the agent may also assist in the evaluation of the relationship of exposure and response in the range of observation and at exposure levels below the range of tumor observation.

A dose–response analysis is generally developed from each study that reports quantitative data on dose and response. A two-step approach distinguishes the analysis of the dose–response data. The first step is an analysis of dose and response in the range of observation of the experimental or epidemiological studies to yield a Point of Departure (POD). The second step is extrapolation to lower doses.

The first step of the dose–response assessment is the evaluation of the data within the range of observation. If there are sufficient quantitative data and adequate understanding of the carcinogenic process, a biologically based model may be developed to relate dose and response data. Otherwise, as a default procedure, a standard model can be used to curve-fit the data. For each tumor response, a POD from the observed data is estimated to mark the beginning of extrapolation to lower doses. The POD is an estimated dose (expressed in human-equivalent terms) near the lower end of the observed range, without significant extrapolation to lower doses.

The POD is used as the starting point for subsequent extrapolations and analyses. For linear extrapolation, the POD is used to calculate a slope factor, and for nonlinear extrapolation the POD is used in the calculation of a Reference Dose (RfD) or Reference Concentration (RfC). In a risk characterization, the POD is part of the determination of an MOE, defined as the ratio of the POD over an exposure estimate ($MOE = POD/Exposure$).

When tumor data are used, a POD is obtained from the modeled tumor incidences. Response levels at or below 10% can often be used as the POD. The POD alone, being a single-point estimate of a single dose–response curve, does not convey all the critical information present in the data from which it is derived. To convey a measure of uncertainty, the POD should be presented as a central estimate with upper and lower bounds. The POD for extrapolating the relationship to environmental exposure levels of interest, when the latter are outside the range of observed data, is generally the lower 95% confidence limit on the lowest dose level that can be supported for modeling by the data.

The second step of the dose–response assessment is an extrapolation to lower dose levels, i.e., below the observable range. The purpose of low-dose extrapolation is to provide as much information as possible about risk in the range of doses below the observed data. The most versatile forms of low-dose extrapolation are dose–response models that characterize risk as a probability over a range of environmental exposure levels. Otherwise, default approaches for extrapolation below the observed data range should take into account considerations about the agent's mode of action at each tumor site. Mode-of-action information can suggest the likely shape of the dose–response curve at these lower doses. Both linear and nonlinear approaches are available.

The linear approach should be used in two distinct circumstances: (1) When there are mode-of-action data to indicate that the dose–response curve is expected to have a linear component below the POD. Agents that are generally considered to be linear in this region include agents that are DNA-reactive and have direct mutagenic activity. (2) As a default when the weight of evidence evaluation of all available data is insufficient to establish the mode of action for a tumor site, because linear extrapolation generally is considered to be a health-protective approach.

For linear extrapolation, a line is drawn from the POD (from observed data), generally as a default, a LED (the 95% lower confidence limit on a dose associated with an extra tumor risk) chosen to be representative of the lower end of the observed range, to the origin (zero dose/zero response), corrected for background incidences. This implies a proportional (linear) relationship between risk and dose at low doses (note that the dose–response curve generally is not linear at higher doses). The slope of this line, known as the slope factor, is an upper-bound estimate of risk per increment of dose that can be used to estimate risk probabilities for different exposure levels. The slope factor is equal to $0.01/LED_{01}$ if the LED_{01} is used as the POD.

Unit risk estimates express the slope in terms of $\mu g/L$ drinking water or $\mu g/m^3$ (or ppm) air. In general, the drinking water unit risk is derived by converting a slope factor from units of mg/kg body weight per day to units of $\mu g/L$, whereas an inhalation unit risk is developed directly from a dose–response analysis using equivalent human concentrations already expressed in units of $\mu g/m^3$. Unit risk estimates often assume a standard intake rate (L/day drinking water or m^3/day air) and body weight (kg), which may need to be reconciled with the exposure factors for the population of interest in an exposure assessment (Section 7.3). Alternatively, when the slope factor for inhalation is in units of ppm, it may sometimes be termed the inhalation unit risk.

Risk-specific doses are derived from the slope factor or unit risk to estimate the dose associated with a specific risk level, for example a one-in-a-million (10^{-6}) increased lifetime risk. Risk below the POD is typically approximated by multiplying the slope factor by an estimate of exposure, i.e., Risk = Slope Factor × Exposure. For exposure levels above the POD, the dose–response model is used instead of this approximation.

A nonlinear approach should be selected when there are sufficient data to ascertain the mode of action and to conclude that it is not linear at low doses, and the agent does not demonstrate mutagenic or other activity consistent with linearity at low doses. The POD is in this case generally a BMDL when incidence data are modeled. A sufficient basis to support this nonlinear procedure is likely to include data on responses that are key events in the carcinogenic process. This means that the POD may be based on these precursor response data, for example hormone levels or mitogenic effects, rather than tumor incidence data. A nonlinear approach can be used to develop an RfD or an RfC. This approach expands such reference values, previously reserved for threshold effects, to include carcinogenic effects determined to have a nonlinear mode of action. A nonlinear approach should generally not be used in cases where the mode of action has not been ascertained.

If a nonlinear dose–response function has been determined, it can be used with the expected exposure to estimate a risk. If an RfD or RfC is calculated, the hazard can be expressed as a Hazard Quotient (HQ), defined as the ratio of an exposure estimate over the RfD or RfC, i.e., HQ = Exposure/(RfD or RfC).

Both linear and nonlinear approaches may be used when there are multiple modes of action. If there are, e.g., multiple tumor sites, one with a linear and another with a nonlinear mode of action, then the corresponding approach is used at each site. If there are, e.g., multiple modes of action at a single tumor site, one linear and another nonlinear, then both approaches are used to consider the respective contributions of each mode of action in different dose ranges. For example, an agent can act predominantly through cytotoxicity at high doses and through mutagenicity at lower doses where cytotoxicity does not occur. Modeling to a low response level can be useful for estimating the response at doses where the high-dose mode of action would be less important.

Where alternative approaches with significant biological support are available for the same tumor response and no scientific consensus favors a single approach, an assessment may present results based on more than one approach.

6.3.3   EU APPROACH: INDUSTRIAL CHEMICALS

Within the EU chemical's regulation, a more simple approach based on the dose descriptor T25 has been proposed as a basis for quantitative risk characterization of non-threshold carcinogens.

The following overview of the T25 approach is primarily based on the papers by Dybing et al. (1997) and Sanner et al. (2001).

The T25 was originally proposed as a simplified carcinogenic potency index as a practical method for potency considerations in carcinogen classification systems (Dybing et al. 1997) and is used within the EU context of classification and labeling of chemical substances (see Section 2.4.1.8) for inclusion of potency considerations in setting specific concentration limits for carcinogens in Annex I of Directive 67/548/EEC (EC 1999).

The T25 is defined as the chronic daily dose (in mg/kg body weight per day), which will give 25% of the animal's tumors at a specific tissue site, after correction for spontaneous incidence, within the standard lifetime of that species. It is a value calculated from a single observed dose–response and based upon the assumption of a linear dose–response relationship over the entire dose range.

According to Sanner et al. (2001), the use of the T25 has advantages in comparison to dose descriptors such as the TD_{50} and LED_{10} (US-EPA 1996). It does not require computer modeling. Also, T25 values, in contrast to TD_{50} values, were considered much more likely to be within the range of the experimental data and the use of data from the lowest dose giving a significant response should in most instances reduce the problem of intercurrent mortality to an acceptable degree.

Available animal carcinogenicity studies are compared and evaluated with regard to their adequacy for analysis of carcinogenic activity. The T25 is calculated from the data set and for a tumor site that is also used in the classification of the carcinogen (see Section 4.9.7.3). Malignant tumors as well as benign tumors that are suspected of possibly progressing to malignant tumors are taken into account in obtaining the T25 value.

The lowest tumorigenic doses showing a significant response (on statistical or biological basis) are generally used for obtaining the T25. However, if the tumor incidence at higher dose levels results in a lower T25 value, this latter value is used unless the higher tumor incidence is likely to be associated with increased general toxicity or local toxicity at the tumor site, which may have interfered with tumor formation, or there is evidence of nonlinearity. If more than one data set is available, data from the study giving the lowest T25 value are normally used, unless the data from a study giving a higher T25 value are judged to be more relevant.

The T25 value may either be incidentally obtained from the experimental study or calculated from other tumor incidences at the selected tumorigenic dose (determined above), using linear extrapolation, i.e., by multiplication of the dose with the factor 25/p where p is the actual tumor incidence (e.g., in case of a net 15% incidence, multiply by 25/15).

The T25 should be expressed in mg/kg body weight per day. To enable a conversion of feed, drinking water, or air concentrations of carcinogens to the T25, physiological parameters should be used, which normally are provided by the study itself; otherwise, default values for lifetime studies are used (see Section 7.4.3). In cases where only inhalation is relevant the unit mg/m^3 may be used directly without the need for conversion to mg/kg body weight per day.

The data for calculating the T25 should preferably be from lifetime oral studies or inhalation studies performed according to accepted test guidelines. If a study has been terminated before the standard life span of the species, the number of tumors found is assumed to be an underestimate of the number that would have been present after lifetime administration, and dose correction should be performed. If, e.g., dosing is terminated at w weeks ($w < 104$ weeks) before the standard life span of 104 weeks and the animals are observed until termination of the study at 104 weeks, the lifetime daily dose d giving the observed tumor incidence is corrected by $w/104$ (if the study was terminated at after 90 weeks ($w = 90$) and the tumor incidence is corrected by the factor "90/104"). Similarly, if animals are dosed 5 days per week, the daily dose giving the observed tumor incidence will be simply corrected by $(5/7) \times d$.

The animal dose descriptor T25 is converted to the corresponding human dose descriptor HT25, by dividing it with the appropriate scaling factor for interspecies dose scaling based on caloric requirement (see Section 5.2.3.2). The lifetime cancer risk due to a specific exposure is then obtained by linear extrapolation by dividing the relevant dose with the coefficient (HT25/0.25), or the daily lifetime dose that would represent a specific lifetime cancer risk is obtained by multiplying this specific lifetime risk with the coefficient (HT25/0.25).

Sanner et al. (2001) have evaluated the proposed T25 method by comparing risk estimates obtained with this method to those obtained by using the LMS method (Section 6.3.1) as well as the LED_{10} method proposed by the US-EPA in 1996 (Section 6.3.2). The comparisons included both genotoxic and non-genotoxic carcinogens, as the main purpose was to compare the methods when the same data set was used, as well as to evaluate the possible effects of different shapes of the dose–response curves.

The comparison of the T25 method with the LMS method showed a good correlation between the two methods (correlation coefficient of 0.85 in a log–log plot) for 33 substances identified in the US-EPA IRIS database. The ratios between the lifetime cancer risks calculated by the T25 method and the LMS method were in the range 0.5–2.0 for 30 out of the 33 substances (calculated for the 10^{-5} lifetime cancer risk). The distribution of the ratios was plotted and the parameters characterizing this distribution were estimated. The mean and the median were both 1.21, the 5th and 95th percentiles were 0.50 and 1.87, respectively, and the minimum and maximum values were 0.45 and 2.31, respectively. For 24 substances, the T25 method gave a higher result than the LMS method, and for the remaining 9 substances a lower result.

In order to compare the T25 method with the LED_{10} method, the T1 and the LED_1 (i.e., the dose expected to increase the tumor frequency by 1%) were obtained by dividing the dose descriptors by 25 and 10, respectively, and the ratio $T1/LED_1$ was calculated. The ratios were in the range 0.5–2.0 for 63 out of the 68 substances. The distribution of the ratios was plotted and the parameters characterizing this distribution were estimated. The mean and the median were both 1.25, the 5th and 95th percentiles were 0.54 and 1.98, respectively, and the minimum and maximum values were 0.5 and 2.3, respectively. The comparison of the T25 method with the LED_{10} method showed a very good correlation between the two methods (correlation coefficient of 0.94 in a log–log plot) for the 68 substances. For 44 substances, the T25 method gave a higher result than the LMS method, and for the remaining 21 substances a lower result.

The authors concluded that the results with the T25 method, which can easily be calculated without the need for computer programs, are in excellent agreement with results from the computer-based extrapolation methods, even though the T25 method only takes into consideration one single dose–response point. In order to overcome possible shortcomings, the authors suggested that the estimated risk figures should be accompanied by a commentary statement giving an overall evaluation of data that may have bearing on the carcinogenic risk, i.e., information from epidemiological studies, dose–response relationships, sites–species–gender activity, mechanistic information relevant to humans, toxicokinetics, structure–activity relationships, and other animal studies. The commentary statement should also indicate whether the real human risk is likely to be higher or lower than the calculated lifetime risk.

The T25 approach was discussed at a workshop organized by the European Centre for the Ecotoxicologicy and Toxicology of Chemicals (ECETOC) (Roberts et al. 2001, ECETOC 2002). It was concluded that the use of the T25 method in risk assessment is problematic due to uncertainties arising from the false assumption of both precision and linearity in the dose–response curves for tumor induction.

6.4 JECFA AND EFSA: NEW APPROACH, MARGIN OF EXPOSURE

The Joint FAO/WHO Expert Committee on Food Additives (JECFA) and the European Food Safety Authority (EFSA) do not establish health-based guidance values for compounds that are both genotoxic and carcinogenic using the threshold approach. Instead it is often advised that the exposure to compounds that are both genotoxic and carcinogenic should be as "low as reasonably achievable" (ALARA). Such advice is of limited value, because it does not take into account either human exposure or carcinogenic potency, and has not allowed risk managers to prioritize different contaminants or to target risk management actions. In addition, ever-increasing analytical sensitivity means that the numbers of chemicals with both genotoxic and carcinogenic potential detected in food will increase.

In order to avoid any quantitative estimate, the JECFA (2005) and EFSA (2005) have recommended to use an MOE approach in the assessment of compounds that are both genotoxic and carcinogenic.

The following overview is primarily based on the following references: JECFA (2005), EFSA (2005), and Larsen (2006).

According to the OECD/IPCS definitions listed in Annex 1 (OECD 2003):

Margin of Exposure is the "Ratio of the no-observed-adverse-effect level (NOAEL) for the critical effect to the theoretical, predicted or estimated exposure dose or concentration."

A related term is the margin of safety. According to the OECD/IPCS definitions listed in Annex 1 (OECD 2003), the margin of safety, for some experts, has the same meaning as the Margin of Exposure, while for others, the margin of safety means the margin between the RfD and the actual exposure dose or concentration.

The Margin of Exposure (MOE) in the context of the assessment of compounds that are both genotoxic and carcinogenic, as defined in EFSA (2005), is different from the OECD/IPCS definition given above: "The Margin of Exposure (MOE) is the ratio between a defined point on the dose–response curve (reference point) for the adverse effect of the compound in the animal carcinogenicity study and the estimated human intake of the compound."

When using the MOE approach, the following steps need to be taken:

- Selection of an appropriate point of comparison from the dose–response curve (reference point)
- Estimation of human dietary exposure
- Calculation of an MOE

To estimate the reference point, JECFA and EFSA recommended using the BMD approach (Section 4.2.5). The BMD is based on mathematical modeling being fitted to the experimental tumor data within the observed range and estimates the dose that causes a low but measurable response. The use of the $BMDL_{10}$ representing the lower bound of a 95% confidence interval on the BMD corresponding to a 10% tumor incidence (BMR_{10}) was recommended as a reference point on the dose–response curve. According to JECFA (2005), a BMR of a 10% incidence is likely to be the most appropriate for modeling of data from cancer bioassays, because the values for different mathematical models show wider divergence at incidences below 10%.

EFSA (2005) noted that in case the dose–response data are unsuitable for deriving a reliable estimate of the BMDL, the use of the T25 (Section 6.3.3) is recommended as the reference point.

Different exposure scenarios should be provided, e.g., for the whole population and for specific groups of the population depending on the compound considered, and its presence in the diet.

The MOEs are calculated by dividing the reference point on the dose–response curve (e.g., $BMDL_{10}$ or T25) by the estimated human intakes ($BMDL_{10}$ or T25/intake).

The following aspects have to be taken into account for the interpretation of an MOE:

- Interspecies and interindividual differences
- Nature of the carcinogenic process
- Reference point on the animal dose–response curve

For interspecies and interindividual differences, EFSA considered that, also for genotoxic and carcinogenic compounds, a default factor of 100 would take account for interspecies and interindividual differences (see Section 5.3 for interspecies differences and Section 5.4 for interindividual differences).

For the nature of the carcinogenic process, EFSA considered it appropriate that an additional factor of 10 would account for human variability in DNA-repair and cell cycle control.

For the reference point on the dose–response curve, EFSA considered that a factor of 10 would take into account that the $BMDL_{10}$ is not a surrogate for a threshold. If the reference point is based on the T25, which is less conservative than the $BMDL_{10}$, EFSA considered that an additional factor of 2.5 would be needed.

The overall MOE of 10,000 is obtained if based on the $BMDL_{10}$, or 25,000 if based on the T25.

The selection of an MOE that would be considered acceptable is a societal judgment and primarily the responsibility of risk managers rather than risk assessors. EFSA considered that an overall MOE of 10,000 for the $BMDL_{10}$, or 25,000 for the T25, would be of low health concern.

JECFA (2005) used this approach to evaluate the carcinogenicity of polycyclic aromatic hydrocarbons (PAH), acrylamide, and ethyl carbamate in food.

REFERENCES

CPDB. 2007: The carcinogenic potency database website. http://potency.berkeley.edu/cpdb.html

DK NFA. 1990. *Quantitative Risk Analysis of Carcinogens.* Søborg, Denmark: National Food Agency of Denmark, Institute of Toxicology, Ministry of Health.

Dybing, E., T. Sanner, H. Roelfzema, K. Kroese, and R.W. Tennant. 1997. T25: A simplified carcinogenic potency index: Description of the system and study of correlations between carcinogenic potency and species/site specificity and mutagenicity. *Pharmacol. Toxicol.* 80:272–279.

EC. 1999. *Guidelines for setting specific concentration limits for carcinogens in Annex I of directive 67/548/EEC. Inclusion of potency considerations. Commission working group on the classification and labeling of dangerous substances.* Brussel, 1999.

EC. 2003. Technical Guidance Document on Risk Assessment in support of Commission Directive 93/67/EEC on Risk Assessment for new notified substances, Commission Regulation (EC) No 1488/94 on Risk Assessment for existing substances and Directive 98/8/EC of the European Parliament and of the Council concerning the placing of biocidal products on the market. http://ecb.jrc.it/tgd

ECETOC. 2002. *The use of T25 estimates and alternative methods in the regulatory risk assessment of non-threshold carcinogens in the European Union.* Technical Report No. 83. Brussels: ECETOC.

EFSA. 2005. *Draft opinion on a harmonized approach for risk assessment of compounds which are both genotoxic and carcinogenic.* Request No EFSA-Q-2004-020, EFSA Scientific Committee, The European Food Safety Authority, 7 April 2005. Brussels: EFSA. http://www.efsa.eu.int/en/

JECFA. 2005. *Summary and conclusions of the Sixty-Fourth Joint FAO/WHO Expert Committee on Food Additives.* 8–17 February 2005. Rome: WHO. http://www.who.int/ipcs/food/jecfa/summaries/summary_report_64_final.pdf

Larsen, J.C. 2006. Risk assessment of chemicals in European traditional foods. *Trends Food Sci. Tech.* 17:471–481.

Lovell, D.P. and G. Thomas. 1996. Quantitative risk assessment and the limitations of the linearized multistage model. *Hum. Exp. Toxicol.* 15:87–104.

Nielsen, E., G. Østergaard, J.C. Larsen, and O. Ladefoged. 2005. *Principles for human health assessments of chemical substances in relation to the establishment of health based quality criteria for ambient air, soil and drinking water.* Environmental Project No. 974/2004. Copenhagen: Danish Environmental Protection Agency, Danish Ministry of the Environment (In Danish with a summary in English).

OECD. 2003. Descriptions of Selected Key Generic Terms Used in Chemical Hazard/Risk Assessment. Joint Project with IPCS on the Harmonisation of Hazard/Risk Assessment Terminology. OECD Series on Testing and Assessment No. 44. Environment Directorate, Joint Meeting of the Chemicals Committee and the Working Party on Chemicals, Pesticides and Biotechnology. ENV/JM/MONO(2003)15. Paris: OECD.

Roberts, R.A., K.S. Crump, W.K. Lutz, et al. 2001. Scientific analysis of the proposed uses of the T25 dose descriptor in chemical carcinogen regulation. An ECETOC Workshop overview. *Arch. Toxicol.* 75:507–512.

Sanner, T., E. Dybing, M.I. Willems, and E.D. Kroese. 2001. A simple method for quantitative risk assessment of non-threshold carcinogens based on the dose descriptor T25. *Pharmacol. Toxicol.* 88:331–341.

US-EPA. 1986. *Guidelines for Carcinogen Risk Assessment.* Federal Register 51(185):33992–34003. Washington, DC: Risk Assessment Forum. U.S. Environmental Protection Agency. http://www.epa.gov/ncea/raf/car2sab/guidelines_1986.pdf

US-EPA. 1996. *Proposed Guidelines for Carcinogen Risk Assessment* (April 23, 1996). Federal Register 61(79):17960–18011. Washington, DC: Office of Research and Development. U.S. Environmental Protection Agency. http://www.epa.gov/ncea/raf/pdfs/propcra_1996.pdf

US-EPA. 1998. *Ambient water quality criteria derivation for the protection of human health—Technical support document.* Final draft. United States Environmental Protection Agency, Office of Water 4304, EPA-822-B-98-005. Washington, DC: Office of Science and Technology. U.S. Environmental Protection Agency. 20460 http://www.epa.gov/waterscience/criteria/humanhealth/awqc-tsd.pdf

US-EPA. 1999. *Guidelines for Carcinogen Risk Assessment.* NCEA-F-0644, July 1999, Reviewed Draft. Washington, DC: Risk Assessment Forum. U.S. Environmental Protection Agency. http://www.epa.gov/iris/cancer_gls.pdf

US-EPA. 2005. *Guidelines for Carcinogen Risk Assessment.* EPA 630/P-03/001F, 2005. Washington, DC: Risk Assessment Forum. U.S. Environmental Protection Agency. http://www.epa.gov/iris/cancer032505-final.pdf

Weil, C.S. 1972. Statistics versus safety factors and scientific judgement in the evaluation for man. *Toxicol. Appl. Pharmacol.* 21:454–463.

WHO. 1996. *12. Chemical and physical aspects: Introduction.* In: *Guidelines for Drinking-Water Quality. Second Edition.* Volume 2: Health Criteria and other Supporting Information. Geneva: WHO.

WHO. 2000. *Criteria for carcinogenic endpoint.* In: *Air Quality Guidelines for Europe. Second Edition.* WHO Regional Publications, European Series, No. 91, 20–29 Copenhagen: WHO Regional Office for Europe. http://www.euro.who.int/document/e71922.pdf

WHO/IPCS. 1994. *Assessing human health risks of chemicals: Derivation of guidance values for health-based exposure limits.* Environmental Health Criteria 170. Geneva: WHO. http://www.inchem.org/documents/ehc/ehc/ehc170.htm

WHO/IPCS. 1999. *Principles for the assessment of risks to human health from exposure to chemicals.* Environmental Health Criteria 210. Geneva: WHO. http://www.inchem.org/documents/ehc/ehc/ehc210.htm

7 Exposure Assessment

The aim of the exposure assessment is to determine the nature and extent of contact with chemical substances experienced or anticipated under different conditions.

7.1 INTRODUCTION

According to the OECD/IPCS definitions listed in Annex 1 (OECD 2003a):

Exposure assessment is "Evaluation of the exposure of an organism, system or (sub) population to an agent (and its derivatives)."

Exposure Assessment is the third step in the process of risk assessment.

The term "human exposure" means contact with a substance and has been defined by US-EPA as taking place at the visible external boundary of the person, i.e., skin and openings into the body such as mouth and nostrils (US-EPA 1992).

An exposure assessment is the quantitative or qualitative evaluation of the amount of a substance that humans come into contact with and includes consideration of the intensity, frequency and duration of contact, the route of exposure (e.g., dermal, oral, or respiratory), rates (chemical intake or uptake rates), the resulting amount that actually crosses the boundary (a dose), and the amount absorbed (internal dose). Depending on the purpose of an exposure assessment, the numerical output may be an estimate of the intensity, rate, duration, and frequency of contact exposure or dose (the resulting amount that actually crosses the boundary). For risk assessments of chemical substances based on dose–response relationships, the output usually includes an estimate of dose (WHO/IPCS 1999).

In most cases, the substance coming into contact with the outer boundary of the body is contained in air, water, soil, or a consumer product. The substance concentration in these media at the point of contact (i.e., the exposure) is the concentration, on which exposure estimates are based. The most accurate exposure assessment would give information on the amount of a specific substance at the target site in the body where toxicity occurs, the biologically effective doses. Figure 7.1 illustrates the relationship between exposure and different types of dose. The applied dose is the amount of a substance at the absorption barrier (skin, lung, gastrointestinal tract) available for absorption. Usually it is very difficult to measure the applied dose directly at the absorption barrier. An approximation of the applied dose can be made using the potential dose, which is the amount of the substance ingested, inhaled, or in material applied to the skin. The applied dose may often be less than the potential dose if the substance is only partly bioavailable. The amount of a substance that has been absorbed and is available for interaction with the biologically target organs and tissues is called the internal dose. The amount transported to an individual organ, tissue, or fluid of interest (the target) is termed the delivered dose; the delivered dose may be only a part of the total internal dose. The biologically effective dose, or the amount that actually reaches cells, sites, or membranes where adverse effects occur, may only be a part of the delivered dose. Doses are often presented as dose rates, or the amount of a chemical dose (applied or internal) per unit time (e.g., mg/day), or per-unit-body-weight basis (e.g., mg/kg body weight per day) (WHO/IPCS 1999).

Currently, most risk assessments on environmental chemicals use dose–response relationships based on potential dose or internal dose, since the toxicokinetics necessary to base relationships on

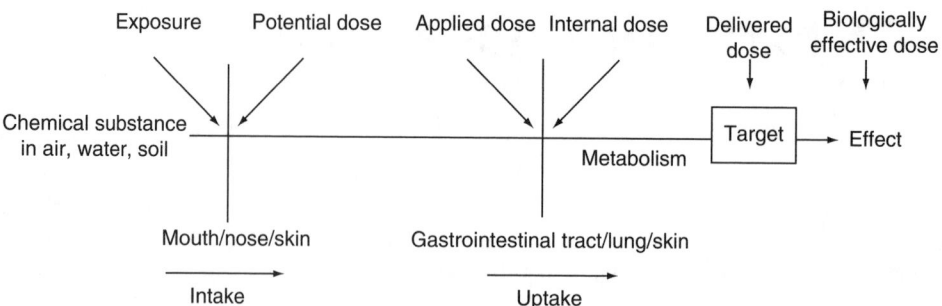

FIGURE 7.1 Inhalation - correlation between exposure concentration and internal dose. (Modified from US-EPA, *Exposure Factors Handbook*, National Center for Environmental Assessment, Washington DC, 1997.) Available at http://www.epa.gov/ncea/efh/

the delivered dose or biologically effective dose are not available. This may change in the future as more knowledge about the toxicokinetics of environmental chemicals becomes available.

The exposure assessment may be performed for a single situation, i.e., a single subgroup of humans exposed for a certain period of time via a particular route. An example of this could be an exposure estimate for children with respect to exposure to a contaminant via ingestion of play-ground dirt on a single incident.

Often, the exposure assessment is more comprehensive, including various subgroups of humans exposed for various lengths of time via several routes of exposure. A common categorization of humans is into the subgroups workers and consumers. Also "everybody" may be looked at as a category. With respect to the length of exposure, the two situations, exposure to a single high dose and exposure to a lower dose repeatedly for a longer time period, are commonly addressed. The routes of exposure commonly include oral ingestion, dermal contact, and via inhalation. Further routes may be included in special cases, such as intravenous, intramuscular, etc. where this is relevant for the risk assessment.

Exposure data can be either measured or calculated. Measured exposure data are preferred, if they are valid. Often measured exposure data are not available, or they are not considered valid, and therefore model-generated data must be used. An exposure model is "a conceptual or mathematical representation of the exposure process" (WHO/IPCS 2004); that is, a tool to calculate an estimate (a figure) to use in the risk characterization, e.g., where a NOAEL is compared with an exposure estimate. The output of an exposure model can be an exposure concentration; in practice, however, exposure often includes estimates of intake (e.g., amount of chemical inhaled or ingested) and the amount of a chemical that is absorbed into the body (e.g., amount of chemical that penetrates the skin or the lining of the lung) (WHO/IPCS 2005). Exposure models can be developed to estimate exposures and doses of individuals, defined population groups, or entire populations. Exposure may be estimated as a continuous variable or integrated over time ranging from minutes to a lifetime. The modeled outputs may include mean or median values, distribution parameters (standard deviations, quartiles, ranges), or entire distributions. Consequently, exposure models vary widely in complexity, approach, inputs, and outputs (WHO/IPCS 2005).

The approach to exposure assessment is not as internationally harmonized as hazard assessment. A synopsis of current activities regarding exposure assessment for industrial chemicals in a number of OECD Member countries has been published (OECD 2006). The executive summary of this document states that while there is a significant level of sharing of approaches used for hazard character-ization for risk assessment, this is not the case for exposure characterization. Although broad consistency in the overall approaches used by different countries in conducting exposure assessment exists, there is variation in policy-related factors, including the regulatory context for assessment and the way that information is applied, as well as in the types of approaches and tools used.

7.2 GUIDELINES AND GUIDANCE DOCUMENTS

7.2.1 WHO

A WHO/IPCS criteria document on human exposure assessment presents the concepts, rationale, and statistical and procedural methodologies for human exposure assessment, but does not give detailed guidance on technical issues regarding instrumental and laboratory methods (WHO/IPCS 2000).

The International Programme on Chemical Safety (IPCS) has undertaken a project to harmonize approaches to the assessment of risk from exposure to chemicals through increased understanding. The project focuses on specific issues and attempts to achieve agreement on basic principles. Among the project's focus areas are exposure assessment and terminology for exposure assessment and risk assessment. The status for the activities of the harmonization project is published in a newsletter (WHO/IPCS 2007).

Under this project, an IPCS Harmonization Project Document on the Principles of Characterizing and Applying Human Exposure has been published (WHO/IPCS 2005). This document sets out the characteristics of exposure assessment models that should be described to aid in model selection by exposure assessors. The document summarizes current practice in exposure modeling and principles for evaluating exposure models, but does not provide a comprehensive list of existing exposure models. The focus of the document is on the discussion of general properties of exposure models and how they should be described. The characteristics of different modeling frameworks are examined, and 10 principles are recommended for characterization, evaluation, and use of exposure models in order to help model users select and apply the most appropriate models. The report also discusses issues such as validation, input data needs, time resolution, and extrapolation of the model results to different populations and scenarios.

Also under the IPCS harmonization project, a working group is preparing a harmonized set of principles for the treatment of uncertainty in exposure assessment. The document will review the types of uncertainty analyses used in exposure assessments, evaluate their effectiveness in giving decision-makers the types of information they need, and derive a set of principles for uncertainty analysis (WHO/IPCS 2006).

7.2.2 OECD

OECD has not published any guidelines for exposure assessment for workers or consumers.

An OECD Guidance Document on Reporting Summary Information on Environmental, Occupational and Consumer Exposure (OECD 2003b) provides guidance for the reporting of summary exposure information (quantitative and qualitative), which can be used in various chemical assessment programs.

The purpose of the guidance document is to:

- Enhance consistency in reporting summary exposure information
- Support reporting of different levels of information (e.g., qualitative, screening level, comprehensive) as appropriate to the purpose of the report and the data available to the submitter
- Harmonize definitions
- Promote transparency of reported exposure information
- Provide clarity on the purpose and coverage (or scope) of information reported
- Provide a consistent approach for describing the reliability of the information

Three formats for reporting summary information on environmental, occupational, and consumer exposure to agents are provided, together with guidance for completing and using the formats. The formats are to be used for the reporting of post Screening Information Data Set (SIDS) exposure information collected in the assessment of high production volume chemicals.

OECD also provides advice on the development of so-called Emission Scenario Documents (ESDs) in its Guidance Document on Emission Scenario Documents (OECD 2000a). An ESD is a document that describes the sources, production processes, pathways, and use patterns with the aim of quantifying the emissions (or releases) of a chemical into water, air, soil, and/or solid waste. ESDs are used in risk assessment of chemicals to establish the conditions on use and releases of the chemicals that are the bases for estimating the concentration of chemicals in the environment. ESDs are already widely used in national and regional contexts; information hereon is compiled in the OECD Database on Use and Releases of Chemicals (OECD 2007). The ESD guidance document is currently being revised. Also, the Task Force on Environmental Exposure Assessment is developing a matrix of emission estimation methods included in existing ESDs.

The OECD Task Force on Environmental Exposure Assessment is developing OECD-wide ESDs, in order to make it possible to reflect conditions on production, use, etc. that are different between countries, and to avoid duplicative efforts by Member countries and industry in gathering exposure information. The first process for developing OECD-wide ESDs is a submission of a project proposal from member countries. The proposal specifies the industry categories and/or use categories that the documents will cover. Once approved by the Task Force, the lead country drafts the document, and the draft will then be circulated to the member countries for their comments. Taking the comments into account, the drafts are amended and published by the OECD.

7.2.3 US-EPA

Various documents have been published by the US-EPA to provide guidance for exposure assessment.

7.2.3.1 US-EPA Guidelines for Exposure Assessment

In 1992, US-EPA published *Guidelines for Exposure Assessment* (US-EPA 1992) that were intended to apply generically to noncancer risk assessments. The Guidelines describe the general concepts of exposure assessment, and provide guidance on the planning and conducting of an exposure assessment. Guidance is also provided on presenting the results of the exposure assessment and characterizing uncertainty. Although these Guidelines focus on exposures of humans to chemical substances, much of the guidance also pertains to assessing wildlife exposure to chemicals, or human exposures to biological, noise, or radiological agents. The Guidelines include a glossary, which helps standardize terminology used by the US-EPA in exposure assessment. They emphasize that exposure assessments performed as part of a risk assessment need to consider the hazard identification and dose–response parts of the risk assessment in the planning stages of the exposure assessment so that these three parts can be smoothly integrated into the risk characterization. The Guidelines discuss a number of approaches and tools for exposure assessment and their appropriate use. The Guidelines also stress that exposure estimates along with supporting information will be fully presented in US-EPA risk assessment documents, and that US-EPA scientists will identify the strengths and weaknesses of each assessment by describing uncertainties, assumptions, and limitations, as well as the scientific basis and rationale for each assessment.

7.2.3.2 US-EPA Guidance for Exposure Assessment

Guidance for assessment of exposure in US-EPA exposure assessments can be found in US-EPA's 1995 "Guidance for Risk Characterization" (US-EPA 1995).

The US-EPA's Office of Pollution Prevention and Toxics (OPPT) uses a tiered approach to exposure assessment (US-EPA 2007a). Exposure assessments may use measured data or model estimates. Representative measured data of known quality are preferred over model estimates and are needed to validate and improve models. The US-EPA *Guidelines for Exposure Assessment* include guidance on the collection and use of monitoring data for exposure assessments (Section 7.2.3.1). The approach should include the development of an estimate having an acceptable amount of uncertainty. In general, estimates based on quality-assured measurement data, gathered to directly answer the questions of the assessment, are likely to have less uncertainty than estimates based on indirect information (e.g., modeling or estimation approaches). For risk assessment purposes, a quantitative exposure assessment approach is needed and exposure information must be clearly linked to the hazard identification and dose–response relationship. The steps in the tiered approach are as follows:

Step 1: Gather basic data and information for a complete and transparent exposure assessment
Step 2: Develop a screening level exposure assessment
Step 3: If needed, develop an advanced exposure assessment

These steps are explained in more detail below.

7.2.3.2.1   Basic Data and Information for a Complete and Transparent Exposure Assessment
7.2.3.2.1.1   Manufacturing/Processing/Use
The first step in assessing exposure for a chemical is to identify all of the manufacturing, processing, and use activities for the chemical. This would include identifying all industrial, commercial, and consumer uses.

7.2.3.2.1.2   Gather Measured Data
Monitoring or measured data may be available in a variety of resources, such as company records or databases, national databases, studies published in the open literature, references, and other resources (e.g., for physico-chemical properties, fate, exposure factors, etc.). When obtaining measured or monitoring data, it is important to obtain all of the needed supporting information. Information on data quality objectives, the sampling plan, use of quality assurance samples, measurement of background levels, establishment and use of quality assurance and quality control measures, and selection and validation of analytical methods are important considerations when evaluating monitoring data or determining a strategy to collect additional monitoring data. The US-EPA *Guidelines for Exposure Assessment* include additional information on these important considerations, see Section 7.2.3.1.

7.2.3.2.1.3   Estimates of Environmental Releases
Environmental release estimates are critical inputs for models that calculate indirect human exposures via the environment such as through ambient air or drinking water. They are also critical to modeling exposures to nonhuman aquatic and terrestrial species. Release estimates may be site-specific or they may be generic for a particular industrial process or industrial use. Releases from consumer and commercial products should also be estimated if applicable.

7.2.3.2.1.4   Potentially Exposed Human Populations
All potentially exposed populations should be identified. The exposed populations should be associated with the activity, task, or source of environmental releases that lead to the exposure. Highly exposed or highly susceptible populations should be addressed whenever possible. All routes of exposure should be included.

7.2.3.2.1.5   Chemical Properties and Fate
Reliable, measured values are preferred, and should be used when available. Measured values or estimates of water solubility and vapor pressure are important in evaluating whether a chemical will dissolve in water or exist as a vapor at ambient temperature, and are used to estimate worker and consumer exposures. Measured data or estimates of biodegradation, sorption, and volatilization potential are used to predict removal in wastewater treatment. Information on decay rates in the

atmosphere, surface water, soil, and groundwater are important in evaluating how long it takes a chemical to break down in the environment, and are used to estimate exposures to the general population and the environment.

7.2.3.2.1.6   Mitigation of Exposures
Process and engineering controls, which are used to control exposures, should be identified. Personal protective equipment (PPE) that will mitigate occupational exposures should be noted and quantitative estimates of exposure with and without the use of PPE should be provided.

7.2.3.2.1.7   Documentation of Basic Data and Information
All measured data, environmental release scenarios, exposure scenarios, assumptions, and estimation techniques must be documented.

7.2.3.2.2   Screening Level Exposure Assessment
7.2.3.2.2.1   Purpose of a Screening Level Exposure Assessment
Screening level exposure assessments should be used to quickly prioritize exposures for further work.

7.2.3.2.2.2   Approach
A screening level exposure assessment will generate a quantitative conservative estimate of exposure. The screening approach generally involves using readily available measured data, existing release and exposure estimates, and other exposure related information. Where conservative estimates of exposure are not available, simple models, which often use generic scenarios and assumptions, may be used to fill in gaps. For example, a screening-level model for ambient air exposure that is using generic assumptions may assume that the exposed populations live near the chemical release locations.

The exposure assessment should include a characterization of the exposure estimates with respect to quality and relevance of exposure data, assumptions, major uncertainties, etc. Guidance for characterizing exposure in US-EPA exposure assessments can be found in "Guidance for Risk Characterization" (US-EPA 1995).

7.2.3.2.3   Advanced Exposure Assessment
Purpose of an Advanced Exposure Assessment
An advanced assessment will develop more accurate estimates of exposure and will generally focus on the higher priority exposures identified in screening activities.

7.2.3.2.3.1   Approach
An advanced exposure assessment should quantify central tendency (e.g., median, arithmetic mean) and high-end (i.e., greater than 90th percentile) exposures. A representative, well-designed monitoring study of known quality is the ideal. Information on data quality objectives, the sampling plan, use of quality assurance samples, measurement of background levels, establishment and use of quality assurance and quality control measures, and selection and validation of analytical methods are important considerations when evaluating monitoring data or determining a strategy to collect additional monitoring data. The US-EPA *Guidelines for Exposure Assessment* include additional information on these important considerations, see Section 7.2.3.1. Higher tier exposure models may also be used in advanced assessments. When they are used, every effort should be made to obtain accurate input data. For example, a higher tier model for ambient air exposure may use facility-specific parameters for emission rates, plant parameters such as stack height and exact location of the exposed populations.

The advanced exposure assessment should include a characterization of the exposure estimates, see Section 7.2.3.2.2.

7.2.3.2.3.2   General Notes
The approach described above is tailored to single chemical exposure assessments, although the general process could also be used for other types of hazards (e.g., biological hazards). Sometimes the focus of an exposure assessment will not be an assessment of human and ecological exposures to

a single chemical across manufacturing, processing, and uses. If the goal of the assessment is to identify safer substitutes for a particular use, the exposure assessment focus will be on all chemicals within that use (e.g., solvents used in a consumer product). In this case the basic data and information collected at the start of the assessment would need to be modified accordingly.

Exposure assessments may use measured data or model estimates. Representative measured data of known quality are preferred over model estimates and are needed to validate and improve models. OPPT encourages the appropriate use of the screening and higher tier models (US-EPA 2007a).

7.2.3.3 Other US-EPA Guidelines for Exposure Assessment

US-EPA guidelines for endpoint-specific risk assessments contain guidance on exposure issues of relevance for the particular endpoint. For example, in the US-EPA *Guidelines for Reproductive Toxicity Risk Assessment* (US-EPA 1996) exposure issues important to reproductive toxicity risk assessment are addressed, and a number of unique considerations regarding the exposure assessment for reproductive toxicity are discussed.

7.2.4 EU

7.2.4.1 Exchange and Assessment of Information on Consumer Exposure

The European Commission's Joint Research Centre (on behalf of DG SANCO) has started a project known as "European Information System on Risks from Chemicals Released from Consumer Products/Articles" (EIS-ChemRisks) (EU 2004), which is designed as a network to collect exposure data, exposure factors, exposure models, and health-related data. The overall objective is to develop tools and reference data to enable harmonized exposure assessment procedures in the EU. A toolbox has been designed to collect exposure information from four reference systems to systematically support exposure assessors in the EU:

- EU-ExpoFactors (European Reference System for Exposure Factors)
- ExpoData (Collection of Reference Exposure Data)
- ExpoHealthData (Reference Exposure-Associated Health Data)
- ExpoScenarios (Reference Exposure Scenarios)

The EIS-ChemRisks project includes sectoral projects that are focused on specific exposure scenarios, e.g., tattoos, textiles, toys, automobiles, etc. The project interfaces with the Consumer Exposure Modelling Task Force (CEM TF 2004), which was set up in November 2002 by the European Commission's Joint Research Centre. CEM TF's main objectives are to make a web-based and -managed inventory of existing consumer exposure modeling tools, to identify harmonization and validation needs for these models, to proceed with the harmonization and validation of an appropriately selected subset of models based on specific scenarios, to prepare a comprehensive overview of modeling approaches to estimate consumer exposure, and to create the Global Net on Consumer Exposure Models, a consortium of expert model developers and users from Europe, America, Canada, and Asia, aiming at harmonizing and validating existing consumer exposure models on the basis of common procedures and protocols.

7.2.4.2 EU Guidance Document for Exposure Assessment

The EU Technical Guidance Document (TGD) for risk assessment of new and existing substances and biocides (EC 2003) contains guidance on exposure assessment (Chapter 2). The core principles of human exposure assessments according to the TGD are: humans may be exposed to substances in the workplace (occupational exposure), from use of consumer products (consumer exposure), and indirectly via the environment. Guidance is given on how to perform an exposure assessment for each of these human populations. The guidance pertains to the general principles that apply, the data evaluation that needs to be performed, and to the way the actual quantitative assessment, based on either measured or modeled data, should be performed.

In a first screening step of the exposure assessment, the likelihood of an exposure of the three populations (workers, consumers, and man indirectly via the environment) to the substance under consideration has to be evaluated. If in the screening step it is indicated that exposure to one or more of the human populations does not occur or when the expected exposure is so low that it can be neglected further in the risk characterization phase, no further assessment is needed and the conclusion can be mentioned in the risk assessment report.

If actual or potential exposure has been identified, a quantitative exposure assessment is necessary. Exposure levels/concentrations for each potentially exposed population need to be derived from the available measured data and/or from modeling. A range of exposure values to characterize different subpopulations and scenarios may result. These results are taken forward to the risk characterization where they are combined with the results of the effects assessment in order to decide whether or not there is concern for the human population exposed to the substance. In some cases all three types of exposure estimates may contribute to an overall exposure value (combined exposure), which should be considered in the risk characterization.

In addition to the quantitative exposure estimates for each of the human populations (workers, consumers, and humans exposed via the environment), it may in some cases also be relevant to assess the combined exposure of humans via two or more routes. Workers may, for instance, be exposed in their private life to consumer products that contain the same substance as the products they are exposed to professionally. In addition, consumers may be exposed to substances via food packaging materials and at the same time be exposed to water and/or air that contain the substance as a result of (diffuse) environmental emissions. In calculating the actual combined exposure value care should be taken of the time scales at which the exposures occur. In general, combined exposure can be of particular relevance when long-term exposure to substances with widespread use and emissions occurs. General guidance on when these situations become relevant cannot be easily given. On a case-by-case basis the assessor needs to decide whether the combined exposure of one or more populations leads to different or additional conclusions regarding the risks of a specific substance under evaluation.

It may often be useful to initially conduct an exposure assessment based on "worst-case" assumptions, and to use default values when model calculations are applied. Such an approach can also be used in the absence of sufficiently detailed data. If the outcome of the risk characterization based on worst-case exposure assumptions is that the substance is "not of concern," the risk assessment for that substance can be stopped with regard to the effect/population considered. If, in contrast, the outcome is that a substance is "of concern," the assessment must, if possible, be refined using a more realistic exposure prediction in order to come to a definitive conclusion.

The following core principles relate to human exposure assessments that need to be carried out for new substances, existing substances, and biocides:

- Exposure assessments should be based upon sound scientific methodologies. The basis for conclusions and assumptions should be made clear and be supportable and any arguments should be developed in a transparent manner.
- Exposure assessment should describe the exposure scenarios of key populations undertaking defined activities. Such scenarios that are representative of the exposure of a particular (sub)population should, where possible, be described using both reasonable worst-case and typical exposures. The reasonable worst-case prediction should also consider upper estimates of the extreme use and reasonably foreseeable other uses. However, the exposure estimate should not be grossly exaggerated as a result of using maximum values that are correlated with each other. Exposure as a result of accidents or from abuse shall not be addressed.

- Actual exposure measurements, provided that they are reliable and representative for the scenario under scrutiny, are preferred to estimates of exposure derived from either analogous data or from the use of exposure models.
- Exposure estimates should be developed by collecting all necessary information (including that obtained from analogous situations or from models), evaluating the information (in terms of its quality, reliability, etc.), thus enabling reasoned estimates of exposure to be derived. These estimates should preferably be supported by a description of any uncertainties relevant to the estimate.
- In carrying out the exposure assessment the risk reduction/control measures that are already in place should be taken into account. Consideration should be given to the possibility that, for one or more of the defined populations, risk reduction/control measures which are required or appropriate in one use scenario may not be required or appropriate in another (i.e., there might be subpopulations legitimately using different patterns of control, which could lead to different exposure levels).

Exposure should normally be understood as external exposure, which can be defined as the amount of substance ingested, the total amount in contact with the skin (which can be calculated from exposure estimates expressed as mg/cm^2 or mg/cm^3), or either the amount inhaled or the concentration of the substance in the atmosphere, as appropriate. In cases where a comparison needs to be made with systemic effects data (e.g., when inhalation or dermal toxicity values are lacking or when exposures due to more than one exposure route need to be combined) the total body burden has to be estimated. Since the assessment of the amount that is absorbed after ingestion, by inhalation or by the skin is usually done in the effects assessment (section on toxicokinetics), this calculation of the total body burden is often placed in the section on risk characterization.

Exposure is considered as single events, or series of repeated events, or as continuous exposure. The duration and frequency of exposure, the routes of exposure, human habits and practices, as well as the technological processes need to be considered. Furthermore, the spatial scale of the exposure (e.g., personal/local/regional level) has to be taken into account.

7.2.4.3 The European Union System for the Evaluation of Substances

The European Union System for the Evaluation of Substances (EUSES) is a decision-support instrument, which enables government authorities, research institutes, and chemical companies to carry out rapid and efficient exposure assessments for chemical substances (ECB 2007).

The system is fully described in the extensive EUSES documentation and is based on the TGD (EC 2003).

The database IUCLID (Section 2.4.1.6) serves as the data source for the calculations to be carried out with EUSES.

As mentioned above, the exposure assessment of new and existing substances is carried out separately for three subgroups of the human population: workers, consumers, and man exposed indirectly via the environment (EC 2003).

Indirect exposure of humans via the environment may occur by consumption of food (fish, crops, meat, and milk) and drinking water, inhalation of ambient air, and ingestion of soil. For existing substance, measured levels in various environmental compartments may be available; however, for new substances, usually no relevant measured data are available and concentrations of a substance in the environment must be estimated.

The indirect exposure is estimated by the use of EUSES. EUSES estimates concentrations in food and the total daily intake of a substance based on predicted environmental concentrations for (surface) water, groundwater, soil, sediment, and ambient air. The indirect exposure is principally assessed on two spatial scales: locally near a point source of the substance, and regionally using averaged concentrations over a larger area. A third spatial scale, the continental scale, is also assessed by EUSES; however, this scale is not included in the estimations of the indirect exposure.

EUSES is intended mainly for the initial (screening) and intermediate (refined) stages of exposure assessments rather than comprehensive assessments (ECB 2007). On the basis of the screening, it can be decided whether more data need to be generated and whether a more refined assessment is necessary. EUSES can also be applied for refined assessments by allowing the replacement of default values, estimated parameter values, or intermediate results by more accurately estimated values or by measured data. EUSES is not specifically designed for site-specific assessments (defaults represent a standard region in EU), but adjustment of parameters may allow for insight into specific local or regional situations.

A new EUSES 2.0.3 version (dated 2005) has been updated according to the revision of the TGD. The documentation and the program can be downloaded (free of charge) from the ECB Web site (ECB 2007).

7.2.4.4 REACH: The New EU Chemicals Regulation

In REACH, the new EU chemicals regulation, which entered into force on 1 June 2007, detailed guidance documents on different REACH elements, including exposure assessment of chemical substances, are currently in preparation (spring 2007). These documents will probably be available on the EU DG Environment REACH Web site (EU 2006) when published.

7.2.5 EXPOSURE FACTORS, SOURCES

In order to determine the exposure of a population, it is necessary to have data about the activities that can lead to an exposure. These data are called exposure factors. They are generally drawn from the scientific literature or governmental statistics. For example, exposure factors may be information about amount of various foodstuffs eaten, breathing rates, or time spent for various activities, e.g., showering or car-driving. The main U.S. and EU sources of exposure factors will be described in the following text, and examples of human exposure factors are addressed in more detail in Section 7.3.

7.2.5.1 United States

The US-EPA *Exposure Factors Handbook* (US-EPA 1997), first published in 1989, provides a summary of the available data on consumption of drinking water; consumption of fruits, vegetables, beef, dairy products, and fish; soil ingestion; inhalation rates; skin surface area; soil adherence; lifetime; activity patterns; body weight; consumer product use; and the reference residence (data that are available on residence characteristics that affect exposure in an indoor environment).

The US-EPA *Child Specific Exposure Factors Handbook* (US-EPA 2006), first published in 2002, consolidates all children's exposure factors data into one document. The document provides a summary of the available and up-to-date statistical data on various factors assessing children's exposures. These factors include drinking water consumption; soil ingestion; inhalation rates; dermal factors including skin area and soil adherence factors; consumption of fruits, vegetables, fish, meats, dairy products, homegrown foods, and breast milk; activity patterns; body weight; consumer products; and life expectancy.

The US-EPA Consolidated Human Activity Database (CHAD) (US-EPA 2007b) contains data obtained from preexisting human activity studies that were collected at city, state, and national levels. CHAD is intended to be an input file for exposure/intake dose modeling and/or statistical analysis. CHAD is a master database providing access to other human activity databases using a consistent format. This facilitates access and retrieval of activity/and questionnaire information from those databases that US-EPA currently has access to and uses in its various regulatory analyses undertaken by program offices.

7.2.5.2 European Union

ECETOC (European Center for Ecotoxicology and Toxicology of Chemicals) has published a report with European exposure factors titled *"Exposure Factors Sourcebook for European Populations, with Focus on UK Data"* (ECETOC 2001). This document was the first broader compilation of European exposure factors and contains methodological information. The ECETOC sourcebook updates and builds upon other collections of exposure factor data including the 1994 *American Industrial Health Council's Exposure Factors Sourcebook* (AIHC 1994) and the US-EPA *Exposure Factors Handbook* (US-EPA 1997); the AIHC no longer exists and the sourcebook is not publicly available. The information in the ECETOC sourcebook includes physiological parameters (body weight, body surface areas, life expectancy), time–activity patterns (work hours, indoor/outdoor time, etc.), and receptor contact rates (soil ingestion rates, food consumption rates, etc.).

In 2002, the European Exposure Factors (ExpoFacts) database started as a 2-year project funded by CEFIC-LRI (European Chemical Industry Council, Long Range Research Initiative) to create a European database of factors affecting exposure to environmental contaminants. The aim was to create a public access data source, similar to the US-EPA *Exposure Factors Handbook* (US-EPA 1997), which has been widely used by European researchers, but with European data. Since 2006, the project is hosted by the European Commission's Joint Research Centre (JRC 2007).

When the project was started in 2002, European exposure factor data were scattered within numerous national and international institutions. ExpoFacts has created no new data, but instead compiled the existing data into one Internet database, where it can be easily found, screened, and downloaded from. Data were collected from the EU countries, candidate countries to EU, and EFTA* countries. As a result, the ExpoFacts database contains data from 30 European countries. In addition to the population time use patterns and exposure route information, e.g., dietary statistics, the database contains socio-demographic and physiologic information to enable database use as a tool for population-wide exposure modeling and risk assessment.

The methodological information, which is provided in the ECETOC sourcebook (ECETOC 2001), is not found in the ExpoFacts database. Therefore ExpoFacts does not replace the ECETOC sourcebook, but adds new information and Internet accessibility.

7.3 HUMAN EXPOSURE FACTORS, EXAMPLES

In the following sections, human exposure factors for ambient air (Section 7.3.1), soil (Section 7.3.2), and drinking water (Section 7.3.3) will be described. These media are used as examples, which serve to illustrate the differences in exposure factors provided by various exposure factor documents. Such differences can have a great impact on the risk characterization (Chapter 8) as well as on the development of regulatory standards and health-based guidance values (Chapter 9), and it is therefore important that the most relevant and reliable values are used for the particular situation.

7.3.1 Air

The general population can be exposed to chemical substances in indoor as well as in outdoor (ambient) air via inhalation of vapors, aerosols, and dusts in the air. The term "inhalation exposure" is defined as the concentration of a substance in inhaled air at the boundary of the body, and is expressed as an average concentration per unit time (e.g., mg/m^3 per day). In order to estimate a daily dose of a substance from the exposure concentration of the substance in the air, the inhalation rate is used. According to US-EPA (1997), the average daily dose (ADD) can be estimated from the exposure concentration by using the following equation:

$$ADD = [C \times IR \times ED]/[BW \times AT]$$

---

* European Free Trade Association, presently comprising Iceland, Liechtenstein, Norway and Switzerland. European countries, that are not EU Member States, can cooperate with EU via EFTA.

where

ADD = average daily dose (mg/kg body weight per day)
C   = concentration of a substance in inhaled air (mg/m^3)
IR   = inhalation rate (m^3/day)
ED   = exposure duration (days)
BW = body weight (kg)
AT   = averaging time (days), for noncarcinogenic effects AT = ED, for carcinogenic or chronic effects AT = 70 years or 25,550 days (lifetime)

The ADD is the dose rate averaged over a pathway-specific period of exposure expressed as a daily dose on a per-unit-body-weight basis (US-EPA 1997). The ADD is used for exposure to chemicals with noncarcinogenic or nonchronic effects. For compounds with carcinogenic or chronic effects, the lifetime average daily dose (LADD) is used. The LADD is the dose rate averaged over a lifetime. The C refers to the concentration of the contaminant in inhaled air. The ED refers to the total time an individual is exposed to an air pollutant.

In the case of inhalation, the situation is complicated by the fact that oxygen exchange with carbon dioxide takes place in the distal portion of the lung. The anatomy and physiology of the respiratory system diminishes the substance concentration in inspired air (potential dose) such that the amount of a substance that actually enters the body through the lung (internal dose) is less than that measured at the boundary of the body; this is illustrated in Figure 7.1 and explained in more detail in the introduction to this chapter. When performing risk assessments that concern the inhalation route of exposure, one must be aware if any adjustments have been employed in the estimation of the substance concentration to account for this reduction in potential dose.

The estimation of the applied dose for a given substance in the air is dependent on the inhalation rate (IR), commonly described as ventilation rate (VR) or breathing rate. Breathing rates are affected by numerous individual characteristics, including age, gender, body weight, health status, and levels of physical activity (working, running, walking, jogging, or resting). VR is usually measured as minute volume, which is the volume in liters of air exhaled per minute (V_E). V_E is the product of the number of breaths per minute and the volume of air respired in each breath, the tidal volume (V_T). Reported values for VR are based upon a number of methods, including direct measurements of respired volume by using a spirometer and a collection system, indirectly from heart rate (HR) measurements, or self-estimates. The direct method is not practical for normal living conditions as it requires that a person wear a device for monitoring airflow. For the indirect HR method, HR and VR are measured for a given individual under a variety of physical activities in a controlled setting; these measurements are used to develop equations relating VR to HR. HR is then recorded during normal activities and VR is calculated based on the VR–HR correlation equations.

7.3.1.1 WHO

WHO standard values for respiratory volumes (average figures) are those recommended by the International Commission on Radiological Protection (ICRP 1974 - cited in WHO/IPCS 1994, 1999). These values are shown in Table 7.1.

7.3.1.2 US-EPA

In the *Exposure Factors Handbook*, US-EPA (1997) has identified five key studies and five other studies on VRs. The results of these studies, which are all from the United States, show the following general tendencies:

- Children at rest inhale relatively larger amounts of air compared with adults
- Men inhale larger amounts of air than women
- Young people inhale larger amounts of air than older people

TABLE 7.1
WHO's Standard Values for Respiratory Volumes

	Respiratory Volume (average)
Long-term exposure (8 h resting, 16 h light/nonoccupational activity	
Child (10 years)	15 m^3/day
Adult man	23 m^3/day
Adult woman	21 m^3/day
Adult, average	22 m^3/day
Short-term exposure **Resting**	
Child (10 years)	2300 L/8 h
Adult man	3600 L/8 h
Adult woman	2900 L/8 h
Light/nonoccupational activity	
Child (10 years)	6240 L/8 h
Adult man	9600 L/8 h
Adult woman	9100 L/8 h

Source: Modified from WHO/IPCS, *Assessing Human Health Risks of Chemicals: Derivation of Guidance Values for Health-Based Exposure Limits*, Geneva, 1994. Available at http://www.inchem.org/documents/ehc/ehc/ehc170.htm; WHO/IPCS, *Principles for the Assessment of Risks to Human Health from Exposure to Chemicals*, Geneva, 1999. Available at http://www.inchem.org/documents/ehc/ehc/ehc210.htm

- Physical activity leads to increased ventilation (the relative difference between the individual groups mentioned in the first three points remains the same as described in these three points, also during physical activity)
- VR is higher outdoor than indoor, also when differences in physical activity are taken into consideration
- Asthmatics (children as well as adults) have an increased VR compared with healthy individuals of the same age groups

Table 7.2 presents selected results from one of the key studies (Layton 1993 - cited in US-EPA 1997). In this key study, three alternative energy-based approaches were used to estimate daily average VRs for different age/gender cohorts of the U.S. population (approximately 50,000 persons) over the entire lifetime: (1) average daily intakes of food energy from dietary surveys with an upward adjustment to account for underreporting of food consumption; (2) average daily energy expenditure based upon the ratio of total daily expenditure to basal metabolism; and (3) daily energy expenditures based upon time–activity data. Consistent results were obtained using these three different approaches.

The recommended VRs in the *Exposure Factors Handbook* (US-EPA 1997) are summarized in Table 7.3; the VR for each age group was calculated as the average of the VRs for the various activity levels, see Table 7.2.

The values for adults (11.3 m^3/day for women and 15.2 m^3/day for men), which are averages of the VRs provided for males and females in each of the three approaches of Layton (1993 - cited in US-EPA 1997), are lower than the default daily average VR of 20 m^3/day employed in IRIS

TABLE 7.2

Daily Inhalation Rates Calculated from Food-Energy Intakes

Age (years)	Body Weight (kg)	VR - average (m³/day)	VR - inactive[a] (m³/day)	VR - active (m³/day)
Children				
Less than 1	7.6	4.5	2.35	6.35
1–2	13	6.8	4.16	9.15
3–5	18	8.3	4.98	10.96
6–8	26	10	5.95	13.09
Men				
9–11	36	14	7.32	18.3
12–14	50	15	8.71	19.16
15–18	66	17	10.31	21.65
19–22	74	16	10.21	19.4
23–34	79	16	10.62	19.12
35–50	82	15	10.25	18.45
51–64	80	15	10.11	17.19
65–74	76	13	8.34	15.01
Over 75	71	13	8.02	15.24
Lifetime average		14		
Women				
9–11	36	13	6.63	16.58
12–14	49	12	7.61	15.20
15–18	56	12	8.14	13.84
19–22	59	11	7.68	12.29
23–34	62	11	7.94	12.7
35–50	66	10	7.80	11.7
51–64	67	10	7.86	11.8
65–74	66	9,7	7.10	10.65
Over 75	62	9,6	6.90	11.04
Lifetime average		10		

Source: Layton (1993, as cited in US-EPA 1997); modified from US-EPA, *Exposure Factors Handbook*, National Center for Environmental Assessment, Washington, DC, 1997. Available at http://www.epa.gov/ncea/efh

[a] Inactive, defined as sleep, was 8 h from 15 years and older, 9 h for children between the ages of 9 and 14 years, 10 h for children between the ages of 3 and 8 years, and 11 h for children less than 3 years old.

(Integrated Risk Information System, the US-EPA's database of human health effects that may result from exposure to various substances found in the environment). The VRs for children are also based on the key study results from Layton (1993 - cited in US-EPA 1997). For short-term exposures, a number of specific VRs are recommended, based on various patterns of activity. On the basis of the existing data it was not possible at the time to recommend upper values (95th percentiles) for the VRs.

7.3.1.3   EU

In the TGD (EC 2003) on risk assessment of new and existing substances and biocides, no exact standard values for VRs are recommended. However, in An appendix (Appendix IV C) regarding

TABLE 7.3
US-EPA's Recommended Standard Values for Ventilation Rate (VR) as Presented in the Exposure Factors Handbook

	VR - average
Long-term exposures	
Infants	
Less than 1 year	4.5 m^3/day
Children	
1–2 years	6.8 m^3/day
3–5 years	8.3 m^3/day
6–8 years	10 m^3/day
9–11 years male/female	14/13 m^3/day
12–14 years male/female	15/12 m^3/day
15–18 years male/female	17/12 m^3/day
Adults (over 19 years)	
Men	15.2 m^3/day
Women	11.3 m^3/day
Short-term exposures	
Adults	
Rest	0.4 m^3/h
Sedentary activities	0.5 m^3/h
Light activities	1.0 m^3/h
Moderate activities	1.6 m^3/h
Heavy activities	3.2 m^3/h
Children	
Rest	0.3 m^3/h
Sedentary activities	0.4 m^3/h
Light activities	1.0 m^3/h
Moderate activities	1.2 m^3/h
Heavy activities	1.9 m^3/h
Outdoor workers	
Average	1.3 m^3/h
Light activities	1.1 m^3/h
Moderate activities	1.5 m^3/h
Heavy activities	2.5 m^3/h

Source: Modified from US-EPA, *Exposure Factors Handbook*, National Center for Environmental Assessment, Washington, DC, 1997. Available at http://www.epa.gov/ncea/efh

physiological factors to be used in the risk assessment, the US-EPA's recommendations in the *Exposure Factors Handbook* (US-EPA 1997) are provided, see Table 7.3.

ECETOC (2001) has collected exposure data in the *Exposure Factors Sourcebook for European Populations*. For VRs, the Sourcebook presents the values, which are recommended by the US-EPA in the *Exposure Factors Handbook* (US-EPA 1997), see Table 7.3. ECETOC states that these values are probably representative of Europeans as well.

7.3.2 SOIL

The ingestion of soil is a potential source of human exposure to chemical substances. The exposure is usually expressed as an average amount of soil ingested per unit time (e.g., mg/day).

The potential for exposure to contaminants via soil is greater for small children because they are naturally curious and often examine new objects by putting them into their mouth, this mouthing behavior is considered to be a normal phase of childhood development. In addition to ingestion of soil through the mouthing of objects or hands, true eating of soil (geophagia) is also a source of oral exposure to chemicals for children. Furthermore, children tend to play on the ground and on the floor, which may also lead to an increased exposure to chemicals in soil and dust particles when compared to adults. Soil ingestion among children may be uneven as most children only ingest relatively small amounts, while a few children may consume larger amounts. Deliberate ingestion of larger amounts of soil is termed *pica* (which is the Latin word for the magpie, refers to "the persistent eating of nonnutritive substances"). Adults may also ingest soil and dust particles that adhere to food, cigarettes, or their hands thereby being exposed to contaminants in soil and dust particles.

Several studies have been conducted to estimate the amount of soil ingested by children. Most of the early studies have attempted to estimate the amount of soil ingested by measuring the amount of dirt present on children's hands and making generalizations based on behavior. More recently, soil intake studies have been conducted using a methodology that measures trace elements (e.g., aluminum, silicon, and titanium) in feces and soil. These trace elements are found in relatively high levels in soil, but not in foodstuffs, and are believed to be poorly absorbed in the gut. In the most recent studies, mass balance analyses have been performed, including measurements of trace elements in feces, soil, and foodstuffs. These measurements are used to estimate the amount of soil ingested over a specified time period.

7.3.2.1 WHO

WHO (1994) has specified the daily amount of soil ingested as 20 mg/day. This is a median value from Health and Welfare Canada (the national Department of Health of the Government of Canada). It has not been specified whether this standard value is valid for adults or for children.

7.3.2.2 US-EPA

In the *Exposure Factors Handbook*, US-EPA (1997) has identified seven key studies which employ methods based on measurements of tracer elements, and nine other studies on soil intake among children. The individual key studies used between three and eight different tracer elements. Not all the tracer elements turned out to be useful for estimation of the daily soil intake; aluminum, silicon, and yttrium appeared to be the most reliable. The results of the studies are summarized in Table 7.4.

A rather large variation in the results is obvious. The estimates of the daily average soil ingestion by children ranged from 39 to 271 mg/day with an average of 146 mg/day for ingestion of soil alone, and of 191 mg/day for combined ingestion of soil and dust. The 95th percentiles ranged from 106 to 1432 mg/day with an average of 383 mg/day for ingestion of soil alone, and of 587 mg/day for combined ingestion of soil and dust.

This variation may be a result of (1) a real difference in children's intake of soil, (2) that the intake of the various tracer elements are not only from soil, but also from food or other objects which children put into their mouth, (3) errors in sampling excreta, primarily feces, from the diaper, (4) tracer elements that are transferred to feces via contact with the diaper, e.g., from certain skin lotions, (5) a difference in absorption of the various tracer elements from the gastrointestinal tract, and (6) the collected soil samples that are usually inhomogeneous and not representative of an average exposure.

The values presented in Table 7.4 may possibly not be representative of long-term exposures as the duration of the studies was not sufficient in any of the studies in order to get a good estimate of the usual intake (sampling periods were between 3 and 8 days). The values presented in Table 7.4

TABLE 7.4

Estimates of Daily Soil Intake in Children

Average (mean) (mg/day)					95th Percentile (mg/day)				
Al	Si	AIR[a]	Ti	Y	Al	Si	TI	Y	Reference
181	184				584	578			Binder et al. (1986)
230		129							Clausing et al. (1987)
39	82		245.5						Davis et al. (1990)
64.5[b]	160[b]	268.4[b]							Davis et al. (1990)
153	154		218		223	276	1432	106	Calabrese et al. (1989)
154[b]	483[b]		170[b]	65[b]	478[b]	653[b]	1059[b]	159[b]	Calabrese et al. (1989)
122	139		271	165	254	224	279	144	Stanek and Calabrese (1995a)
133[c]					217[c]				Stanek and Calabrese (1995b)
69–120[d]									Van Wijnen et al. (1990)

Source: Modified from US-EPA, *Exposure Factors Handbook*, National Center for Environmental Assessment, Washington, DC, 1997. Available at http://www.epa.gov/ncea/efh

Note: All references cited in US-EPA (1997).

[a] AIR: acid-insoluble residue.
[b] Soil and dust combined.
[c] The so-called best tracer method (BTM) was used.
[d] Based on the so-called limiting tracer method (LTM).

are considered as being in the upper end for daily intake since all studies, with the exception of Calabrese et al. (1989 - cited in US-EPA 1997), were conducted during the summer when soil contact is more likely.

According to US-EPA (1997), data on soil ingestion rates for children who deliberately ingest soil (pica) are limited; however, it does not appear to be a very common behavior. This conclusion is based on the seven key studies used by US-EPA to estimate the daily soil intake in children, where only a single child, out of more than 600 examined, exhibited pica behavior directed toward soil. The results for this child are shown in Table 7.5.

TABLE 7.5

Daily Soil Ingestion by a Pica Child

Trace Element	Intake of Soil (mg/day) - week 1	Intake of Soil (mg/day) - week 2
Aluminum	74	13,600
Barium	458	12,088
Manganese	2,221	12,341
Silicon	142	10,955
Titanium	1,543	11,870
Vanadium	1,269	10,071
Yttrium	147	13,325
Zirconium	86	2,695

Source: From Calabrese et al. (1991, as cited in US-EPA 1997); modified from US-EPA, *Exposure Factors Handbook*, National Center for Environmental Assessment, Washington, DC, 1997. Available at http://www.epa.gov/ncea/efh

In the *Exposure Factors Handbook*, US-EPA (1997) has identified three studies where the daily soil intake for adults has been estimated. In one of these studies, an annual average soil intake of 60.5 mg/day was estimated based on assumptions regarding the amount of soil and dust on the hands, so-called "mouthing behavior", and indoor and outdoor activities. In the second study, a daily soil intake of 50 mg/day was estimated based on measurements of urinary arsenic, "mouthing behavior", and information about behavior patterns. In the third study, a daily soil intake of 30–100 mg/day was estimated based on tracer element measurements. The latter study was evaluated by the US-EPA as the most reliable one.

The US-EPA has generally used a standard value for soil intake among children of 200 mg/day. In the *Exposure Factors Handbook* (US-EPA 1997), a standard value of 100 mg/day for children up to 6 years is recommended as a best estimate taking into consideration that the highest values were seen with titanium, which may exhibit greater variability than the other tracers, and the fact that the Calabrese et al. (1989 - cited in US-EPA 1997) study included a pica child, see Table 7.4. However, US-EPA also recommended that a standard value of 200 mg/day can be used as a conservative estimate. Since the children were studied for short periods of time and the prevalence of pica behavior is not known, excluding the pica child from the calculations may underestimate soil intake rates. It is plausible that many children may exhibit some pica behavior if studied for longer periods of time. Over the period of study, upper percentile values ranged from 106 to 1432 mg/day with an average of 383 mg/day for soil ingestion and 587 mg/day for soil and dust ingestion, see Table 7.4. Rounding to one significant figure, the recommended upper percentile soil ingestion rate for children is 400 mg/day.

For pica behavior, an ingestion rate of 10 g/day is considered by US-EPA as a reasonable value for use in acute exposure assessments, based on the available information. It should be noted, however, that this value is based on only one pica child observed in the Calabrese et al. (1989 - cited in US-EPA 1997) study, see Table 7.5.

In the past, many US-EPA risk assessments have assumed an adult soil ingestion rate of 50 mg/day for industrial settings and 100 mg/day for residential and agricultural scenarios. These values are within the range of estimates from the three studies discussed above. Thus, 50 mg/day is considered by US-EPA still to represent a reasonable central estimate of adult soil ingestion and is the recommended value in the *Exposure Factors Handbook* (US-EPA 1997). This recommendation is commented by US-EPA as being clearly highly uncertain and considering the uncertainties in the central estimate, a recommendation for an upper percentile value is considered as being inappropriate.

Table 7.6 summarizes soil ingestion recommendations in the *Exposure Factors Handbook* (US-EPA 1997).

TABLE 7.6
US-EPA Standard Values for Soil Intake as Presented in the Exposure Factors Handbook

Population	Average	95th Percentile
Children	100 mg/day[a]	400 mg/day[b]
Adults	50 mg/day	—
Pica children	10 g/day[c]	—

Source: Modified from US-EPA, *Exposure Factors Handbook*, National Center for Environmental Assessment, Washington, DC, 1997. Available at http://www.epa.gov/ncea/efh

[a] 200 mg/day can be used as a conservative estimate.
[b] Short period of sampling; therefore the value is not an estimate for usual daily intake.
[c] Used for very short-term (acute) exposure. The value is based on a single pica child.

Recently, a U.S. study on daily soil ingestion estimates for children at a Superfund site has been published (Stanek and Calabrese 2000). The estimates are based on the results of an earlier investigation (Calabrese et al. 1997 - cited in ECETOC 2001), but with the use of improved methods. An average (mean) daily soil intake of 31 mg/day and a median of 17 mg/day were estimated. Furthermore 95th percentiles for soil intake during 7, 30, 90, and 365 days of 133, 112, 108, and 106 mg/day, respectively, were estimated.

7.3.2.3 EU

ECETOC (2001) has collected exposure data in the *Exposure Factors Sourcebook for European Populations*. In this Sourcebook, the median soil ingestion rates are estimated as 40 mg/day for children and 1 mg/day for adults. These estimates are based on two U.S. studies (children: Calabrese et al. 1997, adults: Stanek et al. 1997 - both cited in ECETOC 2001), which have been published after the publication of the US-EPA *Exposure Factor Handbook* (US-EPA 1997). ECETOC has decided to base their recommendations on these two more recent studies, as they consider these studies as being more reliable because of improved methods in study design and analysis. However, regarding the child study (Calabrese et al. 1997 - cited in US-EPA 1997), it is noted that the outdoor area was "a grassy Superfund site", and that the physical nature of the site and possible changes in activity patterns associated with Superfund designation could have led to a depression in soil ingestion rates in this study. As upper limits (95th percentiles) for soil ingestion, a value of 200 mg/day for children and 300 mg/day for adults is considered. The higher upper limit for adults compared with children is, according to ECETOC, related to data variability and indicates the uncertainty in current estimates.

7.3.3 DRINKING WATER

Drinking water is a potential source of human exposure to chemical substances, naturally occurring or contaminants. Contamination of drinking water may occur by, e.g., percolation of chemicals through the soil to ground water that is used as a source of drinking water; runoff or discharge to surface water that is used as a source of drinking water; intentional or unintentional addition of substances to treat water (e.g., chlorination); and leaching of materials from plumbing systems (e.g., lead). The exposure is usually expressed as an average amount of drinking water consumed per unit time (e.g., liter/day).

For the estimation of the magnitude of the potential dose of chemicals from drinking water, information on the quantity of water consumed per unit time is required. The intake of drinking water depends on age, level of physical activity (working, running, walking, or resting), and the ambient temperature.

7.3.3.1 WHO

In developing drinking water guideline values for potentially hazardous chemicals, WHO (1996) assumes a daily per capita consumption of 2 L of drinking water by a person weighing 60 kg. This standard value is considered to be on the safe side in most situations. But under certain circumstances, this assumption may underestimate the consumption of water per unit weight, and thus exposure, for those living in hot climates as well as for infants and children who consume more fluid per unit weight than adults. Where it was judged that children were at a particularly high risk from exposure to certain chemicals, the drinking water guideline value was derived on the basis of a 10-kg child consuming 1 L of drinking water per day, or a 5-kg infant consuming 0.75 L per day. The corresponding daily fluid intakes are higher than for adults on a body-weight basis.

7.3.3.2 US-EPA

In the *Exposure Factors Handbook*, US-EPA (1997) has identified three key studies and nine other studies on drinking water intake. The results of these studies, which apart from one are all from the United States, showed a good agreement across studies for the estimate of the mean and 90th percentile. The results also showed an increase in drinking water intake with age, with level of physical activity, and with ambient temperature, and US-EPA notes that based on body weight, children consume more water than adults, and old people consume more than middle-aged people. Table 7.7 presents the results of one central study (Ershow and Cantor 1989 - cited in US-EPA 1997). Table 7.8 summarizes the results of the three key studies and Table 7.9 the results of the nine other studies.

The US-EPA has in most risk assessments used a standard value for daily drinking water consumption of 2 L/day for adults, and 1 L/day for infants (up to 10 kg).

In the *Exposure Factors Handbook* (US-EPA 1997), an average daily drinking water consumption of 1.41 L/day for adults (age 19 years and above) is recommended; this value is the population-weighted mean of the two national studies (Ershow and Cantor 1989 and Canadian Ministry of Health and Welfare 1981 - both cited in US-EPA 1997). The average of the 90th percentile values from the same two studies (2.35 L/day) is recommended as the appropriate upper limit. It is stated that the commonly used standard value of 2 L/day corresponds to the 84th percentile of the intake rate distribution among the adults in the Ershow and Cantor (1989 - cited in US-EPA 1997) study. The standard values provided in the *Exposure Factors Handbook* (US-EPA 1997) are recommended

---

TABLE 7.7
Mean Daily Total Intake of Drinking Water[a]

Age (years)	Mean (ml/day)	Mean (ml/kg/day)	90th Percentile (ml/day)	90th Percentile (ml/kg/day)
Under 0.5	272	52.4	640	128.3
0.5–0.9	328	36.2	688	69.4
1–3	646	46.8	1162	82.1
4–6	742	37.9	1302	69.3
7–10	787	26.9	1338	47.3
11–14	925	20.2	1621	35.7
15–19	999	16.4	1763	29.0
20–44	1255	18.6	2121	32.2
45–64	1546	22.0	2451	35.5
65–74	1500	21.9	2333	35.2
over 75	1381	21.6	2170	33.9
Infants (<1)	302	43.5	649	101.8
Children (1–10)	736	35.5	1294	64.4
Teens (11–19)	965	18.2	1701	32.3
Adults (20–64)	1366	19.9	2268	33.7
Adults (>65)	1459	21.8	2287	34.7
All	1193	22.6	2092	39.8

Source: From Ershow and Cantor (1989, as cited in US-EPA 1997); modified from US-EPA, *Exposure Factors Handbook*, National Center for Environmental Assessment, Washington, DC, 1997. Available at http://www.epa.gov/ncea/efh

[a] Drinking water total is defined as the entire amount of drinking water in the household which is consumed directly as beverages or in the preparation of food and beverages.

TABLE 7.8

Daily Intake of Drinking Water. Results of the Key Studies Identified and Cited in the Exposure Factors Handbook

Age (years)	Mean (liter/day)	90th Percentile (liter/day)	Reference
Under 1	0.30	0.65	Ershow and Cantor (1989)
Under 3	0.61	1.50	Health and Welfare Canada (1981)
3–5	0.87	1.50	Health and Welfare Canada (1981)
1–10	0.74	1.29	Ershow and Cantor (1989)
6–17	1.14	2.21	Health and Welfare Canada (1981)
11–19	0.97	1.70	Ershow and Cantor (1989)
Total	1.38	2.41	Health and Welfare Canada (1981)
	1.41	2.28	Ershow and Cantor (1989)

Source: Modified from US-EPA, *Exposure Factors Handbook*, National Center for Environmental Assessment, Washington, DC, 1997. Available at http://www.epa.gov/ncea/efh

Note: All references cited in US-EPA (1997).

for chronic intake although the available studies have been performed for a shorter term exposure, but it is also stressed that the chronic 90th percentile may very well be larger than the recommended value of 2.35 L/day. It is also stated that, in keeping with the desire to incorporate body weight into exposure assessments without introducing extraneous errors, the values from the Ershow and Cantor (1989 - cited in US-EPA 1997) study expressed as ml/kg body weight per day are recommended in preference to the liter/day units. For adults, the mean and 90th percentile values are 21 ml/kg body weight per day and 34.2 ml/kg body weight per day, respectively.

For children, a general standard value is not recommended by US-EPA. Instead, age-specific standard values are recommended (see Table 7.10); these values are also based on the results of the two key studies.

Finally, specific standard values for pregnant and nursing women are also recommended, as well as for activity at various ambient temperatures (see Table 7.10).

The US-EPA recommendations for drinking water intake are summarized in Table 7.10.

7.3.3.3 EU

ECETOC (2001) has collected exposure data in the *Exposure Factors Sourcebook for European Populations*. In this Sourcebook, average values for drinking water intake of 1.1 L/day for adults and 0.5 L/day for children (1–11 years) are reported. These values are based on a British survey (Hopkins and Ellis 1980 - cited in ECETOC 2001). Selected data from this study are presented in Table 7.11.

7.4 STANDARD INTAKE (FEED, WATER, AIR) FOR EXPERIMENTAL ANIMALS

In most studies with experimental animals, the exposure to a specific substance is often expressed as a concentration of the substance in feed, drinking water, or air.

In order to compare (no) effect levels across studies within a species, or across studies in different species it is often necessary to estimate the average daily dose to which the animals have been exposed. In order to estimate the average daily dose, expressed as mg/kg body weight per day,

TABLE 7.9

Daily Intake of Drinking Water. Results from Other (Nonkey) Studies Cited in the Exposure Factors Handbook

Age (years)	Mean (liter/day)	90th Percentile (liter/day)	Reference
6–11 months	0.20		Pennington (1983)
Under 1	0.19		US-EPA (1984)
Under 1	0.32		Roseberry and Burmaster (1992)
2	0.50		Pennington (1983)
1–4	0.58		US-EPA (1984)
5–9	0.67		US-EPA (1984)
1–10	0.70		Roseberry and Burmaster (1992)
10–14	0.80		US-EPA (1984)
14–16	0.72		Pennington (1983)
15–19	0.90		US-EPA (1984)
11–19	0.91		Roseberry and Burmaster (1992)
Total			
	1.30	2.40	Cantor et al. (1987)
	1.63	—	NAS (1977)
	1.25	1.90	Gillies and Paulin (1983)
25–30	1.04	—	Pennington (1983)
60–65	1.26	—	Pennington (1983)
Over 20	1.04–1.47	—	US-EPA (1984)
20–64	1.37	2.27	Ershow and Cantor (1989)
Over 65	1.46	2.29	Ershow and Cantor (1989)
	1.15		USDA (1995)
	1.07	1.87	Hopkins and Ellis (1980)

Source: Modified from US-EPA, *Exposure Factors Handbook*, National Center for Environmental Assessment, Washington, DC, 1997. Available at http://www.epa.gov/ncea/efh

Note: All references cited in US-EPA (1997).

it is necessary to apply standard values for parameters such as body weight, feed intake, drinking water intake, and VR (inhaled volume of air per unit time) for the most commonly used laboratory animal species.

There are no internationally agreed standard values for these parameters, but various authorities and organizations have suggested standard values for a number of these parameters; this will be presented in the following sections.

7.4.1   OECD

In the "Guidance Notes for Analysis and Evaluation of Repeat-Dose Toxicity Studies" (OECD 2000b), OECD provides standard values for body weight, daily feed intake, and conversion factors which allow the conversion of the concentration of a substance in the feed, expressed as ppm or mg/kg, into a dose, expressed as mg/kg body weight per day. These values originally derive from Lehrman (1954 - cited in WHO/IPCS 1987) and have previously been published by the WHO (WHO/IPCS 1987).

TABLE 7.10
US-EPA's Standard Values for Intake of Drinking Water as Presented in the Exposure Factors Handbook

Age (years)	Average (mean)	50th Percentile	90th Percentile	95th Percentile
Less than 1[a]	0.30 L/day	0.24 L/day	0.65 L/day	0.76 L/day
	44 ml/kg/day	35 ml/kg/day	102 ml/kg/day	127 ml/kg/day
Less than 3[b]	0.61 L/day		1.5 L/day	
3–5[b]	0.87 L/day		1.5 L/day	
1–10[a]	0.74 L/day	0.66 L/day	1.3 L/day	1.5 L/day
	35 ml/kg/day	31 ml/kg/day	64 ml/kg/day	79.4 ml/kg/day
11–19[a]	0.97 L/day	0.87 L/day	1.7 L/day	2.0 L/day
	18 ml/kg/day	16 ml/kg/day	32 ml/kg/day	40 ml/kg/day
Adults (above 19 years)[a]	1.4 L/day	1.3 L/day	2.3 L/day	
	21 ml/kg/day	19 ml/kg/day	34 ml/kg/day	
Pregnant women[c]	1.2 L/day	1.1 L/day	2.2 L/day	2.4 L/day
	18.3 ml/kg/day	16 ml/kg/day	35 ml/kg/day	40 ml/kg/day
Nursing women[c]	1.3 L/day	1.3 L/day	1.9 L/day	2.2 L/day
	21.4 ml/kg/day	21 ml/kg/day	35 ml/kg/day	37 ml/kg/day

Adults[d], high activity, warm climate: 0.21–0.65 L/h depending on activity and temperature
Adults[e], active: 6 L/day (temperate climate) - 11 L/day (warm climate)

Source: Modified from US-EPA, *Exposure Factors Handbook*, National Center for Environmental Assessment, Washington, DC, 1997. Available at http://www.epa.gov/ncea/efh

[a] From Ershow and Cantor (1989 - cited in US-EPA 1997).
[b] From Canadian Ministry of Health and Welfare (1992 - cited in US-EPA 1997).
[c] From Ershow et al. (1991 - cited in US-EPA 1997).
[d] From McNall and Schlegal (1968 - cited in US-EPA 1997).
[e] From U.S. Army (1983 - cited in US-EPA 1997).

TABLE 7.11
Daily Intake of Drinking Water

Age (years)	Mean (ml/day)		10th Percentile (ml/day)		90th Percentile (ml/day)	
	Men	Women	Men	Women	Men	Women
1–4	477	464	170	150	850	890
5–11	550	533	220	220	900	930
12–17	805	725	290	310	1350	1160
18–30	1006	991	450	500	1620	1550
31–54	1201	1091	640	620	1880	1680
Over 54	1133	1027	620	540	1720	1570

Source: From Hopkins and Ellis (1980, as cited in ECETOC 2001); modified from ECETOC, *Exposure Factors. Sourcebook for European Populations, with Focus on UK Data*, Technical Report No. 79, 2001. Available at http://ecb.jrc.it/tgdoc/

TABLE 7.12

Approximate Relation between the Concentration of a Substance in Feed (ppm or mg/kg) and Dose (mg/kg Body Weight Per Day)

Animal Species	Body Weight (kg)	Grams Feed Consumed Per Day[a]	Type of Diet	Conversion Factor[b]
Mouse	0.02	3	Dry laboratory	0.150
Rat, young	0.10	10	Chow diets	0.100
Rat, older	0.40	20		0.050
Guinea pig	0.75	30		0.040
Rabbit	2.0	60		0.030
Dog	10	250		0.025
Dog	10	750	Moist semisolid diets	0.075
Cat	2	100		0.050
Monkey	5	250		0.050
Human	60	1500		0.025

Source: Modified from OECD, *Approaches to Exposure Assessment in OECD Member Countries: Report from the Policy Dialogue on Exposure Assessment in June 2005.* OECD Environment, Health and Safety Publications Series on Testing and Assessment No. 51, Paris, 2006.

[a] Excluding liquid.

[b] The figures in the column are the factors, which the feed intake (expressed as ppm or mg/kg) should be multiplied with in order to obtain the dose (expressed as mg/kg body weight per day). For example, for a rat (weight: 400 g), 1 ppm in the feed is equivalent to 0.05 mg/kg body weight per day.

The OECD conversion factors, which are summarized in Table 7.12, are used by, e.g., the EU's Scientific Committee for Food (SCF), the Joint FAO/WHO Expert Committee on Food Additives (JECFA), and the Joint FAO/WHO Meeting on Pesticide Residues (JMPR) in their evaluations of food additives, contaminants, and pesticides.

The OECD states, as also outlined in the Environmental Health Criteria Monograph No. 104 (WHO/IPCS 1990), if dietary intake is measured, then JMPR evaluations indicate that X ppm (mg/kg) in feed is equal to Y mg/kg body weight per day but if there is inadequate feed intake data and the tabulated conversion factors are used, then it is reported that X ppm (mg/kg) in the feed is equivalent to Y mg/kg body weight per day.

With respect to water consumption, the OECD (2000b) only provides a standard value for the rat: 24–35 ml/day. Assuming an average intake of 30 ml/day and a body weight of 400 g, the average daily intake corresponds to 75 ml/kg body weight per day. In order to estimate the average daily dose, expressed as mg/kg body weight per day, from a concentration of a substance in the drinking water, expressed as ppm or mg/liter, in long-term studies (90 days and more), a conversion factor of 0.075 (75 ml/kg body weight per day divided by 1 mg/ L) is estimated. This means that 1 ppm (mg/ L) in the drinking water corresponds to 0.075 mg/kg body weight per day.

7.4.2 UNITED STATES

The US-EPA's Recommendations for and Documentation of Biological Values for Use in Risk Assessment (US-EPA 1988) consists of an extensive compilation of values taken from published literature sources for life span, body weight, food consumption, water consumption and VRs for a

TABLE 7.13

Default Reference Values for Body Weights in Oral and Inhalation Toxicity Studies

| Type of Study | Sex | Body Weight | | | | |
		28 Days	90 Days	Chronic	Pregnant	Nursing
Rat (g)	M	250	325	475		
	F	175	200	275	300	300
	M + F	200	275	375		
Mouse (g)	M	30	35	45		
	F	25	30	35	35	35
	M + F	30	35	40		
Dog (kg)	M	11	12			
	F	9	10			
	M + F	10	11			

Source: Modified from EC, 2003. Technical guidance document in support of Commission Directive 93/67/EEC on Risk Assessment for new notified substances, Commission Regulation (EC) No 1488/94 on Risk Assessment for existing substances and Directive 98/8/EC of the European Parliament and of the Council concerning the placing of biocidal products on the market. http://ecb.jrc.it/tgd.

Proposed body weight defaults for the 28- and 90-day studies in the mouse, and the 28-, 90-day, and chronic studies in the rat are based on data from more than one strain.

wide range of predominately mammalian species. Synthesis of the information to provide recommendations concerning parameters for conversion of exposure data (concentrations of a substance in drinking water, food, and air) from mammalian toxicological studies to dose estimates, expressed as mg/kg body weight per day, is provided. The report is not available from the US-EPA in electronic format but can be obtained as a hard copy from the US-EPA.

7.4.3 EU

The TGD (EC 2003) provides default reference values for certain biological parameters. The default reference values for body weights (oral/inhalation studies and dermal studies) and VRs, which are based on a report from TNO (the Netherlands Organisation for Applied Scientific Research) (Paulussen et al. 1998), are presented in Tables 7.13 through 7.15.

The TGD does not suggest default reference values for feed or water intake with the explanation that experimental conditions (e.g., type of diet) may affect the water and feed consumption. Instead allometric equations are provided in order to derive default values on a case-by-case basis. The allometric equations have been divided into species-specific equations, see Table 7.16, and general equations, see Table 7.17. The allometric equations were originally presented in a report published by the US-EPA (1988), see Section 7.4.2. The TGD recommends that where possible the equations that are species-specific should be applied as they represent the most realistic approach, although they do not always correlate well.

7.5 PROBABILISTIC METHODS

A description of a population's exposure should include an estimation of the variability. Individuals within the population are likely to be exposed to different doses or concentrations because of

TABLE 7.14
Default Reference Values for Body Weights
in Dermal Toxicity Studies

| Species | | Body Weights | |
Type of Study	Sex	28 Days	90 Days
Rabbit (kg)[a]	M	3	
	F	3	
	M + F	3	
Rat (g)[b]	M	350	400
	F	250	275
	M + F	300	350
Guinea pig (g)[c]	M + F	475	

Source: Modified from EC, 2003. Technical guidance document in support of Commission Directive 93/67/EEC on Risk Assessment for new notified substances, Commission Regulation (EC) No 1488/94 on Risk Assessment for existing substances and Directive 98/8/EC of the European Parliament and of the Council concerning the placing of biocidal products on the market. http://ecb.jrc.it/tgd.

[a] Defaults based on New Zealand white rabbit strain.
[b] Defaults based on Sprague–Dawley rat strain.
[c] Defaults based on strain 13 guinea pigs where only joint male and female data were available.

TABLE 7.15
Default Reference Values for Ventilation Rates (Volume of Inhaled Air Per Unit Time)

| Species | | Ventilation[a] | | | | | |
| | | 28 Days | | 90 Days | | Chronic | |
Type of Study	Sex	ml/minute	liter/day	ml/minute	liter/day	ml/minute	liter/day
Rat	M	175	252	200	288	300	432
	F	125	180	150	216	200	288
	M + F	150	216	175	252	250	360
Mouse	M	30	43	35	50	50	72
	F	25	36	30	43	40	58
	M + F	25	36	30	43	45	65
Rabbit	M	750	1080				
	F	750	1080				
	M + F	750	1080				
Guinea pig	M + F	200	288	225	324		
Dog	M	2550	3672	2700	3888		
	F	2250	3240	2350	3384		
	M + F	2400	3456	2500	3600		

Source: Modified from EC, 2003. Technical guidance document in support of Commission Directive 93/67/EEC on Risk Assessment for new notified substances, Commission Regulation (EC) No 1488/94 on Risk Assessment for existing substances and Directive 98/8/EC of the European Parliament and of the Council concerning the placing of biocidal products on the market. http://ecb.jrc.it/tgd.

[a] The values are based on species-specific allometric relationships.

TABLE 7.16
Species-Specific Allometric Equations for Feed Consumption

Animal Group	Allometric Equation
Rat	$F = 0.040\ W^{0.479}$
Mouse	$F = 0.064\ W^{0.7242}$
Dog	$F = 5.13\ W^{-0.918}$
Guinea pig	$F = 0.041\ W^{0.3308}$
Rabbit	$F = 0.041\ W^{0.7898}$
Hamster	$F = 0.082\ W^{0.9285}$
Laboratory animals [a]	$F = 0.056\ W^{0.6611}$
All species combined [b]	$F = 0.065\ W^{0.7919}$

Source: From US-EPA, Recommendations for and Documentation of Biological Values for Use in Risk Assessment. Available at http://cfpub.epa.gov/ncea/cfm/recordisplay.cfm?deid = 34855; modified from EC, 2003. Technical guidance document in support of Commission Directive 93/67/EEC on Risk Assessment for new notified substances, Commission Regulation (EC) No 1488/94 on Risk Assessment for existing substances and Directive 98/8/EC of the European Parliament and of the Council concerning the placing of biocidal products on the market. http://ecb.jrc.it/tgd.

F = feed consumption in kg/day; W = body weight in kg.

[a] Includes gerbils, guinea pigs, hamsters, mice, rats, cats, dogs, and rabbits.
[b] Includes chickens in addition to laboratory animals.

differences, e.g., in physiology or behavior. These differences lead to variation among individuals. Variability is an unavoidable component of a biological sample and cannot be reduced by additional study or measurement. In addition, the description of the population's exposure should also include uncertainty about the true levels of exposure arising from limitations in the available measured data, and from limitations in the methods and techniques used to obtain the data. Uncertainty is a measure of the incompleteness of one's knowledge about an unknown parameter. Uncertainty can be reduced by improving the accuracy and precision of the exposure measurements.

Traditional risk assessments involve the use of point estimates of exposure with no regard for variability or uncertainty as input to the exposure models, and this may not be acceptable, as the exposure assessments have great impact on the outcome of the risk assessment.

Probabilistic exposure models attempt to provide inputs to exposure models by representing variability or uncertainty via frequency or probability distributions. Probabilistic methods can be used in the exposure assessment because pertinent variables (e.g., concentration, intake rate, exposure duration, and body weight) have been identified, their distributions can be observed, and the formula for combining the variables to estimate the exposure is well defined.

Regulatory agencies are accepting and actively encouraging probabilistic approaches. An international workshop recommended that guidance on the appropriate occasions for use of a probabilistic approach in exposure assessment, and guidance to facilitate and interpret probabilistic exposure assessment should be developed on an international level to ensure harmonization of regulation (van Drooge and van Haelst 2001).

TABLE 7.17

General Allometric Equations for Feed and Drinking-Water Consumption

Animal Group	Allometric Equation	Equation Number
Feed and water consumption[a]		
Dry diet: all species	$F = 0.31\ L^{0.7923}$ (kg)	1a
	$L = 3.59\ F^{1.2041}$ (l)	1b
Wet diet: all species	$F = 2.09\ L^{0.7389}$ (kg)	2a
	$L = 0.39\ F^{1.2447}$ (l)	2b
Laboratory mammals: dry diet	$F = 0.28\ L^{0.7613}$ (kg)	3a
	$L = 0.31\ F^{1.2226}$ (l)	3b
Laboratory rodents: dry diet	$F = 0.16\ L^{0.6426}$ (kg)	4a
	$L = 0.25\ F^{1.2943}$ (l)	4b
Body weight to feed or water consumption		
Dry diet: all species	$F = 0.049\ W^{0.6087}$ (kg)	5
	$L = 0.093\ W^{0.7584}$ (l)	6
Unknown diet: all species	$F = 0.065\ W^{0.7919}$ (kg)	7
	$L = 0.11\ W^{0.7872}$ (l)	8

Source: Modified from EC, 2003. Technical guidance document in support of Commission Directive 93/67/EEC on Risk Assessment for new notified substances, Commission Regulation (EC) No 1488/94 on Risk Assessment for existing substances and Directive 98/8/EC of the European Parliament and of the Council concerning the placing of biocidal products on the market. http://ecb.jrc.it/tgd.

The equations that describe the feed and water consumption data cannot be applied to pregnant females used in reproductive multigeneration studies.

When either feed or water consumption is known for animals on a wet or dry diet, Equations 1–4 should be used to estimate the missing value. If diets are specified or can be reasonably assumed to have been dry or moist, Equations 1 and 2 are recommended. If body weight is known or can be estimated, Equations 5–8 are recommended for estimating feed and water consumption.

[a] F = feed consumption in kg/day; L = liquid (water) consumption in liter/day; W = body weight in kg.

This work has resulted in the establishment of an IPCS harmonization project working group. A Guidance Document has been developed and is presently being finalized (WHO/IPCS 2006).

For further information on probabilistic methods in exposure assessment, the reader is recommended to consult the textbook by Cullen and Frey (1999).

REFERENCES

CEM TF. 2004. Consumer Exposure Modelling Tools. Ispra, European Commission, Joint Research Centre, Institute for Health and Consumer Protection, Physical and Chemical Exposure Unit, Exposure Modelling Sector. http://cem.jrc.it/cemdb/qstart.php

Cullen, A.C. and H.C. Frey. 1999. *Probabilistic Techniques in Exposure Assessment: A Handbook for Dealing with Variability and Uncertainty in Models and Inputs.* New York: Plenum Publishing Corporation.

EC. 2003. Technical Guidance Document in support of Commission Directive 93/67/EEC on Risk Assessment for new notified substances, Commission Regulation (EC) No 1488/94 on Risk Assessment for existing substances and Directive 98/8/EC of the European Parliament and of the Council concerning the placing of biocidal products on the market. http://ecb.jrc.it/tgd

ECB. 2007. The European Union system for the evaluation of substances (EUSES). http://ecb.jrc.it/euses/

ECETOC. 2001. *Exposure Factors Sourcebook for European Populations, with Focus on UK Data*. Technical Report No. 79. Brussels: and of the Council Concerning the Placing of Biocidal Products on the Market. http://ecb.jrc.it/tgdoc/

EU. 2004. European information system on risks from chemicals released from consumer products/articles. Consumer exposure modelling task force. Brussels, European Commission, Joint Research Centre. http://www.jrc.cec.eu.int/eis-chemrisks/

EU. 2006. EU DG Environment REACH website. http://ec.europa.eu/environment/chemicals/reach/reach_intro.htm

JRC. 2007. The ExpoFacts Homepage. http://cem.jrc.it/expofacts/

OECD. 2000a. *Guidance Document on Emission Scenario Documents*. OECD Series On Emission Scenario Documents No 1. Environment Directorate, Joint Meeting of the Chemicals Committee and the Working Party on Chemicals, Pesticides and Biotechnology. ENV/JM/MONO(2000)12. Paris: OECD.

OECD. 2000b. *Guidance Notes for Analysis and Evaluation of Repeat-Dose Toxicity Studies*. OECD Series on Testing and Assessment No. 32 and OECD Series on Pesticides No. 10. Environment Directorate, Joint Meeting of the Chemicals Committee and the Working Party on Chemicals, Pesticides and Biotechnology. ENV/JM/MONO(2000)18. Paris: OECD.

OECD. 2003a. *Descriptions of Selected Key Generic Terms Used in Chemical Hazard/Risk Assessment. Joint Project with IPCS on the Harmonisation of Hazard/Risk Assessment Terminology*. OECD Series on Testing and Assessment No. 44. Environment Directorate, Joint Meeting of the Chemicals Committee and the Working Party on Chemicals, Pesticides and Biotechnology. ENV/JM/MONO(2003)15. Paris: OECD.

OECD. 2003b. *Guidance Document on Reporting Summary Information on Environmental, Occupational and Consumer Exposure*. OECD Series on Testing and Assessment No. 42. Environment Directorate, Joint Meeting of the Chemicals Committee and the Working Party on Chemicals, Pesticides and Biotechnology. ENV/JM/MONO(2003)16. Paris: OECD.

OECD. 2006. *Approaches to Exposure Assessment in OECD Member Countries: Report from the Policy Dialogue on Exposure Assessment in June 2005*. OECD Environment, Health and Safety Publications Series on Testing and Assessment No. 51. Paris: OECD.

OECD. 2007. OECD's Database on Use and Release of Industrial Chemicals. http://appli1.oecd.org/ehs/urchem.nsf

Paulussen, J.J.C., C.M. Mahieu, and P.M.J. Bos. 1998. *Default Values in Occupational Risk Assessment*. TNO Report V98.380. Zeist, The Netherlands: TNO Nutrition and Food Research Institute.

Stanek III, E.J. and E.J. Calabrese. 2000. Daily Soil Ingestion Estimates for Children at a Superfund Site. *Risk Anal*. 20:627–635.

US-EPA. 1988. Recommendations for and Documentation of Biological Values for Use in Risk Assessment. http://cfpub.epa.gov/ncea/cfm/recordisplay.cfm?deid = 34855

US-EPA. 1992. *Guidelines for Exposure Assessment*. 600Z-92/001, 1992. Washington, DC: U.S. Environmental Protection Agency, Risk Assessment Forum. http://www.epa.gov/nceawww1/pdfs/guidline.pdf

US-EPA. 1995. U.S. Environmental Protection Agency Science Policy Council. *Guidance for Risk Characterization*. http://www.epa.gov/OSA/spc/pdfs/rcguide.pdf

US-EPA. 1996. *Guidelines for Reproductive Toxicity Risk Assessment*. 630/R-96/009, 1996. Washington, DC: U.S. Environmental Protection Agency, Risk Assessment Forum. http://www.epa.gov/ncea/raf/pdfs/repro51.pdf

US-EPA. 1997. *Exposure Factors Handbook*. Washington, DC: National Center for Environmental Assessment. Office of Research and Development. http://www.epa.gov/ncea/efh/

US-EPA. 2006. *Child Specific Exposure Factors Handbook 2006 (External Review Draft)*. Washington, DC: National Center for Environmental Assessment. Office of Research and Development. Presently undergoing update (In press). http://cfpub.epa.gov/ncea/cfm/recordisplay.cfm?deid = 56747

US-EPA. 2007a. EPA/OPPT Exposure Assessment Tools and Models website. http://www.epa.gov/opptintr/exposure/pubs/opptexpo.htm

US-EPA. 2007b. Consolidated Human Activity Database. http://www.epa.gov/chadnet1/

van Drooge, H.L. and A.G. van Haelst. 2001. Probabilistic Exposure Assessment is Essential for Assessing Risks - Summary of Discussions. *Ann. Occup. Hyg*. 45:S159–S162.

WHO/IPCS. 1987. *Principles for the Safety Assessment of Food Additives and Contaminants in Food.* Environmental Health Criteria 70. Geneva: WHO. http://www.inchem.org/documents/ehc/ehc/ehc70.htm

WHO/IPCS. 1990. *Principles for the Toxicological Assessment of Pesticide Residues in Food.* Environmental Health Criteria 104. Geneva: WHO. http://www.inchem.org/documents/ehc/ehc/ehc104.htm

WHO/IPCS. 1994. *Assessing Human Health Risks of Chemicals: Derivation of Guidance Values for Health-based Exposure Limits.* Environmental Health Criteria 170. Geneva: WHO. http://www.inchem.org/documents/ehc/ehc/ehc170.htm

WHO. 1996. *12. Chemical and physical aspects: Introduction.* In: *Guidelines for Drinking-Water Quality, Second Edition.* Volume 2: Health Criteria and other Supporting Information. Geneva: WHO. http://www.who.int/water_sanitation_health/dwq/gdwq2v1/en/index1.html

WHO/IPCS. 1999. *Principles for the Assessment of Risks to Human Health from Exposure to Chemicals.* Environmental Health Criteria 210. Geneva: WHO. http://www.inchem.org/documents/ehc/ehc/ehc210.htm

WHO/IPCS. 2000. *Human Exposure Assessment.* Environmental Health Criteria 214. Geneva: WHO. http://www.inchem.org/documents/ehc/ehc/ehc214.htm

WHO/IPCS. 2004. *IPCS glossary of key exposure assessment terminology.* In: *IPCS Risk Assessment Terminology.* Harmonization Project Document No. 1. Geneva: WHO. http://www.inchem.org/documents/sids/sids/risk_assess.pdf

WHO/IPCS. 2005. *Principles of Characterizing and Applying Human Exposure Models.* IPCS Harmonization Project Document No. 3. Geneva: WHO. http://whqlibdoc.who.int/publications/2005/9241563117_eng.pdf

WHO/IPCS. 2006. *Draft Guidance Document on Characterizing and Communicating Uncertainty in Exposure Assessment.* Geneva: WHO. http://www.who.int/ipcs/methods/harmonization/areas/exposure_assessment/en/index.html

WHO/IPCS. 2007. IPCS Harmonization Project. Harmonization of approaches to the assessment of risk from exposure to chemicals website. http://www.who.int/ipcs/methods/harmonization/en/

8 Risk Characterization

The aim of the risk characterization of a chemical substance under evaluation is to integrate the hazard assessment and exposure assessment in order to evaluate the qualitative and quantitative probability for a health risk likely to occur in a given human population due to actual or predicted exposure to that specific chemical as well as the seriousness of any health risk.

The process of risk characterization has been described and discussed extensively in the scientific literature. It is beyond the scope of this book to review all these references. This chapter is limited to give a very short overview of some general aspects related to the risk characterization process, as well as of the currently used approaches in the WHO, the US-EPA, and the EU, and is thus not meant to be exhaustive.

8.1 INTRODUCTION

According to the OECD/IPCS definitions listed in Annexure 1 of Chapter 1 (OECD 2003):

Hazard assessment is "A process designed to determine the possible adverse effects of an agent or situation to which an organism, system or (sub) population could be exposed. The process includes hazard identification and hazard characterization. The process focuses on the hazard in contrast to risk assessment where exposure assessment is a distinct additional step."

Hazard identification is the first step in the process of risk assessment, and hazard characterization is the second step. The hazard assessment is also known as "effect assessment" and is addressed in detail in Chapter 4.

Exposure Assessment is "Evaluation of the exposure of an organism, system or (sub) population to an agent (and its derivatives)."

Exposure assessment is the third step in the process of risk assessment, and is addressed in detail in Chapter 7.

Risk characterization is "The qualitative and, wherever possible, quantitative determination, including attendant uncertainties, of the probability of occurrence of known and potential adverse effects of an agent in a given organism, system or (sub)population, under defined exposure conditions."

Risk characterization is the fourth step in the process of risk assessment.

Hazard assessment and exposure assessment of chemicals are generally common stages performed similarly independent of the chemical use category (industrial chemical, pesticide, biocide, food additive, food contact material, etc.). However, variation occurs in the way in which the exposure assessment and hazard assessment information are integrated in the risk characterization step, depending on the regulation involved and the goal of the risk assessment. This will be addressed in more detail in the next section.

8.2 RISK CHARACTERIZATION: GENERAL ASPECTS

In the risk characterization step, the health risk likely to occur in a given human population due to actual or predicted exposure to the particular chemical under evaluation is evaluated, and all the assumptions, uncertainties, and scientific judgments from the preceding three steps are considered.

Risk characterization is thus the step in the risk assessment process where the outcome of the exposure assessment (e.g., daily intake via food and drinking water, or via inhalation of airborne substances) and the hazard (effects) assessment (e.g., NOAEL and tolerable intake) are compared. If possible, an uncertainty analysis should be carried out, which produces an estimation of the risk.

Several questions should be answered before comparison of hazard and exposure is made:

- What is the target population to protect?
- What is the timescale of exposure?
- What is the spatial scale of exposure?
- Which route(s) of exposure is or are relevant?
- Are sufficient toxicity data available to derive a meaningful toxicological parameter, i.e., point of departure (POD), corresponding to the timescale of exposure and the route(s) of exposure as established in the exposure assessment?
- What degree of uncertainty is acceptable?

Exposure of the general population to chemicals present in the environment is an example of long-term exposure on a local or regional spatial scale. The general population is mainly exposed to environmental chemicals via oral exposure through food and drinking water and via inhalation from ambient and indoor air. The total body burden can, e.g., be expressed as a total oral intake (the outcome of the exposure assessment). This intake should be compared with a POD derived from preferably long-term studies or at least subchronic studies (outcome of the hazard (effects) assessment).

The most frequently used POD for threshold effects (Section 4.2) is the NOAEL (Section 4.2.4). This NOAEL is generally obtained from studies in experimental animals. If reliable human data are available to derive the NOAEL, this value is preferable to the NOAEL from experimental animals. Where a NOAEL cannot be derived, a LOAEL, if available, can be used. An alternative POD to the NOAEL/LOAEL is the benchmark dose (BMD) (Section 4.2.5). The tolerable intake can also, in some cases, form the basis as the POD. In this chapter, the POD will be denoted as a "derived no-effect level" (DNEL) in order to provide a general term for the various types of PODs that can form the basis for the risk characterization.

Risk characterization for non-threshold effects, e.g., for chemicals that are both genotoxic and carcinogenic, generally proceeds by comparing the acceptable risk level (Section 6.2.4) with the actual or estimated total daily intake. An alternative, new approach is the margin of exposure approach (Section 6.4).

Consumers and workers may be exposed to a variety of chemicals via different exposure routes. This exposure can be judged to be acute, subchronic, or chronic by analogy with the exposure schedules in the various experimental animal studies. The spatial scale of exposure is at the personal level. At the risk characterization step, the actual or estimated acute, subchronic, and chronic exposure level can be compared with a suitable DNEL derived from the acute, subchronic, and chronic studies, respectively.

For risk assessment of chemicals based on a DNEL as described above, the output of the exposure assessment (Chapter 7) is usually an estimate of dose or concentration. Exposure data can either be measured or predicted. Measured exposure data are preferred if they are valid, i.e., an actual exposure estimate. However, in most cases, measured data are not available and therefore model-generated data must be used for the risk characterization, i.e., a predicted exposure estimate.

Predictive methods of exposure assessment often rely on single values for input parameters to the exposure model that represent one point on the distribution curve of all possible values for this parameter. This point value can range from a 50th percentile, mean, median, or "typical" value to a "worst-case" estimate. In the predictive exposure assessment, a number of parameters are integrated through an algorithm to produce an output such as the predicted environmental concentration (PEC). If many worst-case values are involved, this integration can result in a PEC that has a very low probability of occurring. Instead of generating a single-point estimate, distributions of parameters can be stochastically combined to generate a distribution of the output value.

The exposure assessment could be performed for a single exposure scenario, or be more comprehensive including several exposure scenarios. In some situations, the estimated exposure from a single scenario is taken forward to the risk characterization while for other purposes, the estimated exposures from various scenarios form the basis for an estimation of a combined exposure to the chemical under evaluation from all characterized exposure scenarios.

8.2.1 Toxicity Exposure Ratio Approach

In the "toxicity exposure ratio" approach, the output of the hazard (effects) assessment is compared with the output of the exposure assessment.

The direct comparison of a POD (DNEL) with the estimated exposure (E) leads to the establishment of a ratio (DNEL/E), often denoted as the margin of safety (MOS) or margin of exposure (MOE).

According to the OECD/IPCS definitions listed in Annex 1 (OECD 2003):

Margin of Exposure is "Ratio of the no-observed-adverse-effect level (NOAEL) for the critical effect to the theoretical, predicted or estimated exposure dose or concentration."

A related term is the margin of safety, which, according to the OECD/IPCS definitions listed in Annex 1 (OECD 2003), for some experts, has the same meaning as the margin of Exposure, while for others, the margin of Safety means the margin between the reference dose and the actual exposure dose or concentration.

The MOS (DNEL/E or E/DNEL) gives an indication of the degree of risk. In many international regulatory frameworks, the risk is often expressed as such a ratio, often denoted as a risk quotient. The interpretation of the significance of the magnitude of the derived ratio in terms of the possible risk of adverse consequences for the health of the exposed population is the key aspect of this approach. The magnitude of the ratio will determine the ultimate regulatory position that is adopted: either regulatory action or acceptance of the existing situation.

In this approach, the toxicological uncertainties, which are essentially similar to those involved in standard setting for threshold effects (Chapter 5), are addressed as part of the risk characterization step, i.e., considering whether the ratio is sufficiently large to give the degree of confidence that the exposure situation will not result in adverse human health consequences.

Depending on the biological species for which the DNEL has been derived (humans or experimental animals), the exposure duration and frequency, the incidence, type and severity of the effects, the dose–effect and dose–response relationships observed, and the availability of other data on the toxicological profile of the chemical under evaluation, the risk assessor can evaluate the uncertainty in the hazard (effects) assessment and then judge whether the resulting margin is sufficient or not. This margin should also take into account uncertainties due to interindividual variation, i.e., variations in sensitivity among individuals of one species (generally humans), uncertainties resulting from interspecies variation, i.e., variations between the species human and the test species animal, as well as uncertainties resulting from differences in the exposure scenario, e.g., short-term exposure versus long-term exposure and continuous versus intermittent exposure.

Generally, local effects can only be evaluated qualitatively, taking into account the airborne concentration of the chemical or the concentration of the chemical in consumer products, and effects such as irritation and corrosion (Section 4.5), and sensitization (Section 4.6). This is because the standard guideline tests for these endpoints are generally aimed at providing the evidence that a particular chemical has the potential to induce such effects, rather than to evaluate dose/concentration–response relationships, i.e., they provide a yes/no type of result (Section 4.2.3). For that purpose, substances are administered as a single exposure according to the standard test guidelines for these endpoints. Therefore, it is generally not possible to incorporate such properties in a quantitative risk characterization framework based on the toxicity exposure ratio approach. In some cases, however, it might be possible to establish a NOAEC for humans for irritative effects, particularly for respiratory tract irritation, primarily based on human data from the working environment (Section 3.2.4), or based on longer term studies in experimental animals, e.g., repeated dose toxicity studies (Section 4.7).

Finally, the MOS should also take into account the uncertainties in the estimated exposure. For predicted exposure estimates, this requires an uncertainty analysis (Section 8.2.3) involving the determination of the uncertainty in the model output value, based on the collective uncertainty of the model input parameters. General sources of variability and uncertainty in exposure assessments are measurement errors, sampling errors, variability in natural systems and human behavior, limitations in model description, limitations in generic or indirect data, and professional judgment.

The toxicity exposure ratio approach, rather than a more rigid standard setting approach (Section 8.2.2), allows greater room for expert judgment because the size of an overall assessment factor is not fixed. Furthermore, this approach can be readily applied to substances for which limited data are available. The risk assessor can decide how wide the MOS should be in the light of the data available.

It should be noted that MOS ratios are no absolute measure of risks. Nobody knows the real risks of chemicals where the exposure exceeds the derived no-effect level (DNEL). The risk assessor only knows that the likelihood of adverse effects increases when the DNEL/E ratios decrease or the E/DNEL ratios increase. Thus, such ratios are internationally accepted only as substitutes for risks.

8.2.2 STANDARD SETTING APPROACH

An alternative approach to the toxicity exposure ratio approach described in the previous section is the "standard setting" approach.

Often, the output of the hazard (effects) assessment (e.g., the NOAEL) leads directly to the establishment of a regulatory standard, for example the derivation of an acceptable or tolerable daily intake (ADI/TDI) (Section 5.12) for a chemical in relation to a specific use category such as, e.g., pesticide, biocide, food additive, food contact material, etc.

The risk characterization is conducted by comparing this regulatory standard with the outcome of the exposure assessment, i.e., the exposure estimate. Regulatory decisions on the need for further risk management action are then made on the basis of this comparison.

This form of risk assessment is based on the concept of defining an exposure level, the derived standard, expressed usually on a temporal basis (e.g., daily, weekly), which is considered to offer sufficient reassurance of protection of human health, and then comparing this with an estimated level of exposure. If the estimated exposure is higher than the standard, then further regulatory intervention may be needed. Please see Section 5.12 for a discussion of the health implications of exceeding the tolerable intake.

In this approach, the toxicological uncertainties are addressed as part of the hazard assessment and standard setting (Chapter 5), and are thus incorporated before considerations of exposure.

8.2.3 UNCERTAINTY ANALYSIS

If possible, as a last step in the risk characterization process, an uncertainty analysis should be carried out, which, if it results in a quantifiable overall uncertainty, produces an estimation of the risk.

The same uncertainties exist in moving from hazard assessment to the development of a regulatory standard (e.g., ADI/TDI) as in the standard setting approach, or in applying the hazard information to assessing the significance of a derived ratio (e.g., MOS/MOE) as in the toxicity exposure ratio approach, i.e., the uncertainties inherent in the hazard assessment.

In both approaches, allowance is often made for these uncertainties by the application of numerical factors (assessment factors, extrapolation factors, uncertainty factors).

In the case of the standard setting approach, assessment factors are directly applied to the hazard assessment output (i.e., the NOAEL/LOAEL or BMD) in order to derive a standard (Chapter 5).

In the case of the toxicity exposure ratio approach, it is part of the consideration of the magnitude of the ratio (i.e., MOS/MOE) between the hazard assessment output and the exposure assessment output, i.e., by considering whether the ratio is large enough to accommodate the numerical factors that are used to allow for uncertainty. In should be recognized that, in this approach, the toxicological uncertainties are essentially similar to those involved in the standard setting approach.

As described in detail in this book, the use of assessment factors is an established practice in chemical risk assessment to account for uncertainties inherent in the hazard (effects) assessment and consequently, inherent in the risk assessment. The use of assessment factors to address this uncertainty is part of the conventional approach that has developed over the years. According to the current risk assessment paradigm, the usual approach is simply to multiply these individual assessment factors in order to establish an overall composite numerical assessment factor (Section 5.10). An alternative to the traditional assessment factor approach is to combine estimates of the ranges that these factors may encompass through a probabilistic assessment; this is essentially a variation of the standard paradigm.

Furthermore, uncertainties in the exposure assessment should also be taken into account. However, no generally, internationally accepted principles for addressing these uncertainties have been developed. For predicted exposure estimates, an uncertainty analysis involving the determination of the uncertainty in the model output value, based on the collective uncertainty of the model input parameters, can be performed. The usual approach for assessing this uncertainty is the Monte Carlo simulation. This method starts with an analysis of the probability distribution of each of the variables in the uncertainty analysis. In the simulation, one random value from each distribution curve is drawn to produce an output value. This process is repeated many times to produce a complete distribution curve for the output parameter.

Other, more sophisticated approaches for addressing uncertainties in the risk assessment process may be adopted in future risk assessments. For example, the use of probabilistic approaches in risk assessment will probably have an impact in future risk assessments. For more information on probabilistic approaches in risk assessment, the reader is kindly referred to the short overviews presented in Sections 4.14 and 5.11 as well as scientific papers, for example Baird et al. (1996), Slob and Pieters (1998), Edler et al. (2002), Renwick et al. (2004), and van der Voet and Slob (2007).

It should be recognized that precise risk assessments do not exist and risk assessors will often differ in the conclusions they draw from the same set of data, particularly if the conclusions contain some implicit value judgments.

At present, risk assessors cannot adequately predict what part of the human population will be affected by exposure to a particular chemical under evaluation. We are only able to assess risks in a very general and simplified manner. Therefore, the best we can do is relative risk ranking. Risk ranking enables us to compare single chemicals or groups of chemicals once the risks of the respective chemicals have been assessed in a consistent and simplified manner. Relative risk ranking then allows us to replace dangerous chemicals, processes, or techniques by safer alternatives in the risk management phase, without knowing the precise risks.

It should also be recognized that, according to the currently used practices, health risk assessments of exposure to chemicals and the subsequent regulatory measures, e.g., health-based guidance values (Chapter 9), are generally based upon data from studies on the individual chemical, i.e., risk assessments are generally chemical-specific. However, humans are simultaneously exposed to a large number of chemicals that potentially possess a number of similar or different adverse effects. Therefore, the aspect of combined actions of chemicals in mixture needs to be addressed to a greater extent in the risk assessment process (Chapter 10).

8.3 RISK CHARACTERIZATION: CURRENTLY USED APPROACHES

This section gives a short overview of the currently used risk characterization approaches in the WHO, the US-EPA, and the EU.

8.3.1 WHO

The WHO and its bodies, e.g., IPCS, JECFA, and JMPR, are primarily involved in standard settings rather than risk characterization in its strict meaning, i.e., comparing the outcome of the hazard (effects) assessment and the outcome of the exposure assessment.

Examples of standard settings developed by the WHO include "air quality guidelines" (Section 9.2.1.1) and "drinking water guidelines" (Section 9.2.1.2), and (in collaboration with the FAO) "maximum residue limits" (MRLs) for pesticides and veterinary drugs and "maximum levels" for food additives (Section 9.2.1.3).

8.3.2 US-EPA

In 1995, the US-EPA updated and issued the current "Agency-wide Risk Characterization Policy" (US-EPA 1995). The Policy called for all risk assessments performed at US-EPA to include a risk characterization to ensure that the risk assessment process is transparent. It also emphasized that risk assessments should be clear, reasonable, and consistent with other risk assessments of similar scope prepared by programs across the Agency.

The US-EPA's "*Risk Characterization Handbook*" (US-EPA 2000) was developed to implement the "Risk Characterization Policy." This Handbook provides a single, centralized body of risk characterization implementation guidance for the US-EPA's risk assessors and risk managers to help make the risk characterization process Transparent and the risk characterization products, Clear, Consistent, and Reasonable (TCCR). TCCR became the underlying principle for a good risk characterization. The Handbook has two parts.

The first part is the Risk Characterization guidance itself. The Risk Characterization Guide is designed to provide risk assessors, risk managers, and other decision-makers an understanding of the goals and principles of risk characterization, the importance of planning and scoping for a risk assessment, the essential elements to address in a risk characterization, the factors that are considered in decision making by risk managers, and the forms that risk characterization takes for different audiences. A discussion of the various administrative details regarding risk characterization completes the guide.

The second part comprises the Appendices, which contain the Risk Characterization Policy, the risk characterization case studies, and references. The case studies contain examples of risk characterizations from risk assessments that apply the principles described in the Risk Characterization Guide.

The Handbook emphasized that other aspects than science influence risk characterization, and that science policy choices must be made to deal with uncertainties. Many choices are usually made during the course of the risk assessment process, resulting in a particular outcome. Therefore, it is possible to perform parallel risk assessments of the same data, but reach different results.

The Handbook also pointed out that risk characterization does not stand alone, as it is one of the four steps in risk assessment. There is only a single technical characterization of risk as a final product of the risk assessment. This technical characterization must be written with enough detailed technical information to allow another expert (e.g., other risk assessors, peer reviewers) to reasonably reconstruct what was done in the risk assessment, including to be able to identify the assumptions made during the risk assessment. Since the risk characterization is a part of the risk assessment itself, it should be kept in mind that the goal of the risk characterization is not to repeat the entire risk assessment, but just to identify the key elements from the risk assessment that really make a difference in its outcome.

The major elements to be considered in the risk characterization part include: key information, context, sensitive subpopulations, scientific assumptions, policy choices, variability, uncertainty, bias and perspective, strengths and weaknesses, key conclusions, alternatives considered, and research needs. Whether every element is actually written into the risk characterization or not, depends upon the purpose of the risk assessment and the detail necessary to adequately characterize it. By the time the risk assessment is completed, the universe of policy choices, management decisions, and uncertainties should have been identified, as well as the conclusions of the risk assessment. Because key findings differ for each risk assessment, it is not possible to define exactly what they are generically. Professional judgment is necessary to define them.

The *Risk Characterization Handbook* (US-EPA 2000) is thus a practical guide in how to perform the risk characterization. However, the Handbook does not include any detailed information on the practices employed in the risk assessment itself, including use of uncertainty factors and use of default and extrapolation assumptions in the risk characterization step. This information and practices are provided in the US-EPA staff paper from 2004 titled "An Examination of EPA Risk Assessment Principles and Practices" (US-EPA 2004).

The US-EPA is the federal agency responsible for standard settings in relation to contaminants in ambient air (Section 9.2.2.1), drinking water (Section 9.2.2.2), and soil (Section 9.2.2.3) while the U.S. Food and Drug Administration (US-FDA) is the federal agency responsible for standard settings in relation to contaminants in food (Section 9.2.2.4).

8.3.3 EU: Industrial Chemicals

The process of human health risk characterization has been extensively addressed within the EU framework of risk assessment of new and existing chemical substances (industrial chemicals). Detailed guidance is provided in the Technical Guidance Documents on risk assessment for new and existing Substances (TGD). In the first edition of the TGD (EC 1996), guidance was provided in relation to risk characterization in general. The TGD has been revised and the second edition was published in 2003 (EC 2003); however, the human health risk characterization part was not included in this second edition. The general guidance presented below is cited from both the first and second editions of the TGD.

The definition of risk characterization in terms of the TGD differs somewhat from that of the OECD/IPCS given in Section 8.1. According to the TGD (EC 2003):

Risk characterization is "Estimation of the incidence and severity of the adverse effects likely to occur in a human population or environmental compartment due to actual or predicted exposure to a substance, and may include "risk estimation", i.e., the quantification of that likelihood."

The risk characterization is carried out by quantitatively comparing the outcome of the hazard (effects assessment) to the outcome of the exposure assessment, i.e., a comparison of the NOAEL, or LOAEL, and the exposure estimate. The ratio resulting from this comparison is called the Margin of Safety (MOS) (MOS = N(L)OAEL/Exposure). This is done separately for each potentially exposed population, i.e., workers, consumers, and man exposed via the environment, and for each toxicological endpoint, i.e., acute toxicity, irritation and corrosion, sensitization, repeated dose toxicity, mutagenicity, carcinogenicity, and toxicity to reproduction.

Unless a N/LOAEL is available from human data, the N/LOAEL are those derived from the animal studies without any modifications. When a human N/LOAEL is available, the approaches described below can still be used in principle but some of the factors which need to be considered when using data from animal studies will obviously not be applied.

Where a N/LOAEL has been identified for any of the toxicological endpoints mentioned above, it will be compared with the exposure estimate for the exposed human population. Where more than one N/LOAEL has been identified for a specific endpoint, then the most relevant N/LOAEL will be used.

Where it is not possible to determine a N/LOAEL, the likelihood that the effect will occur is evaluated on the basis of the quantitative and/or qualitative information on exposure relevant to the human populations under consideration. Where a N/LOAEL has not been determined, the test results nevertheless demonstrate a relationship between dose (or concentration) and the severity of an adverse effect or where, in connection with a test method which entails the use of only one dose or concentration, it is possible to evaluate the relative severity of the effect, such information shall also be taken into account in evaluating the likelihood of the effect occurring.

Where the exposure estimate is higher than or equal to the N/LOAEL (MOS $\leq$ 1), this indicates that the substance is "of concern" with regard to the exposure of the human population considered. The risk assessor will then need to decide whether additional data either on exposure or toxicity would allow a refinement of the exposure or N/LOAEL values and, subsequently, a refinement of the comparison which would influence the risk characterization result.

Where the exposure estimate is less than the N/LOAEL (MOS > 1), the risk assessor will need to evaluate the magnitude of the MOS taking account of the following factors.

From Regulation 1488/94 (EC 1994) and Directive 93/67/EEC (EEC 1993):

- Uncertainty arising, among other factors, from the variability in the experimental data and intra- and interspecies variation
- Nature and severity of the effect
- Human population to which the quantitative and/or qualitative information on exposure applies

Other factors:

- Differences in exposure (route, duration, frequency, and pattern)
- Dose–response relationship observed
- Overall confidence in the database

Expert judgment is required to weigh these individual parameters on a case-by-case basis. The approach used should be transparent and a justification should be provided by the risk assessor for the conclusion reached. It should be recognized that these parameters are parallel to those being considered in the evaluation of the assessment factors to be applied in the establishment of a tolerable intake (Chapter 5). It should be noted that the first edition of the TGD (EC 1996) did not provide any quantitative guidance on the minimal size of the MOS.

A final draft version of the human health risk characterization part was released in 2005 with a detailed guidance on, among others, the main issues to be included in derivation of the reference MOS (MOSref), which is analogous to an overall assessment factor. The individual factors contributing to the MOSref are described separately and guidance is given on how to combine these into the MOSref. The guidance provided in this draft version has been extensively used in relation to the risk assessment of prioritized substances carried out since the draft version was released; however, this version is not publicly available.

In REACH, the new chemicals regulation, which entered into force on 1 June 2007, detailed guidance documents on different REACH elements, including risk characterization and the use of assessment factors, are currently in preparation (spring 2007). These documents will probably be available on the EU DG Environment REACH Web site (EU 2006) when published.

The EU is also involved in standard settings in relation to contaminants in ambient air (Section 9.2.3.1), drinking water (Section 9.2.3.2), soil (Section 9.2.3.3), and food (Section 9.2.3.4).

REFERENCES

Baird, S.J.S, J.T. Cohen, J.D. Graham, A.I. Shlyakhter, and J.S. Evans. 1996. Noncancer risk assessment: A probabilistic alternative to current practice. *Hum. Ecol. Risk Assess.* 2:79–102.

EC. 1994. Commission Regulation (EC) No 1488/94 of 28 June 1994 laying down the principles for the assessment of risks to man and the environment of existing substances in accordance with Council Regulation (EEC) No 793/93. *Off. J. Eur. Communities* L 161, 29.6.1994, pp. 3–11.

EC. 1996. Technical Guidance Document in support of Commission Directive 93/67/EEC on Risk Assessment for new notified substances and Commission Regulation (EC) No 1488/94 on Risk Assessment for existing substances. http://ecb.jrc.it/tgd

EC. 2003. Technical Guidance Document in support of Commission Directive 93/67/EEC on Risk Assessment for new notified substances, Commission Regulation (EC) No 1488/94 on Risk Assessment for existing substances and Directive 98/8/EC of the European Parliament and of the Council concerning the placing of biocidal products on the market. http://ecb.jrc.it/tgd

Edler, L., K. Poirier, M. Dourson, J. Kleiner, B. Mileson, H. Nordmann, A. Renwick, W. Slob, K. Walton, and G. Würtzen. 2002. Mathematical modeling and quantitative methods. *Food Chem. Toxicol.* 40:283–326.

EEC. 1993. Commission Directive 93/67/EEC of 20 July 1993, laying down the principles for the assessment of risks to man and the environment of substances notified in accordance with Council Directive 67/548/67. *Off. J. Eur. Communities* No L 227, 8.9.1993, pp. 9–18.

EU. 2006. The DG Environment REACH website. http://ec.europa.eu/environment/chemicals/reach/reach_intro.htm

OECD. 2003. *Descriptions of Selected Key Generic Terms Used in Chemical Hazard/Risk Assessment.* Joint Project with IPCS on the Harmonisation of Hazard/Risk Assessment Terminology. OECD Series on Testing and Assessment No. 44. Environment Directorate, Joint Meeting of the Chemicals Committee and the Working Party on Chemicals, Pesticides and Biotechnology. ENV/JM/MONO(2003)15. Paris: OECD.

Renwick, A.G., A. Flynn, R.J. Fletcher, D.J.G. Müller, S. Tuijelaars, and H. Verhagen. 2004. Risk–benefit analysis of micronutrients. *Food Chem. Toxicol.* 42:1903–1922.

Slob, W. and M.N. Pieter. 1998. A probabilistic approach for deriving acceptable human intake limits and human risks from toxicological studies: General framework. *Risk Anal.* 18:787–798.

US-EPA. 1995. Policy for risk characterization. Washington, DC: Science Policy Council. http://www.epa.gov/osp/spc/rcpolicy.htm

US-EPA. 2000. *Science Policy Council Handbook: Risk Characterization Handbook.* EPA 100-B00-002. Washington, DC: Science Policy Council. December. http://www.epa.gov/osp/spc/rchandbk.pdf

US-EPA. 2004. *An Examination of EPA Risk Assessment Principles and Practices.* EPA/100/B-04/001 March 2004. Washington, DC: Office of the Science Advisor. U.S. Environmental Protection Agency. http://www.epa.gov/osa/ratf.htm

van der Voet, H., and W. Slob. 2007. Integration of probabilistic exposure assessment and probabilistic hazard characterization. *Risk Anal.* 27:351–371.

9 Regulatory Standards Set by Various Bodies

Regulatory standards, or health-based guidance values, in this chapter denoted "guidance values," for exposure to chemicals in various media such as air, drinking water, soil, and food are set by various international, federal, and national bodies. This chapter will give an overview of the development of guidance values in general terms and present some examples.

9.1 GUIDANCE VALUES: DEVELOPMENT

This section gives an overview of the development of guidance values in general terms.

According to the OECD/IPCS definitions listed in Annex 1 (OECD 2003):

Guidance Value is "Value, such as concentration in air or water, which is derived after allocation of the reference dose among the different possible media (routes) of exposure."

Combined exposures from all media at their respective guidance values over a lifetime would be expected to be without appreciable health risk. The aim of a guidance value is to provide quantitative information from risk assessment for risk managers to enable them to make decisions concerning the protection of human health.

Guidance values are developed from a standard such as, e.g., an Acceptable/Tolerable Daily Intake (ADI/TDI), and Reference Dose/Concentration (RfD/RfC). For threshold effects, the standard is derived by dividing the No-Observed-Adverse-Effect Level (NOAEL) or Lowest-Observed-Adverse-Effect Level (LOAEL), or alternatively a Benchmark Dose (BMD) for the critical effect (s) by an overall assessment factor, described in detail in Chapter 5. For non-threshold effects, the standard is derived by a quantitative assessment, described in detail in Chapter 6.

To the extent possible, guidance values should reflect consideration of total exposure to the substance whether present in air, drinking water, soil, food, or other media. Guidance values should be derived for a clearly defined exposure scenario, e.g., for ambient air, drinking water, and soil based on the human exposure factors presented in Section 7.3. It should be recognized that there are no internationally agreed standard values for human exposure factors, and the examples presented in Section 7.3 serve to illustrate the differences in exposure factors provided by various exposure factor documents. Such differences can have a great impact on the guidance values, and it is therefore important that the most relevant and reliable exposure factors are used for the particular situation.

The development of guidance values for chemicals present in more than one environmental medium will require the allocation of proportions of the tolerable intake to various media based on information on relative proportion of total exposure from each of the media. If possible, estimation of exposure should be based on concentrations in the environmental media including ambient air, drinking water, soil, and food as well as consumer products. Unless there are age groups being more sensitive or having widely differing exposure profiles, the intake from each of the media should be estimated for adults, e.g., for ambient air, drinking water, and soil based on the human exposure factors presented in Section 7.3. If the data on concentrations of a substance in environmental media are inconsistent or inadequate, exposure could be estimated based on models, which incorporate as much data as possible on, e.g., production, use patterns, and physico-chemical properties.

An example of an exposure model to predict distribution in environmental media and estimation of the proportion of total exposure by various routes from consumer products is the EUSES (Section 7.2.4.3). It is important to recognize that the proportions of total intake from various media may vary, based on circumstances.

It should be noted that a source in one medium may lead to additional intake from other routes, e.g., for drinking water, dermal and inhalation exposure may occur during a shower, and for soil, exposure would often be both via ingestion and dermal contact; when possible, such intake should also be considered in the derivation of guidance values.

Total allocations of less than 100% of the tolerable intake are recommended to account for, e.g., those media for which exposure has not been characterized, and cross-route exposure. The proportion of the total intake, which is not allocated, should vary according to the adequacy of the exposure characterization from all media.

The precision of the guidance value is dependent upon the validity and reliability of the available data. In addition to the impact on the guidance value from the choice of an exposure factor as mentioned above, there are often sources of uncertainty in the derivation of the tolerable intake (Section 5.12) and in their allocation as a basis for the guidance values. The resulting guidance values thus represent a best estimate based on the available data at the time of development. The numerical value of a guidance value should reflect the precision in its derivation, and should usually be given to only one significant figure.

Guidance values can be set for the general population, occupationally exposed population, as well as for susceptible subgroups. The approach for setting guidance values in the ambient environment, i.e., ambient air, drinking water, soil, food, and other media relates primarily to long-term exposure of the general population. Some degree of human variability is taken into account in the assessment factors applied in the derivation of the tolerable intake (Chapter 5). Where a uniquely sensitive group forms a significant proportion of the population, the tolerable intake could be derived based on that group. In cases where the exposure profiles of this subgroup and the general population are similar, the guidance values should be based on the tolerable intake for the sensitive subgroup. If the exposure profiles differ, guidance values should be calculated separately for the subgroup and general population based on their respective tolerable intakes and exposure profiles, and the more conservative values adopted. The approach for setting guidance values relating to intermittent, short-term (e.g., accidental), and occupational exposures is basically similar to that for long-term exposure although there might be other considerations to take into account.

9.2   GUIDANCE VALUES: EXAMPLES

Various international, federal, and national bodies set guidance values for exposure to chemicals in various media such as air, drinking water, soil, and food. This section will present some examples of guidance values set by the WHO representing an international body, the US-EPA and the EU representing federal bodies, and Denmark representing a national body.

9.2.1   WHO

Examples of guidance values developed by the WHO include "air quality guidelines" and "drinking water guidelines," and (in collaboration with the FAO) "maximum residue limits" (MRLs) for pesticides and veterinary drugs and "maximum levels" for food additives.

9.2.1.1   Air Quality Guidelines

Recognizing the need of humans for clean air, the WHO Regional Office for Europe in 1987 published the first edition of the "*Air Quality Guidelines for Europe*" containing health risk assessments of 28 chemical air contaminants (WHO 1987). In 1993, air pollutants of special environmental and health significance to countries of the European Region were identified by a

WHO planning group, and 35 air pollutants were selected to be included in a second edition of the *Air Quality Guidelines*, which was published in 2000 (WHO 2000). It is noted that relevant EHC documents (Section 3.6.1.1) were of great value with respect to the selection of pollutions to be included in the second edition of the *Air Quality Guidelines* (WHO 2000). The publication is available via the WHO Regional Office for Europe's Web site (WHO 2007b).

The second edition of the *Air Quality Guidelines for Europe* (WHO 2000) comprises four introductory chapters plus sections on health risk evaluation and guidelines of the various pollutants.

The introductory Chapter 1 of the *Air Quality Guidelines* sets the scene regarding air quality issues, states the nature of the air quality guidelines, and describes the procedures used in the updating and revision process. In the following text, the most essential information in the context of this book is presented.

The second edition is limited to summaries of the data on which the guidelines are based; the full background evaluation should become progressively available on the WHO Regional Office for Europe's Web site (WHO 2007b). As in the first edition, detailed referencing of the relevant literature has been provided with indications of the periods covered by the reviews of individual pollutants.

The primary aim of the guidelines is to provide a basis for protecting public health from adverse effects of air pollution and for eliminating, or reducing to a minimum, those contaminants of air that are known or likely to be hazardous to human health and well-being. The guidelines are intended to provide background information and guidance to governments in making risk management decisions, particularly as a basis for setting standards or limit values for air contaminants.

The guidelines are not restricted to a numerical value below which exposure for a given period of time does not constitute a significant health risk; they also include any kind of recommendation or guidance in the relevant field. Numerical values either indicate the airborne concentration of a chemical combined with exposure times at which no adverse effect is expected in terms of noncarcinogenic endpoints, or they provide an estimate of lifetime cancer risk arising from those substances that are proven human carcinogens or carcinogens with at least limited evidence of human carcinogenicity. It is pointed out that the risk estimates for carcinogens do not indicate a safe level, but they are presented so that the carcinogenic potencies of different carcinogens can be compared and an assessment of overall risk made. The guidelines, including numerical values, are further addressed in the introductory Chapter 3 of the *Air Quality Guidelines* (WHO 2000).

When numerical guideline values are given, these values are not standards in themselves. Before transforming them into legally binding standards, the guideline values must be considered in the context of prevailing exposure levels, technical feasibility, source control measures, abatement strategies, and social, economic, and cultural conditions; this is further addressed in the introductory Chapter 4 of the *Air Quality Guidelines* (WHO 2000).

It is stated that inhalation of an air pollutant in concentrations and for exposure times below a guideline value will not have adverse effects on health; however, compliance with recommendations regarding guideline values does not guarantee the absolute exclusion of effects at levels below such values. As an example it is mentioned that highly sensitive groups such as those impaired by concurrent disease or other physiological limitations may be affected at or near concentrations referred to in the guideline values. Health effects at or below guideline values may also result from combined exposure to various chemicals or from exposure to the same chemical by multiple routes.

It has been agreed with the EU Commission that the final drafts of the revised WHO guideline documents would provide a starting point for discussions by the Commission's working groups aiming at setting legally binding limit values for air quality in the EU, see Section 9.2.4.1.

The introductory Chapter 2 of the *Air Quality Guidelines* (WHO 2000) gives a very detailed and comprehensive description of the criteria used in establishing the guideline values including criteria for selection of NOAEL/LOAEL, adverse effect, benchmark approach, and uncertainty factors. These criteria are comparable to the principles outlined in Chapters 4 and 5 in this book. There are also criteria for selection of averaging times and for consideration of sensory effects (malodorous properties at concentrations far below those at which toxic effects occur). In addition, there are criteria for carcinogenic endpoint including qualitative assessment of carcinogenicity, quantitative assessment of carcinogenic potency, quantitative assessment of carcinogenicity based on human data, risk estimates from animal cancer bioassays, and interpretation of risk estimates; these criteria are comparable to the principles outlined in Chapter 6 in this book. It is specifically pointed out that during the preparation of this second edition of the guidelines, attention was paid to defining specific sensitive subgroups in the population.

WHO global air quality guidelines are available for the following air pollutants: particulate matter, ozone, nitrogen dioxide, and sulfur dioxide. The latest revision, which was published in 2005, substantially lowered the recommended limits of particulate matter, ozone, and sulfur dioxide. The publication is available via the WHO Regional Office for Europe's Web site (WHO 2007b).

9.2.1.2 Drinking-Water Guidelines

The WHO published the first and second editions of the *"Guidelines for Drinking-water Quality,"* in three volumes, in 1984–1985 and in 1993–1997, respectively. The first edition contains health risk assessments of 36 inorganic constituents and physical parameters, 27 organic compounds, 30 pesticides, 4 disinfectants, 5 disinfectant by-products, and 6 other chlorination by-products. The second edition contains health risk assessments of 6 inorganic constituents (updated), 3 organic compounds (1 new, 2 updated), 10 pesticides (7 new, 3 updated), and 1 disinfectant by-product (new). A third edition of Volume 1 was published in 2004. The Guidelines are kept up to date by a "rolling revision" and a fourth edition is scheduled for 2008. The publications are available to the public via the WHO Water Sanitation and Health Web site (WHO 2007a).

Volume 1 "Recommendations" addresses requirements to ensure drinking-water safety, including minimum procedures and specific guideline values, and how those requirements are intended to be used. The volume also describes the approaches used in deriving the guidelines, including guideline values. It includes fact sheets on significant microbial and chemical hazards. Chapter 12 provides the fact sheets for the individual chemical contaminants evaluated. For those contaminants for which a guideline value has been established, the fact sheets include a brief toxicological overview of the chemical, the basis for guideline derivation, treatment achievability, and analytical limit of detection.

Volume 2, *"Health Criteria and other Supporting Information,"* explains how guideline values for the contaminants are to be used, defines the criteria used to select the various chemical, physical, microbiological, and radiological contaminants included, describes the approaches used in deriving guideline values, and presents, in the form of brief monographs, critical reviews and evaluations of the effects on human health of the substances or contaminants examined. The introductory chapter of Volume 2, second edition (WHO 1996), sets the scene regarding drinking-water quality issues, gives some general considerations, states the nature of the drinking-water quality guidelines, and describes the criteria for the selection of health-related drinking-water contaminants. In the following text, the most essential information in the context of this book is presented.

The primary aim of the *Guidelines for Drinking-water Quality* is the protection of public health. The guidelines are addressed primarily to water and health regulators, policymakers and their advisors and intended to be used as a basis for the development of national standards that may ensure the safety of drinking-water supplies through the elimination, or reduction to a minimum concentration, of constituents of water that are known to be hazardous to health. It is pointed out that the guideline values recommended are not mandatory limits as, in order to define such limits, it is necessary to consider the guideline values in the context of local or national environmental, social, economic, and cultural conditions.

It is noted that the problems associated with chemical constituents of drinking water arise primarily from their ability to cause adverse health effects after prolonged periods of exposure, and it is pointed out that of particular concern are contaminants that have cumulative toxic properties such as heavy metals and carcinogenic substances. It is also pointed out that the use of chemical disinfectants in water treatment usually results in the formation of chemical by-products, some of which are potentially hazardous.

Guideline values have been set for potentially hazardous water constituents and provide a basis for assessing drinking-water quality. A guideline value represents the concentration of a constituent that does not result in any significant risk to the health of the consumer over a lifetime of consumption. However, it is emphasized that the guideline values should not be regarded as implying that the quality of drinking water may be degraded to the recommended level. It is pointed out that short-term deviations above the guideline values do not necessarily mean that the water is unsuitable for consumption. The amount by which, and the period for which, any guideline value can be exceeded without affecting public health depends upon the specific substance involved.

In some instances, provisional guideline values have been set for constituents for which there is some evidence of a potential hazard but where the available information on health effects is limited. Provisional guideline values have also been set for substances for which the calculated guideline value would be below the practical quantification level, or below the level that can be achieved through practical treatment methods, as well as for certain substances when it is likely that guideline values will be exceeded as a result of disinfection procedures.

It is stated that thousands of chemicals have been identified in drinking-water supplies around the world, many in extremely low concentrations. The chemicals selected for the development of guideline values include those considered potentially hazardous to human health, those detected relatively frequently in drinking water, and those detected in relatively high concentrations.

Chapter 12 of Volume 2, second edition (WHO 1996), gives a detailed and comprehensive description of the principles used in establishing the guideline values.

The assessment of the toxicity of drinking-water contaminants has been made on the basis of published reports from the open literature, information submitted by governments and other interested parties, and unpublished proprietary data. In the development of the guideline values, existing international approaches to developing guidelines were carefully considered. Previous risk assessments developed by the WHO/IPCS in EHC monographs (see Section 3.6.1.1), IARC (see Section 3.6.1.2), JMPR (see Section 3.6.1.3), and JECFA (see Section 3.6.1.3) were reviewed. These assessments were relied upon except where new information justified a reassessment. The quality of new data was critically evaluated prior to their use in risk assessment.

The section on derivation of guideline values using a tolerable daily intake (TDI) includes criteria for derivation of a TDI from a NOAEL or LOAEL by application of uncertainty factors; these criteria are comparable to the principles outlined in Chapters 4 and 5 in this book. The guideline value is then derived from the TDI as follows:

$$GV = [TDI \times bw \times P]/C$$

where
   bw = body weight (60 kg for adults, 10 kg for children, 5 kg for infants)
   P = proportion of the TDI allocated to drinking water
   C = daily drinking-water consumption (2 L for adults, 1 L for children, 0.75 L for infants)

In many cases, the intake of a chemical from drinking water is small in comparison with that from other sources such as food and air. Guideline values derived using the TDI approach take into account exposure from all sources by allocating a percentage of the TDI to drinking water. When possible, data concerning the proportion of total intake normally ingested in drinking water (based on mean levels in food, air, and drinking water) or intakes estimated on the basis of consideration of physical and chemical properties were used in the derivation of the guideline values. Where such information was not available, an arbitrary (default) value of 10% for drinking water was used.

This default value was considered, in most cases, to be sufficient to account for additional routes of intake, i.e., inhalation and dermal absorption of contaminants in water.

There is also a section on derivation of guideline values for potential carcinogens; the criteria in this section are comparable to the principles outlined in Chapter 6 in this book.

Volume 3 "Surveillance and control of community supplies" contains recommendations and information concerning what needs to be done in small communities, particularly in developing countries, to safeguard their water supplies.

9.2.1.3 Food

The Codex Alimentarius Commission was created in 1963 by FAO and WHO to develop food standards, guidelines, and related texts such as codes of practice under the Joint FAO/WHO Food Standards Program. The main purposes of this Program are to protect the health of the consumers and to ensure fair trade practices in the food trade, and to promote the coordination of all food standards work undertaken by international governmental and nongovernmental organizations (CA 2007).

The Codex Alimentarius is a collection of standards, codes of practice, guidelines, and other recommendations. Some of these texts are very general, and some are very specific. Some deal with detailed requirements related to a food or group of foods; others deal with the operation and management of production processes or the operation of government regulatory systems for food safety and consumer protection.

The main FAO/WHO expert bodies include the Joint FAO/WHO Expert Committee on Food Additives (JECFA), the Joint FAO/WHO Meetings on Pesticide Residues (JMPR), and the Joint FAO/ WHO Expert Meetings on Microbiological Risk Assessment (JEMRA). Codex Alimentarius provides lists of MRLs for pesticides and veterinary drugs, and maximum levels for food additives.

The Joint FAO/WHO Expert Committee on Food Additives (JECFA) was established in 1955 to consider chemical, toxicological, and other aspects of contaminants and residues of veterinary drugs in foods for human consumption. The Codex Committee on Food Additives and Contaminants and the Codex Committee on Residues of Veterinary Drugs in Foods identify food additives, contaminants, and veterinary drug residues that should receive priority evaluation and refer them to JECFA for assessment before incorporating them into Codex standards.

The Joint FAO/WHO Meetings on Pesticide Residues (JMPR) began work in 1963 following a decision that the Codex Alimentarius Commission should recommend MRLs for pesticides and environmental contaminants in specific food products to ensure the safety of foods containing residues. It was also decided that JMPR should recommend methods of sampling and analysis. There is close cooperation between JMPR and the Codex Committee on Pesticide Residues (CCPR). CCPR identifies those substances requiring priority evaluation. After JMPR evaluation, CCPR discusses the recommended MRLs and, if they are acceptable, forwards them to the Commission for adoption as Codex MRLs.

The Joint FAO/WHO Expert Meetings on Microbiological Risk Assessment (JEMRA) began work in 2000 to develop and provide advice to the Codex Alimentarius Commission on microbiological aspects of food safety. In addition to providing risk assessments, JEMRA develops guidance on related areas such as data collection and the application of risk assessment. JEMRA works most closely with the Codex Committee on Food Hygiene, but has also provided advice to other Codex committees, such as the Committee on Fish and Fishery Products.

9.2.2 UNITED STATES

The U.S. Environmental Protection Agency (US-EPA) is the federal agency responsible for regulating the level of contaminants in ambient air (Section 9.2.2.1), drinking water (Section 9.2.2.2), and soil (Section 9.2.2.3) while the U.S. Food and Drug Administration (US-FDA) is the federal agency responsible for regulating the level of contaminants in food (Section 9.2.2.4).

9.2.2.1 Air

The Clean Air Act (CAA), amended in 1970 and in 1990 (US-EPA 1990), is the federal law under which the US-EPA sets limits for air pollutants anywhere in the United States. This ensures uniform basic health and environmental protection across the country. The law allows individual states to have stronger pollution controls, but states are not allowed to have weaker pollution controls than those set for the whole country. The states do much of the work to carry out the Act (US-EPA 2007a).

States have to develop state implementation plans (SIPs) that explain how each state will do its job under the Clean Air Act. A state implementation plan is a collection of the regulations a state will use to clean up polluted areas. The states must involve the public, through hearings and opportunities to comment, in the development of each state implementation plan.

US-EPA must approve each SIP, and if a SIP is not acceptable, US-EPA can take over enforcing the Clean Air Act in that state.

The U.S. government, through US-EPA, assists the states by providing scientific research, expert studies, engineering designs, and money to support clean air programs.

US-EPA refers to chemicals that cause serious health and environmental hazards as hazardous air pollutants (HAPs) or air toxics. The 1970 Clean Air Act gave US-EPA authority to list air toxics for regulation and then to regulate the chemicals. The agency listed and regulated seven chemicals through 1990. The 1990 Act includes a list of 189 hazardous air pollutants selected by Congress on the basis of potential health and/or environmental hazard; US-EPA must regulate these listed air toxics. The 1990 Act allows US-EPA to add new chemicals to the list as necessary.

US-EPA develops regulations, MACT (Maximum Achievable Control Technology) standards, requiring sources to meet specific emission limits that are based on emission levels already being achieved by many similar sources in the country. Then US-EPA applies a risk-based approach to assess how these technology-based emission limits are reducing health and environmental risks. Based on this assessment, US-EPA may implement additional standards to address any significant remaining, or residual, health or environmental risks.

US-EPA is developing an air toxics risk assessment (ATRA) reference library for conducting air toxics analyses at the facility and community-scale. This library provides information on the fundamental principles of risk-based assessment for air toxics and how to apply those principles in different settings as well as strategies for reducing risk at the local level.

A Technical Resource Manual is available, which gives comprehensive guidance on the risk assessment process (US-EPA 2004).

9.2.2.2 Drinking Water

The Safe Drinking Water Act (SDWA) is the main federal law that ensures the quality of the drinking water (US-EPA 2007b).

The SDWA was originally passed by Congress in 1974 to protect public health by regulating the nation's public drinking-water supply. The law was amended in 1986 and 1996 and requires many actions to protect drinking water and its sources: rivers, lakes, reservoirs, springs, and groundwater wells. SDWA does not regulate private wells, which serve fewer than 25 individuals. SDWA authorizes the US-EPA to set national health-based standards for drinking water to protect against both naturally occurring and man-made contaminants that may be found in drinking water. US-EPA, States, and water systems then work together to make sure that these standards are met.

US-EPA sets national standards for tap water, which help ensure consistent quality in the water supply. US-EPA prioritizes contaminants for potential regulation based on risk and how often they occur in water supplies. Certain water systems are monitored for the presence of contaminants for which no national standards currently exist and collect information on their occurrence. US-EPA sets a health goal based on risk, including risks to the most sensitive people, e.g., infants, children, pregnant women, the elderly, and the immuno-compromised. US-EPA then sets a legal limit for the contaminant in drinking water or a required Treatment Technique (TT). This limit or TT is set to be as close to the health goal as feasible. US-EPA also performs a cost–benefit analysis and obtains input from interested parties when setting standards.

US-EPA uses the following steps to set enforceable, health-based drinking-water standards:

- Determine whether a contaminant should be regulated based on peer-reviewed science, including data on: how often the contaminant occurs in the environment; how humans are exposed to it; and the health effects of exposure (particularly to vulnerable subpopulations).
- Set a Maximum Contaminant Level Goal (MCLG) (the level of a contaminant in drinking water below which there is no known or expected health risk. MCLGs allow for a margin of safety). These goals take into account the risks of exposure for certain sensitive populations, such as infants, the elderly, and persons with compromised immune systems. These goals are not enforceable levels because they do not take available technology into consideration, and therefore are sometimes set at levels which public water systems cannot meet.
- Propose an enforceable standard in the form of a Maximum Contaminant Level (MCL) (the maximum amount of a contaminant allowed in water delivered to a user of any public water system) or a treatment technique (TT) (required procedure or level of technological performance set when there is no reliable method to measure a contaminant at very low levels). MCLs are set as close to MCLGs as feasible, considering available technology and cost. Examples of rules requiring TTs are the Surface Water Treatment Rule (requires disinfection and filtration) and the Lead and Copper Rule (requires optimized corrosion control). Water samples that contain lead or copper exceeding the action level trigger additional treatment or other requirements that a water system must follow. Required testing (monitoring) schedules are part of the enforceable standard. After determining a proposed MCL or TT that is as close to the MCLG as possible based on affordable technology, US-EPA must complete an economic analysis to determine whether the benefits of that standard justify the costs. If not, US-EPA may adjust the MCL for a particular class or group of systems to a level that maximizes health risk reduction benefits at a cost that is justified by the benefits. US-EPA may not adjust the MCL if the benefits justify the costs to large systems and small systems that are unlikely to receive variances.
- Set an enforceable MCL or TT. After considering comments on the proposed standard and other relevant information, US-EPA makes final an enforceable MCL or TT, including required testing and reporting schedules. States are authorized to grant variances from standards for systems serving up to 3,300 people if the systems cannot afford to comply with a rule (through treatment, an alternative source of water, or other restructuring) and the systems install US-EPA approved variance technology. States can grant variances to systems serving 3,301–10,000 people with US-EPA approval. SDWA does not allow small systems to have variances for microbial contaminants. Under certain circumstances exemptions from standards may be granted to allow extra time to seek other compliance options or financial assistance. After the exemption period expires, the public water system must be in compliance. The terms of variances and exemptions must ensure no unreasonable risk to public health.

9.2.2.2.1   The Contaminant Candidate List

The 1996 Amendments to SDWA require that every 5 years US-EPA establish a list of contaminants which are known or anticipated to occur in public water systems and may require future regulations under SDWA. The list is developed with significant input from the scientific community and other interested parties. After establishing this contaminant candidate list, US-EPA identifies contaminants, which are priorities for additional research and data gathering. US-EPA uses this information to determine whether or not a regulation is appropriate and this process is repeated for each list, every 5 years.

In order to support this decision making, US-EPA has also established a National Contaminant Occurrence Database (NCOD), which stores data on the occurrence of both regulated and unregulated contaminants. US-EPA is also required to list and develop regulations for monitoring certain unregulated contaminants. This monitoring data will provide the basis for identifying contaminants that may be placed on future Contaminant Candidate Lists and support the US-EPA Administrator's decisions to regulate contaminants in the future.

9.2.2.3 Soil

The U.S. National Environmental Policy Act of 1969 required careful analysis of the consequences of any federally funded project. The Resource Conservation and Recovery Act (RCRA) of 1976 established guidelines for handling, transport, and hauling of hazardous materials, such as required in cleanup of soil contaminants. The Comprehensive Environmental Response, Compensation, and Liability Act (CERCLA) of 1980 established, for the first time, strict rules on legal liability for soil contamination. CERCLA stimulated identification and cleanup of thousands of contaminated land sites, and consequently raised awareness of property buyers and sellers to make soil contamination a focal issue of land use and management practices (US-EPA 2007c).

9.2.2.3.1   The Comprehensive Environmental Response, Compensation, and Liability Act

The Comprehensive Environmental Response, Compensation, and Liability Act (CERCLA), commonly known as Superfund, was enacted by the U.S. Congress on 11 December 1980. CERCLA created a tax on the chemical and petroleum industries and provided broad Federal authority to respond directly to releases or threatened releases of hazardous substances that may endanger public health or the environment. Over 5 years, $1.6 billion was collected and the tax went to a trust fund for cleaning up abandoned or uncontrolled hazardous waste sites (US-EPA 2007c).

CERCLA established prohibitions and requirements concerning closed and abandoned hazardous waste sites; provided for liability of persons responsible for releases of hazardous waste at these sites; and established a trust fund to provide for cleanup when no responsible party could be identified.

CERCLA authorizes two kinds of response actions: Short-term Removals, where actions may be taken to address releases or threatened releases requiring prompt response; and Long-term Remedial Response Actions, that permanently and significantly reduce the dangers associated with releases or threats of releases of hazardous substances that are serious, but not immediately life threatening. These actions can be conducted only at sites listed on US-EPA's National Priorities List (NPL).

CERCLA also enabled the revision of the U.S. National Contingency Plan (NCP). The NCP provides the guidelines and procedures needed to respond to releases and threatened releases of hazardous substances, pollutants, or contaminants. The NCP also established the NPL.

CERCLA was amended by the Superfund Amendments and Reauthorization Act (SARA) on 17 October 1986.

9.2.2.3.2   The Superfund Program

The goal of the Superfund program is to clean up uncontrolled hazardous waste sites that pose unacceptable risks to human health and environment in a manner that restores these sites to uses appropriate for nearby communities. As already mentioned, the program was authorized under the CERCLA of 1980 (US-EPA 2007c).

The Soil Screening Guidance, Technical Background Document (US-EPA 1996a) provides the technical background for the development of methodologies described in the Soil Screening Guidance: User's Guide (US-EPA 1996b), along with additional information useful for soil screening. Together, these documents define the framework and methodology to develop soil screening levels (SSLs) for chemicals commonly found at Superfund sites.

SSLs are risk-based concentrations derived from standardized equations combining exposure information assumptions with US-EPA toxicity data. For the ingestion, dermal, and inhalation pathways, toxicity criteria are used to define an acceptable level of contamination in soil, based on a one-in-a-million (10^{-6}) individual excess cancer risk for carcinogens and a Hazard Quotient (HQ) of 1 for noncarcinogens. The hazard quotient is defined as the ratio of an exposure estimate over the Reference Dose or Concentration (Section 5.1), i.e., HQ = Exposure/(RfD or RfC).

9.2.2.4   Food

The US-FDA is the federal agency responsible for regulating the level of contaminants in food (US-FDA 2007). The US-FDA establishes "action levels" for poisonous or deleterious substances to control levels of contaminants in human food and animal feed.

Action levels and tolerances are established based on the unavoidability of the poisonous or deleterious substances and do not represent permissible levels of contamination where it is avoidable. The blending of a food or feed containing a substance in excess of an action level or tolerance with another food or feed in order to lower the concentration of a contaminant is not permitted, and the final product resulting from blending is unlawful, regardless of the level of the contaminant.

Action levels and tolerances represent limits at or above which the US-FDA will take legal action to remove products from the market. Where no established action level or tolerance exists, the US-FDA may take legal action against the product at the minimal detectable level of the contaminant.

The action levels are established and revised according to criteria specified in Title 21, Code of Federal Regulations, Parts 109 and 509 and are revoked when a regulation establishing a tolerance for the same substance and use becomes effective.

With respect to veterinary medicines, the US-FDA establishes tolerances to include a safety factor to assure that the drug will have no harmful effects on consumers of the food product. The US-FDA first determines the level at which the drug does not produce any measurable effect in laboratory animals. From this, the US-FDA determines an acceptable daily intake (ADI), and the drug tolerance and withdrawal times are then determined so that the concentrations of drug residues in edible tissues are below the ADI. Depending on the drug, "safety factors" of between 100-fold to 2000-fold are included in the calculations used to set the tolerances.

9.2.3   EUROPEAN UNION

9.2.3.1   Air

Since the early 1970s, the European Union (EU) has been working to improve air quality by controlling emissions of harmful substances into the atmosphere, improving fuel quality, and by integrating environmental protection requirements into the transport and energy sectors.

Through EU legislation, major air pollutants have been regulated. For example, through an EU Directive (EC 1999), the EU has established limit values for concentrations of sulfur dioxide, nitrogen dioxide and nitrogen oxides, particulate matter and lead, as well as alert thresholds for concentrations of sulfur dioxide and nitrogen oxide, in ambient air. Member States must take the measures necessary to ensure that concentrations of the pollutants in ambient air do not exceed the limit values.

The EU's Sixth Environment Action Programme (EAP), "Environment 2010: Our future, Our choice," includes Environment and Health as one of the four main target areas requiring greater effort. Air pollution is one of the issues highlighted in this area. The Sixth EAP aims to achieve levels of air quality that do not result in unacceptable impacts on, and risks to, human health and the environment.

The EU is acting at many levels to reduce exposure to air pollution: through EC legislation, through work at international level to reduce cross-border pollution, through cooperation with sectors responsible for air pollution, through national, regional authorities and NGOs, and through research. The Clean Air for Europe (CAFE) initiative has led to a thematic strategy setting out the objectives and measures for the next phase of European air quality policy.

9.2.3.1.1 Clean Air for Europe

Clean Air for Europe (CAFE) was launched in March 2001 (CAFE 2007). CAFE is a program of technical analysis and policy development, which supported the development of the Thematic Strategy on Air Pollution under the Sixth EAP. The EU Commission adopted the Thematic Strategy in September 2005.

The press release stated

> The European Commission today proposed an ambitious strategy for achieving further significant improvements in air quality across Europe. The Thematic Strategy on air pollution aims by 2020 to cut the annual number of premature deaths from air pollution-related diseases by almost 40% from the 2000 level. It also aims to substantially reduce the area of forests and other ecosystems suffering damage from airborne pollutants. While covering all major air pollutants, the Strategy pays special attention to fine dust, also known as particulates, and ground-level ozone pollution because these pose the greatest danger to human health. Under the Strategy the Commission is proposing to start regulating fine airborne particulates, known as $PM_{2.5}$, which penetrate deep into human lungs. The Commission also proposes to streamline air quality legislation by merging existing legal instruments into a single Ambient Air Quality Directive, a move that will contribute to better regulation.

The proposed Directive (the CAFE Directive) is the Directive on Ambient Air Quality and Cleaner Air for Europe (EC 2005).

9.2.3.2 Drinking Water

The Drinking Water Directive (EC 1998), concerns the quality of water intended for human consumption. The objective of the Drinking Water Directive is to protect the health of the consumers in the EU and to make sure the water is wholesome and clean (free of unacceptable taste, odor, color) and that it has a pleasant appearance (EC 2007a).

The Drinking Water Directive sets standards for the most common substances (so-called parameters) that can be found in drinking water. In the Drinking Water Directive a total of 48 microbiological and chemical parameters must be monitored and tested regularly. In principle WHO guidelines for drinking water are used as a basis for the standards in the Drinking Water Directive.

EU Member States must transpose the Drinking Water Directive into their own national legislation. The Member States can include additional requirements, e.g., regulate additional substances that are relevant within their territory or set higher standards. But Member States are not allowed to set lower standards as the level of protection of human health should be the same within the whole EU.

Member States have to monitor the quality of the drinking water supplied to their citizens and this has to be done mainly at the tap inside private and public premises. Also the quality of drinking water used in the food production industry has to be monitored to make sure it complies with the EU standards. Member States report at three yearly intervals the monitoring results to the European Commission.

The Commission assesses the results of water quality monitoring against the standards in the Drinking Water Directive. After each reporting cycle the Commission produces a synthesis report, which summarizes the quality of drinking water and its improvement at a European level. The synthesis reports are available to the public (EU 2007).

9.2.3.3 Soil

An EU Soil Framework Directive is presently being developed. The EU has decided to adopt a Thematic Strategy on Soil Protection as part of its aim of protection and preservation of natural resources. A proposal for this is being finalized in 2006 (EC 2007b).

The Strategy will comprise three elements:

- A Communication laying down the principles of Community Soil Protection Policy
- A Legislative proposal for the protection of soil - A Soil Framework Directive
- An analysis of the environmental, economic, and social impacts of the proposals

The proposed EU Soil Framework Directive will include the following main objectives:

- Implementation of common principles
- Future prevention of soil degradation
- Conservation of soil functions
- Protection of sustainable land use

9.2.3.4 Food

9.2.3.4.1 The European Food Safety Authority

The European Food Safety Authority (EFSA), based in Parma, Italy, since 2005 is the keystone of EU risk assessment regarding food and feed safety.

EFSA was legally established by a European Parliament and Council Regulation adopted in 2002 following a series of food scares in the 1990s including the Bovine Spongiform Encephalopathy and dioxins scandals, which undermined consumer confidence in the safety of the food chain (EFSA 2007).

The responsibility for risk assessment is clearly separated from that of risk management. While EFSA advises on possible risk related to food safety, the responsibility for risk management lies with the EU political institutions (European Commission, European Parliament, and the Council, i.e., EU Member States). It is the role of the EU institutions, to propose and adopt legislation as well as regulatory and control measures when and where required, taking into account EFSA's advice as well as other considerations. With respect to chemicals, EFSA assesses food additives, flavorings, processing aids and materials in contact with food, additives and products or substances used in animal feed, plant health, plant protection products and their residues, and contaminants in the food chain (EFSA 2007). The basic principles of EU legislation on contaminants in food are that food containing a contaminant to an amount unacceptable from the public health viewpoint and in particular at a toxicological level, shall not be placed on the market, contaminant levels shall be kept as low as can reasonably be achieved following recommended good working practices, and maximum levels must be set for certain contaminants in order to protect public health.

Maximum residue levels in certain foods are set for the following contaminants: nitrate, mycotoxins (aflatoxins, ochratoxin A, patulin, deoxynivalenol, zearalenone, fumonisins, T'-2 and HT-2-toxin), metals (lead, cadmium, mercury, inorganic tin), 3-MCPD (3-monochloro-propane-1,2-diol), dioxins and PCBs, and polycyclic aromatic hydrocarbons (benzo(*a*)pyrene) (EC 2007c).

9.2.3.4.2 Residues of Veterinary Medicines in Food

The European Medicines Agency (EMEA) is a decentralized body of the EU with headquarters in London. The EMEA began its activities in 1995, when the European system for authorizing medicinal products was introduced, providing for a centralized and a mutual recognition procedure. The EMEA has a role in both, but is primarily involved in the centralized procedure. Where the centralized procedure is used, companies submit one single marketing authorization application to the EMEA. A single evaluation is carried out through the Committee for Medicinal Products for Human Use (CHMP) or Committee for Medicinal Products for Veterinary Use (CVMP). If the relevant Committee concludes that quality, safety, and efficacy of the medicinal product are sufficiently proven, it adopts a positive opinion. This is sent to the Commission to be transformed into a single market authorization valid for the whole of the EU (EMEA 2007).

The CVMP is responsible for preparing the Agency's opinions on all questions concerning veterinary medicinal products, in accordance with Regulation (EC) No. 726/2004 (EC 2004).

A core activity of the CVMP is the establishment of MRLs of veterinary medicinal products permissible in food produced by or from animals for human consumption, including dairy products, meat, honey, etc. These limits must be established for all pharmacologically active substances contained in a medicinal product before the product can be granted a marketing authorization.

9.2.4 Denmark

The procedure for setting health-based quality criteria in Denmark is described briefly, to illustrate the use of international principles in national chemicals regulation.

In Denmark, health-based quality criteria are set for chemical substances in ambient air, drinking water, and soil according to principles laid down in a guidance document from the Danish Environmental Protection Agency (D-EPA 2006). These principles are for threshold effects comparable to the principles outlined in Chapters 4 and 5 in this book, and for non-threshold effects in Chapter 6.

The health-based quality criteria are derived for the relevant media (ambient air, drinking water, soil) by dividing the TDI with a standard exposure factor for this media.

The following general calculation method is used:

$$QC_{asw} = \frac{TDI \cdot V \cdot f}{e_{asw}}$$

where

QC_{asw} is the quality criterion for air, soil, or drinking water
TDI is the tolerable daily intake expressed as mg/kg bw/day
V is the body weight (bw) in kg
f is the allocation factor, i.e., the fraction of the TDI, which is allocated to exposure from air, soil, or drinking water (Section 9.1)
e_{asw} is the standard exposure factor

The standard exposure factors applied are based on children's exposures. The following factors are used:

- Air: 0.5 m^3/kg body weight per day (children 1–5 years old)
- Soil: 0.0002 kg/day if the major part of the TDI is allocated to intake of the contaminant from soil
  Soil: 0.0001 kg/day if only a minor part of the TDI is allocated to intake of the contaminant from soil, or if the basis for the soil quality criterion is the 10^{-6} lifetime risk for a carcinogenic substance
- Drinking water: 0.08 L/day if the major part of the TDI is allocated to intake of the contaminant from drinking water
  Drinking water: 0.03 L/day if only a minor part of the TDI is allocated to intake of the contaminant from drinking water, or if the basis for the drinking-water quality criterion is the 10^{-6} lifetime risk for a carcinogenic substance

The health-based quality criteria derived as described above are used as the basis for the setting of quality criteria for chemical substances in soil and drinking water, and of C-values (Contribution values, the maximum amount of any pollutant a company is allowed to emit in the air) in ambient air. In this step, other than health-based viewpoints may be taken into account, including aesthetic factors such as odor (all media), discoloration (soil, drinking water), taste (drinking water), and microbial growth (drinking water).

REFERENCES

CA. 2007. The CODEX Alimentarius website. http://www.codexalimentarius.net/web/index_en.jsp

CAFE. 2007. The CAFE Programme website. http://ec.europa.eu/environment/air/cafe/index.htm

D-EPA. 2006. Principles for establishment of health based quality criteria for ambient air, soil and drinking water. Vejledning fra Miljøstyrelsen Nr. 5 2006. Danish Environmental Protection Agency, Danish Ministry of the Environment (in Danish). http://www2.mst.dk/Udgiv/publikationer/2006/87-7052-182-4/pdf/87-7052-182-4.pdfEC. 1998. Council Directive 98/83/EC of 3 November 1998 on the quality of water intended for human consumption. *Off. J. Eur. Communities* L 330, 5.12.1998, 32–54.

EC. 1999. Council Directive 1999/30/EC of 22 April 1999 relating to limit values for sulphur dioxide, nitrogen dioxide and oxides of nitrogen, particulate matter and lead in ambient air. *Off. J. Eur. Communities* L 163, 29.6.1999, 41–60.

EC. 2004. Regulation (EC) No. 726/2004 of the European Parliament And of the Council of 31 March 2004 laying down Community procedures for the authorisation and supervision of medicinal products for human and veterinary use and establishing a European Medicines Agency. *Off. J. Eur. Communities* L 136, 30.4.2004, 1–33.

EC. 2005. Proposal for a Directive of the European Parliament and of the Council on ambient air quality and cleaner air for Europe. http://ec.europa.eu/environment/air/cafe/pdf/cafe_dir_en.pdf

EC. 2007a. European Commission. Environment Policies Water website. http://europa.eu.int/comm/environment/water/index.html

EC. 2007b. European Commission. Environment Policies Soil website. http://europa.eu.int/comm/environment/soil/index.htm

EC. 2007c. European Commission. Food and feed - Chemical contaminants website. http://ec.europa.eu/food/food/chemicalsafety/contaminants/legisl_en.htm

EFSA. 2007. European Food Safety Authority website. http://www.efsa.europa.eu/en.html

EMEA. 2007. The European Medicines Agency website. http://www.emea.europa.eu/

EU. 2007. DG Environment website. http://ec.europa.eu/environment/water/water-drink/index_en.html

OECD. 2003. *Descriptions of Selected Key Generic Terms Used in Chemical Hazard/Risk Assessment. Joint Project with IPCS on the Harmonisation of Hazard/Risk Assessment Terminology.* OECD Series on Testing and Assessment No. 44. Environment Directorate, Joint Meeting of the Chemicals Committee and the Working Party on Chemicals, Pesticides and Biotechnology. ENV/JM/MONO(2003)15. Paris: OECD.

US-EPA. 1990. U.S. EPA Clean Air Act. http://www.epa.gov/air/caa/index.html

US-EPA. 1996a. U.S. EPA Soil Screening Guidance (SSG). Soil Screening Guidance: Technical Background Document. EPA Document Number: EPA/540/R-95/128. July 1996. Washington, DC:Office of Emergency and Remedial Response. U.S. Environmental Protection Agency http://www.epa.gov/superfund/resources/soil/toc.htm

US-EPA. 1996b. *Soil Screening Guidance: User's Guide.* EPA Document Number: EPA540/R-96/018. July 1996. Washington, DC: Office of Emergency and Remedial Response.U.S. Environmental Protection Agency. http://www.epa.gov/superfund/resources/soil/ssg496.pdf

US-EPA. 2004. *U.S. EPA Air Toxics Risk Assessment Reference Library. Technical Resource Manual.* EPA-453-K-04-001A. April 2004. Research Triangle Park, NC: Office of Air Quality Planning and Standards. http://www.epa.gov/ttn/fera/risk_atra_vol1.html

US-EPA. 2007a. U.S. EPA Air and Radiation website. http://www.epa.gov/air/

US-EPA. 2007b. U.S. EPA Ground Water and Drinking Water website. http://www.epa.gov/safewater/

US-EPA. 2007c. U.S. EPA Superfund website http://www.epa.gov/superfund/index.htm

US-FDA. 2007. U.S. Food and Drug Administration website. http://www.fda.gov/

WHO. 1987. *Air Quality Guidelines for Europe.* Copenhagen, WHO Regional Office for Europe. WHO Regional Publications, European Series, No. 23. Copenhagen: WHO.

WHO. 1996. *Guidelines for Drinking-Water Quality. Second Edition.* Volume 2: Health Criteria and other Supporting Information. Geneva:WHO.

WHO. 2000. *Air Quality Guidelines for Europe. Second Edition.* Copenhagen, WHO Regional Office for Europe. WHO Regional Publications, European Series, No. 91. Copenhagen: WHO.

WHO. 2007a. WHO Water Sanitation and Health website. http://www.who.int/water_sanitation_health/dwq/guidelines/en/

WHO. 2007b. WHO Regional Office for Europe's website. http://www.euro.who.int/air

10 Combined Actions of Chemicals in Mixture

10.1 INTRODUCTION

Following the current practice, health risk assessments of exposure to chemicals and the subsequent regulatory measures, e.g., classification and labeling, establishment of regulatory standards such as Maximum Residue Limits (MRLs), etc. are generally based upon data from studies on the individual substances. However, humans are simultaneously exposed to a large number of chemicals that potentially possess a number of similar or different toxic effects. Consequently, not only opponents against the use of chemicals but also the consumers at large are increasingly challenging the authorities to consider that this "chemical cocktail" or "total chemical load" does not produce unforeseen health effects.

This question was even more highlighted in 1996 when the U.S. Congress passed the U.S. Food Quality Protection Act (FQPA) (US-EPA 1996). This act requires that the US-EPA consider the effects of exposure to all pesticides and other chemicals that act by a common mechanism of toxicity when tolerances for pesticide use in crops are derived. Therefore, the aspect of combined actions of chemicals needs to be addressed to a greater extent in the risk assessment process. A major obstacle in doing so is the lack of data from studies on chemical mixtures employing generally accepted toxicological methods, such as short- and long-term animal studies. This is because most of the limited resources in experimental toxicology are used to study single chemicals or the effects of pretreatment with one chemical on the effects of another. In addition, data on human exposures to chemical mixtures are in general very inadequate. Therefore, regulatory agencies are faced with the situation that they cannot always reliably predict whether the simultaneous exposure to foreign chemicals in the environment and food constitutes a real health problem. As the possible combinations of chemicals are innumerable and experimental testing of all such mixtures is not feasible for obvious reasons, there is a need for science-based advice on how exposure to mixtures of chemicals can be dealt with in the risk assessment.

Interactions between chemicals administered to humans at high doses have been known for many years in the field of pharmacology. However, these experiences are not directly useful for predicting toxic effects of mixtures of environmental chemicals because the exposure levels for the general human population are relatively low and interactions occurring at high doses may not be representative for low-dose exposures (Könemann and Pieters 1996).

Toxicity studies with mixtures have been performed for several decades. Initially, most studies were done with binary mixtures. Later, studies with defined mixtures of more than two compounds have been reported. Studies have also been performed with complex mixtures of environmental chemicals, such as exhaust condensates, in order to gain insight into the toxic effects of such a particular mixture. However, the interpretation of the toxicity seen in these latter studies is complicated because the exact composition of the mixtures is normally not known, and the "real life" mixtures may vary considerably in composition. Therefore, extrapolation to other situations may be difficult. This fact is often ignored for the sake of simplicity.

TABLE 10.1

Classification of Combined (Joint) Toxic Actions of Two Compounds in Mixture

Interaction	Combined Action	
	Similar Action	Dissimilar Action
Absent (No interaction)	Simple similar action (Dose addition)	Independent action (Response addition)
Present (Interaction)	Complex similar action (Antagonism or synergism)	Dependent action or complex dissimilar action (Antagonism or synergism)

Source: Modified from Placket, R.L. and Hewlett, P.S., *J. Royal Statistical Society, Series B* 14, 143, 1952.

A major issue in the assessment of the combined toxicological effect of chemicals in a mixture is the type of combined action to be expected. What kind of toxicity may be expected, given the toxicity profiles of the individual components? Bliss (1939) was the first to provide a conceptual framework for the combined action of chemicals and later contributions were made by Finney (1942), Hewlett and Placket (1959, 1964), Placket and Hewlett (1952, 1963, 1967), Ashford and Cobby (1974), and Ashford (1981).

Placket and Hewlett (1952) provided a scheme of possibilities of combined (joint) actions, see Table 10.1. A major clue that can be taken from this scheme is that, in the initial assessment, it is important to evaluate whether interactions are actually occurring (present) or not (absent).

Interaction was defined as the influence of one chemical on the biological action of another, either qualitatively or quantitatively. The scheme represents the extremes of combined actions.

In many cases, adequate information about the underlying mechanisms of combined actions is not available. This led Berenbaum (1985, 1989) to propose three classes of combined action: zero interaction, synergism, and antagonism.

Current knowledge about the combined toxicological effects that may occur from exposures to different chemicals in mixtures is outlined in this chapter. Special attention is paid to the low levels of exposures normally encountered from the unintended, indirect exposure to chemical mixtures through food and environment. It should be recognized that it has not been possible to cover all possible combined exposures to chemicals in this book.

The main emphasis is paid to the identification of the basic principles for combined actions and interactions of chemicals (Section 10.2), and to the current knowledge on effects of exposures to mixtures of industrial chemicals, including pesticides and environmental contaminants. Test strategies to assess combined actions and interactions of chemicals in mixtures (Section 10.3) as well as toxicological test methods (Section 10.4) are addressed, approaches used in the assessment of chemical mixtures are presented (Section 10.5), and examples of experimental studies using simple, well-defined mixtures are given (Section 10.6).

10.2 BASIC CONCEPTS AND TERMINOLOGY USED TO DESCRIBE THE COMBINED ACTION OF CHEMICALS IN MIXTURES

The major objective in the risk assessment of exposure to mixtures of chemicals is to establish or predict how the resulting toxicological effect might turn out. Will the toxic effect be determined by simple additivity of dose or effect, or will it deviate from additivity, either by an effect stronger or less than expected on the basis of additivity?

The prediction of the toxicological properties of a chemical mixture requires detailed information on the composition of the mixture and the mechanism of action of each of the individual compounds. In order to perform a risk assessment, proper exposure data are also needed. Most often such detailed information is not available. Complex chemical mixtures may contain hundreds, or even thousands of compounds, and their composition is qualitatively and quantitatively not fully known and may change over time. Adequate testing of such mixtures is most often impossible because they are either virtually unavailable for testing or only available in such a limited amount that a sufficient number of dose levels cannot be applied. In addition, high-dose levels of a chemical mixture may have different types of effects than low dose levels and high- to low-dose extrapolation may be meaningless.

In the following, several terms used to describe interactions between chemicals are mentioned as well as basic concepts used in the hazard and risk assessment of chemical mixtures. The description of these basic concepts, first outlined by Bliss (1939) and Placket and Hewlett (1952), are based on the publications by Könemann and Pieters (1996), Cassee et al. (1998), and Groten et al. (2001). The definitions of additivity, synergism, antagonism, and potentiation are those of Klaassen (1995) and Seed et al. (1995).

As has already been outlined in the introduction, one of the main points to consider is whether there will be no interaction or interaction in the form of either synergism or antagonism. These three basic principles of combined actions of chemical mixtures are purely theoretical and one often has to deal with two or all three concepts at the same time, especially when mixtures consist of more than two compounds and when the toxicity targets are more complex.

Interactions between chemicals may be of a physico-chemical and/or biological nature. Examples of physico-chemical interactions are the reaction of nitrite with alkylamines to produce carcinogenic nitrosamines, and the binding of toxic chemicals to active charcoal resulting in a decreased absorption from the gastrointestinal tract. It is held that physico-chemical interactions will normally only occur at high doses and therefore are of lesser importance for low-dose scenarios. Physico-chemical interactions will therefore not be considered in any detail in this book.

10.2.1 No Interaction

According to Placket and Hewlett, there are two types of combined action without interaction (Table 10.1): simple similar action (dose addition, Loewe additivity) and simple dissimilar action. This latter type contains two concepts: effect or response additivity and Bliss independence. The independence criterion seems not to be widely used in toxicology (Groten et al. 2001).

The response to a mixture of compounds depends not only on the dose, but also on the correlation of tolerances between the effects of the chemicals in the mixture, which can vary between -1 and $+1$ (Bliss 1937). There is a complete negative correlation ($r = -1$) between the effects of two chemicals if the individuals that are most susceptible to one toxicant are least susceptible to the other, while a complete positive correlation ($r = +1$) exists if the individuals most susceptible to one toxicant are also most susceptible to the other.

10.2.1.1 Simple Similar Action (Dose Addition, Loewe Additivity)

Simple similar action (simple joint action or concentration/dose addition) is a noninteractive process in which the chemicals in the mixture do not affect the toxicity of one another. All the chemicals of concern in the mixture act on the same biological site, by the same mechanism of action, and differ only in their potencies. The correlation of tolerances is completely positive ($r = +1$) and each chemical contributes to the toxicity of the mixture in proportion to its dose, expressed as the percentage of the dose of that chemical alone that would be required to obtain the given effect of the mixture. Thus, the individual components of the mixture act as if they were dilutions of the same toxic compound and their relative potencies are assumed to be constant throughout all dose levels. An important implication is that, in principle, no threshold exists for dose additivity.

Simple similar action serves as the basis for the use of toxic equivalency factors (TEF, Section 10.5.1.4) often used to describe the combined toxicity of isomers or structural analogues. Additive effects are described mathematically using summation of doses of the individual compounds in a mixture adjusted for differences in potencies. This method is assumed to be only valid for compounds that produce linear dose–response curves. Probably, the best validated example of a group of compounds that obey the principles of simple similar actions are the dioxins (polychlorinated dibenzo-*p*-dioxins and dibenzofurans) that produce most (if not all) of their toxicities through interaction with the Ah-receptor.

10.2.1.2 Simple Dissimilar Action (Response or Effect Additivity, Bliss Independence)

Simple dissimilar action (simple independent action, independent joint action, Bliss independence, and effect addition or response addition) is also a noninteractive process where the toxic effect of each chemical in the mixture is not affected by the other chemicals present. However, the modes of action of the constituents in the mixture will always differ and possibly, but not necessarily, the nature and site of action also differs among the constituents. Response addition is referred to when each individual of a population (e.g., a group of experimental animals or humans) has a certain tolerance to each of the chemicals in a mixture and will only exhibit a response to a toxicant if the concentration exceeds the tolerance dose. In such a case, the number of responders within the group will be recorded rather than the average effect of a mixture on a group of individuals. By definition, response addition is determined by summing the responses of the animals to each toxic chemical in the mixture.

Three different concepts have been developed for effect/response additivity depending on the correlation of susceptibility of individuals to the toxic agents:

- Complete Negative Correlation
  There is a complete negative correlation between the effects of two chemicals if the individuals that are most susceptible to one toxicant are least susceptible to the other. This is the simplest form of response additivity. The proportion (*P*) of individuals responding to the mixture is equal to the sum of the responses to each of the components:

$$P_{\text{mixture A,B}} = P_A + P_B \text{ less than or equal to } 1$$

- Complete Positive Correlation
  There is a complete positive correlation between the effects of two chemicals if the individuals most susceptible to one toxicant are also most susceptible to the other. The proportion (*P*) of individuals responding to the mixture is equal to the response to the most toxic compound in the mixture:

$$P_{\text{mixture A,B}} = P_A \text{ if toxicity A} \geq \text{B}$$

- No Correlation
  This situation is equal to Bliss independence. There is no correlation if the proportion of individuals responding to the mixture is equal to the sum of proportions of individuals responding to each of the toxicants taking into account that those individuals that respond to constituent A cannot react to B as well:

$$P_{\text{mixture A,B}} = P_A + P_B \cdot (1 - P_A)$$

Although this type of correlation seems to be similar to complete negative correlation, the difference is that, in this case, an individual can respond to both compounds A and B but not to both at the same time.

The approach of response addition can be easily applied to simple problems, such as acute toxicity of pesticides. However, more complex effects are not always easy to summate. Experimental animals are usually obtained from inbred strains, while human populations are more heterogeneous. In addition, various effects on different organ systems may occur within different time frames in experimental animals.

The US-EPA (1986) applied the concept of response addition to the determination of cancer risks, assuming a complete negative correlation of tolerance. This assumption is considered to contribute to a conservative estimation of risk, since the correlation of tolerances may not be strictly negative in inbred homogenous experimental animals. There is a major difference between the concepts of response addition and dose addition when the human situation of low exposure levels is assessed. Response addition implies that when doses of chemicals are below the no-effect levels of the individual compounds (i.e., the response of each chemical equals zero), the combined action of all compounds together will also be zero. In contrast, dose addition can also occur below the no-effect level and the combined toxicity of a mixture of compounds at individual levels below the no-effect level may lead to a response.

For compounds with presumed linear dose–response curves, such as genotoxic and carcinogenic compounds for which it is assumed that a no-effect level does not exist and for which the mechanism of action may be regarded as similar, response addition and dose addition will provide identical results (Könemann and Pieters 1996).

10.2.2 Interactions: Complex Similar Action and Complex Dissimilar Action

Chemicals in mixtures may interact with one another and modify the magnitude and sometimes also the nature of the toxic effect. As illustrated in Table 10.1, the combined action of chemicals that interacts can be divided into two categories: complex similar action and complex dissimilar action (dependent action).

Interactions may take place in the toxicokinetic phase and/or in the toxicodynamic phase. The interactions may result in either a weaker (antagonistic) or stronger (potentiated, synergistic) combined effect than would be expected from knowledge about the toxicity and mode of action of each individual compound.

- Antagonism
  An antagonistic effect occurs when the combined effect of two chemicals is less than the sum of each chemical given alone. Synonyms sometimes used for antagonism are interaction, depotentiation, desensitization, infra-addition, negative synergy, less than additive, subaddition, inhibition, antergism, competitive antagonism, noncompetitive antagonism, uncompetitive antagonism, or acompetitive antagonism.

- Synergism
  A synergistic effect occurs when the combined effect of two chemicals is greater than the sum of the effects of each chemical given alone. Synonyms sometimes used to describe synergism are: coalitivity, interaction, uni-synergism, augmentation, sensitization, supra-addition, independent synergism, dependent synergism, degradative synergism, greater than additive, co-synergism, super-addition, conditional independence, or potentiation.

- Potentiation
  Potentiation, being a form of synergism, occurs when the toxicity of a chemical on a certain tissue or organ system is enhanced when given together with another chemical that does not have toxic effects on the same tissue or organ system. This form of interaction is especially well described in mutagenesis and carcinogenesis where a number of compounds have been identified as co-mutagens or co-carcinogens.

The ultimate toxicological response following exposure to a chemical substance is most commonly the result of the action of this substance on a definite site or receptor. For a given concentration of the agent at the target site, the intensity of the response will depend on the quality of the action (the intrinsic activity) and the affinity of the compound for the receptor.

When two compounds exert the same action by acting at different sites, their interaction will often result in a synergistic effect but a simple additive effect is also a possibility (the synergism between smoking and asbestos exposure is the classical example).

10.2.2.1 Complex Similar Action

In the case of complex similar action, two compounds acting on the same target receptor do not produce an additive effect as would be expected from simplicity, but either an antagonistic or synergistic effect. This phenomenon is well known for substances competing for the same hormonal or enzymatic receptor sites. In such cases, lower than additive effects are often observed. An example could be two chemicals that exert the same action (e.g., accumulation of acetylcholine) by acting in the same manner (e.g., by inhibition of acetylcholine esterase). An additive effect may occur if the intrinsic activities and affinities of the two substances are identical but most often an antagonistic effect is observed as both compounds compete for the same receptor. A maximal antagonism is found when the substance with the lowest intrinsic activity possesses the higher affinity for the receptor or has been the first to get into contact with the target.

In order to predict the effect of a mixture of chemicals with the same target receptor, but with different nonlinear dose–effect relationships, either physiological or mathematical modeling can be applied. For interactions between chemicals and a target receptor or enzyme, the Michaelis–Menten kinetics (first order kinetics but with saturation) are often applicable. This kind of action can then be considered a special case of similar combined action (dose addition).

It is highly likely that, for compounds thought to have complex similar actions, the observed deviations from the expected additivity in some cases are due to the fact that the compounds are actually not acting at the very same site at the target receptor. This means that the compounds actually have complex dissimilar actions and the combined action is misclassified as a complex similar action due to insufficient knowledge about the exact mechanisms of action.

10.2.2.2 Complex Dissimilar Actions

Complex dissimilar actions are probably the most frequently occurring interactions operating in experimental studies on mixtures applying high doses. The most obvious cases in the toxicokinetic phase involve enzyme induction or inhibition. Enzyme induction could result in a synergistic effect if more reactive (and toxic) intermediates are formed or in an antagonistic effect if the toxic agent is removed by detoxification. Compounds which influence the amount of biotransformation enzymes can have paramount effect on the toxicity of other chemicals. Uptake and excretion are often active processes which may also be affected by other chemicals. Interaction between substrates for the same membrane receptors or pumps, as well as for biotransformation enzymes could result in synergism and antagonism, too.

10.3 TEST STRATEGIES TO ASSESS COMBINED ACTIONS AND INTERACTIONS OF CHEMICALS IN MIXTURES

Ideally, all chemicals in a mixture should be identified and the toxicity profile of each of the constituents as well as their potential combined actions and/or interactions should be determined over a wide range of exposure levels. For complex environmental mixtures, this approach is not realistic and therefore a number of approaches and test scenarios have been presented to obtain toxicological information on mixtures with a limited number of test groups (Cassee et al. 1998).

10.3.1 TESTING OF WHOLE MIXTURES

Although testing of the whole mixture as such seems to be the proper way to approach the risk assessment of exposure to that mixture, it will not provide data on combined actions and/or interactions between the individual components of the mixture. Even if the effect of the mixture is compared with the effects of each individual component at comparable concentrations, this will not allow a description of potential synergism, potentiation, or antagonism, and it is even doubtful that deviations from additivity can be concluded. This can only be achieved if dose–response curves are obtained for each of the single compounds.

Testing of the whole mixture as such has been recommended for mixtures that are not well characterized (Mumtaz et al. 1993), and has successfully been applied for assessing the combined toxicity of simple, defined chemical mixtures where the toxicological properties of the individual components were also investigated, see Section 10.6.

10.3.2 PHYSIOLOGICALLY BASED TOXICOKINETIC MODELING

For many chemicals, their metabolism is the major determinant of the risk and for a number of hazardous compounds, there is a considerable knowledge from experimental studies on the relationship between metabolism and toxicity. In particular, *in vitro* studies using cell cultures, subcellular fractions, or pure enzymes have provided information on the nature of reactive intermediates as well as on detoxification pathways. Moreover, the significance of these processes has been demonstrated in several species of experimental animals and humans.

Physiologically Based Toxicokinetic (PBTK) models are derived similarly to Physiologically Based Pharmacokinetic (PBPK) models, which have been used for a number of years in the development of medicinal drugs. They describe the rat or man as a set of tissue compartments, i.e., liver, adipose tissues, poorly perfused tissues, and richly perfused tissues along with a description of metabolism in the liver. In case of volatile organic compounds a description of gas exchange at the level of the lung is included, see also Section 4.3.6.

In principle, the *in vivo* human metabolism can be predicted by using *in vitro* enzyme kinetic data and can thus be compared with the *in vitro* and *in vivo* data from experimental animals. For example, experiments using microsomes or hepatocytes may predict the *in vivo* velocity of metabolism for a single metabolic pathway. Such data may be incorporated in PBTK modeling (Andersen et al. 1995, Leung and Paustenbach 1995, Yang et al. 1995). As a rule, the description of the rate constants such as V_{max} and K_m for the individual (iso)enzymes follows Michaelis–Menten kinetics. Therefore, interindividual differences in expression levels of enzymes and genetic polymorphism can also be modeled. Ploemen et al. (1997) have presented a strategy to combine PBPK modeling with human *in vitro* metabolic data to explore the relative and overall contribution of critical metabolic pathways in man.

In order to use PBTK modeling in the assessment of mixtures, Cassee et al. (1998) suggest that one of the components is first modeled and regarded as the prime toxicant being modified by the other components. Based on *in vitro* data on the other components, effects of, e.g., inhibition or induction of specific biotransformation isoenzymes can be incorporated in the model. Effects of competition between chemicals in a mixture for the same biotransformation enzymes may also be incorporated by translating the effects into effects on the Michaelis–Menten parameters that are then incorporated into the model.

PBTK models can potentially be extended to include the toxicodynamic phase (PBTK/TD model) if a direct relationship exists between the concentration of the active metabolite (or parent compound) and the toxic effect (Yang et al. 1995).

10.3.3 ISOBOLE METHODS

An isobole is a contour line representing equi-effective quantities of two agents or their mixtures (Loewe and Muischnek 1926).

The theoretical line of additivity is the straight line that connects the individual doses of each of the single agents that produce a predetermined, fixed effect alone, for example an ED_{50} (50% response) of a given toxic or biochemical effect.

The isobole method is widely used to evaluate the effects of binary mixtures. However, a large number of different mixtures of the two compounds have to be tested in order to identify combinations that produce the fixed effect.

If the graphical representation (isobologram) of the combinations that produce the fixed effect shows a straight line, the two compounds behave in a dose-additive manner and subsequently, can be regarded as compounds that have a similar mode of action, see Figure 10.1. In case of an antagonistic interaction all the equi-effect concentrations in the mixtures represent an upward concave line in the isobologram (Figure 10.2), whereas a synergistic interaction would produce a downward curve in the isobologram (Figure 10.3).

In practice, the interpretation of test results strongly depends on the accuracy of the estimated intercepts of the theoretical isobole with the axis, which represents the doses of the single compounds that induce the desired effect. In fact, large standard deviations of these intercepts prevent a reliable conclusion as to the deviation from additivity.

Berenbaum (1981) introduced an equation to calculate an interaction index (CI). This enables the effects of noninteractive combinations to be calculated directly from dose–effect relationships of the individual compounds, regardless of the particular types of dose–effect relations involved.

$$CI = d_1/D_1 + d_2/D_2 + \ldots + L \, d_n/D_n$$

$d_1, d_2, \ldots, d_n$ are the doses of the agents in the mixture
$D_1, D_2, \ldots, D_n$ are the doses of the individual agents producing the same effect as the mixture

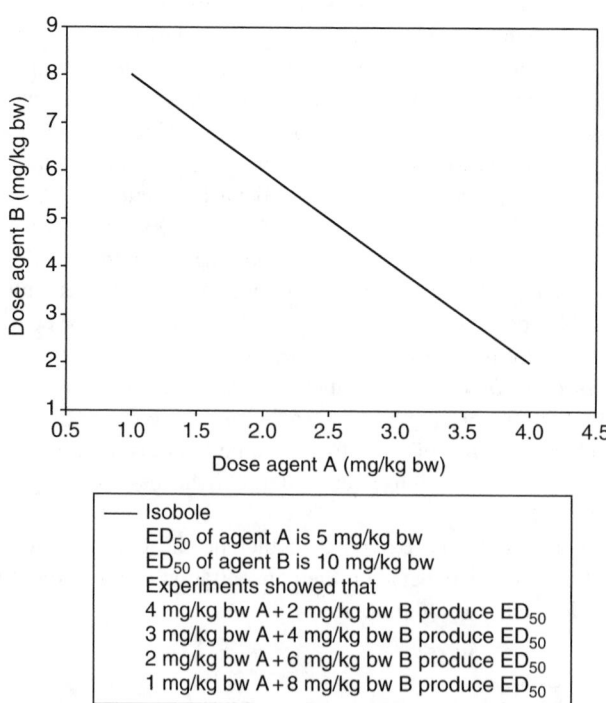

FIGURE 10.1 Isobologram of two agents A and B that act additively.

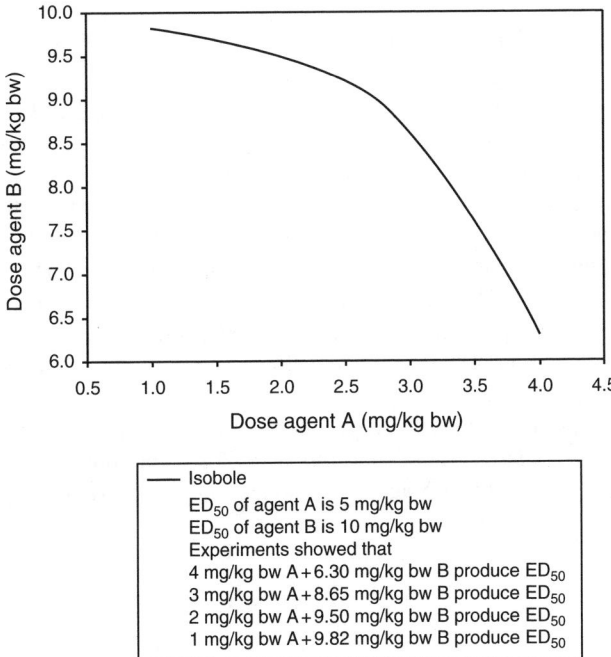

FIGURE 10.2 Isobologram of two agents that act antagonistically.

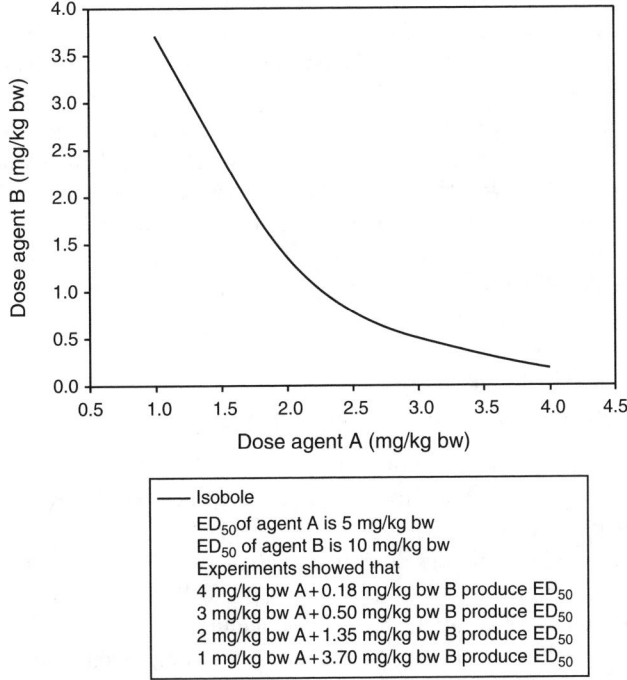

FIGURE 10.3 Isobologram of two agents A and B that act synergistically.

For binary mixtures, a straight line (isobole) is produced joining D_1 and D_2 and passing through (d_1, d_2). The interaction index (CI) is 1, <1, or >1 when the combinations show zero interaction, synergism, or antagonism using dose addition, respectively.

In cases of departure from additivity, the magnitude of CI depends on the ratio of the concentrations of the constituents of the mixture. Thus the CI is not a general figure but depends on the specific concentrations of the chemicals in the mixture.

One difficulty in using this approach is to determine when a specific CI actually deviates from 1 (additivity), as the method of isoboles as developed does not include measures to decide whether deviations from the line of additivity are systematic or simply due to chance or experimental error (Cassee et al. 1998). One way of dealing with this problem is to calculate confidence intervals for the iso-effective doses of the single compounds and to add a confidence belt to the line of additivity (Kortenkamp and Altenburger 1998). This envelope of additivity is an area in which those combinations of two compounds are lying that has a specific effect and may reasonably be considered as showing no interaction (for details see Cassee et al. 1998).

The isobole method can also be applied to mixtures where only one of the two agents produces the effect under consideration. In case agent A produces an effect, whereas agent B does not, the equation is reduced to

$$CI = d_1/D_1 = 1$$

In this case the iso-effective dose, D_2, of the agent lacking the effect of interest can be regarded as infinitely large, so that the resulting additivity isobole runs parallel to the respective dose axis.

Combination of three agents can be analyzed by constructing three-dimensional isobolar surfaces, and combinations of more than three compounds can be assessed more easily by using a generalization of the above-mentioned equation. However, new procedures using a polynomial model have been proposed to evaluate more complex mixtures (Cassee et al. 1998).

One of the strengths of the isobole method is that it can be used to analyze combined effects of compounds irrespective of the shape of their dose–response curves. It is possible to assess mixtures of agents with dissimilar dose–response relationships, even when the maximal effects are not identical (Kortenkamp and Altenburger 1998).

Although isoboles are very illustrative, a complete construction requires a large amount of data sets both on the single compounds and mixtures and large standard deviations may limit the interpretation (Cassee et al. 1998). However, even if it is desirable to test combinations of agents at several mixture ratios so that the isoboles can be constructed reliably, Kortenkamp and Altenburger (1998) are of the opinion that this is not always a necessary prerequisite. Valid conclusions about the combination effect of mixtures can often be drawn on the basis of surprisingly few data.

10.3.4 COMPARISON OF INDIVIDUAL DOSE–RESPONSE CURVES

Comparison of dose–response curves of one chemical (A) in the absence and presence of a second chemical (B) has been proposed as a tool to predict whether the combined action of the two chemicals is either additive or independent (Cassee et al. 1998).

In the case of dose additivity, the dose–response curve of A is determined on a linear- or log-dose scale, and an equi-effective dose of A ($d_{A,equi}$) and B (d_B) resulting in the same effect is estimated. Using the fixed dose d_B of chemical B and adding various doses ($d_A - d_{A,equi}$) of A, the dose–response curve should shift to the left and reach the same maximum as the maximum for the dose–response curve of A alone when the effect of B is smaller than A_{max}. However, in case of competitive agonism, the effect of B does not affect the effect of A + B at higher dose of A.

In the case of an independent effect, the addition of a fixed dose of B will produce an upward shifted dose–response curve. An independent effect can be calculated from the equation

$$E_{AB} = E_A + E_B - (E_A \times E_B) \text{ (Bliss independence)}$$

This method is based on the idea of response addition and was developed to accommodate the observation that compounds may act on different subsystems within an organism, which may well involve different sites and modes of action. Individual mixture components are not assumed to contribute to the overall mixture effect if they are present at subthreshold levels.

10.3.5 RESPONSE SURFACE ANALYSIS

Response or effect surface analysis (RSA) uses multiple linear regressions to produce a statistically based mathematical relationship between the doses of each of the chemicals in a mixture and the effect parameter. The equation for a mixture containing three compounds would be

$$E = \alpha + \beta_1 \cdot d_1 + \beta_2 \cdot d_2 + \beta_3 \cdot d_3 + \gamma_1 \cdot d_1 \cdot d_2 + \gamma_2 \cdot d_1 \cdot d_3 + \gamma_3 \cdot d_2 \cdot d_3 + \delta_1 \cdot d_1 \cdot d_2 \cdot d_3$$

where
d_n represents the dose of a chemical in the mixture
the coefficient α represents the control situation
the constants β are associated with the main effects of each of the compounds
the coefficients γ and δ indicate two- or three-factor interactions, respectively

In the case of negative values for β, positive values for γ and δ indicate a less than additive interaction between two or three compounds. Zero values of γ or δ indicate absence of a particular interaction. The p values of these coefficients are estimated using t test.

Cassee et al. (1998) stressed that it should be avoided to use high-effect levels in this method because saturation and competition will play a major role for the combined effect and might lead to erroneous conclusions about combined action at lower (and more realistic) dose levels. They advocated using a concentration range not exceeding a 50% effect level, or even less.

10.3.6 STATISTICAL DESIGNS

When a mixture contains more than two compounds, all kinds of two- or more (three)-factor interactions are possible. In order to determine these in experiments, the number of possible test combinations increases exponentially with increasing numbers of compounds. In addition, the number of experimental groups will increase with the number of doses of each compound.

Factorial designs, in which n chemicals are tested at x dose levels (x^n treatment groups) have been suggested by the US-EPA (US-EPA 1986) as a statistical approach for risk assessment of chemical mixtures. A 2^5 factorial design has been used to describe interactions between the carcinogenic activity of five polycyclic aromatic hydrocarbons at two dose levels (Nesnow 1994) and a 5^3 design to identify nonadditive effects of three chemicals on developmental toxicity at five dose levels (Narotsky et al. 1995).

However, full factorial designs using conventional toxicity testing are very costly, even if only two dose levels are used. This would require $2^n - 1$ test groups to identify interactions between all chemicals of interest. Therefore, the use of fractionated factorial designs have been suggested (Plackett and Burman 1946, Svengaard and Hertzberg 1994).

Fractionated factorial designs have been used to identify interactive effects between seven trace elements and cadmium accumulation in the body (Groten et al. 1991). Groten et al. (1997) also used a fractionated factorial design to study the interactions of nine chemicals in mixture in subacute rat studies, see Section 10.6.1. They used two dose levels of each compound, whereas a full factorial design would have required 511 different test groups. The study applied a 1/32 fraction of the complete design involving 16 experimental groups. The design was based on a general balance of groups with and without one of the compounds and required prior knowledge on the chemicals in the mixture. For a discussion of advantages and limitations in using this approach, see Groten et al. (1997) and Cassee et al. (1998).

10.4 TOXICOLOGICAL TEST METHODS

When epidemiological studies form the basis for the risk assessment of a single chemical or even complex mixtures, such as various combustion emissions, it may be stated that in those cases the effects of combined action of chemicals have been incorporated. Examples can, for instance, be found in the updated WHO Air Quality guidelines (WHO 2000). Thus, the guideline value for, e.g., ozone was derived from epidemiological studies of persons exposed to ozone as part of the total mixture of chemicals in polluted ambient air. In addition, the risk estimate for exposure to polycyclic aromatic hydrocarbons was derived from studies on coke-oven workers heavily exposed to benzo[a]pyrene as a component of a mixture of PAH and possibly many other chemicals at the workplace. Therefore, in some instances the derivation of a tolerable intake for a single compound can be based on studies where the compound was part of a complex chemical mixture.

However, for most compounds the risk assessment has to be based on results from *in vitro* and *in vivo* studies.

The *in vitro* and *in vivo* test methods available to study combined actions and toxicological and biochemical interactions of chemicals in mixtures are essentially the same as those used for the study of single chemicals in order to examine their potential general toxicity and special effects such as mutagenicity, carcinogenicity, and reproductive toxicity.

One especially successful method of testing complex mixtures is bioassay-directed fractionation followed by chemical identification of active compounds. Until now this method has mainly been used for the testing and identification of genotoxic compounds in environmental mixtures such as extracts of air particulates, exhaust condensates, and cooked foods. In this approach, each fraction is bioassayed until the major class of specific chemical(s) responsible for the activity can be isolated and chemically characterized, which make a risk assessment of the mixture possible.

The advantage of fractionation includes the separation of active constituents from inactive or otherwise toxic components. Disadvantages include the limited amount of sample available for testing following processing, the likelihood of "spillover" of chemical classes between fractions, and the possible loss or modification of components with fractionation.

Alternatively (or initially) the mixture is treated as a whole and tested in its crude state. The advantage of this strategy includes the relevancy of the tested sample to its environmental counterpart, decreased potential for artefact formation, and inclusion of combined effects of chemicals in the mixture. Moreover if the mixture is representative of others in its class (e.g., diesel emissions from different sources would share certain characteristics), it may be possible to extrapolate results across samples. This method also circumvents the labor-intensive process of individual testing of multiple chemicals. But sometimes a complex mixture is too cytotoxic to be tested directly in a bioassay. Furthermore, it may be incompatible with the test system because of the physical matrix. Other disadvantages include the inability to specify the constituent of the mixture responsible for the toxicity, as well as potential masking effects (e.g., the masking of mutagenicity by cytotoxicity).

10.5 APPROACHES USED IN THE HAZARD ASSESSMENT OF CHEMICAL MIXTURES

Various approaches have been suggested in the scientific literature for use in the evaluation of the health risks from exposure to mixtures of chemicals. Most attention and effort has been devoted to procedures to assess cumulative effects of exposure to chemicals that act by a similar mechanism/mode of action (US-EPA 2000a), in which case the concept of dose addition applies.

However, the Dutch group around Victor Feron (Groten et al. 2001, Feron and Groten 2002, Jonker et al. 2004), as well as US-EPA (1999, 2000b) and the Agency for Toxic Substances and Disease Registry (ATSDR 2004) have suggested approaches that cover also chemicals that differ in their modes of action.

10.5.1 Procedures Used to Assess Cumulative Effects of Chemicals That Act by a Common Mechanism of Action: Cumulative Risk Assessment by Dose Addition

A number of approaches have been suggested to combine the exposure to chemicals that act by a similar mechanism of action but have different potencies and exposure characteristics (US-EPA 1999, 2000a,b). Wilkinson et al. (2000) have critically evaluated these approaches.

The approaches discussed are the hazard index (HI) (Section 10.5.1.1) and the weight-of-evidence (WOE) modification to the HI (Section 10.5.1.2), the point of departure index (PODI) (Section 10.5.1.3), toxicity equivalency factors (TEFs) (Section 10.5.1.4), the margin of exposure (MOE) procedures (Section 10.5.1.5), and the cumulative risk index (CRI) method (Section 10.5.1.6).

An ILSI Working Group (Mileson et al. 1998) has also addressed the "common mechanism" issue. The ILSI Working Group concluded that a common mechanism might exist if two compounds:

- Cause the same critical effect
- Act on the same molecular target at the same target tissue
- Act by the same pharmacological mechanism of action and may share a common toxic intermediate

It should be realized that with the exception of a few groups of chemicals (such as some organophosphorous and carbamate pesticides as well as some polychlorinated dibenzo-p-dioxins (PCDDs), polychlorinated dibenzofurans (PCDFs), and polychlorinated biphenyls (PCBs), precise mechanistic information on their toxic effects are scarce. In realizing that the exact molecular mechanism is not known for most chemicals the term "mode of action" is used to describe toxicities that appear to be similar albeit the mechanism is not known in detail, see also Section 4.2.6. For several groups of endocrine disrupters this terminology seems appropriate.

Another critical issue is the question of concurrent exposure. This refers to co-exposure to more than one chemical able to interact with a defined target in a specific target tissue during a particular time frame of interest (Wilkinson et al. 2000). It is important to distinguish between concurrent or simultaneously "external" exposure, referring to the timing of oral, dermal, or inhalation exposure, from concurrent "internal" exposure that relates to the dose actually attained at a given biological target in a given time frame. For the risk assessment, it is the "internal" exposure that is of toxicological significance; however, it is seldom known. The factors that determine whether a cumulative effect is likely from exposure to several different common mechanism compounds, are the timing and duration of external exposure, the persistence (biological half-life) of the chemicals in the body, and the duration of the effect.

The effect of any chemical at a biological target depends on its ability to attain a target site concentration that exceeds the threshold required to elicit the response. The intensity and duration of the response depends on the toxicokinetic properties of the compound (absorption, distribution, metabolism, and excretion) and the nature of the target site interaction (reversible, irreversible). If recovery is complete between successive exposures, no cumulative toxicity is to be expected. However, a short-term acute exposure could potentially add to the long-term burden of a persistent chemical and be relevant for the magnitude of the chronic effect.

For acute and short-term exposures difference in the toxicokinetic properties, which will result in different times to maximum effect for the individual compounds, are critical in determining concurrency at the target site. Therefore, exposure intervals and the sequence of exposures to different chemicals may have significant impact on the potential cumulative effect.

The risk assessment of exposure to mixtures of defined chemicals should make optimal use of the toxicological databases. Ideally, the point of departure (POD) for the assessment should be a dose associated with a particular biological response (BMD_{10} or $BMDL_{10}$, Section 4.2.5) since this takes into account all of the dose–response data available. A POD based on doses causing a particular response should always take preference over the NOAEL. This is because the NOAEL is a single point value and not a measure of a biological response, and is largely an artefact of experimental design. The POD should also ideally be based on studies with the same animal species using the same route of administration. However, the data available for most chemicals will not permit an estimation of, for instance, BMD_{10} and relative potencies may have to be based on NOAELs as PODs.

In describing the various procedures proposed to evaluate the risk associated with combined exposure to a group of chemicals with a common mechanism of action, the biggest problem associated with all methods of cumulative risk assessment is how to accommodate the different assessment factors (AFs) that are applied to derive regulatory standards such as ADIs or RfDs (Chapter 5). If the assessment factors applied are the same for all the chemicals, all the methods will give the same result. However, this is most often not the case and the different assessment factors applied to the various chemicals will dominate the result of the risk assessment. In order to illustrate this, exposure to a hypothetical group of four common mechanism chemicals, differing in potency by 100-fold and having exposures ranging from 0.01 to 0.5 mg/kg bw/day, was assessed by Wilkinson et al. (2000) assuming they had either the same AF of 100 (Table 10.2, scenario A) or AFs ranging from 10 to 1000 (Table 10.2, scenario B).

10.5.1.1 Hazard Index

The HI is the sum of the Hazard Quotients (HQ, defined as the ratio of an exposure estimate over the RfD, i.e., HQ = Exposure/RfD) of the individual chemicals, i.e., the sum of exposure to each chemical expressed as a fraction of its RfD/ADI/TDI (for definitions, see Section 5.1). The HI

TABLE 10.2

Hypothetical Example for Cumulative Risk Assessment

Compound	BMD_{10} (mg/kg/d)	Uncertainty Factor (UF)	RfD (mg/kg/d)	Exposure (mg/kg/d)
Scenario A: Chemicals with the same AF				
I	100	100	1	0.5
II	500	100	5	0.5
III	25	100	0.25	0.01
IV	5	100	0.05	0.01
Scenario B: Chemicals with different AF				
I	100	10	10	0.5
II	500	100	5	0.5
III	25	1000	0.025	0.01
IV	5	100	0.05	0.01

Source: Adapted from Wilkinson, C.F., Christoph, G.R., Jolien, E., et al., *Reg. Toxicol. Pharmacol.* 31, 30, 2000.

should not exceed 1 since this indicates that the FQPA (US-EPA 1996) "risk cup," a kind of a combined RfD for the common mechanism group, is full.

$$HI = HQ_I + HQ_{II} + HQ_{III} + HQ_{IV} \text{ or}$$
$$HI = Exp_I/RfD_I + Exp_{II}/RfD_{II} + Exp_{III}/RfD_{III} + Exp_{IV}/RfD_{IV}$$

Although the HI method is transparent, easily understandable, and directly relates to the RfD, the major disadvantage is that the RfD is not an appropriate metric to use as a POD for cumulative risk assessment, since the RfD is normally derived by using NOAELs and uncertainty factors, which are not data based, but may incorporate significant policy-driven assumptions. This issue is addressed in detail in Chapter 5.

Use of the information in Table 10.2, scenario A, where the AF values for each compound is the same gives the following result:

$$HI = 0.5 + 0.1 + 0.04 + 0.2 = 0.84 \text{ Risk units}$$

Whereas use of the information in Table 10.2, scenario B, where the AF values differ gives

$$HI = 0.05 + 0.1 + 0.4 + 0.2 = 0.75 \text{ Risk units}$$

Although the overall HI is quite similar, this example illustrates that the contribution of each chemical is highly dependent on the AF. Moreover, the method does not reflect that the components of the mixture do not all have the same critical effect.

The HI method has been refined by the introduction of the target-organ toxicity dose (TTD) method. This method suggests that separate HIs should be estimated for all endpoints of concern. This implies that a TTD should be established for all relevant endpoints for each chemical using the same principles as used in the "normal" derivation of the RfD/ADI/TDI and that HQs should be calculated for the relevant effects of each chemical (for details see ATSDR 2004).

10.5.1.2 Weight-of-Evidence Modification to the Hazard Index

The HI method does not incorporate information on interactions among components of the mixture (ATSDR 2004). Mumtaz and Durkin (1992) proposed a weight-of-evidence (WOE) method to systematically address this need. The method was designed to modify the HI to account for interactions, using the weight of evidence for interactions among pairs of mixture components. Thus, the basic assumption is that pairwise interactions will dominate in the mixture and adequately represent all the interactions. For example, if chemicals A and B interact in a certain way, the presence of chemical C will not cause the interaction to be substantially different.

It should be noted that experience with the method has revealed that it is mainly useful for a qualitative prediction as to whether the hazard may be greater or less than indicated by the HI (ATSDR 2004).

The method evaluates the data relevant to joint actions for each possible pair of chemicals in the mixture in order to make qualitative binary weight-of-evidence (BINWOE) determinations for the effect of each chemical on the toxicity of every other chemical. Two BINWOEs are needed for each pair: one for the effect of chemical A on the toxicity of chemical B, and another for the effect of chemical B on the toxicity of chemical A.

The BINWOE determination indicates the expected direction of the interaction, such as greater than additive, less than additive, additive, or intermediate. It scores the data qualitatively by using an alphanumeric scheme that takes into account mechanistic understanding, toxicological significance, and relevance of the exposure duration, sequence, bioassay, and route of exposure. The alphanumeric terms are finally converted into a single numeric score. The BINWOE evaluations should be target organ specific. A more detailed description and discussion has been provided by ATSDR (2004).

10.5.1.3 Point of Departure Index

A scientifically more appropriate method of addition is summing the exposures of each compound expressed as a fraction of their respective POD instead of the ADIs or RfDs. These POD fractions (PODF) are reciprocals of the individual MOEs (see Section 10.5.1.5) of each compound. This approach sums the exposures to the compounds in terms of their relative potencies. In this example, the BMD_{10} (Table 10.2) was used as POD:

$$PODI = 0.005 + 0.001 + 0.0004 + 0.002 = 0.0084 \text{ Risk units}$$

The PODI can be converted into a risk cup unit by multiplying by an appropriate group AF. For example, a group AF of 100 would result in a combined risk of 0.84 risk units.

10.5.1.4 Toxicity Equivalency Factors

The TEF approach normalizes exposures to common mechanism chemicals with different potencies to yield a total equivalent exposure (TEQ) to one of the chemicals, the "index compound." TEFs are derived as the ratio of the POD of the index compound to that of each member in the group. The exposure to each chemical is then multiplied by the respective TEF value to express exposure in terms of the index compound. Summation of these values result in the total combined exposure (TEQ) expressed in terms of the index compound.

This approach was initially developed to estimate the potential toxicity of mixtures of polychlorinated dibenzo-*p*-dioxins (PCDDs), polychlorinated dibenzofurans (PCDFs), and polychlorinated dioxin-like biphenyls (PCBs). Over the years, a number of different TEF systems for PCDDs, PCDFs and PCBs have been used. A system was internationally agreed upon at a WHO Consultation in 1997 (WHO–TEF) as published by Van den Berg et al. (1998). A WHO update has been published recently (Van den Berg et al. 2006) (Table 10.3).

WHO only assigned TEFs for compounds that:

- Show a structural relationship to the PCDDs and PCDFs
- Bind to the aryl hydrocarbon (Ah) receptor
- Elicit Ah receptor-mediated biochemical and toxic responses
- Are persistent and accumulate in the food chain

A TEF for a compound is determined as the toxicity of the compound relative to the toxicity of the index compound 2,3,7,8-TCDD based on available *in vitro* and *in vivo* data (Van den Berg et al. 1998, 2006).

The majority of studies assessing the combined effects of PCDDs, PCDFs, and dioxin-like PCB congeners in complex mixtures have supported the hypothesis that the toxic effects of combinations of congeners follow dose additivity. Therefore, the concentrations and TEFs of individual congeners in a mixture may be converted into a toxic equivalent concentration (TEQ) by multiplying the analytically determined amounts of each congener by the corresponding TEF and summing the contribution from each congener using the following equation:

$$TEQ = \sum (PCDD_i \times TEF_i) + \sum (PCDF_i \times TEF_i) + \sum (PCB_i \times TEF_i)$$

TEFs were also used by the NRC Committee on Pesticides in the Diet of Infants and Children to estimate the aggregate risk to children from dietary exposure to a mixture of pesticides (NRC 1993).

TABLE 10.3
Toxicity Equivalency Factors (WHO–TEFs) for Dioxins and Dioxin-Like PCBs

PCDDs and PCDFs	TEF
2,3,7,8-TCDD	1
1,2,3,7,8-PnCDD	1
1,2,3,4,7,8-HxCDD	0.1
1,2,3,6,7,8-HxCDD	0.1
1,2,3,7,8,9-HxCDD	0.1
1,2,3,4,6,7,8-HpCDD	0.01
OCDD	0.0003
2,3,7,8-TCDF	0.1
1,2,3,7,8-PnCDF	0.03
2,3,4,7,8-PnCDF	0.3
1,2,3,4,7,8-HxCDF	0.1
1,2,3,6,7,8-HxCDF	0.1
1,2,3,7,8,9-HxCDF	0.1
2,3,4,6,7,8-HxCDF	0.1
1,2,3,4,6,7,8-HpCDF	0.01
1,2,3,4,7,8,9-HpCDF	0.01
OCDF	0.0003
PCBs (IUPAC number)	**TEF**
Non-*ortho PCBs*	
3,3′,4,4′-TCB (77)	0.0001
3,4,4′,5-TCB (81)	0.00003
3,3′,4,4′,5-PnCB (126)	0.1
3,3′,4,4′,5,5′-HxCB (169)	0.03
Mono-*ortho PCBs*	
2,3,3′,4,4′-PnCB (105)	0.00003
2,3,4,4′,5-PnCB (114)	0.00003
2,3′,4,4′,5-PnCB (118)	0.00003
2,3,4,4′5-PnCB (123)	0.00003
2,3,3′,4,4′,5-HxCB (156)	0.00003
2,3,3′,4,4′,5′-HxCB (157)	0.00003
2,3′,4,4′,5,5′-HxCB (167)	0.00003
2,3,3′,4,4′,5,5′-HpCB (189)	0.00003

Abbreviations: PnCDD, pentachlorodibenzo-*p*-dioxin; HxCDD, hexachlorodibenzo-*p*-dioxin; HpCDD, heptachlorodibenzo-*p*-dioxin; OCDD, octachlorodibenzo-*p*-dioxin; PnCDF, pentachlorodibenzofuran; HxCDF, hexachlorodibenzofuran; HpCDF, heptachlorodibenzofuran; OCDF, octachlorodibenzofuran; TCB, tetrachlorobiphenyl; PnCB, pentachlorobiphenyl; HxCB, hexachlorobiphenyl; HpCB, heptachlorobiphenyl.

Source: Adapted from Van den Berg, M., Birnbaum, L.S., Denison, M., De Vita, M., et al., *Toxicol. Sci.* 93, 223, 2006.

The Committee examined five organophosphate pesticides (acephate, chlorpyrifos, dimethoate, disulfoton, and ethion), which are all cholinesterase inhibitors and may be present as residues in fruits and vegetables. Chlorpyrifos was used as the index compound. The TEF was defined as the ratio of the NOAEL or LOAEL for each pesticide to the NOAEL or LOAEL for chlorpyrifos. TEFs based on LOAELs were used when a NOAEL could mot be established for two of the compounds.

Based on US-FDA residue data on the five pesticides, total chlorpyrifos equivalents concentrations were estimated for each food item included.

The UK Pesticide Safety Directorate (PSD) has decided to use the TEF approach for assessment of combined risk from exposure to mixtures of acetyl cholinesterase inhibitors (organophosphate (OP) compounds and carbamates) (PSD 1999). Despite clear differences in the action of carbamates and OP compounds, the index compounds selected for all acetyl cholinesterase inhibitors were either aldicarb (carbamate) or chlorpyrifos (OP). The POD for determining relative potency was predetermined as the dose level that produced 20% inhibition of red blood cell cholinesterase in a 90-day dietary study in rats.

Another well-known example within the pesticide area is the group ADI of 0–0.03 mg/kg bw/day allocated to dithiocarbamate fungicides. Thus, the Joint FAO/WHO Meeting on Pesticide Residues (JMPR) in 1993, in its evaluation of mancozeb, concluded that

> the data on mancozeb would support an ADI of 0–0.05 mg/kg bw, based on the NOAEL of 4.8 mg/kg bw/day for the thyroid effects in rats using a 100 fold safety factor. However, the Meeting established a group ADI of 0–0.03 mg/kg bw for mancozeb, alone or in combination with maneb, metiram, and/or zineb, because of the similarity of the chemical structure of these compounds, the comparable toxicological profiles of the dithiocarbamates based on the toxic effect of ethylenethiourea (ETU), the main common metabolite, and the fact that parent dithiocarbamate residues cannot be differentiated using the presently-available analytical procedures.

Wilkinson et al. (2000) used the information in Table 10.2, chose compound IV as the index compound (TEF = 1), assigned TEF values to compounds I (0.05), II (0.01), and III (0.2) and calculated the total compound IV equivalent exposure (TEQ) to 0.042 mg/kg bw/day. When this TEQ was compared to the RfD of compound IV (0.05 mg/kg bw/day), a value of 0.84 was obtained, representing a kind of combined HQs that indicates that 84% of the risk cup was filled. This risk estimate will be the same regardless which compound is selected as the index compound, provided that the AF for each member in the group is the same (Table 10.2, scenario A).

There are no specific guidance criteria available for the selection of the index compound. US-EPA (1986) has suggested that the index compound should be the member of the group that is the best studied and has the largest body of scientific data of acceptable quality. This will be associated with a low AF and lead to the lowest combined risk. However, this has been criticized for using data on well-studied compounds to improve the acceptability of compounds that have poor toxicological databases.

By using the information in Table 10.2, scenario B, Wilkinson et al. (2000) illustrated that if the AF for each compound is different, the selection of the index compound is critical. If compound 3 (AF of 1000) was selected as index compound, the TEQ exposure would be 0.21 and the combined risk estimate 8.4-fold higher than considered acceptable.

10.5.1.5 Margin of Exposure

The margin of exposure (MOE) is the ratio of the POD (e.g., NOAEL, BMD_{10}) to the level of exposure.

$$MOE = \frac{BMD_{10}}{Exposure}$$

The MOE approach is often used to determine the acceptability of acute risks for single chemicals and MOEs of >100 or >10 are usually considered acceptable when derived from toxicological data from animal and human studies, respectively. The US-EPA favors this concept for performing aggregate and cumulative risk assessments (Whalan and Pettigrew 1997).

The combined MOE (MOE_T) is the reciprocal of the MOEs of each compound in the mixture.

$$MOE_T = \frac{1}{(1/MOE_1) + (1/MOE_2) + (MOE_3) + (MOE_4)}$$

Using the hypothetical data in Table 10.2:

$$MOE_T = \frac{1}{0.005 + 0.001 + 0.0004 + 0.002} = 119$$

There are no established criteria to define the magnitude of an acceptable MOE_T for exposure to mixture of chemicals. If the compounds act through a common mechanism of toxicity, then a MOE_T of 100 may by intuition be considered acceptable as this value is considered acceptable for single compounds. However, as the number of compounds in the mixture increases the MOE_Ts decreases and combinations of two, three, and four compounds, each having acceptable MOEs of 100, will yield MOE_Ts of 75, 33, and 25, respectively. In such cases, to obtain a MOE_T of >100 for mixtures containing two, three, or four compounds, the individual MOEs have to be greater than 200, 300, and 400, respectively. Alternatively, the exposure level to each compound should be reduced by 2, 3, or 4 times, respectively. In the example, the MOE_T of 100 results from summation of compounds that have MOEs of 200 (I), 1000 (II), 2500 (III), and 500 (IV). This shows the pronounced influence of the compound (I) that has the lowest MOE. In particular, the $MOE_T > 100$ approach seems inappropriate when the individual MOEs originate from data (NOAELs, BMD_{10}s) that would relate to application of different AF (e.g., data from animals and humans). Therefore, a stepwise reduction in the magnitude of the acceptable MOE_T has to be considered as the size of the group increases.

10.5.1.6   Cumulative Risk Index

The cumulative risk index (CRI), also referred to as the aggregate risk index (ARI) has been suggested by the US-EPA (Whalan and Pettigrew 1997) to combine MOEs for chemicals with different AFs. The risk index (RI) of a chemical is the MOE divided by the AF or simply the reference dose divided by exposure, and is the reciprocal of the HQ:

$$RI = \frac{POD}{Exposure \times UF} = \frac{RfD}{Exposure} = \frac{1}{HQ}$$

The CRI is thus defined as

$$CRI = \frac{1}{1/RI_I + 1/RI_{II} + 1/RI_{III} + 1/RI_{IV}} \text{ or}$$
$$= \frac{1}{Exp_I/RfD_I + Exp_{II}/RfD_{II} + Exp_{III}/RfD_{III} + Exp_{IV}/RfD_{IV}}$$

The CRI has the same disadvantages as described for the HI and in addition, since it is derived from the MOE approach, the CRI is not as transparent and understandable as the HI. It also involves more complex calculations.

10.5.2   Procedures Used to Assess Cumulative Effects of Chemicals That Do Not Act by a Common Mechanism of Action

In this case two situations may be defined:

- Compounds in a mixture that do not interact, for which the concept of response/effect addition may apply
- Compounds in a mixture that do interact, producing either synergism, potentiation, or antagonism, in which case special considerations should be done

For compounds operating by simple dissimilar action, available studies have not shown interaction leading to toxic effects when exposure is below the NOAEL for each of the compounds, but interactions may possibly occur when exposure is at the LOAEL for each of the compounds considered. Therefore, interaction of compounds with simple dissimilar action is not of concern at levels below the ADI for all these compounds.

Occurrence of complex dissimilar actions is thought to be rare at low exposure (ADI) levels but it should always be considered whether a plausible hypothesis exists for effect interactions of two or more compounds. Interactions can occur both in the toxicodynamic phase (e.g., endocrine disruptors) and in the toxicokinetic phase (e.g., interference with transport, metabolism (activation, deactivation), distribution, and elimination of another compound).

Like for compounds with a common mechanism of action, the HI, combined MOE procedures, the PODI, and CRI methods can be applied. The TEF concept is based on a common mechanism of action for the compounds involved, and therefore the TEF approach is not applicable for the evaluation of a mixture of compounds with a dissimilar mode of action.

10.5.2.1 Interactions in Toxicokinetics

Toxicokinetic interactions occur when the disposition of a toxic compound, i.e., its absorption, distribution (including localization at the target site), biotransformation, or excretion is altered by exposure, either simultaneously or displaced in time, to another compound. The toxicological net-outcome of a toxicokinetic interaction depends on whether a higher or lower level of the biologically active species is achieved at the target site and/or whether the target site is exposed for a shorter or longer duration.

10.5.2.1.1 Interference with Absorption

Absorption of chemicals from the gastrointestinal tract is usually a passive diffusion-driven process. Interactions are mainly to be expected when an active transport process or a specific transporter is involved (Feron et al. 1995c). For example, iron is known to decrease the gastrointestinal absorption of cadmium presumably by competing for the proteins involved in the transport of cadmium, and thus protects against cadmium accumulation and toxicity in experimental animals (Groten et al. 1991). This makes iron deficient women a particular risk group for cadmium toxicity due to increased uptake from the gastrointestinal tract.

As regards absorption through the skin, it is well known that surface-active compounds and skin irritants can enhance the absorption of other chemicals.

10.5.2.1.2 Interference with Distribution

Chemicals are distributed throughout the body via the bloodstream (or the lymph in special cases). Lipophilic compounds are to a large extent bound to proteins in the blood instead of just dissolved in water. A more lipophilic compound may remove a less lipophilic substance from the binding site and thus severely increase the concentration of unbound compound available for toxicological effect. This situation is well known for medical drugs administered simultaneously (Feron et al. 1995c).

10.5.2.1.3 Interference with Biotransformation

The majority of compounds that enter the organism require metabolism in order to be excreted. If the parent compound is responsible for the toxicity and its metabolites are less toxic, an increased biotransformation rate will reduce the toxicity, and conversely. However, if the chemical's toxicity is mainly due to its metabolite, stimulating the biotransformation will enhance the toxicity.

There are numerous possibilities for interactions among chemicals at the level of the enzymes involved in the biotransformation processes. Such interactions may in principle be due to competition for a given enzyme or cofactor. Well known examples are the detoxification of different alkylating agents by conjugation with glutathione, which may be reduced by compounds that compete for the glutathione-*S*-transferases and/or glutathione (Feron et al. 1995c).

Another important possibility for interactions is induction or inhibition of the drug metabolizing enzymes. Inducers or inhibitors of the microsomal cytochrome P450 oxidative systems may either potentiate (via increased production of active metabolites) or reduce (via increased detoxification) the toxicity of other chemicals. Thus, ketones like acetone and methyl *n*-butyl ketone and methyl isobutyl ketone can potentiate the hepatotoxicity of carbon tetrachloride and 1,2-dichlorobenzene by induction of cytochrome P450. On the other hand, inhibition of cytochrome P450 by disulfiram strongly enhances the carcinogenicity of ethylene dichloride and ethylene dibromide by forcing their biotransformation through the glutathione pathway, leading to enhanced formation of the ultimate carcinogenic glutathione conjugate. The principle of enhancing the toxicity of some pesticides by adding an inhibitor of cytochrome P450 (e.g., piperonyl butoxide) in the formulation is well known (Feron et al. 1995c).

A review of the literature (Krishnan and Brodeur 1991) demonstrated that the majority of toxicokinetic interaction results from metabolic induction or inhibition caused by some components of the mixture. These interactions may alter tissue dosimetry and thereby the toxicity of components in the mixture. The tissue doses of chemicals in mixture can be predicted with PBTK models when the binary interactions between all of the components in the mixture are known (Haddad et al. 1999a,b, 2000a,b). However, the quantitative characteristics of each of these binary interactions have to be determined by experimentation. Given the complexity of the mixtures, to which humans are exposed, this would obviously require an unrealistic large number of experiments in order to characterize the qualitative and quantitative nature of the possible interactions.

Haddad et al. (2000b) addressed this problem by using the theoretical limits of the PBTK modulation of the tissue dose that would arise from hypothetical metabolic interactions between 10 volatile organic compounds (VOCs) in the male rat. The VOCs used were dichloromethane, benzene, trichloroethylene, toluene, tetrachloroethylene, ethylbenzene, styrene, and *para*-, *ortho*-, and *meta*-xylene. All rat physiological parameters and physico-chemical (partition coefficient) and biochemical (metabolic constants) parameters used in the PBTK models were taken from the vast literature on these compounds. All model equations, except those describing metabolism, were taken from Ramsey and Andersen (1984). PBTK models predicting the blood concentrations of each mixture component were simulated using either the description of saturable metabolism or the description using the hepatic extraction ratio (E). In the latter case the numerical value of E was set to either 1 (maximal enzyme induction) or 0 (maximal enzyme inhibition). Data on blood concentration kinetics following exposure to binary, quaternary, quinternary, octernary, and decernary mixtures of the VOCs were obtained in rats exposed for 4 h by inhalation (50–100 ppm each). For all chemicals the simulation lines obtained using $E = 1$ and $E = 0$ formed the boundary lines, whereas the one obtained using V_{max} and K_m values was in between. The kinetic data from mixture exposures were within the simulated boundaries of blood concentrations. However, with increasing complexity of the mixtures, the impact on the blood kinetics of the single components became progressively more important, i.e., blood concentrations of unchanged parent chemicals increased with mixture complexity. This is consistent with the occurrence of metabolic inhibition among the chemicals in the mixture.

In a second experiment rats were pre-exposed to the mixture of all 10 chemicals (50 ppm each) 4 h a day for 3 consecutive days. On day four the rats were once more exposed and the kinetics of the compounds followed in blood. There seemed to be a systematic decrease (although not statistically significant) in blood concentrations indicative of greater metabolism due to enzyme induction.

Chaturvedi et al. (1991) studied the effects of mixtures of parathion, toxaphene, and/or 2,4-D on the hepatic mixed-function oxygenase in ICR male mice. They found that a 7-day toxaphene pretreatment enhanced the hepatic biotransformation of parathion and its metabolite paraoxon, both in the presence and absence of NADP. However, in the absence of NADP the enhancement was minor. The authors suggested that toxaphene induced the metabolic pathways of parathion and paraoxon involving the mixed-function oxygenase and that paraoxonase is not involved in the toxaphene-induced decreases of the two compounds. Toxaphene is enhancing the NADP-dependent metabolism of parathion and paraoxon and thereby decreasing their toxicity. Carboxyl esterase is involved in decreasing the toxicity of parathion and paraoxon by acting as a pool of noncritical enzymes, which compete for the binding of paraoxon thereby preventing an inhibition of choline-sterase. The increase in the level of carboxyl esterase and cholinesterase has the potential to enhance further the ability of toxaphene to limit the toxicity of parathion. The authors therefore anticipated the toxicity of a mixture of parathion and toxaphene to be lower than that of parathion. Thus the results of the study could indicate an antagonistic effect of toxaphene on parathion and on paraoxon.

Chaturvedi (1993) also examined the effect of mixtures of 10 pesticides (alachlor, aldrin, atrazine, 2,4-D, DDT, dieldrin, endosulfan, lindane, parathion, and toxaphene) administered by oral intubations or by drinking water on the xenobiotic-metabolizing enzymes in male mice. He concluded, "The pesticide mixtures have the capability to induce the xenobiotic-metabolizing enzymes, which possibly would not have been observed with individual pesticides at the doses and experimental conditions used in the study."

However, it is not possible to categorize the type of combined action because Chaturvedi (1993) only examined the combined effects of the 10 compounds in the mixture and did not consider the effect of the individual pesticides.

10.5.2.1.4 Interference with Excretion

For excretion processes, the same reasoning may be used as for absorption. Cases of interaction are only to be expected when active processes are involved. Increased excretion of a chemical following administration of an osmotic diuretic or alteration of the pH of the urine are well known examples of dispositional interaction.

10.5.3 USE OF RESPONSE/EFFECT ADDITION IN THE RISK ASSESSMENT OF MIXTURES OF CARCINOGENIC POLYCYCLIC AROMATIC HYDROCARBONS

The application of the concept of response addition has been suggested by the US-EPA (1986) to determine the cancer risk from mixtures containing carcinogenic compounds. The assumption was that such compounds show simple dissimilar action with a complete negative correlation of tolerance. However, as pointed out by Könemann and Pieters (1996) for compounds with presumed linear dose–response curves, such as genotoxic and carcinogenic compounds for which it is assumed that a no-effect level does not exist and for which the mechanism of action may be regarded as similar, response addition and dose addition will provide identical results. Therefore, various authors have used different terminology in the assessment of PAH, relative response factors, relative potency factors, or TEFs.

A number of PAH as well as coal tar and some occupational exposures to combustion emissions containing these compounds have shown carcinogenicity in experimental animals and genotoxicity and mutagenicity *in vitro* and *in vivo* (WHO/IPCS 1998). Several attempts have been made to derive relative potency factors, often expressed as TEFs for individual PAH (relative to benzo[*a*] pyrene, the best studied PAH) with the purpose of summarizing the contributions from individual PAH in a mixture into a total benzo[*a*]pyrene equivalent dose, assuming additivity in their carcinogenic effects (Krewski et al. 1989, Rugen et al. 1989, Thorslund and Farrar 1990, Nisbet and LaGoy 1992, Larsen and Larsen 1998). Because there is a total lack of adequate data from oral carcinogenicity studies on PAH others than benzo[*a*]pyrene, TEF values for PAH in food have been suggested based on studies using skin application, pulmonary instillation, and subcutaneous or intraperitoneal injections.

There are several problems in using the TEF approach in the risk assessment of PAH in food. The use of the TEF approach requires that the compounds in question exert the toxicological effect by the same mechanism of action, such as is the case for the polychlorinated dibenzo-*p*-dioxins and polychlorinated dibenzofurans, which act through binding to the Ah-receptor. Although a number of

PAH bind to the Ah receptor, this effect is not the only effect that determines the carcinogenic potency of PAH. DNA binding and induction of mutations are other significant effects in the carcinogenesis of PAH, and there is no indication that different PAH are activated via the same metabolic route, binds DNA in the same positions, and induce the same types of mutations in the same organs or tissues. In fact, the study by Culp et al. (1998) showed that a coal-tar mixture of PAH also produced tumors in other tissues and organs than those affected by benzo[*a*]pyrene alone, and that the additional PAH in the mixture did not significantly contribute to the incidence of stomach tumors observed after benz[*a*]pyrene alone.

The limitations in using the TEF approach for the assessment of PAH carcinogenicity following oral administration was illustrated when it was used on the carcinogenicity data and the analytical data on the PAH composition in the coal tars used in the study by Culp et al. (1998). When the TEF values derived by Larsen and Larsen (1998) were used (Table 10.4), the carcinogenic potency of both coal-tar mixtures was predicted to be only approximately 1.5 times that of the benzo[*a*]pyrene content. However, the observed potencies of the coal-tar mixtures were up to 5 times that accounted for by the benzo[*a*]pyrene content. In this case, the use of the TEF approach for PAH carcinogenicity would underestimate it.

Schneider et al. (2002) also examined the use of the TEF approach on the data from the Culp et al. (1998) study, and from several other studies using dermal or lung application of PAH mixtures of known composition. They used the TEF derived by Brown and Mittelman (1993) (Table 10.4) and concluded that the benzo[*a*]pyrene equivalency factors do not adequately describe the potency of PAH mixtures and lead to underestimation of the carcinogenic potencies in most cases.

10.5.4 APPROACH TO ASSESS SIMPLE AND COMPLEX MIXTURES SUGGESTED BY THE DUTCH GROUP

The Dutch group around Victor Feron initiated their research program in order to test the hypothesis that exposure to chemicals at (low) nontoxic doses of the individual chemicals, as a rule, would be of no health concern. One reason being that most test guidelines from national and international organizations often suggest the use of simple "dose addition" or "response addition" models for the assessment of chemical mixtures totally ignoring any knowledge on the mode of action of the chemicals. Clearly, such an approach would greatly overestimate the risk in case of chemicals that act by mechanisms where the additivity assumptions are invalid. The group clearly recognizes that for mixtures of compounds known to act by the same mechanism and therefore not showing interactions, a cumulative approach is the valid choice using dose or response addition.

They considered it important to distinguish between simple and complex mixtures. According to Feron et al. (1998) a simple mixture consists of a relatively small number of chemicals (e.g., 10 or less) and the composition of the mixture is known, both qualitatively and quantitatively. An example would be a cocktail of pesticide residues in food. A complex mixture comprises tens, hundreds, or thousands of chemicals, and the qualitative and quantitative composition is not fully known. They also emphasized to distinguish between whole–mixture analysis (top–down approach) and component–interaction analysis (bottom–up approach), the latter requiring an understanding of the basic concepts of combined action of chemicals.

10.5.4.1 Simple Mixtures

A general scheme for the safety evaluation of simple mixtures has been proposed (Groten et al. 2001). The most pragmatic and perhaps simplest approach is to test the toxicity of the mixture without identifying the type of interactions between the individual components. However, the results of the testing can only be used for hazard characterization following exposure to that particular mixture. A more detailed approach is to assess the combined action of the individual components in the mixture. Several experimental designs can be used, primarily depending on the complexity and number of compounds in the mixture. The major concern in the analysis of the data

TABLE 10.4

Estimates of Carcinogenic Potencies of Various PAH, Relative to Benzo[a]pyrene (BaP)

Compound	Studies Using Rat Lung Installation		Studies Using Mouse Skin Painting			Combined Estimates from Different Types of Studies				
	Calc.[a]	Publ.[b]	Calc.[c]	Calc.[d]	Publ.[e]	Publ.[f]	Publ.[g]	Publ.[h]	Publ.[i]	Publ.[j]
Anthracene				<0.0046		0.32		0.01	0.0005	0.01
Fluorene							0.001		0.0005	0
Phenanthrene	0.0004						0.001		0.0005	0
Benz[a]anthracene					0.0039–0.0055	0.145			0.005	0.1
Chrysene	0.030			0.013		0.0044	0.0044	0.01	0.03	0.01
Cyclopenteno[cd]pyrene			0.0084				0.023		0.02	
Fluoranthene				<0.105				0.001	0.05	0.01
Pyrene				<0.0046		0.081	0.081	0.001	0.001	0
Benzo[b]fluoranthene	0.089	0.123	0.18	0.037	0.023	0.140	0.140	0.1	0.1	1
Benzo[j]fluoranthene	0.053	0.052	0.022	0.040	0.075		0.061		0.05	0.1
Benzo[k]fluoranthene	0.052	0.053	4×10^{-8}	0.0004		0.066	0.066	0.1	0.05	0.1
Benzo[ghi]fluoranthene	0.012	0.021					0.022		0.01	0.01
Benzo[a]pyrene	1	1	1	1	1	1	1	1	1	1
Benzo[e]pyrene	0.0019	0.007		0.0039			0.004		0.002	
Dibenz[a,h]anthracene	1.23			0.65	0.59	1.1	1.1		1.1	1
Anthanthrene	0.340	0.316					0.320		0.3	
Benzo[ghi]perylene							0.022	0.01	0.02	0.01
Dibenzo[a,e]pyrene				0.221					0.2	0.1
Dibenzo[a,h]pyrene				0.843					1	1
Dibenzo[a,i]pyrene				0.082					0.1	1
Dibenzo[a,l]pyrene				1.27					1	1
Indeno[1,2,3-cd]pyrene	0.102	0.278	4×10^{-8}	0.035	0.0059	0.232	0.234	0.1	0.1	0.1
Coronene			0.007						0.01	

[a] Calculated by Nielsen et al. (1995).
[b] Thorslund and Farrar (1990).
[c] and [d] Calculated by Nielsen et al. (1995).
[e] Rugen et al. (1989).
[f] Clement (1988, as cited by Nielsen et al. 1995).
[g] Krewski et al. (1989).
[h] Nisbet and LaGoy (1992).
[i] Larsen and Larsen (1998).
[j] Brown and Mittelman (1993) (US-EPA OPPTS).

is whether the components act via similar toxicological processes, by the same mode of action or their modes of action are functionally independent (Figure 10.4).

10.5.4.2 Complex Mixtures

As regards complex mixtures, the Dutch group initially recommended a two-step approach (see Figure 10.5): first to identify the "*n*" (e.g., 10) most risky chemicals in the mixture, and then to perform hazard identification and risk assessment of the defined mixture of the (10) priority chemicals using procedures appropriate for simple, defined mixtures (Feron et al. 1995a,b, 1998; Cassee et al. 1998.

Selection of the top-ten chemicals in the first step should be based on the level of exposure and level of toxicity of the individual chemicals. The higher the value of the risk quotient (RQ) the higher the probability of adverse health effect in humans (e.g., higher risk) and the higher the

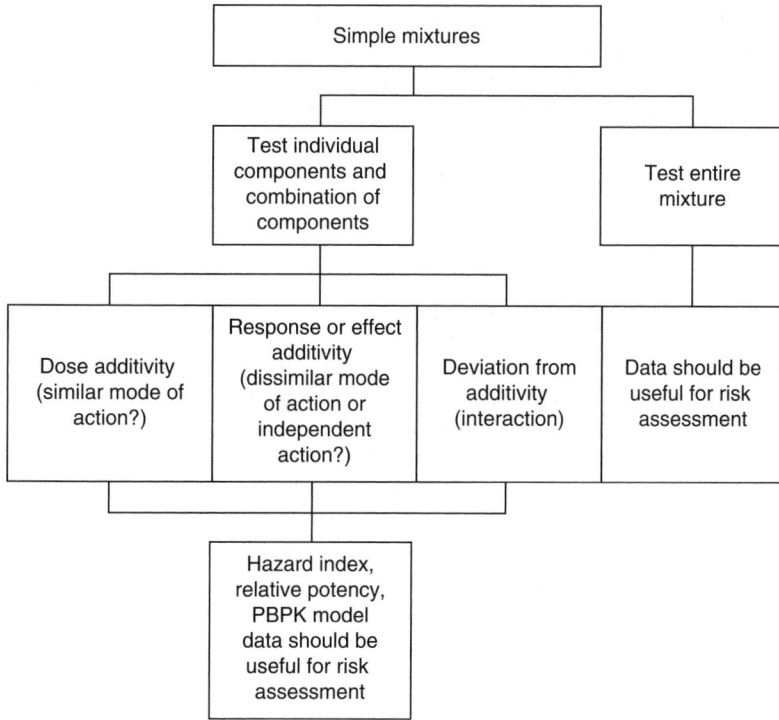

FIGURE 10.4 Scheme for safety evaluation of simple mixtures.

chemical should rank on the list of priority chemicals. The hazard identification and risk assessment of the mixture of selected chemicals (the top-ten chemicals) should be based on toxicity data and on the mechanism of action of the individual compounds and on the prediction of presence or absence

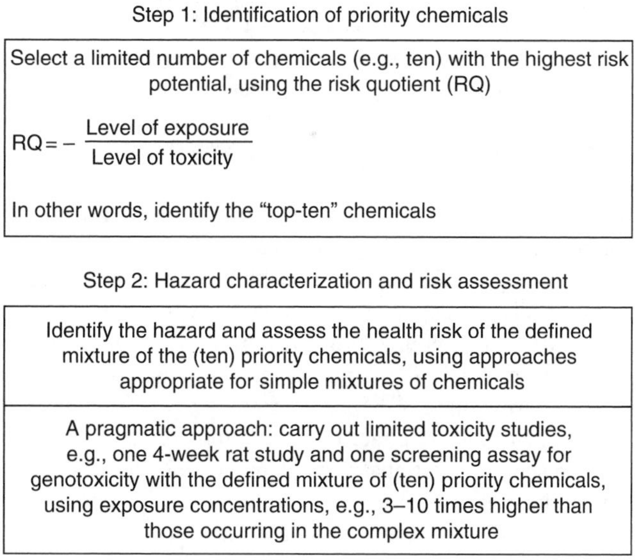

Step 1: Identification of priority chemicals

Select a limited number of chemicals (e.g., ten) with the highest risk potential, using the risk quotient (RQ)

$$RQ = - \frac{\text{Level of exposure}}{\text{Level of toxicity}}$$

In other words, identify the "top-ten" chemicals

Step 2: Hazard characterization and risk assessment

Identify the hazard and assess the health risk of the defined mixture of the (ten) priority chemicals, using approaches appropriate for simple mixtures of chemicals

A pragmatic approach: carry out limited toxicity studies, e.g., one 4-week rat study and one screening assay for genotoxicity with the defined mixture of (ten) priority chemicals, using exposure concentrations, e.g., 3–10 times higher than those occurring in the complex mixture

FIGURE 10.5 Two-step procedure for the safety evaluation of complex mixtures. (Adapted from Feron, V.J., Groten, J.P., van Zorge, J.A., Cassee, F.R., Jonker, D., and van Bladeren, P.J., *Toxicology* 105, 415, 1999b.)

of additive or potentiating interactive effects. In order to predict combined action (additivity) or interactions between the selected chemicals, knowledge about the presumed mechanism of action is necessary. Therefore, a classification system of chemicals on the basis of their mechanism of action would be extremely helpful. A classification could be based on:

- Similar or identical biotransformation pathways, including ability to induce or inhibit biotransformation enzymes
- Similar or identical receptors for the compounds or their active metabolites
- Structural similarities pointing to either of the above

The group recognizes that a major practical problem is lack of information of biotransformation and relevant receptor or target site of many chemicals. In such cases, the chemicals should be classified using computer-based structure–toxicity relationships and expert judgment and experience.

The evaluation assumes that the hazard and possible risk of the defined (top-ten) mixture of chemicals is representative for the hazard and risk of the entire complex mixture. For some mixtures that are relatively easily available (e.g., combustion fumes, food products, pesticide mixtures), this assumption could be validated by comparing the toxicity of the top-ten mixture with the toxicity of the original complex mixture in short-term test.

The group has elaborated on their approach and provided a scheme (decision tree) for hazard identification and risk assessment of complex mixtures (Feron et al. 1998, Groten et al. 2001) (Figure 10.6). In this scheme it is suggested that:

- For complex mixtures that are virtually unavailable for testing as a whole (such as workplace atmospheres, coke oven emissions, atmospheres at waste sites) the top-ten approach as mentioned above is suggested.
- For complex mixtures that are readily available for testing as a whole (such as drinking water, diesel exhaust, welding fumes, tobacco smoke, pesticide mixtures, food products) three possible approaches are suggested:
  1. Testing as a whole: This may characterize the toxicity profile of the mixture and eventually verify the (presumed) safety or hazard from exposure to mixture. One problem may be that incorporation of the test material in the diet at a sufficiently high dose may result in an unbalanced diet and nutritional deficiencies. Another problem is that most mixtures may change in chemical composition over time.
  2. Identification of the top-ten chemicals to be treated as a simple mixture: This should be considered as the primary option if the available data on the composition and the toxicity profile of the mixture indicate that the hazard is driven by a small number of the constituents.
  3. The "pseudo top-ten" approach: This should be considered for mixtures which consist of a large number of widely varying chemicals with no obvious ranking of individual constituents according to their potential health risks and the top-ten chemicals of the mixture are not easily identified. This approach involves identification of the top-ten classes of chemicals to be lumped together by class to the top-ten chemicals to be treated as a simple mixture. The lumping technique is based on grouping chemicals with relevant similarity such as the same target organ or similar mode of action. The selected top-ten chemicals are either chemicals representative of each class or pseudo components representing a fictional average of a certain class. This technique has been described by Verhaar et al. (1997) who proposed it to be combined with QSAR and PBPK/PD (physiologically based pharmacokinetic/pharmacodynamic) modeling in predicting the toxicity of complex mixtures of petroleum products.

Groten et al. (2000) used the principle that joint actions and/or interactions could possibly occur if chemicals shared a common target organ and produced similar adverse effects to analyze all

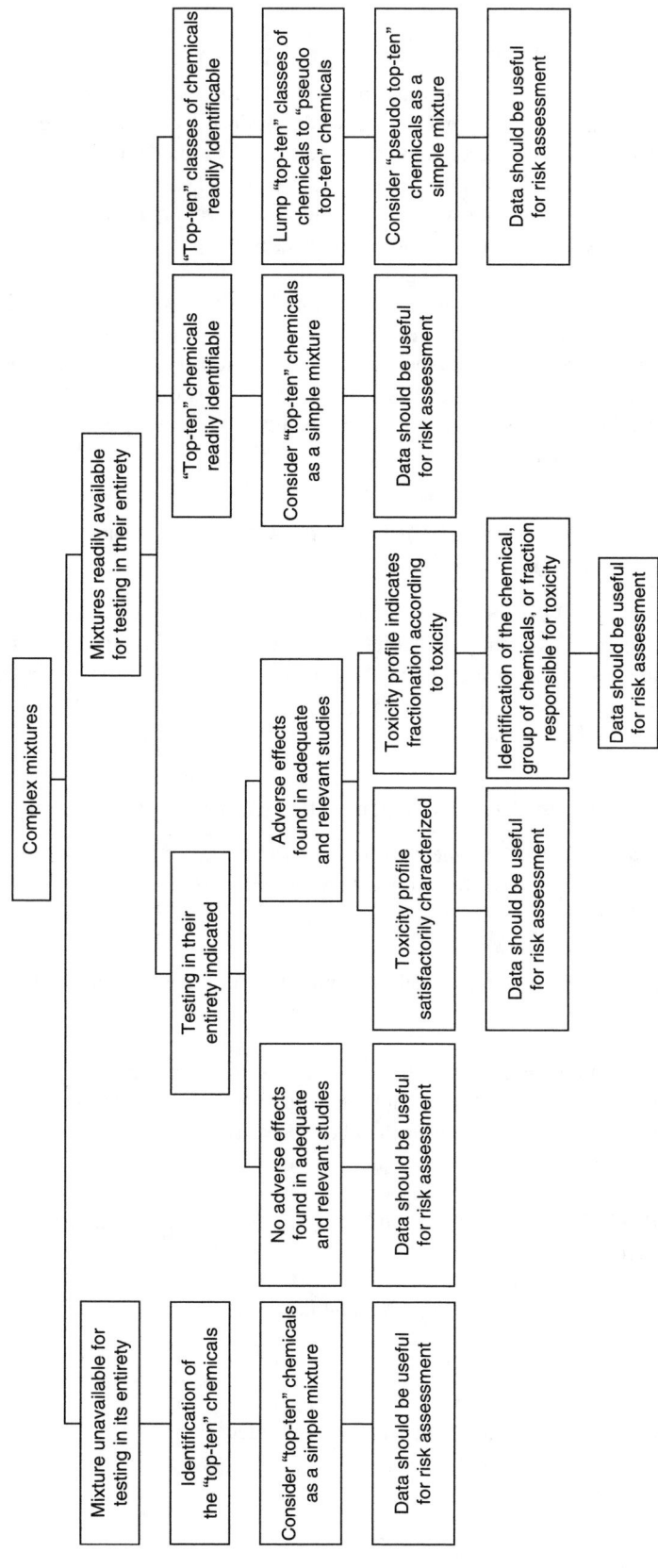

FIGURE 10.6 Scheme for safety evaluation of complex chemical mixtures. (Adapted from Feron, V.J., Groten, J.P., and van Bladeren, P.J., *Arch. Toxicol. Suppl.* 20, 363, 1998; Groten, J.P., Feron, V.J., and Sühnel, J., *Trends Pharmacol. Sci.* 22, 316, 2001.)

approved food additives allocated a numerical ADI (a total of 65 additives). Target organs were identified based on the adverse effects reported at dose levels above the NOAELs in animal or human studies. Description of pathological and other changes found were used to assess whether different additives, sharing the same target organ, would produce a common toxic effect. In many cases the adverse effects were considered to be nonspecific and/or related to nutritional/palatability problems, such that a clear target toxicity could not be identified. Twelve different target organs were identified, and each group of additives, sharing the same target organ, was studied in detail for possible joint actions and/or interactions. In all but four cases, the possibility of joint actions and/or interactions could be excluded on scientific grounds. The exceptions were groups of additives with critical effects on the liver (curcumin, thiabendazole, propyl gallate, and butyl hydroxyl toluene), the kidneys (diphenyl, o-phenylphenol, and ferrocyanide salts), the blood (azorubine and propyl gallate), and the thyroid (erythrosine, thiabendazole, and nitrate). However, an in-depth analysis of both the specific use and the intake levels of these four groups of additives led the authors to conclude that joint actions or interactions among the additives in a group was a theoretical rather than a practical concern (Groten et al. 2000).

10.5.5 Approaches for Assessment of Joint Toxic Action of Chemical Mixtures Suggested by ATSDR

The American Agency for Toxic Substances and Disease Registry (ATSDR) in 2004 issued a "Guidance Manual for the Assessment of Joint Toxic Action of Chemical Mixtures" (ATSDR 2004) in which strategies for exposure-based assessments are proposed for noncarcinogenic and carcinogenic effects, respectively. The strategies are based on a number of sequential questions asked and are outlined in the following.

10.5.5.1 ATSDR Strategy for Noncarcinogenic Effects of Chemical Mixtures

Steps 1 and 2 ask questions as to whether the mixture has already been either regulated or scientifically evaluated in which case this information should be used. If this is not the case, further steps are proposed:

Step 3
   Are the HQs (see Section 10.5.1.1) equal to or greater than 0.1 for at least two of the mixture components?

1. If no, additivity and interactions are unlikely to result in health hazard.
2. If yes, further evaluation of additivity and interactions is necessary for the components of concern having HQ equal to or greater than 0.1. Go to step 4.

Step 4
   Is a relevant PBPK/PD model or studies on joint toxic action available for mixture of components of concern?

1. If yes, use model or study to evaluate potential health hazard.
2. If no, go to step 5.

Step 5
   Do the components of concern have the same critical effects?

1. If yes, go to step 6a.
2. If no, go to step 6b.

Step 6a

Apply the HI method (see Section 10.5.1.1).

Step 6b

Apply target-organ toxicity dose (TTD, see Section 10.5.1.1) modification of the HI method for overlapping targets of toxicity or access any unique critical effect with separate HQ.

1. If HI or separate HQ is equal to or less than 1 go to step 7a.
2. If HI or separate HQ is greater than 1 (potential health hazard due to additivity potential health hazard due to unique critical effect) go to step 7b.

Step 7a

Apply qualitative WOE (see Section 10.5.1.2 and ATSDR 2004): is the combined score positive?

1. If no, health hazard is unlikely.
2. If yes, there may be potential health hazard due to interactions and/or additivity.

Step 7b

Apply qualitative WOE: is combined score positive?

1. If no, health hazard likely to be less than indicated by HI or by separate HQ for unique critical effect.
2. If yes, there may be potential health hazard due to interactions and/or additivity.

10.5.5.2 ATSDR Strategy for Carcinogenic Effects of Chemical Mixtures

Steps 1 and 2 ask questions as to whether the mixture has already been either regulated or scientifically evaluated in which case this information should be used. If this is not the case further steps are proposed:

Step 3

Are the estimated risks equal to or greater than 1×10^{-6} for at least two of the individual mixture components?

1. If no, additivity and interactions are unlikely to result in health hazard.
2. If yes, further evaluation of additivity and interactions is necessary for the components of concern (having risk estimates equal to or greater than 1×10^{-6}). Go to step 4.

Step 4

Is a relevant PBPK/PD model or studies on joint toxic action available for mixture of components of concern?

1. If yes, use model or study to evaluate potential health hazard.
2. If no, go to step 5.

Step 5

Is the sum of the risks for the components equal to or greater than 1×10^{-4}?

1. If no, go to step 6a.
2. If yes (potential health hazard due to additivity; evaluate interactions), go to step 6b.

Step 6a

Apply qualitative WOE (see Section 10.5.1.2 and ATSDR 2004): is combined score positive?

1. If no, health hazard is unlikely.
2. If yes, there may be potential health hazard due to interactions and/or additivity.

Step 6b

Apply qualitative WOE: is combined score positive?

1. If no, health hazard likely to be less than indicated by the sum of risks.
2. If yes, there may be potential health hazard due to interactions and/or additivity.

10.6 EXPERIMENTAL STUDIES USING SIMPLE, WELL-DEFINED MIXTURES

In recognizing the difficulties in the risk assessment of chemical mixtures, the group around Victor Feron at the TNO Nutrition and Food Research Institute in Zeist, the Netherlands, initiated a research program to obtain some basic information on the toxicological interactions between toxicologically well-characterized chemicals in well-defined mixtures. The objective was to establish knowledge about some general principles for the interaction of chemicals in mixtures that would be useful in the risk assessment of complex mixtures. The group used relatively simple mixtures (not more than 10 different compounds), which were tested in short-term repeated-dose toxicity studies in order to examine the concepts of simple similar action or simple dissimilar action and its implication for the risk assessment of chemical mixtures.

10.6.1 CHEMICALS WITH DIFFERENT TARGET ORGANS AND/OR DIFFERENT MODES OF ACTION

10.6.1.1 Dutch Group

Two 4-week studies of the toxicity (clinical chemistry, hematology, biochemistry, and pathology) in rats were performed on combinations of compounds with different target organs and/or different modes of actions. The study designs are given in Tables 10.5 and 10.6, respectively. NOAELs and LOAELs had been previously established for each single compound in the same laboratory using the same strain of rats and comparable experimental conditions.

In the first study (Jonker et al. 1990, Feron et al. 1995a) (Table 10.5), no treatment related adverse effects were found below the NOAEL, and only slight effect (slightly decreased hemoglobin concentration and slightly increased kidney weight in male rats) at the NOAEL. As was expected from the studies of the toxicity of the individual compounds, a wide range of adverse effects was seen at the LOAEL, a number of which had not been seen at all or had been found at doses higher than the LOAEL of the individual compounds, which suggested some kind of interaction resulting in a slightly more severe and maybe broader range of toxic response. The authors concluded that the study demonstrates absence of a simple additive effect and synergism, and provides some, but no convincing evidence for an increased risk from exposure to a combination of chemicals when each chemical is administered at its own individual NOAEL. At lower dose levels no increased risk appeared to exist.

The second experiment (Feron et al. 1995a, Groten et al. 1997) was a 4-week oral/inhalatory study in male Wistar rats in which the toxicity of combinations of nine compounds was examined (Table 10.6) in a complex design. The study comprised a main study and a satellite study. In the main study, the rats were simultaneously exposed to mixtures of all nine chemicals. The satellite study was designed as a fractionated two-level factorial study in which the rats were simultaneously exposed to combinations of maximally five compounds at their LOAEL; these 16 combinations of 9 factors (9 chemicals) jointly comprise a 1/32 fraction of a complete study.

TABLE 10.5
Toxicological Studies of Chemicals with Different Target Organs and/or Modes of Action

Compound	NOAEL/10	NOAEL/3	NOAEL	LOAEL	Target Organ
		mg/kg Body Weight Per Day			
Potassium nitrate[a]	10	33	100	300	Adrenals
Stannous chloride[b]	100	330	1,000	3,000	Body weight, blood, liver
Sodium metabisulphite[b]	500	1,670	5,000	20,000	Blood, stomach, liver
Metaldehyde[b]	20	70	200	1,000	Liver
Loperamide[b]	0.5	1.7	5	25	Body weight
Mirex[b]	0.5	1.7	5	80	Body weight, liver, blood
Lysinoalanine[b]	3	10	30	100	Kidneys
Di-*n*-octyltin dichloride[b]	0.6	2	6	30	Thymus, liver

Sources: Jonker, D., Woutersen, R.A., van Bladeren, P.J., Til, H.P., and Feron, V.J., *Food Chem. Ttoxicol.* 28, 623, 1990; Feron, V.J., Groten, J.P., van Zorge, J.A., Cassee, F.R., Jonker, D., and van Bladeren, P.J., *Toxicol. Lett.* 82/83, 505, 1995a.

[a] Oral administration in drinking water.
[b] Oral administration in diet.

In the main study, a number of effects were observed at the LOAEL and, based on the toxicity data of the individual compounds, most of these effects were expected. A few effects seen in the toxicity studies with the individual compounds had disappeared in the combination, whereas some effects not seen in the range-finding studies with the individual compounds appeared in the combination. Only minor treatment related effects were observed in the NOAEL.

For most of the endpoints studied in the satellite groups, the factorial analysis revealed main effects of the individual compounds and interactions (cases of nonadditivity) between the compounds.

TABLE 10.6
Toxicological Studies of Chemicals with Different Target Organs and/or Modes of Action

Compound	NOAEL/3	NOAEL	LOAEL	Target Organ
	mg/kg Body Weight Per Day			
Aspirin[a]	330	1000	5000	Liver, stomach
Cadmium chloride[a]	3	10	50	Red blood cell, liver
Stannous chloride[a]	260	800	3000	Red blood cell
Spermine[a]	130	400	2000	Heart, liver
Loperamide[a]	2	6	30	Liver
BHA[a]	330	1000	3000	Stomach
DEHP[a]	65	200	1000	Liver
Dichloromethane[b]	30	100	500	Blood
Formaldehyde[b]	0.3	1	3	Nose

Sources: Feron, V.J., Groten, J.P., van Zorge, J.A., Cassee, F.R., Jonker, D., and van Bladeren, P.J., *Toxicol. Lett.* 82/83, 505, 1995a; Groten, J.P., Schoen, E.D., van Bladeren, P.J., Kuper, F.C.F., van Zorge, J.A., and Feron, V.J., *Fund. Appl. Toxicol.* 36, 15, 1997.

[a] Oral administration in diet.
[b] Inhalation.

In comparison with the adverse effects produced by the individual compounds, both more severe and less severe effects were observed at the LOAEL of the combined compounds, indicating interactions of effects at this exposure level. Several of the effects were more severe than seen with the individual compounds.

It was concluded that simultaneous exposure to these nine chemicals did not constitute an evidently increased hazard compared to exposure to each of the chemicals separately, provided the exposure level of each chemical in the mixture was at most similar to or lower than its own NOAEL (Groten et al. 1997).

10.6.1.2   Other Studies

Ito et al. (1995) examined the combined dietary administration to rats of 19 organophosphate pesticides and 1 organochlorine pesticide, all permitted for use in Japan, each at its ADI level. The dietary exposure at this level did not enhance the development of diethyl nitrosamine initiated pre-neoplastic lesions whereas at 100 times the ADI, the number and area of lesions were increased. The authors concluded that the study provided direct support for the present use of the safety factor approach in the quantitative hazard evaluation of pesticides.

In rats treated orally by gavage with 100- and 1000-fold the ADI of three pesticides, indication of immune suppression was found at 1000-fold the ADIs. No immunological effects were reported at the 100-fold ADI level as well as after treatment with the individual pesticides. The authors suggest that one of the pesticides had a slight antagonistic effect on the two others at high doses (Akay et al. 1999).

Wade et al. (2002) studied effects (systemic, immune, and reproductive) of a 70-day exposure to a complex mixture of persistent contaminants in sexually mature rats. Each chemical was included in the mixture at the MRL, RfD, or TDI as determined by ATSDR or US-EPA, and the rats were exposed to the mixture at 1, 10, 100, and 1000 times the estimated safe levels daily for 70 days (see Table 10.7).

TABLE 10.7
Composition of Contaminant Mixture Administered to Male Rats

Contaminant	1 × MRL/RfD/TDI (xg/kg bw/d)	10 × MRL/RfD/TDI (xg/kg bw/d)	100 × MRL/RfD/TDI (xg/kg bw/d)	1000 × MRL/RfD/TDI (xg/kg bw/d)
Aldrin	30 ng	0.3 μg	3 μg	30 μg
p,p'-DDT	30 ng	0.3 μg	3 μg	30 μg
p,p'-DDE	570 ng	5.7 μg	57 μg	570 μg
Dieldrin	50 ng	0.5 μg	5 μg	50 μg
Endosulfan	50 ng	0.5 μg	5 μg	50 μg
Heptachlor	0.5 μg	5 μg	50 μg	500 μg
Hexachlorbenzene	0.3 μg	3 μg	30 μg	300 μg
Hexachlorocyclohexane	0.3 μg	3 μg	30 μg	300 μg
Mirex	0.8 μg	8 μg	80 μg	800 μg
Methoxychlor	2 μg	20 μg	200 μg	2000 μg
1,2,3-Trichlorobenzene	0.77 μg	7.7 μg	77 μg	770 μg
1,2,4-Trichlorobenzene	2.3 μg	23 μg	230 μg	2300 μg
1,2,3,4-Tetrachlorobenzene	0.2 μg	2 μg	20 μg	200 μg
Pentachlorobenzene	0.5 μg	5 μg	50 μg	500 μg
TCDD	1 ng	10 ng	0.1 μg	1 μg
PCB (Aroclor 1254)	1 μg	10 μg	100 μg	1000 μg
Cadmium chloride	0.7 μg	7 μg	70 μg	700 μg
Lead chloride	0.1 ng	1 ng	10 ng	0.1 μg

Source: Wade, M.G., Foster, W.G., Younglai, E.V., et al., *Toxicol. Sci.* 67, 137, 2002.

The authors concluded that the mixture induced effects on the liver and the kidney, and on the general metabolism at high doses but caused only minor effects on immune function, reproductive hormone levels, or general indices of reproductive function measures. The results suggested that additive or synergistic effects of exposure to contaminants resulting in residue levels representative of contemporary human tissue levels are unlikely to result in adverse effects on immune function or reproductive physiology in male rats.

10.6.2 SAME TARGET ORGAN WITH DISSIMILAR OR SIMILAR MODES OF ACTION

10.6.2.1 Nephrotoxicants with Dissimilar Modes of Action

The toxicity of mixtures of chemicals with the same target organ was examined in rats using nephrotoxicants with similar or dissimilar modes of action.

In a 4-week feeding study (Jonker et al. 1993, Feron et al. 1995a), rats were administered lysinoalanine, mercuric chloride, hexachloro-1,3-butadiene (HCBD), and d-limonene, each affecting renal proximal tubular cells but through different modes of action. The compounds were administered simultaneously at their individual lowest-observed-nephrotoxic-effect level (LONEL), no-observed-nephrotoxic-effect level (NONEL), and one-quarter of the NONEL (see Table 10.8). Nephrotoxicity was more severe in males fed the combination than in males given the nephrotoxins alone whereas in females, renal changes induced by the combination were not more severe than those observed with individual compounds. No adverse changes attributable to treatment were observed in rats fed the combination at one-quarter of the NONEL. The authors concluded that combined exposure to the four nephrotoxins at their individual NONEL did not constitute an obviously increased hazard, indicating absence of clear additivity and synergistic interaction, whereas at the LONEL clearly enhanced renal toxicity occurred in males, although not in females.

10.6.2.2 Nephrotoxicants with Similar Modes of Action

In a subsequent study, the additivity assumption (dose addition) was tested, using the similarly acting nephrotoxicants tetrachloroethylene, trichloroethylene, hexachloro-1,3-butadiene (HCBD), and 1,1,2-trichloro-3,3,3-trifluoropropene (Jonker et al. 1996). The compounds were given to female rats by daily oral gavage for 32 days either alone, at the LONEL and NONEL (= LONEL/4), or in combinations of four (at the NONEL and LONEL/2) or three (at the LONEL/3) (see Table 10.9).

TABLE 10.8

Toxicological Studies of Chemicals with the Same Target Organ, the Kidney, but Different Modes of Action

Compound	NONEL/4 (ppm in diet)	NONEL (ppm in diet)	LONEL (ppm in diet)	Mode of Action
Lysino-alanine	7.5	30	240	Metal ion chelator
Mercuric chloride	3.75	15	120	Mitochondrial dysfunction
Hexachloro-1,3-butadiene	5	20	100	β-Lyase mediated activation
d-Limonene	125	500	4000	$\alpha_{2\mu}$-Globulin accumulation

Sources: Jonker, D., Woutersen, R.A., van Bladeren, P.J., Til, H.P., and Feron, V.J., *Food Chem. Toxicol.* 31, 125, 1993; Feron, V.J., Groten, J.P. van Zorge, J.A., Cassee, F.R., Jonker, D., and van Bladeren, P.J., *Toxicol. Lett.* 82/83, 505, 1995a.

TABLE 10.9

Four-Week Oral Toxicity Study in Female Rats with Mixtures of Nephrotoxicants Having Similar Modes of Action

Treatment (mg/kg bw/d)		Total Doses in Toxicity Units
Control: corn oil 10 ml/kg bw/d		
Individual compounds at NONEL[a]		
Tetra[b]	600 mg/kg bw/d	¼
Tri[c]	500 mg/kg bw/d	¼
TCTFP[d]	1.5 mg/kg bw/d	¼
HCBD[e]	1.0 mg/kg bw/d	¼
Individual compounds at LONEL[f]		
Tetra	2400 mg/kg bw/d	1
Tri	2000 mg/kg bw/d	1
TCTFP	6.0 mg/kg bw/d	1
HCBD	4.0 mg/kg bw/d	1
Combination of all four compounds		
At NONEL		1
At LONEL/2		2
Combination of three compounds		
Tetra + Tri + TCTFP at LONEL/3		1
Tetra + Tri + HCBD at LONEL/3		1
Tetra + TCTFP + HCBD at LONEL/3		1
Tri + TCTFP + HCBD at LONEL/3		1

Sources: From Feron, V.J., Groten, J.P., van Zorge, J.A., Cassee, F.R., Jonker, D., and van Bladeren, P.J., *Toxicol. Lett.* 82/83, 505, 1995a; Jonker, D., Woutersen, R.A., and Feron, V.J., *Food Chem. Toxicol.* 34, 1075, 1996.

[a] NONEL = no-observed-nephrotoxic-effect level (= LONEL/4).
[b] Tetra = tetrachloroethylene.
[c] Tri = trichloroethylene.
[d] TCTFP = 1,1,2-trichloro-3,3,3-trifluoropropene.
[e] HCBD = hexachloro-1,3-butadiene.
[f] LONEL = lowest-observed-nephrotoxic-effect level.

Relative kidney weight was increased on exposure to the individual compounds at their LONEL and, to about the same extent, on combined exposure at the NONEL or the LONEL/3. The other endpoints studied (histopathology, concentrating ability, urinary excretion of glucose, protein and marker enzymes, and plasma creatinine and urea) were not or only scarcely affected upon combined exposure at the NONEL or LONEL/3. As assessed by the effect on kidney weight, the renal toxicity of the mixtures corresponded to the effect expected on the basis of the additivity assumption (Feron et al. 1995a, Jonker et al. 1996).

10.6.2.3 Mixtures of Chemicals Affecting the Same Target Organ but with Different Target Sites

A number of 3-day inhalation studies (Feron et al. 1995a, Cassee et al. 1996, 1998) were carried out in male rats with formaldehyde, acetaldehyde, and acrolein and mixtures of two or three of these toxicants (Table 10.10). They all produce the same type of adverse effect (nasal cytotoxicity) but with different target sites (different regions of the nasal mucosa). The nasal changes seen after

TABLE 10.10

Exposure Concentrations (ppm) Used in 3-Day (6 h/day) Inhalation Toxicity Studies in Male Rats with Nasal Toxicants

Groups[a]	Formaldehyde (ppm)	Acetaldehyde (ppm)	Acrolein (ppm)
Formaldehyde/low	1.0		
Formaldehyde/high	3.2		
Acetaldehyde/low		750	
Acetaldehyde/high		1500	
Acrolein/low			0.25
Acrolein/high			0.67
Mix 1	1.0		0.25
Mix 2	1.0	750	0.25
Mix 3	3.2	1500	0.67

Sources: From Feron, V.J., Groten, J.P., van Zorge, J.A., *Toxicol. Lett.* 82/83, 505, 1995a; Cassee, F. R., Groten, J.P. and Feron, V.J., *Fund. Appl. Toxicol.* 29, 208, 1996; Cassee, F.R., Groten, J. P., van Bladeren, P.J., and Feron, V.J., *Crit. Rev. Toxicol.* 28, 73, 1998.

[a] Each separate study included a control group exposed to clean air only.

exposure to mixtures 1 and 2 were very similar in site, type, degree, and incidence to those induced by 0.25 ppm acrolein alone. These changes were therefore considered to be induced by acrolein and not influenced by co-exposure to 1.0 ppm formaldehyde or to 1.0 ppm formaldehyde +750 ppm acetaldehyde. Mixture 3 induced pronounced changes that were more severe than those found after exposure to the individual compounds at comparable concentrations. The changes indicated that the combined effect of formaldehyde and acrolein on the nasal respiratory epithelium was at least additive and that formaldehyde and/or acrolein probably potentiated the effect of acetaldehyde on the olfactory epithelium.

The authors concluded that neither effect addition nor potentiating interactions occurred, providing the exposure concentrations of the aldehydes are at their NOAECs. They also stated that the type of combined action or interaction found at clearly toxic effect levels was not very helpful in predicting what would happen at levels that are not toxic.

10.6.3 CONCLUSIONS OF THE DUTCH STUDIES

The overall conclusion that the Dutch group drew from their experiments was that combined exposure to arbitrarily chosen chemicals clearly demonstrated the absence of full additivity, and provided some evidence of partial additivity when all chemicals in the mixture were administrated at their own individual NOAELs. At slightly lower dose levels, no clear evidence of toxicity was found. This conclusion was found valid for combinations of chemicals that have either different target organs and/or different target sites within the same organ (i.e., differ in the mode of action). Therefore, exposure to such mixtures is not associated with a greater hazard than exposure to the individual chemicals, provided that the exposure levels are at or below the individual NOAELs. At exposure levels higher than the NOAELs, both synergistic and antagonistic effects may be seen, dependent on the compounds. When the exposure levels are at the ADI/TDI levels, no greater hazard is to be expected. The group is of the opinion that the use of the "dose addition" approach to the risk assessment of chemical mixtures is only scientifically justifiable when all the chemicals in the mixture act in the same way, by the same mechanism, and thus differ only in their potencies. Application of the "dose addition" model to mixtures of chemicals that act by mechanisms for which the additivity assumptions are invalid would greatly overestimate the risk (Feron et al. 1995b,c; Cassee et al. 1998).

REFERENCES

Akay M.T., G. Ozmen, and E.A. Elcüman. 1999. Effects of combinations of endosulfan, dimethoate and carbaryl on immune and hematological parmeters of rats. *Vet. Hum. Toxicol.* 41:296–299.

Andersen, M.E., H.J. Clewell III, and C.G. Fredrick. 1995. Applying simulation modeling to problems in toxicology and risk assessment - A short perspective. *Toxicol. Appl. Pharmacol.* 133:181–187.

Ashford, J.R. 1981. General models for the joint action of drugs. *Biometrics* 37:457–474.

Ashford, J.R. and J.M. Cobby. 1974. A system of models for the action of drugs applied singly or jointly to biological organisms. *Biometrics* 30:11–31.

ATSDR. 2004. *Guidance manual for the assessment of joint toxic action of chemical mixtures.* Atlanta, U.S.A.: U.S. Department of Health and Human Services, Public Health Service, Agency of Toxic Substances and Disease Registry, Division of Toxicology. http://www.atsdr.cdc.gov/interactionprofiles/ipga.html

Berenbaum, M.C. 1981. Criteria for analyzing interactions between biologically active agents. *Adv. Cancer Res.* 35:269–335.

Berenbaum, M.C. 1985. The expected effect of a combination of agents: The general solution. *J. Theoret. Biol.* 114:413–431.

Berenbaum, M.C. 1989. What is synergy? *Pharmacol. Rev.* 41:93–141.

Bliss, C.I. 1939. The toxicity of poisons applied jointly. *Ann. Appl. Biol.* 26:585–615.

Brown, R. and A. Mittelman. 1993. *Evaluation of existing methods to rank the relative carcinogenicity of polycyclic aromatic compounds (PAH).* Draft. Technical Resources, Inc., Contract No. 68-01-0022, for Office of Emergency and Remedial Response, Office of Solid Waste and Emergency Response, U.S. Environmental Protection Agency, Office of Pesticides, Pollution Prevention and Toxic Substances (OPPTS).

Cassee, F.R., J.P. Groten, and V.J. Feron. 1996. Changes in the nasal epithelium of rats exposed by inhalation to mixtures of formaldehyde, acetaldehyde, and acrolein. *Fund. Appl. Toxicol.* 29:208–218.

Cassee, F.R., J.P. Groten, P.J. van Bladeren, and V.J. Feron. 1998. Toxicological evaluation and risk assessment of chemical mixtures. *Crit. Rev. Toxicol.* 28:73–101.

Chaturvedi, A.K. 1993. Toxicological evaluation of mixtures of ten widely used pesticides. *J. Appl. Toxicol.* 13:183–188.

Chaturvedi, A.K., D.J. Kuntz, and N.G. Rao. 1991. Metabolic aspects of the toxicology of mixtures of parathion, toxaphene and/or 2,4-D in mice. *J. Appl. Toxicol.* 11:245–251.

Culp, S.J., D.W. Gaylor, W.G. Sheldon, L.S. Goldstein, and F.A. Beland. 1998. A comparison of the tumours induced by coal tar and benzo[a]pyrene in a 2-year bioassay. *Carcinogenesis* 19:117–124.

Feron, V.J., J.P. Groten, J.A. van Zorge, F.R. Cassee, D. Jonker, and P.J. van Bladeren. 1995a. Toxicity studies in rats of simple mixtures of chemicals with the same or different target organs. *Toxicol. Lett.* 82/83: 505–512.

Feron, V.J., J.P. Groten, F.R. Cassee, D. Jonker, and P.J. van Bladeren. 1995b. Toxicology of chemical mixtures: Challenges for today and the future. *Toxicology* 105:415–417.

Feron, V.J., R.A. Woutersen, J.H.E. Arts, F.R. Cassee, F. de Vrijer, and P.J. van Bladeren. 1995c. Safety evaluation of the mixture of chemicals at a specific workplace: Theoretical considerations and a suggested two-step procedure. *Toxicol. Lett.* 76:47–55.

Feron, V.J., J.P. Groten, and P.J. van Bladeren. 1998. Exposure of humans to complex chemical mixtures: Hazard identification and risk assessment. *Arch. Toxicol. Suppl.* 20:363–373.

Feron, V.J. and J.P. Groten. 2002. Toxicological evaluation of chemical mixtures. *Food Chem Toxicol.* 40:825–839.

Finney, D.J. 1942. The analysis of toxicity tests on mixtures of poisons. *Ann. Appl. Biol.* 29:82–93.

Groten, J.P., E.J. Sinkeldam, T. Muys, J.B. Luten, and P.J. van Bladeren. 1991. Interaction of dietary Ca, P, Mg, Mn, Cu, Fe, Zn, and Se with the accumulation and oral toxicity of cadmium in rats. *Food Chem. Toxicol.* 29:249–258.

Groten, J.P., E.D. Schoen, P.J. van Bladeren, F.C.F. Kuper, J.A. van Zorge, and V.J. Feron. 1997. Subacute toxicity of a combination of nine chemicals in rats: Detecting interactive effects with a two-level factorial design. *Fund. Appl. Toxicol.* 36:15–29.

Groten, J.P., W. Butler, V.J. Feron, G. Kozianowski, A.G. Renwick, and R. Walker. 2000. An analysis of the possibility for health implications of joint actions and interactions between food additives. *Reg. Toxicol. Pharmacol.* 31:77–91.

Groten, J.P., V.J. Feron, and J. Sühnel. 2001. Toxicology of simple and complex mixtures. *Trends Pharmacol. Sci.* 22:316–322.

Haddad, S., R. Tardif, C. Viau, and K. Krishnan. 1999a. A modelling approach to account for toxicokinetic interaction in the calculation of biological hazard index for chemical mixtures. *Toxicol. Lett.* 108:303–308.

Haddad, S., G. Charest-Tardif, R. Tardif, and K. Krishnan. 1999b. Physiological modelling of the toxicokinetic interactions in a quarternary mixture of aromatic hydrocarbons. *Toxicol. Appl. Pharmacol.* 161:249–257.

Haddad, S., G. Charest-Tardif, R. Tardif, and K. Krishnan. 2000a. Validation of a physiological modelling framework for simulating the toxicokinetics of chemicals in mixtures. *Toxicol. Appl. Pharmacol.* 167:199–209.

Haddad, S., G. Charest-Tardif, and K. Krishnan. 2000b. Physiologically based modelling of the maximal effect of metabolic interactions on the kinetics of components of complex chemical mixtures. *J. Toxicol. Environ. Health* 61 PartA:209–223.

Hewlett, P.S. and R.L. Placket. 1959. An unified theory for quantal responses to mixtures of drugs: Non-interactive action. *Biometrics* 15:591–610.

Hewlett, P.S. and R.L. Placket. 1964. An unified theory for quantal responses to mixtures of drugs: Competitive interactions. *Biometrics* 20:566–575.

Ito, N., R. Hasegawa. K. Imaida, Y. Kurata, A. Hagiwara, and T. Shirai. 1995. Effect of ingestion of 20 pesticides in combination at acceptable daily intake levels on rat liver carcinogenesis. *Food Chem. Toxicol.* 33:159–163.

Jonker, D., R.A. Woutersen, P.J. van Bladeren, H.P. Til, and V.J. Feron. 1990. 4-Week oral toxicity study of a combination of eight chemicals in rats: Comparison with the toxicity of the individual compounds. *Food Chem. Toxicol.* 28:623–631.

Jonker, D., R.A. Woutersen, P.J. van Bladeren, H.P. Til, and V.J. Feron. 1993. Subacute (4-wk) oral toxicity of a combination of four nephrotoxicants in rats: Comparison with the toxicity of the individual compounds. *Food Chem. Toxicol.* 31:125–136.

Jonker, D., R.A. Woutersen, and V.J. Feron, 1996. Toxicity of mixtures of nephrotoxicants with similar or dissimilar mode of action. *Food Chem. Toxicol.* 34:1075–1082.

Jonker, D., A.P. Freidig, J.P. Groten, A.E. de Hollander, R.H. Stierum, R.A. Woutersen, and V.J. Feron. 2004. Safety evaluation of chemical mixtures and combinations of chemical and non-chemical stressors. *Rev. Environ. Health* 19:83–139.

Klaassen, C.D. 1995, ed: *Casarett and Doull's Toxicology: The Basic Science of Poisons. Fifth Edition*, 1995. New York: McGraw-Hill.

Kortenkamp, A. and R. Altenburger. 1998. Synergisms with mixtures of xenoestrogens: A reevalution using the method of isoboles. *Sci. Total Environ.* 221:59–73.

Krewski, D., T. Thorslund, and J. Withey. 1989. Carcinogenic risk assessment of complex mixtures. *Toxicol. Ind. Health* 5:851–867.

Krishnan, K. and J. Brodeur. 1991. Toxicological consequences of combined exposure to environmental pollutants. *Arch. Complex Environ. Stud.* 3:1–106.

Könemann, W.H. and M.N. Pieters. 1996. Confusion of concepts in mixture toxicology. *Food Chem. Toxicol.* 34:1025–1031.

Larsen, J.C. and P.B. Larsen. 1998. Chemical carcinogens. In: *Air Pollution and Health* (Hester, R.E., and Harrison, R.M., eds.), Issues in Environmental Sciences and Technology, 10, The Royal Society of Chemical, Cambridge.

Leung, W.W. and D.J. Paustenbach. 1995. Physiologically based pharmacokinetic and pharmacodynamic modeling in health risk assessment and characterization of hazardous substances. *Toxicol. Lett.* 79:55–65.

Loewe, S. and H. Muischnek. 1926. Über kombinationswirkungen. *Arch. Exp. Pathol. Pharmak.* 114:313–326.

Mileson, B.E., J.E. Chambers, W.L. Chen, et al. 1998. Common mechanism of toxicity: A case study of organophosphorous pesticides. *Toxicol. Sci.* 41:8–20.

Mumtaz, M.M. and P.R. Durkin. 1992. A weight-of-evidence approach for assessing interactions in chemical mixtures. *Toxicol. Ind. Health* 8:377–406.

Mumtaz, M.M., I.G. Sipes, H.J. Clewell, and R.S.H. Yang. 1993. Risk assessment of chemical mixtures: Biological and toxicological issues. *Fund. Appl. Toxicol.* 21:258–269.

Narotsky, M.G., E.A. Weller, V.M. Chinchilli, and R.J. Kavlock. 1995. Nonadditive developmental toxicity in mixtures of trichloroethylene, Di(2-ethylhexyl)phthalate, and heptachlor in a $5 \times 5 \times 5$ design. *Fund. Appl. Toxicol.* 27:203–216.

Nesnow S. 1994. Mechanistic linkage between DNA adducts, mutations in oncogenes, and tumorigenicity of carcinogenic aromatic hydrocarbons in strain A/J mice. In *Chemical mixtures and quantitative risk assessment*. Abstract of the second annual HERL symposium, Nov. 7–10., Raleigh, North Carolina: Health Effects Research Laboratory, U.S. Environmental Protection Agency.

Nielsen, T., Jørgensen, H.E., Poulsen, M., Jensen, F.P., Larsen, J.C., Jensen, A.B., Schramm, J., and Tønnesen, J. 1995. *Traffic PAH and other mutagens in air in Denmark.* Environmental Project No. 285/1995. Copenhagen: Danish Environmental Protection Agency, Ministry of the Environment and Energy, Denmark.

Nisbet, I.C. and P.K. LaGoy. 1992. Toxic equivalency factors (TEFs) for polycyclic aromatic hydrocarbons (PAH). *Regul. Toxicol. Pharmacol.* 16:290–300.

NRC. 1993. Pesticides in the diets of infants and children. National Research Council (NRC). Washington, DC: Natl. Acad. Press.

Plackett, R.L. and J.P. Burman. 1946. The design of optimum multifactorial experiments. *Biomatrika* 33:305–339.

Placket, R.L. and P.S. Hewlett. 1952. Quantal responses to mixtures of poisons. J. *Royal Statistical Society, Series B* 14:143–163.

Placket, R.L. and P.S. Hewlett. 1963. A unified theory for quantal responses to mixtures of drugs: The fitting of data of certain models for two non-interactive drugs with complete positive correlation of tolerances. *Biometrics* 19:517–531.

Placket, R.L. and P.S. Hewlett. 1967. A comparison of two approaches to the construction of models for quantal responses in mixtures of drugs. *Biometrics* 23:27–44.

Ploemen, J.H.T.M., L.W. Wormhoudt, G.R.M.M. Haenen, et al. 1997. The use of human *in vitro* metabolic parameters to explore the risk of hazardous compounds: the case of ethylene dibromide. *Toxicol. Appl. Pharmacol.* 193:56–64.

PSD. 1999. *Methodology for the toxicological assessment of exposures from combinations of cholinesterase inhibiting compounds.* Draft document. Medical and Toxicological Panel, Advisory Committee on Pesticides, Pesticides Safety Directorate, UK Ministry of Agriculture, Fisheries and Food, April 19, 1999.

Rugen, P.J., C.D. Stern, and S.H. Lamm. 1989. Comparative carcinogenicity of the PAH as a basis for acceptable exposure levels (AELs) in drinking water. *Regul. Toxicol. Pharmacol.* 9:273.

Schneider, K., R. Roller, F. Kalberlah, and U. Schuhmacher-Wolz. 2002. Cancer risk assessment for oral exposure to PAH mixtures. *J. Appl. Toxicol.* 22:73–83.

Seed, J., R.P. Brown, S.S. Olin, and J.A. Foran. 1995. Chemical mixtures: Current risk assessment methodologies and future directions. *Reg. Toxicol. Pharmacol.* 22:76–94.

Svengaard, D.J. and R.C. Hertzberg 1994. *Statistical methods for toxicological evaluation.* In Yang, R.S.H. (ed). *Toxicology of Chemical Mixtures,* San Diego: Academic Press.

Thorslund, T.W. and D. Farrar. 1990. *Development of relative potency estimates for PAH and hydrocarbon combustion product fractions compared to benzo[a]pyrene and their use in carcinogenic risk assessment.* EPA/600/R-92/134, Dept. Commerce, NTIS.

US-EPA. 1986. *Guidance for health risk from exposure to chemical mixtures.* U.S. Environmental Protection Agency, *Fed Reg* 51:34014.

US-EPA. 1999. *Guidance for conducting health risk assessment of chemical mixtures* (External scientific peer review draft). April 1999. NCEA-C-0148. Washington, DC: Risk assessment Forum, U.S. Environmental Protection Agency.

US-EPA. 2000a. *Proposed guidance on cumulative risk assessment of pesticide chemicals that has a common mechanism of toxicity.* Public comment draft., June 22. Washington, DC: U.S. Environmental Protection Agency, Office of Pesticide Programs.

US-EPA 2000b. *Supplementary guidance for conducting health risk assessment of chemical mixtures.* EPA/630/R-00/002. Washington, DC: Risk Assessment Forum. U.S. Environmental Protection Agency. http://www.epa.gov/ncea/raf/pdfs/chem_mix/chem_mix_08_2001.pdf

Van den Berg, M., L. Birnbaum, A.T.C. Bosveld, et al. 1998. Toxic Equivalency Factors (TEFs) for PCBs, PCDDs, PCDFs for Humans and for Wildlife. *Environ. Health Perspect.* 106:775–792.

Van den Berg, M., L.S. Birnbaum, M. Denison, M. De Vito, et al. 2006. The 2005 World Health Organization reevaluation of human and mammalian toxic equivalency factors for dioxins and dioxin-like compounds. *Toxicol. Sci.* 93:223–241.

Wade, M.G., W.G. Foster, E.V. Younglai, et al. 2002. Effects of subchronic exposure to a complex mixture of persistent contaminants in male rats: Systemic, immune, and reproductive effects. *Toxicol. Sci.* 67:131–143.

Whalan J.E. and H.M. Pettigrew. 1997. *Inhalation risk assessments and the combining of margins of exposure.* Washington, DC: U.S. Environmental Protection Agency, Office of Pesticide Programs.

WHO/IPCS. 1998. *Selected non-heterocyclic policyclic aromatic hydrocarbons.* Environmental Health Criteria No. 202. Geneva: WHO.

WHO. 2000. *Air Quality Guidelines for Europe, Second Edition.* WHO Regional Publications, European Series, No. 91. World Health Organization Regional office for Europe. Copenhagen: WHO.

Wilkinson, C.F., G.R. Christoph, E. Julien, et al. 2000. Assessing the risks of exposure to multiple chemicals with a common mechanism of toxicity: How to cumulate? *Reg. Toxicol. Pharmacol.* 31 30–43.

Yang, R.S., H.A. El-Masri, R.S. Thomas, A.A. Constan, and J.D. Tessari. 1995. The application of physiologically based pharmacokinetic/pharmacodynamic (PBPK/PD) modeling for exploring risk assessment approaches for chemical mixtures. *Toxicol. Lett.* 79:193–200.

Index

A

Acceptable Daily Intake, 214, 238, 241, 279, 291–292
  and acceptable risk, definition, 4, 305
  definition of, 212
Acceptable lifetime cancer risk, 305
ACGIH, *see* American Conference on Government
    Industrial Hygienists
Acids influence, on ionic substances absorption, 104
Acute Reference Dose (ARfD), 212
Acute toxic class method, 109–110
Acute toxicity
  assessment objectives for, 108
  definition, 107–108
  guideline documents on, 110
  and hazard assessment, 111
  types of, 108
  in vivo test, guidelines, 108–110
ADI, *see* Acceptable Daily Intake
Adverse effect, definition, 4; *see also* Hazard assessment
Agency for Toxic Substances and Disease Registry (U.S.),
    22, 28, 69
Air quality guidelines for Europe, WHO guidelines, 68–69
Alarie test, 117
Allometric equation, for extrapolation, 229
Allometric scaling, 222, 227, 233, 235, 240, 243
American Conference on Government Industrial Hygienists,
    72–73
Ames test, *see* Bacterial reverse mutation test
Animal acute toxicity data, for toxicity study,
    56–57, 111
Animal data, for hazard assessment, 100–101
Area Under the Plasma Curve, 97, 100–101, 107, 255
Armitage and doll multistage mathematical
    model, 301
Aspergillus nidulans, gene mutation test on, 154, 163
Assessment endpoint, definition, 4
Assessment factors
  definition, 4, 213
  various approaches of
    Calabrese and Gilbert approach, 217
    Chemical-Specific assessment factors, 225–226
    Children-Specific assessment factors, 226–227
    Danish EPA's approach, 225
    Dutch approaches, 221–222
    ECETOC approach, 220–221
    EU TGD approach, 219–220
    Kalberlah and Schneider approach, 222–223
    Lewis–Lynch–Nikiforov approach, 217–219
    Renwick approach, 217
    Swedish National Chemicals Inspectorate's
        approach, 223–224
    UK approach, 223
    US-EPA approach, 216

ATSDR, *see* Agency for Toxic Substances
    and Disease Registry
AUC, *see* Area Under the plasma Curve
Average relative risk model, 307

B

Bacterial DNA damage or repair tests, 153
Bacterial reverse mutation test, 153
Bases influence, on ionic substances absorption, 104
BEIs, *see* Biological Exposure Indices
Benchmark dose concept, 211, 273, 277–281, 289–290
  advantage and disadvantage of, 93
  determination, 92
  goals, 91
  technical guidance document (TGD) on, 93
Benchmark Dose Lower Limit, for health risk
    assessments, 92
Benign tumors, 163–164
Biocidal products, 39
Biological effects of low level exposures
    (BELLE), 196
Biological Exposure Indices, 72–73
BMD, *see* Benchmark Dose
BMDL, *see* Benchmark Dose Lower Limit
Body size and extrapolation
  body surface area approach, 231–232
  body weight approach, 230–231
  caloric requirement approach, 232–234
  exposure route, 234–235
  PBPK models, 235–237
Body surface area (BSA) approach, 231–232
Body weight and scaling factor, 230
Bone marrow chromosome aberration test, 147, 160
Buehler test, 118

C

CAAT, *see* Center for Alternatives to Animal
    Testing
Calabrese and Gilbert approach, 217
Caloric requirement approach, 232–234
Canadian Centre for Occupational Health
    and Safety (CCOHS), 74
Carcinogenicity, 163
  assessment objectives for, 165
  categories of, 164–165
  categorization for, 176–179
  chemical, effect, 164
  definition, 165